Stochastic Mechanics
Random Media
Signal Processing and Image Synthesis
Mathematical Economics and Finance
Stochastic Optimization
Stochastic Control
Stochastic Models in Life Sciences

Stochastic Modelling
and Applied Probability

(Formerly:
Applications of Mathematics)

60

Edited by   B. Rozovskiĭ
G. Grimmett

Advisory Board   D. Dawson
D. Geman
I. Karatzas
F. Kelly
Y. Le Jan
B. Øksendal
G. Papanicolaou
E. Pardoux

For other titles published in this series, go to
www.springer.com/series/602

Alan Bain · Dan Crisan

Fundamentals of Stochastic Filtering

Alan Bain
BNP Paribas
10 Harewood Avenue
London NW1 6AA
United Kingdom
alan.bain@bnpparibas.com

Dan Crisan
Department of Mathematics
Imperial College London
180 Queen's Gate
London SW7 2AZ
United Kingdom
d.crisan@imperial.ac.uk

Managing Editors
B. Rozovskiĭ
Division of Applied Mathematics
Brown University
182 George St.
Providence, RI 02912
USA
rozovski@dam.brown.edu

G. Grimmett
Centre for Mathematical Sciences
University of Cambridge
Wilberforce Road
Cambridge CB3 0WB
UK
g.r.grimmett@statslab.cam.ac.uk

ISSN: 0172-4568 Stochastic Modelling and Applied Probability
ISBN: 978-0-387-76895-3   e-ISBN: 978-0-387-76896-0
DOI 10.1007/978-0-387-76896-0

Library of Congress Control Number: 2008938477

Mathematics Subject Classification (2000): 93E10, 93E11, 60G35, 62M20, 60H15

© Springer Science+Business Media, LLC 2009
All rights reserved. This work may not be translated or copied in whole or in part without the written permission of the publisher (Springer Science+Business Media, LLC, 233 Spring Street, New York, NY 10013, USA), except for brief excerpts in connection with reviews or scholarly analysis. Use in connection with any form of information storage and retrieval, electronic adaptation, computer software, or by similar or dissimilar methodology now known or hereafter developed is forbidden.
The use in this publication of trade names, trademarks, service marks, and similar terms, even if they are not identified as such, is not to be taken as an expression of opinion as to whether or not they are subject to proprietary rights.

Printed on acid-free paper

springer.com

Preface

Many aspects of phenomena critical to our lives can not be measured directly. Fortunately models of these phenomena, together with more limited observations frequently allow us to make reasonable inferences about the state of the systems that affect us. The process of using partial observations and a stochastic model to make inferences about an evolving system is known as *stochastic filtering*.

The objective of this text is to assist anyone who would like to become familiar with the theory of stochastic filtering, whether graduate student or more experienced scientist. The majority of the fundamental results of the subject are presented using modern methods making them readily available for reference. The book may also be of interest to practitioners of stochastic filtering, who wish to gain a better understanding of the underlying theory.

Stochastic filtering in continuous time relies heavily on measure theory, stochastic processes and stochastic calculus. While knowledge of basic measure theory and probability is assumed, the text is largely self-contained in that the majority of the results needed are stated in two appendices. This should make it easy for the book to be used as a graduate teaching text. With this in mind, each chapter contains a number of exercises, with solutions detailed at the end of the chapter.

The book is divided into two parts: The first covers four basic topics within the theory of filtering: the filtering equations (Chapters 3 and 4), Clark's representation formula (Chapter 5), finite-dimensional filters, in particular, the Beneš and the Kalman–Bucy filter (Chapter 6) and the smoothness of the solution of the filtering equations (Chapter 7). These chapters could be used as the basis of a one- or two-term graduate lecture course.

The second part of the book is dedicated to numerical schemes for the approximation of the solution of the filtering problem. After a short survey of the existing numerical schemes (Chapter 8), the bulk of the material is dedicated to particle approximations. Chapters 9 and 10 describe various particle filtering methods in continuous and discrete time and prove associated convergence results. The material in Chapter 10 does not require knowledge of stochastic integration and could form the basis of a short introductory course.

We should like to thank the publishers, in particular the senior editor, Achi Dosanjh, for her understanding and patience. Thanks are also due to various people who offered their support and advice during the project, in particular Martin Clark, Mark Davis and Boris Rozovsky. One of the authors (D.C.) would like to thank Robert Piché for the invitation to give a series of lectures on the subject in August 2006.

Part of the book grew out of notes on lectures given at Imperial College London, University of Cambridge and Tampere University of Technology. Special thanks are due to Kari Heine from Tampere University of Technology and Olasunkanmi Obanubi from Imperial College London who read large portions of the first draft and suggested many corrections and improvements.

Finally we would like to thank our families for their support, without which this project would have never happened.

London
December 2007

Alan Bain
Dan Crisan

Contents

Preface v

Notation xi

1 Introduction 1
   1.1 Foreword 1
   1.2 The Contents of the Book 3
   1.3 Historical Account 5

Part I Filtering Theory

2 The Stochastic Process π 13
   2.1 The Observation σ-algebra $\mathcal{Y}_t$ 16
   2.2 The Optional Projection of a Measurable Process 17
   2.3 Probability Measures on Metric Spaces 19
       2.3.1 The Weak Topology on $\mathcal{P}(\mathbb{S})$ 21
   2.4 The Stochastic Process π 27
       2.4.1 Regular Conditional Probabilities 32
   2.5 Right Continuity of Observation Filtration 33
   2.6 Solutions to Exercises 41
   2.7 Bibliographical Notes 45

3 The Filtering Equations 47
   3.1 The Filtering Framework 47
   3.2 Two Particular Cases 49
       3.2.1 X a Diffusion Process 49
       3.2.2 X a Markov Process with a Finite Number of States ... 51
   3.3 The Change of Probability Measure Method 52
   3.4 Unnormalised Conditional Distribution 57
   3.5 The Zakai Equation 61

	3.6 The Kushner–Stratonovich Equation	67
	3.7 The Innovation Process Approach	70
	3.8 The Correlated Noise Framework	73
	3.9 Solutions to Exercises	75
	3.10 Bibliographical Notes	93
4	**Uniqueness of the Solution to the Zakai and the Kushner–Stratonovich Equations**	**95**
	4.1 The PDE Approach to Uniqueness	96
	4.2 The Functional Analytic Approach	110
	4.3 Solutions to Exercises	116
	4.4 Bibliographical Notes	125
5	**The Robust Representation Formula**	**127**
	5.1 The Framework	127
	5.2 The Importance of a Robust Representation	128
	5.3 Preliminary Bounds	129
	5.4 Clark's Robustness Result	133
	5.5 Solutions to Exercises	139
	5.6 Bibliographic Note	139
6	**Finite-Dimensional Filters**	**141**
	6.1 The Beneš Filter	141
	6.1.1 Another Change of Probability Measure	142
	6.1.2 The Explicit Formula for the Beneš Filter	144
	6.2 The Kalman–Bucy Filter	148
	6.2.1 The First and Second Moments of the Conditional Distribution of the Signal	150
	6.2.2 The Explicit Formula for the Kalman–Bucy Filter	154
	6.3 Solutions to Exercises	155
7	**The Density of the Conditional Distribution of the Signal**	**165**
	7.1 An Embedding Theorem	166
	7.2 The Existence of the Density of ρ_t	168
	7.3 The Smoothness of the Density of ρ_t	174
	7.4 The Dual of ρ_t	180
	7.5 Solutions to Exercises	182

Part II Numerical Algorithms

8	**Numerical Methods for Solving the Filtering Problem**	**191**
	8.1 The Extended Kalman Filter	191
	8.2 Finite-Dimensional Non-linear Filters	196
	8.3 The Projection Filter and Moments Methods	199
	8.4 The Spectral Approach	202

	Contents	ix
	8.5 Partial Differential Equations Methods	206
	8.6 Particle Methods	209
	8.7 Solutions to Exercises	217

9 A Continuous Time Particle Filter 221
  9.1 Introduction 221
  9.2 The Approximating Particle System 223
    9.2.1 The Branching Algorithm 225
  9.3 Preliminary Results 230
  9.4 The Convergence Results 241
  9.5 Other Results 249
  9.6 The Implementation of the Particle Approximation for π_t 250
  9.7 Solutions to Exercises 252

10 Particle Filters in Discrete Time 257
  10.1 The Framework 257
  10.2 The Recurrence Formula for π_t 259
  10.3 Convergence of Approximations to π_t 264
    10.3.1 The Fixed Observation Case 264
    10.3.2 The Random Observation Case 269
  10.4 Particle Filters in Discrete Time 272
  10.5 Offspring Distributions 275
  10.6 Convergence of the Algorithm 281
  10.7 Final Discussion 285
  10.8 Solutions to Exercises 286

Part III Appendices

A Measure Theory 293
  A.1 Monotone Class Theorem 293
  A.2 Conditional Expectation 293
  A.3 Topological Results 296
  A.4 Tulcea's Theorem 298
    A.4.1 The Daniell–Kolmogorov–Tulcea Theorem 301
  A.5 Càdlàg Paths 303
    A.5.1 Discontinuities of Càdlàg Paths 303
    A.5.2 Skorohod Topology 304
  A.6 Stopping Times 306
  A.7 The Optional Projection 311
    A.7.1 Path Regularity 312
  A.8 The Previsible Projection 317
  A.9 The Optional Projection Without the Usual Conditions 319
  A.10 Convergence of Measure-valued Random Variables 322
  A.11 Gronwall's Lemma 325

         A.12 Explicit Construction of the Underlying
               Sample Space for the Stochastic Filtering Problem 326

B Stochastic Analysis 329
     B.1 Martingale Theory in Continuous Time 329
     B.2 Itô Integral 330
          B.2.1 Quadratic Variation 332
          B.2.2 Continuous Integrator 338
          B.2.3 Integration by Parts Formula 341
          B.2.4 Itô's Formula 343
          B.2.5 Localization 343
     B.3 Stochastic Calculus 344
          B.3.1 Girsanov's Theorem 345
          B.3.2 Martingale Representation Theorem 348
          B.3.3 Novikov's Condition 350
          B.3.4 Stochastic Fubini Theorem 351
          B.3.5 Burkholder–Davis–Gundy Inequalities 353
     B.4 Stochastic Differential Equations 355
     B.5 Total Sets in L^1 355
     B.6 Limits of Stochastic Integrals 358
     B.7 An Exponential Functional of Brownian motion 360

References 367

Author Name Index 383

Subject Index 387

Notation

Spaces

- $\mathbb{R}^d$ – the d-dimensional Euclidean space.
- $\overline{\mathbb{R}^d}$ – the one-point compactification of $\mathbb{R}^d$ formed by adjoining a single point at infinity to $\mathbb{R}^d$.
- $\mathcal{B}(\mathbb{S})$ – the Borel σ-field on $\mathbb{S}$. That is the σ-field generated by the open sets in $\mathbb{S}$. If $\mathbb{S} = \mathbb{R}^d$ for some d, then this σ-field is countably generated.
- $(\mathbb{S}, \mathcal{S})$ – the state space for the signal. Unless otherwise stated, $\mathbb{S}$ is a complete separable metric space and $\mathcal{S}$ is the associated Borel σ-field $\mathcal{B}(\mathbb{S})$.
- $C(\mathbb{S})$ – the space of real-valued continuous functions defined on $\mathbb{S}$.
- $M(\mathbb{S})$ – the space of $\mathcal{B}(\mathbb{S})$-measurable functions $\mathbb{S} \to \mathbb{R}$.
- $B(\mathbb{S})$ – the space of bounded $\mathcal{B}(\mathbb{S})$-measurable functions $\mathbb{S} \to \mathbb{R}$.
- $C_b(\mathbb{S})$ – the space of bounded continuous functions $\mathbb{S} \to \mathbb{R}$.
- $C_k(\mathbb{S})$ – the space of compactly supported continuous functions $\mathbb{S} \to \mathbb{R}$.
- $C_k^m(\mathbb{S})$ – the space of compactly supported continuous functions $\mathbb{S} \to \mathbb{R}$ whose first m derivatives are continuous.
- $C_b^m(\mathbb{R}^d)$ – the space of all bounded, continuous functions with bounded partial derivatives up to order m. The norm $\|\cdot\|_{m,\infty}$ is frequently used with this space.
- $C_b^\infty(\mathbb{R}^d) = \bigcap_{m=0}^\infty C_b^m(\mathbb{R}^d)$.
- $D_\mathbb{S}[0, \infty)$ – the space of càdlàg functions from $[0, \infty) \to \mathbb{S}$.
- $C_b^{1,2}$ the space of bounded continuous real-valued funtions $u(t, x)$ with domain $[0, \infty) \times \mathbb{R}$, which are differentiable with respect to t and twice differentiable with respect to x. These derivatives are bounded and continuous with respect to (t, x).
- $C^l(\mathbb{R}^d)$ the subspace of $C(\mathbb{R}^d)$ containing functions φ such that $\varphi/\psi \in C_b(\mathbb{R}^d)$, where $\psi(x) = 1 + \|x\|$.
- $W_p^m(\mathbb{R}^d)$ – the Sobolev space of all functions with generalized partial derivatives up to order m with both the function and all its partial derivatives being L^p-integrable. This space is usually endowed with the norm $\|\cdot\|_{m,p}$.

xii   Notation

- $SL(\mathbb{R}^d) = \{\varphi \in C_b(\mathbb{R}^d) : \exists\, M \text{ such that } \varphi(x) \leq M/(1 + \|x\|),\ \forall x \in \mathbb{R}^d\}$
- $\mathcal{M}(\mathbb{S})$ – the space of finite measures over $(\mathbb{S}, \mathcal{S})$.
- $\mathcal{P}(\mathbb{S})$ – the space of probability measures over $(\mathbb{S}, \mathcal{S})$, i.e the subspace of $\mathcal{M}(\mathbb{S})$ such that $\mu \in \mathcal{P}(\mathbb{S})$ satisfies $\mu(\mathbb{S}) = 1$.
- $D_{M_F(\mathbb{R}^d)}[0,\infty)$ – the space of right continuous functions with left limits $a : [0,\infty) \to M_F(\mathbb{R}^d)$ endowed with the Skorohod topology.
- I – an arbitrary finite set $\{a_1, a_2, \ldots\}$.
- $P(I)$ – the power set of I, i.e. the set of all subsets of I.
- $\mathcal{M}(I)$ – the space of finite positive measures over $(I, P(I))$.
- $\mathcal{P}(I)$ – the space of probability measures over $(I, P(I))$, i.e. the subspace of $\mathcal{M}(I)$ such that $\mu \in P(I)$ satisfies $\mu(I) = 1$.

Other notations

- $\|\cdot\|$ – the Euclidean norm, for $x = (x_i)_{i=1}^m \in \mathbb{R}^m$, $\|x\| = \sqrt{x_1^2 + \cdots + x_m^2}$. It is also applied to $d \times p$-matrices by considering them as $d \times p$ vectors, viz
$$\|a\| = \sqrt{\sum_{i=1}^d \sum_{j=1}^p a_{ij}^2}.$$
- $\|\cdot\|_\infty$ – the supremum norm; for $\varphi \colon \mathbb{R}^d \to \mathbb{R}$, $\|\varphi\|_\infty = \sup_{x \in \mathbb{R}^d} |\varphi(x)|$. In general if $\varphi : \mathbb{R}^d \to \mathbb{R}^m$ then
$$\|\varphi\|_\infty = \max_{i=1,\ldots m} \sup_{x \in \mathbb{R}^d} |\varphi^i(x)|.$$
The notation $\|\cdot\|_\infty$ is equivalent to $\|\cdot\|_{0,\infty}$. This norm is especially useful on spaces such as $C_b(\mathbb{R}^d)$, or $C_k(\mathbb{R}^d)$, which only contain functions of bounded supremum norm; in other words, $\|\varphi\|_\infty < \infty$.
- $\|\cdot\|_{m,p}$ – the norm used on the space W_p^m defined by
$$\|\varphi\|_{m,p} = \left(\sum_{|\alpha| \leq m} \|D_\alpha \varphi(x)\|_p^p\right)^{1/p}$$
where $\alpha = (\alpha^1, \ldots, \alpha^d)$ is a multi-index and $D_\alpha \varphi = (\partial_1)^{\alpha^1} \ldots (\partial_d)^{\alpha^d} \varphi$.
- $\|\cdot\|_{m,\infty}$ is the special case of the above norm when $p = \infty$, defined by
$$\|\varphi\|_{m,\infty} = \sum_{|\alpha| \leq m} \sup_{x \in \mathbb{R}^d} |D_\alpha \varphi(x)|.$$
- δ_a – the Dirac measure concentrated at $a \in \mathbb{S}$, $\delta_x(A) \equiv \mathbf{1}_A(x)$.
- $\mathbf{1}$ – the constant function 1.
- $\Rightarrow$ – used to denote weak convergence of probability measures in $\mathcal{P}(\mathbb{S})$; see Definition 2.14.

- μf, $\mu(f)$ – the integral of $f \in B(\mathbb{S})$ with respect to $\mu \in \mathcal{M}(\mathbb{S})$, i.e. $\mu f \triangleq \int_{\mathbb{S}} f(x) \mu(\mathrm{d}x)$.
- $a^\top$ is the transpose of the matrix a.
- $\mathbb{I}_d$ – the $d \times d$ identity matrix.
- $\mathbb{O}_{d,m}$ – the $d \times m$ zero matrix.
- $\mathrm{tr}(A)$ – the trace of the matrix A, i.e. if $A = (a_{ij})$, then $\mathrm{tr}(A) = \sum_i a_{ii}$.
- $[x]$ – the integer part of $x \in \mathbb{R}$.
- $\{x\}$ – the fractional part of $x \in \mathbb{R}$, i.e. $x - [x]$.
- $\langle M \rangle_t$ – the quadratic variation of the semi martingale M.
- $s \wedge t$ – for $s, t \in \mathbb{R}$, $s \wedge t = \min(s, t)$.
- $s \vee t$ – for $s, t \in \mathbb{R}$, $t \vee s = \max(s, t)$.
- $A \vee B$ – the σ-algebra generated by the union $A \cup B$.
- $A \triangle B$ – the symmetric difference of sets A and B, i.e. all elements that are in one of A or B but not both, formally $A \triangle B = (A \setminus B) \cup (B \setminus A)$.
- $\mathcal{N}$ – the collection of null sets in the probability space $(\Omega, \mathcal{F}, \mathbb{P})$.

1
Introduction

1.1 Foreword

The development of mathematics since the 1950s has gone through many radical changes both in scope and in depth. Practical applications are being found for an increasing number of theoretical results and practical problems have also stimulated the development of theory. In the case of stochastic filtering, it is not clear whether this first arose as an application found for general theory, or as the solution of a practical problem.

Stochastic filtering now covers so many areas that it would be futile to attempt to write a comprehensive book on the subject. The purpose of this text is not to be exhaustive, but to provide a modern, solid and accessible starting point for studying the subject.

The aim of stochastic filtering is to estimate an evolving dynamical system, the signal, customarily modelled by a stochastic process. Throughout the book the signal process is denoted by $X = \{X_t,\ t \geq 0\}$, where t is the temporal parameter. Alternatively, one could choose a discrete time process, i.e. a process $X = \{X_t,\ t \in \mathbb{N}\}$ where t takes values in the (discrete) set $\{0, 1, 2, \ldots\}$. The former continuous time description of the process has the benefit that use can be made of the power of stochastic calculus. A discrete time process may be viewed as a continuous time process with jumps at fixed times. Thus a discrete time process can be viewed as a special case of a continuous time process. However, it is not necessarily effective to do so since it is much easier and more transparent to study the discrete case directly. Unless otherwise stated, the process X and all other processes are defined on a probability space $(\Omega, \mathcal{F}, \mathbb{P})$.

The signal process X can not be measured directly. However, a partial measurement of the signal can be obtained. This measurement is modelled by another continuous time process $Y = \{Y_t,\ t \geq 0\}$ which is called the *observation* process. This observation process is a function of X and a measurement noise. The measurement noise is modelled by a stochastic process $W = \{W_t,\ t \geq 0\}$. Hence,

$$Y_t = f_t(X_t, W_t), \qquad t \in [0, \infty).$$

Let $\mathcal{Y} = \{\mathcal{Y}_t,\ t \geq 0\}$ be the filtration generated by the observation process Y; namely,
$$\mathcal{Y}_t = \sigma\left(Y_s,\ s \in [0, t]\right), \qquad t \geq 0.$$
This σ-algebra $\mathcal{Y}_t$ can be interpreted as the information available from observations up to time t. This information can be used to make various inferences about X, for example:

- What is the best estimate (denoted by $\hat{X}_t$) of the value of the signal at time t, given the observations up to time t? If *best estimate* means the best mean square estimate, then this translates into computing $\mathbb{E}[X_t \mid \mathcal{Y}_t]$, the conditional mean of X_t given $\mathcal{Y}_t$.
- Given the observations up to time t, what is the estimate of the difference $X_t - \hat{X}_t$? For example, if the signal is real-valued, we may want to compute $\mathbb{E}[(X_t - \hat{X}_t)^2 \mid \mathcal{Y}_t] = \mathbb{E}[X_t^2 \mid \mathcal{Y}_t] - \mathbb{E}[X_t \mid \mathcal{Y}_t]^2$.
- What is the probability that the signal at time t can be found within a certain set A, again given the observations up to time t? This means computing $\mathbb{P}(X_t \in A \mid \mathcal{Y}_t)$, the conditional probability of the event $\{X_t \in A\}$ given $\mathcal{Y}_t$.

The typical form of such an inference requires the computation or approximation of one or more quantities of the form $\mathbb{E}[\varphi(X_t) \mid \mathcal{Y}_t]$, where φ is a real-valued function defined on the state space of the signal. Each of these statistics will provide fragments of information about X_t. But what if all information about X_t which is contained in $\mathcal{Y}_t$ is required? Mathematically, this means computing π_t, the conditional distribution of X_t given $\mathcal{Y}_t$. This π_t is defined as a random probability measure which is measurable with respect to $\mathcal{Y}_t$ so that[†]

$$\mathbb{E}\left[\varphi(X_t) \mid \mathcal{Y}_t\right] = \int_S \varphi(x) \pi_t(\mathrm{d}x), \tag{1.1}$$

for all statistics φ for which both terms of the above identity make sense. Knowing π_t will enable us, at least theoretically, to compute any inference of X_t given $\mathcal{Y}_t$ which is of interest, by integrating a suitable function φ with respect to π_t.

The measurability of π_t with respect to $\mathcal{Y}_t$ is crucial. However, this condition is sometimes overlooked and treated as a rather meaningless theoretical requirement. The following theorem illustrates the significance of the condition (for a proof see, e.g. Proposition 4.9 page 69 in [23]).

Theorem 1.1. *Let Ω be a probability space and $a, b \colon \Omega \to \mathbb{R}$ be two arbitrary functions. Let $\mathcal{A}$ be the σ-algebra generated by a, that is the smallest σ-algebra*

[†] The identity (1.1) holds $\mathbb{P}$-almost surely, i.e. there can be a subset of Ω of probability zero where (1.1) does not hold. The formal definition of the process π_t can be found in Chapter 2.

such that a is $\mathcal{A}/\mathcal{B}(\mathbb{R})$-measurable. Then if b is also $\mathcal{A}/\mathcal{B}(\mathbb{R})$-measurable there exists a $\mathcal{B}(\mathbb{R})/\mathcal{B}(\mathbb{R})$-measurable function $f \colon \mathbb{R} \to \mathbb{R}$ such that $b = f \circ a$, where $\circ$ denotes function composition.

Hence if b is "a-measurable", then b is determined by a. If we know the value of a then (theoretically) we will know the value of b. In practice however, it is often impossible to obtain an explicit formula for the connecting function f and this is the main difficulty in solving the filtering problem. Translating this concept into the context of filtering tells us that the random probability π_t is a function of Y_s for $s \in [0,t]$. Thus π_t is determined by the values of the observation process in the time interval $[0,t]$.

1.2 The Contents of the Book

The book is divided into two parts. The first part deals with the theoretical aspects of the problem of stochastic filtering and the second describes numerical methods for solving the filtering problem with emphasis on the class of particle approximations.

In Chapter 2 a fundamental measure-theoretic result related to π is proved: that the conditional distribution of the signal can be viewed as a stochastic process with values in the space of probability measures.

The filtering problem is stated formally in Chapter 3 for a class of problem where the signal X takes values in a state space $\mathbb{S}$ and is the solution of a martingale problem associated with an operator A. Two examples of filtering problems which can be considered in this fashion are:

1. The state space $\mathbb{S} = \mathbb{R}^d$ and $X = (X^i)_{i=1}^d$ is the solution of a d-dimensional stochastic differential equation driven by an m-dimensional Brownian motion process $V = (V^j)_{j=1}^m$,

$$X_t^i = X_0^i + \int_0^t f^i(X_s)\,\mathrm{d}s + \sum_{j=1}^m \int_0^t \sigma^{ij}(X_s)\,\mathrm{d}V_s^j, \quad i=1,\ldots,d. \quad (1.2)$$

In this case, the signal process is the solution of a martingale problem associated with the second-order differential operator

$$A = \sum_{i=1}^d f^i \frac{\partial}{\partial x_i} + \frac{1}{2} \sum_{i,j=1}^d \left(\sum_{k=1}^m \sigma^{ik} \sigma^{jk} \right) \frac{\partial^2}{\partial x_i \partial x_j}.$$

2. The state space $\mathbb{S} = I$ and X is a continuous time Markov chain with finite state space I. In this case, the corresponding operator is given by the Q-matrix of the chain.

The observation process Y is required to satisfy a stochastic evolution equation of the form

$$Y_t = Y_0 + \int_0^t h(X_s)\,\mathrm{d}s + W_t, \tag{1.3}$$

where $W = (W^i)_{i=1}^n$ is an n-dimensional Brownian motion independent of X and $h = (h_i)_{i=1}^n : \mathbb{S} \to \mathbb{R}^n$ is called the *sensor function*.

The filtering equations for a problem of this class are then deduced. In particular, it is proved that for any test function φ in the domain of A we have[†]

$$\mathrm{d}\pi_t(\varphi) = \pi_t(A\varphi)\,\mathrm{d}t + \sum_{i=1}^n \left(\pi_t\left(h^i\varphi\right) - \pi_t\left(h^i\right)\pi_t(\varphi)\right)$$
$$\times \left(\mathrm{d}Y_t^i - \pi_t(h^i\varphi)\mathrm{d}t\right). \tag{1.4}$$

Also, π_t has an unnormalized version, denoted by ρ_t, which satisfies the linear equation

$$\mathrm{d}\rho_t(\varphi) = \rho_t(A\varphi)\,\mathrm{d}t + \sum_{i=1}^n \rho_t(h^i\varphi)\,\mathrm{d}Y_t^i. \tag{1.5}$$

The identity

$$\pi_t(\varphi) = \frac{\rho_t(\varphi)}{\rho_t(1)}$$

is called the Kallianpur–Striebel formula.

The first term of (1.5) describes the evolution of the signal and the accumulation of observations is reflected in the second term. The same terms (with the same interpretations) can be found in (1.4) and the additional terms are due to the normalization procedure.

In Chapter 3 we present two approaches to deducing the filtering equations (1.4) and (1.5): the change of measure approach and the innovation approach. An extension is also described to the case where the noise driving the observation process is no longer independent of the signal. This feature is quite common, for example, in financial applications.

Chapter 4 contains a detailed study of the uniqueness of the solution of the filtering equations (1.4) and (1.5). The uniqueness can be shown by following a partial differential equations approach. The solution of certain partial differential equations with final condition is proved to be a partial dual for the filtering equations which leads to a proof of uniqueness. The second approach to proving uniqueness of the solution of the filtering equations follows the recent work of Heunis and Lucic.

In Chapter 5, we study the robust representation formula for the conditional expectation of the signal. The representation is robust in the sense that its dependence on the observation process Y is continuous. The result has important practical and theoretical consequences.

[†] If a is a measure on a space $\mathbb{S}$ and f is an a-integrable function then $a(f) \triangleq \int_\mathbb{S} f(x)a(\mathrm{d}x)$.

Chapter 6 is devoted to finite-dimensional filters. Two classes of filter are described: the Kalman–Bucy filter and the Beneš filter. Explicit formulae are deduced for both π_t and ρ_t and the finite-dimensionality of the filters is emphasized. The analysis of the Beneš filter uses the robust representation result presented in Chapter 5.

Among practitioners, it is generally accepted that the state space for π_t is that of densities with respect to the Lebesgue measure. Inherent in this is the (often unproved) assumption that π_t will always be absolutely continuous with respect to the Lebesgue measure. This is not always the case, although usually practitioners assume the correct conditions to ensure this. We discuss this issue in Chapter 7 and we look at the stochastic PDEs satisfied by the density of π_t and the density of ρ_t.

Chapter 8 gives an overview of the main computational methods currently available for solving the filtering problem. As expected of a topic with such a diversity of applications, numerous algorithms for solving the filtering problem have been developed.

Six classes of numerical method are presented: linearization methods (the extended Kalman filter), approximations by (exact) finite-dimensional filters, the projection filter/moment methods, spectral methods, PDE methods and particle methods.

Chapter 9 contains a detailed study of a continuous time particle filter. Particle filters (also known as *sequential Monte Carlo methods*) are some of the most successful methods for the numerical approximations of the solution of the filtering problem.

Chapter 10 is a self-contained, elementary treatment of particle approximations to the solution of the stochastic filtering problem in the discrete time case.

Finally, two appendices contain an assortment of measure theory, probability theory and stochastic analysis results included in order to make the text as self-contained as possible.

1.3 Historical Account

The origins of the filtering problem in discrete time can be traced back to the work of Kolmogorov [152, 153] and Krein [155, 156]. In the continuous time case Wiener [270] was the first to discuss the optimal estimation of dynamical systems in the presence of noise. The Wiener filter consists of a signal X which is a stationary process and an associated measurement process $Y = X + V$ where V is some independent noise. The object is to use the values of Y to estimate X, where the estimation is required to have the following three properties.

- *Causal*: X_t is to be estimated using Y_s for $s \leq t$.
- *Optimal*: The estimate, say $\hat{X}_t$, should minimise the mean square error $\mathbb{E}[(X - \hat{X}_t)^2]$.

- *Online*: At any (arbitrary) time t, the estimate $\hat{X}_t$ should be available.

The Wiener filter gives a linear, time-invariant causal estimate of the form

$$\hat{X}_t = \int_{-\infty}^{t} h(t-s) Y(s) \, \mathrm{d}s,$$

where $h(s)$ is called the transfer function. Wiener studied and solved this problem using the spectral theory of stationary processes. The results were included in a classified National Defense Research Council report issued in January/February 1942. The report, nicknamed "The Yellow Peril" (according to Wiener [271] this was because of the yellow paper in which it was bound) was widely circulated among defence engineers. Subsequently declassified, it appeared as a book, [270], in 1949.

It is important to note that all consequent advances in the theory and practical implementation of stochastic filtering always adhered to the three precepts enumerated above: causality, optimality and online estimation.

The next major development in stochastic filtering was the introduction of the linear filter. In this case, the signal satisfies a stochastic differential equation of the form (1.2) with linear coefficients and Gaussian initial condition and the observation equation satisfies an evolution equation of the form (1.3) with a linear sensor function. The linear filter can be solved explicitly; in other words, π_t is given by a closed formula. The solution is a finite-dimensional one: π_t is Gaussian, hence completely determined by its mean and its covariance matrix. Moreover it is quite easy to estimate the two parameters. The covariance matrix does not depend on Y and it satisfies a deterministic Riccati equation. Hence it can be solved in advance, before the filter is applied online. The mean satisfies a linear stochastic differential equation driven by Y, whose solution can be easily computed. These were the reasons for the linear filter's widespread success in the 1960s; for example it was used by NASA to get the Apollo missions off the ground and to the moon.[†] Bucy and Kalman were the pioneers in this field. Kalman was the first to publish in a wide circulation journal. In [146], he solved the discrete time version of the linear filter. Bucy obtained similar results independently.

Following the success of the linear filter, scientists started to explore different avenues. Firstly they extended the application of the Kalman filter beyond the linear/Gaussian framework. The basis of this extension is the fact that, locally, all systems behave linearly. So, at least locally, one can apply the Kalman filter equation. This gave rise to a class of algorithm called the extended Kalman filter. At the time of writing these algorithms, most of which are empirical and without theoretical foundation, are still widely used in a variety of applications.[‡]

[†] For an account of the linear filter's applications to aerospace engineering and further references see Cipra [54].

[‡] We study the extended Kalman filter in some detail in Chapter 6.

Stratonovich's work in non-linear filtering theory took place at the same time as the work of Bucy and Kalman. Stratonovich† presented his first results in the theory of conditional Markov processes and the related optimal non-linear filtering at the All-Union Conference on Statistical Radiophysics in Gorki (1958) and in a seminar [257]; they were published as [259].

Nevertheless, there was considerable unease about the methods used by Stratonovich to deduce the continuous time filtering equation. The paper [259] appeared with an editorial footnote indicating that part of the exposition was not wholly convincing. Writing in *Mathematical Reviews*, Bharucha-Reid [17] indicated that he was inclined to agree with the editor's comment concerning the author's arguments in the continuous case.

Part of the problem was that Stratonovich was using the stochastic integral which today bears his name. Stratonovich himself mentions this misunderstanding in [260, page 42]. He also points out (ibid., page 227) that the linear filtering equations were published by him in [258].

On the other side of the Atlantic in the mid-1960s Kushner [175, 176, 178] derived and analysed equation (1.4) using Itô (and not Stratonovich) calculus. Shiryaev [255] provided the first rigorous derivation in the case of a general observation process where the signal and observation noises may be correlated. The equation (1.4) was also obtained in various forms by other authors, namely: Bucy [30] and Wonham [273]. In 1968, Kailath [137] introduced the innovation approach to linear filtering. This new method for deducing the filtering equations was extended in the early 1970s by Frost and Kailath [103] and by Fujisaki, Kallianpur and Kunita [104]. The equation (1.4) is now commonly referred to as either the Fujisaki–Kallianpur–Kunita equation or the Kushner–Stratonovich equation.

Similarly, the filtering equation (1.5) was introduced in the same period by Duncan [85], [84], Mortensen [222] and Zakai [281], and is consequently referred to as the Zakai or the Duncan–Mortensen–Zakai equation.

The stochastic partial differential equations‡ associated with the filtering equations were rigorously analysed and extended in the late 1970s by Pardoux [236, 237, 238] and Krylov and Rozovskii [159, 160, 161, 162]. Pardoux adopted a functional analytic approach in analysing these SPDEs, whereas Krylov and Rozovskii examined the filtering equations using methods inherited from classical PDE theory. See Rozovskii [250] and the references therein for an analysis of the filtering equations using these methods.

Another important development in filtering theory was initiated by Clark [56] and continued by Davis [72, 74, 75]. In the late 1970s, Clark introduced the concept of *robust* or *pathwise* filtering; that is, $\pi_t(\varphi)$ is a function of the observation path $\{Y_s,\ s \in [0, T]\}$,

---

† We thank Gregorii Milstein and Michael Tretyakov for drawing our attention to Stratonovitch's historical account [260].

‡ Here we refer to the strong version of the filtering equations (1.4) and (1.5) as described in Chapter 7.

$$\pi_t(\varphi) = \Phi(Y_s;\ s \in [0,T]),$$

where Φ is a function defined on the corresponding space of trajectories. But Φ is not uniquely defined. Any other function Φ' equal to Φ on a set of measure one would be an equally acceptable version of $\pi_t(\varphi)$. From a computational point of view, we need to identify a continuous version of Φ.[†]

Given the success of the linear/Gaussian filter, scientists tried to find other classes of filtering problem where the solution was finite-dimensional and/or had a closed form. Beneš [9] succeeded in doing this. The class of filter which he studied had a linearly evolving observation process. However the signal was allowed to have a non-linear drift as long as it satisfied a certain (quite restrictive) condition, thenceforth known as the Beneš condition. The linear filter satisfies the Beneš condition.

Brockett and Clark [26, 27, 28] initiated a Lie algebraic approach to the filtering problem. From the linearized form of the Zakai equation one can deduce that ρ_t lies on a surface "generated" by two differential operators. One is the infinitesimal generator of X, generally a second-order differential operator and the other is a linear zero-order operator. From a Lie algebraic point of view the Kalman filter and the Beneš filter are isomorphic, where the isomorphism is given by a state space transformation. Beneš continued his work in [10] where he found a larger class of exact filter for which the corresponding Lie algebra is no longer isomorphic with that associated with the Kalman–Bucy filter. Following Beneš, Daum derived new classes of exact filters in [69] and [70]. A number of other classes of finite-dimensional filter have been discovered; see the series of papers by Chiou, Chen, Hu, Leung, Wu, Yau and Yau [48, 49, 50, 131, 274, 277, 276, 278]. See also the papers by Maybank [203, 204] and Schwartz and Dickinson [254].

In contrast to these finite-dimensional filters, results have been discovered which prove that generically the filtering problem is infinite-dimensional (Chaleyat-Maurel and Michel [42]). Hazewinkel, Marcus and Sussmann [121, 122] and Mitter [210] have contributed to this area. The general consensus is now that finite-dimensional filters are the exceptions and not the rule.

The work of Kallianpur has been influential in the field. The papers which contain the derivation of the Kallianpur–Striebel formula [144] and the derivation of the filtering equation [104] are of particular interest. Jointly with Karandikar in the papers [138, 139, 140, 141, 142, 143], Kallianpur extended the theory of stochastic filtering to finitely additive measures in place of countably additive measures.

The area expanded rapidly in the 1980s and 1990s. Among the topics developed in this period were: stability of the solution of the filtering problem, the uniqueness and Feynman–Kac representations of the solutions of the filtering equations, Malliavin calculus applied to the qualitative analysis of π_t and connections were discovered between filtering and information theory. In addition to the scientists already mentioned Bensoussan

---

[†] We analyze the pathwise approach to stochastic filtering in Chapter 5.

[12, 14, 15], Budhiraja [32, 33, 34, 35], Chaleyat-Maurel [40, 41, 44, 45], Duncan [86, 87, 88, 89], Elliott [90, 91, 92, 94], Grigelionis [107, 108, 109, 111], Gyöngy [112, 113, 115, 116, 117], Hazewinkel [124, 123, 125, 126], Heunis [127, 128, 129, 130], Kunita [165, 166, 167, 168], Kurtz [170, 172, 173, 174], Liptser [52, 190, 191], Michel [46, 47, 207, 20], Mikulevicius [109, 110, 208, 209], Mitter [98, 211, 212, 213], Newton [212, 225, 226], Picard [240, 241, 242, 243], Ocone [57, 228, 229, 230, 232, 233] Runggaldier [80, 96, 154, 191] and Zeitouni [4, 5, 282, 283, 284] contributed during this period. In addition to these papers, monographs were written by Bensoussan [13], Liptser and Shiryaev [192, 193] and Rozovskii [250] and Pardoux published lecture notes [238].

Much of the work carried out in the 1990s has focussed on the numerical solution of the filtering problem. The advent of fast computers has encouraged research in this area beyond the linear/Gaussian filter. Development in this area continues today. In Chapter 8 some historical comments are given for each of the six classes of numerical method discussed. Kushner (see e.g. [177, 179, 180, 181]) worked in particular on approximations of the solution of the filtering problem by means of finite Markov chain approximations (which are classified in Chapter 8 as PDE methods). Among others he introduced the important idea of a robust discrete state approximation, the finite difference method. Le Gland and his collaborators (see [25, 24, 100, 101, 136, 187, 188, 223]) have contributed to the development of several classes of approximation including the projection filter, PDE methods and particle methods.

Rapid progress continues to be made in both the theory and applications of stochastic filtering. In addition to work on the classical filtering problem, there is ongoing work on the analysis of the filtering problem for infinite-dimensional problems and problems where the Brownian motion noise is replaced by either 'coloured' noise, or fractional Brownian motion. Applications of stochastic filtering have been found within mathematical finance. There is continuing work for developing both generic/universal numerical methods for solving the filtering problem and problem specific ones.

At a Cambridge conference on stochastic processes in July 2001, Moshe Zakai was asked what he thought of stochastic filtering as a subject for future research students. He replied that he always advised his students 'to have an alternative subject on the side, just in case!' We hope that this book will assist anyone interested in learning about this challenging subject!

Part I

Filtering Theory

2
The Stochastic Process π

The principal aim of this chapter is to familiarize the reader with the fact that the conditional distribution of the signal can be viewed as a stochastic process with values in the space of probability measures. While it is true that this chapter sets the scene for the subsequent chapters, it can be skipped by those readers whose interests are biased towards the applied aspects of the subject. The gist of the chapter can be summarized by the following.

The principal aim of solving a filtering problem is to determine the conditional distribution of the signal X given the observation σ-algebra $\mathcal{Y}_t$, where

$$\mathcal{Y}_t \triangleq \sigma(Y_s,\ 0 \le s \le t) \vee \mathcal{N},$$

where $\mathcal{N}$ is the collection of all null sets of the complete probability space $(\Omega, \mathcal{F}, \mathbb{P})$ (see Remark 2.3 for comments on what is possible without the addition of these null sets to $\mathcal{Y}_t$). We wish to formalise this by defining a stochastic process describing this conditional distribution. Let the signal process X take values in a measurable space $(\mathbb{S}, \mathcal{S})$. Suppose we naïvely define a stochastic process $(\omega, t) \to \pi_t^\omega$ taking values in the space of functions from $\mathcal{S}$ into $[0, 1]$ by

$$\pi_t^\omega(A) = \mathbb{P}\left[X_t \in A \mid \mathcal{Y}_t\right](\omega), \tag{2.1}$$

where A is an arbitrary set in the σ-algebra $\mathcal{S}$. Recalling Kolmogorov's definition of conditional expectation[†], $\pi_t^\omega(A)$ is not uniquely defined for all $\omega \in \Omega$, but only for ω outside a $\mathbb{P}$-null set, which may depend upon the set A. It would be natural to think of this π_t as a probability measure on $(\mathbb{S}, \mathcal{S})$. However, this is not straightforward. For example consider the countable additivity property which any measure must satisfy. Let $A_1, A_2, \ldots \in \mathcal{S}$ be a sequence of pairwise disjoint sets, then by properties a. and c. of conditional expectation (see Section A.2), $\pi_t(\cdot)(\omega)$ satisfies the expected σ-additivity condition

---

[†] See Section A.2 in the appendix for a brief review of the properties of conditional expectation and conditional probability.

$$\pi_t^\omega\left(\bigcup_n A_n\right) = \sum_n \pi_t^\omega(A_n)$$

for every $\omega \in \Omega \backslash \mathcal{N}(A_n,\ n \geq 1)$, where $\mathcal{N}(A_n,\ n \geq 1)$ is a $\mathbb{P}$-null set which depends on the choice of the disjoint sets $A_n,\ n \geq 1$. Then we define

$$\bar{\mathcal{N}} = \bigcup \mathcal{N}(A_n,\ n \geq 1),$$

where the union is taken over all sequences of disjoint sets $(A_n)_{n \geq 1}$, such that for all $n > 0$, $A_n \in \mathcal{S}$. Then π_t^ω satisfies the σ-additivity property for arbitrary sets $\{A_n,\ n \geq 1\}$ only if $\omega \notin \bar{\mathcal{N}}$. Although the $\mathbb{P}$-measure of $\mathcal{N}(A_n,\ n \geq 1)$ is zero, the set $\bar{\mathcal{N}}$ need not even be measurable because it is defined in terms of an uncountable union, and furthermore, $\bar{\mathcal{N}}$ need not be contained in a $\mathbb{P}$-null set. This would imply that π_t cannot be a probability measure.

To solve this difficulty we require that the state space of the signal $\mathbb{S}$ be a complete separable metric space and $\mathcal{S}$ be the Borel σ-algebra $\mathcal{B}(\mathbb{S})$. This enables us to define π_t as the *regular* conditional distribution (in the sense of Definition A.2) of X_t given $\mathcal{Y}_t$. Defined in this manner, the process $\pi = \{\pi_t, t \geq 0\}$ will be a $\mathcal{P}(\mathbb{S})$-valued $\mathcal{Y}_t$-adapted process which satisfies (2.1) for any $t \geq 0$.

Unfortunately this is not enough. A second requirement must be satisfied by the process π. One of the results established in Chapter 3 is an evolution equation (1.4) for π, which is called the filtering equation. This evolution equation involves a stochastic integral with respect to the observation process Y whose integrand is described in terms of π.

Since the integrator process Y is continuous, it follows from Theorem B.19 that the stochastic integral with respect to Y is defined if π is a progressively measurable process, that is, if the function

$$(t, \omega) \to \pi_t : ([0, T] \times \Omega, \mathcal{B}([0, T]) \otimes \mathcal{Y}_t) \to (\mathcal{P}(\mathbb{S}), \mathcal{B}(\mathcal{P}(\mathbb{S}))),$$

is measurable for any $T > 0$. It is necessary to show that π has a version which is progressively measurable. We construct such a version for a signal process X which has càdlàg paths. In general, such a version is no longer adapted with respect to $\mathcal{Y}_t$, but with respect to a right continuous enlargement of $\mathcal{Y}_t$. In the case of the problems considered within this book $\mathcal{Y}_t$ itself is right continuous (see Section 2.5) so no enlargement is required.

Theorem 2.1. *Let $\mathbb{S}$ be a complete separable metric space and $\mathcal{S}$ be the associated Borel σ-algebra. Then there exists a $\mathcal{P}(\mathbb{S})$-valued $\mathcal{Y}_t$-adapted process $\pi = \{\pi_t, t \geq 0\}$ such that for any $f \in B(\mathbb{S})$*

$$\pi_t f = \mathbb{E}[f(X_t) \mid \mathcal{Y}_t] \qquad \mathbb{P}\text{-}a.s.$$

In particular, identity (2.1) holds true for any $A \in \mathcal{B}(\mathbb{S})$. Moreover, if Y satisfies the evolution equation

$$Y_t = Y_0 + \int_0^t h(X_s)\,\mathrm{d}s + W_t, \qquad t \geq 0, \tag{2.2}$$

where $W = \{W_t,\ t \geq 0\}$ is a standard $\mathcal{F}_t$-adapted m-dimensional Brownian motion and $h = (h_i)_{i=1}^m : \mathbb{S} \to \mathbb{R}^m$ is a measurable function such that

$$\mathbb{E}\left[\int_0^t \|h(X_s)\|\,\mathrm{d}s\right] < \infty \tag{2.3}$$

and

$$\mathbb{P}\left(\int_0^t \|\pi_s(h)\|^2\,\mathrm{d}s < \infty\right) = 1. \tag{2.4}$$

for all $t \geq 0$, then π has a $\mathcal{Y}_t$-adapted progressively measurable modification. Furthermore, if X is càdlàg then π_t can be chosen to have càdlàg paths.

The conditions (2.3) and (2.4) are frequently difficult to check (particularly (2.4)). They are implied by the stronger, but simpler condition

$$\mathbb{E}\left[\int_0^t \|h(X_s)\|^2\,\mathrm{d}s\right] < \infty. \tag{2.5}$$

To prove Theorem 2.1 we prove first a more general result (Theorem 2.24) which justifies the existence of a version of π adapted with respect to a right continuous enlargement of the observation filtration $\mathcal{Y}_t$. This result is proved without imposing any additional constraints on the observation process Y. However, under the additional constraints (2.2)–(2.4) as a consequence of Theorem 2.35, the filtration $\mathcal{Y}_t$ is right continuous, so no enlargement is required. Theorem 2.1 then follows.

In order to prove Theorem 2.24, we must introduce the *optional projection* of a stochastic process with respect to a filtration which satisfies the usual conditions. The standard construction of the optional projection requires the filtration to be right continuous and a priori the filtration $\mathcal{Y}_t$ may not have this property. Therefore choose a right continuous enlargement of the filtration $\mathcal{Y}_t$ defined by $\{\mathcal{Y}_{t+},\ t \geq 0\}$, where $\mathcal{Y}_{t+} = \cap_{s>t}\mathcal{Y}_s$. The existence of such an optional projection is established in Section 2.2.

Remark 2.2. The construction of the optional projection is valid without requiring that the filtration satisfy the usual conditions (see Section A.9). However such conditions are too weak for the proof of Theorem 2.24.

Remark 2.3. We always assume that the process π is this progressively measurable version and consequently $\{\mathcal{Y}_t, t \geq 0\}$ always denotes the augmented observation filtration. However, for any $t \geq 0$, the random probability measure π_t has a $\sigma(Y_s, s \in [0,t])$-measurable version, which can be used whenever the progressive measurability property is not required (see Exercise 2.36). Such a version of π_t, being $\sigma(Y_s, s \in [0,t])$-adapted, is a function of the observation path and thus is completely determined by the observation data. It turns out that π_t is a continuous function of the observation path. This is known as the path-robustness of filtering theory and it is discussed in Chapter 5.

2.1 The Observation σ-algebra $\mathcal{Y}_t$

Let $(\Omega, \mathcal{F}, \mathbb{P})$ be a probability space together with a filtration $(\mathcal{F}_t)_{t \geq 0}$ which satisfies the usual conditions:

1. $\mathcal{F}$ is complete i.e. $A \subset B$, $B \in \mathcal{F}$ and $\mathbb{P}(B) = 0$ implies that $A \in \mathcal{F}$ and $\mathbb{P}(A) = 0$.
2. The filtration $\mathcal{F}_t$ is right continuous i.e. $\mathcal{F}_t = \mathcal{F}_{t+}$.
3. $\mathcal{F}_0$ (and consequently all $\mathcal{F}_t$ for $t \geq 0$) contains all the $\mathbb{P}$-null sets.

On $(\Omega, \mathcal{F}, \mathbb{P})$ we consider a stochastic process $X = \{X_t, t \geq 0\}$ which takes values in a complete separable metric space $\mathbb{S}$ (the state space). Let $\mathcal{S}$ be the associated Borel σ-algebra. We assume that X is measurable. That is, X has the property that the mapping

$$(t, \omega) \to X_t(\omega) : ([0, \infty) \times \Omega, \mathcal{B}([0, \infty)) \otimes \mathcal{F}) \to (\mathbb{S}, \mathcal{S})$$

is measurable. Moreover we assume that X is $\mathcal{F}_t$-adapted.

Also let $Y = \{Y_t, t \geq 0\}$ be another $\mathcal{F}_t$-adapted process. The σ-algebra $\mathcal{Y}_t$ has already been mentioned in the introductory chapter. We now make a formal definition

$$\mathcal{Y}_t \triangleq \sigma(Y_s, \ 0 \leq s \leq t) \vee \mathcal{N}, \qquad (2.6)$$

where $\mathcal{N}$ is the set of $\mathbb{P}$-null sets in $\mathcal{F}$ and the notation $A \vee B$ is the standard notation for the σ-algebra generated by A and B, i.e. $\sigma(A, B)$.

The addition of the null sets $\mathcal{N}$ to the observation σ-algebra considerably increases the complexity of the proofs in the derivation of the filtering equations via the innovations approach in Chapter 3, so we should be clear why it is necessary. It is important that we can modify $\mathcal{Y}_t$-adapted processes. Suppose N_t is a such a process, then we need to be able to construct a process $\tilde{N}_t$ so that for $\omega \in G$ we change the values of the process, and for all $\omega \notin G$, $\tilde{N}_t(\omega) = N_t(\omega)$, where G is a $\mathbb{P}$-null set. In order that $\tilde{N}_t$ be $\mathcal{Y}_t$-adapted, the set G must be in $\mathcal{Y}_t$, which is assured by the augmentation of $\mathcal{Y}_t$ with the $\mathbb{P}$-null sets $\mathcal{N}$.

The following exercise gives a straightforward characterization of the σ-algebra $\mathcal{Y}_t$ and the relation between the expectation conditional upon the augmented filtration $\mathcal{Y}_t$ and that conditional upon the unaugmented filtration $\mathcal{Y}_t^o$.

Exercise 2.4. Let $\mathcal{Y}_t^o = \sigma(Y_s, \ 0 \leq s \leq t)$.

i. Prove that

$$\mathcal{Y}_t = \{F \subset \Omega : F = (G \backslash N_1) \cup N_2, \ G \in \mathcal{Y}_t^o, \ N_1, N_2 \in \mathcal{N}\}. \qquad (2.7)$$

ii. Deduce from part (i) that if ξ is $\mathcal{Y}_t$-measurable, then there exists a $\mathcal{Y}_t^o$-measurable random variable η, such that $\xi = \eta$ $\mathbb{P}$-almost surely. In particular, for any integrable random variable ξ, the identity

$$\mathbb{E}[\xi \mid \mathcal{Y}_t] = \mathbb{E}[\xi \mid \mathcal{Y}_t^o]$$

holds $\mathbb{P}$-almost surely.

As already stated, we consider a right continuous enlargement of the filtration $\mathcal{Y}_t$ defined by $\{\mathcal{Y}_{t+},\ t \geq 0\}$, where $\mathcal{Y}_{t+} = \cap_{s>t}\mathcal{Y}_s$. We do not wish a priori to impose the requirement that this observation σ-algebra be right continuous and satisfy $\mathcal{Y}_{t+} = \mathcal{Y}_t$, because verifying the right continuity of a σ-algebra which depends upon observations might not be possible before the observations have been made! We note, however, that the σ-algebra $\mathcal{Y}_{t+}$ satisfies the usual conditions; it is right continuous and complete.

Finally we note that no path regularity is assumed on either X or Y. Also no explicit connection exists between the processes X and Y.

2.2 The Optional Projection of a Measurable Process

From the perspective of measure theory, the filtering problem is associated with the construction of the optional projection of a process. The results in this section are standard in the theory of continuous time stochastic processes; but since they are often not mentioned in elementary treatments we consider the results which we require in detail.

Definition 2.5. *The optional σ-algebra $\mathcal{O}$ is defined as the σ-algebra on $[0, \infty) \times \Omega$ generated by $\mathcal{F}_t$-adapted processes with càdlàg paths. A process is said to be optional if it is $\mathcal{O}$-measurable.*

There is a well-known inclusion result: the set of previsible processes is contained in the set of optional processes, which is contained in the set of progressively measurable processes. We only require the second part of this inclusion; for a proof of the first part see Rogers and Williams [249].

Lemma 2.6. *Every optional process is progressively measurable.*

Proof. As the optional processes are generated by the adapted processes with càdlàg paths; it is sufficient to show that any such process X is progressively measurable.

For fixed $T > 0$, define an approximation process

$$Y^{(n)}(s,\omega) \triangleq \sum_{k=0}^{\infty} X_{T2^{-n}(k+1)}(\omega) 1_{[Tk2^{-n}, T(k+1)2^{-n})}(s) + X_T(\omega) 1_{[T,\infty)}(s).$$

It is immediate that $Y^{(n)}(s,\omega)$ restricted to $s \in [0,T]$ is $\mathcal{B}([0,T]) \otimes \mathcal{F}_T$-measurable and progressive. Since X has right continuous paths as does $Y^{(n)}$, it follows that $\lim_n Y_t^{(n)} = \lim_{s \downarrow t} X_s = X_t$ as $n \to \infty$. Since the limit exists, $X = \liminf_{n \to \infty} Y^{(n)}$, and is therefore progressively measurable. $\square$

The following theorem is only important in the case of a process X which is not adapted to the filtration $\mathcal{F}_t$. It allows us to construct from X an $\mathcal{F}_t$-adapted process. Unlike in the case of discrete time, we can not simply use the process defined by the conditional expectation $\mathbb{E}[X_t \mid \mathcal{F}_t]$, since this would not be uniquely defined for ω in a null set which depends upon t; thus the process would be unspecified on the uncountable union of these null sets over $t \in [0, \infty)$, which need not be null, therefore this definition could result in a process unspecified on a set of strictly positive measure which is unacceptable.

Theorem 2.7. *Let X be a bounded measurable process, then there exists an optional process ${}^oX$ called the* optional projection[†] *of X such that for every stopping time T*

$$ {}^oX_T 1_{\{T<\infty\}} = \mathbb{E}\left[X_T 1_{\{T<\infty\}} \mid \mathcal{F}_T\right]. \tag{2.8}$$

This process is unique up to indistinguishability, i.e. any processes which satisfy these conditions will be indistinguishable.

As we have assumed that the filtration $\mathcal{F}_t$ satisfies the usual conditions, this result can be established using Doob's result on the regularization of the trajectories of martingales. The proof is given in Section A.7 of the Appendix.

Remark 2.8. A simple consequence of the uniqueness part of this result is the fact that if X is itself optional then ${}^oX = X$. The definition can be extended to unbounded non-negative measurable processes by applying Theorem 2.7 to $X \wedge n$ and taking the limit as $n \to \infty$.

While Theorem 2.7 establishes the existence of the optional projection process, it does not provide us with any information about the trajectories of this process; for example, if the process X has continuous paths, does ${}^oX$ also have continuous paths? This turns out not to be true; see Remark 2.10. We must establish some kind of path regularity in order to apply many of the standard techniques of continuous time processes to the optional projection process.

The following theorem establishes the regularity which we need, however, its proof is fairly long and uses multiple applications of the optional section theorem; therefore the proof is not given here, but can be found in Section A.7.1 of the appendix.

Theorem 2.9. *If Y is a bounded càdlàg process then the optional projection ${}^oY$ is also càdlàg.*

Since the optional projection is only unique up to indistinguishability, the theorem is in fact stating that ${}^oY_t$ is indistinguishable from a càdlàg

---

[†] In some older French literature relevant to the subject, this projection is called the *projection bien-measurable*, although more recently, *projection optionelle* has superseded this.

process. As may be expected, this result depends upon $\mathcal{F}_t$ satisfying the usual conditions.

The restriction to bounded processes in the statement of the theorem is not essential, but is natural since our definition of optional projection was for a bounded process. The theorem can be extended to a process Y in the class D (i.e. the class of processes such that the set $\{X_T : T \text{ is a stopping time and } \mathbb{P}(T < \infty) = 1\}$ is uniformly integrable). As a uniformly integrable martingale is of class D, it follows that the theorem applies to uniformly integrable martingales.

Remark 2.10. The optional projection of a bounded continuous process need not itself be continuous. As an example, consider the process whose value at any time t is given by the same integrable random variable A; that is $X_t(\omega) = A(\omega)$. The optional projection of such a process is clearly the càdlàg modification of the martingale $\mathbb{E}[A \mid \mathcal{F}_t]$, however, clearly this modification need not be continuous.

2.3 Probability Measures on Metric Spaces

This section presents some results on probability measures on metric spaces which are needed in order to construct the process π, and which are used throughout the book. The reader familiar with these topics can skip this section to proceed with the construction of π. Let $\mathcal{P}(\mathbb{S})$ denote the space of probability measures on the space $\mathbb{S}$, that is, the subspace of $\mu \in \mathcal{M}(\mathbb{S})$ such that $\mu(\mathbb{S}) = 1$. Let $B(\mathbb{S})$ be the space of bounded $\mathcal{B}(\mathbb{S})$-measurable functions $\mathbb{S} \to \mathbb{R}$. If $\nu \in \mathcal{P}(\mathbb{S})$ and $f \in B(\mathbb{S})$ we write

$$\nu f = \int_{\mathbb{S}} f(x) \nu(\mathrm{d}x).$$

The following standard results about probability measures are necessary. For more details on these subjects, the reader should consult one of the many references, such as Billingsley [19] and Parthasarathy [239].

Theorem 2.11. *Any probability measure μ on a metric space $\mathbb{S}$ endowed with the associated Borel σ-algebra $\mathcal{B}(\mathbb{S})$ is regular. That is, if $A \in \mathcal{B}(\mathbb{S})$, given $\varepsilon > 0$ we can find an open set G and a closed set F such that $F \subseteq A \subseteq G$ and $\mu(G \setminus F) < \varepsilon$.*

Proof. Let d be the metric on $\mathbb{S}$. If A is closed then we can take $F = A$ and $G = \{x : d(x, A) < \delta\}$; as $\delta \downarrow 0$ the set G decreases to A. So if we let $\mathcal{H}$ be the class of sets A with the property of regularity then all the closed sets are contained in $\mathcal{H}$. As the closed sets are the complements of the open sets they also generate the Borel σ-algebra. So if we show that $\mathcal{H}$ is a σ-algebra then we shall have established the result. As $\mathcal{H}$ is obviously closed under complementation we only need to prove that it is closed under the formation of countable unions. Let $A_n \in \mathcal{H}$ and let F_n and G_n be closed and open sets such that $F_n \subseteq A_n \subseteq G_n$. By the definition of $\mathcal{H}$ we can choose these sets such that $\mathbb{P}(G_n \setminus F_n) < \varepsilon/2^{n+1}$. If we define $G = \bigcup_{n=1}^\infty G_n$ this is clearly an open set. Choose n_0 such that $\mathbb{P}(\bigcup_{n_0+1}^\infty F_n) < \varepsilon/2$ and then define $F = \bigcup_{n=1}^{n_0} F_n$ which, by virtue of the finite union is a closed set. Thus $F \subseteq A \subseteq G$ and $\mathbb{P}(G \setminus F) < \varepsilon$, establishing that $\mathcal{H}$ is closed under countable unions. Hence $\mathcal{H}$ contains the Borel sets. □

The main consequence for us of this theorem is that if two probability measures on $(\mathbb{S}, \mathcal{B}(\mathbb{S}))$ agree on the closed sets then they are equal.

Definition 2.12. *A subset $A \subset B(\mathbb{S})$ is said to be separating if for $\nu, \mu \in \mathcal{P}(\mathbb{S})$, the condition $\nu f = \mu f$ for all $f \in A$ implies that $\mu = \nu$.*

The following result determines a very important separating class which motivates the definition of weak convergence. However, it should be noted that the conclusion of Theorem 2.13 does follow from the more general Portmanteau theorem (Theorem 2.17).

Theorem 2.13. *Let $(\mathbb{S}, d)$ be a metric space and $U^d(\mathbb{S})$ be the space of all continuous bounded functions $\mathbb{S} \to \mathbb{R}$ which are uniformly continuous with respect to the metric d on $\mathbb{S}$. If μ, ν are elements of $\mathcal{P}(\mathbb{S})$, and*

$$\int_\mathbb{S} f(x)\mu(\mathrm{d}x) = \int_\mathbb{S} f(x)\nu(\mathrm{d}x) \qquad \forall f \in U^d(\mathbb{S}),$$

then this implies that $\mu = \nu$. That is, the space $U^d(\mathbb{S})$ is separating.

Proof. By Theorem 2.11 it is sufficient to show that ν and μ agree on closed subsets of $\mathbb{S}$. Let F be a closed set and define $F^\varepsilon = \{x : d(x, F) < \varepsilon\}$ which is clearly open and $\bigcap_{n=1}^\infty F^{1/n} = F$. The sets F and $(F^{1/n})^c$ are disjoint closed sets and $d(F, (F^{1/n})^c) \geq 1/n$. Define

$$f_n = (1 - nd(x, F))^+.$$

It is clear that $f_n \in U^d(\mathbb{S})$ and $0 \leq f_n \leq 1$. For $x \in F$ it follows that $f_n(x) = 1$, and $f_n(x) = 0$ for $x \in (F^{1/n})^c$. Hence

$$\mu(F) = \int_\mathbb{S} 1_F(x)\mu(\mathrm{d}x) \leq \int_\mathbb{S} f_n(x)\mu(\mathrm{d}x)$$

and

$$\nu(F^{1/n}) = \int_\mathbb{S} 1_{F^{1/n}}(x)\nu(\mathrm{d}x) \geq \int_\mathbb{S} f_n(x)\nu(\mathrm{d}x),$$

but by assumption as $f_n \in U^d(\mathbb{S})$ the right-hand sides of these two equations are equal; therefore $\mu(F) \leq \nu(F^{1/n})$ and letting n tend to infinity we obtain $\mu(F) \leq \nu(F)$. By symmetry we obtain the opposite inequality; hence $\mu(F) = \nu(F)$ for all closed sets F. □

2.3.1 The Weak Topology on $\mathcal{P}(\mathbb{S})$

Let us endow the space $\mathcal{P}(\mathbb{S})$ with the weak topology. Familiarity with basic results of general topology is assumed here, but some less elementary results which are required are proved in Appendix A.3.

Definition 2.14. *A sequence of probability measures $\mu_n \in \mathcal{P}(\mathbb{S})$, converges weakly to $\mu \in \mathcal{P}(\mathbb{S})$ if and only if $\mu_n \varphi$ converges to $\mu \varphi$ as $n \to \infty$ for all $\varphi \in C_b(\mathbb{S})$. Weak convergence of μ_n to μ is denoted $\mu_n \Rightarrow \mu$.*

No restriction is implied in this definition by the assertion that the limit μ is a probability measure. Since $\mathbf{1} \in C_b(\mathbb{S})$ it follows that $\mu_n \mathbf{1} = 1$ for all n; hence $\mu \mathbf{1} = 1$. We now exhibit a topology which engenders this form of convergence.

The reader with an interest in functional analysis should be aware that the concept of weak topology in the following definition is really the weak*-topology on the dual of $C_b(\mathbb{S})$ which is the space $\mathcal{M}(\mathbb{S})$, but the terminology weak convergence for this concept has become standard within probability theory. Recall that for a space S, if $\mathcal{T}_1$ and $\mathcal{T}_2$ are *topologies* (collections of subsets of S satisfying the axioms of closure under finite intersections, closure under all unions, and containing the S and $\emptyset$), then we say that $\mathcal{T}_1$ is *weaker* (*coarser*) than $\mathcal{T}_2$ if $\mathcal{T}_1 \subset \mathcal{T}_2$, in which case $\mathcal{T}_2$ is said to be *finer* than $\mathcal{T}_1$.

Definition 2.15. *The weak topology on the space $\mathcal{P}(\mathbb{S})$ is defined to be the weakest topology such that for all $f \in C_b(\mathbb{S})$, the function $\mu \mapsto \mu f$ is continuous.*

A *basis* for the neighbourhoods of a measure μ is defined to be a collection of open sets which contain μ, such that if V is another open set containing μ then there exists an element of the basis which is a subset of V.

It is clear that for $f \in C_b(\mathbb{S})$ the required continuity of the real-valued function on $\mathcal{P}(\mathbb{S})$ given by $\nu \mapsto \nu f$ implies that the set $\{\nu : |\nu f - \mu f| < \epsilon\}$ is open and contains μ. As the axioms of a topology require closure under finite intersections, we can construct a neighbourhood basis from these sets by taking finite intersections of them; thus in the weak topology on $\mathcal{P}(\mathbb{S})$ a basis for the neighbourhood of μ (a local basis) is provided by the sets of the form

$$\{\nu \in \mathcal{P}(\mathbb{S}) : |\mu f_i - \nu f_i| < \varepsilon, 1 \leq i \leq m\} \tag{2.9}$$

for $m \in \mathbb{N}$, $\varepsilon > 0$ and where $f_1, \ldots, f_m$ are elements of $C_b(\mathbb{S})$.

Theorem 2.16. *A sequence of probability measures $\mu_n \in \mathcal{P}(\mathbb{S})$ converges weakly to $\mu \in \mathcal{P}(\mathbb{S})$ if and only if μ_n converges to μ in the weak topology.*

Proof. If μ_n converges to μ in the weak topology then for any set A in the neighbourhood base of μ, there exists n_0 such that for $n \geq n_0$, $\mu_n \in A$. For any $f \in C_b(\mathbb{S})$, and $\varepsilon > 0$, the set $\{\nu : |\mu f - \nu f| < \varepsilon\}$ is in such a neighbourhood basis; thus $\mu_n f \to \mu f$ for all $f \in C_b(\mathbb{S})$, which implies that $\mu_n \Rightarrow \mu$. Conversely suppose $\mu_n \Rightarrow \mu$, and let A be the element of the neighbourhood basis for the weak topology given by (2.9). By the definition of weak convergence, it follows that $\mu_n f_i \to \mu_n f$, for $i = 1, \ldots, m$, so there exists n_i such that for $n \geq n_i$, $|\mu_n f_i - \mu f_i| < \varepsilon$; thus for $n \geq \max_{i=1,\ldots,m} n_i$, μ_n is in A and thus μ_n converges to μ in the weak topology. $\square$

We do not a priori know that this topology is metrizable; therefore we are forced to consider convergence of nets instead of sequences until such point as we prove that the space is metrizable. Consequently we make this proof our first priority. Recall that a net in E is a set of elements in E indexed by $\alpha \in D$, where D is an index set (i.e. a set with a partial ordering). Let x_α be a net in E. Define
$$\limsup_\alpha x_\alpha \triangleq \inf_{\alpha_0 \in D} \left\{ \sup_{\alpha \geq \alpha_0} x_\alpha \right\}$$
and
$$\liminf_\alpha x_\alpha \triangleq \sup_{\alpha_0 \in D} \left\{ \inf_{\alpha \geq \alpha_0} x_\alpha \right\}.$$
The net is said to converge to x if and only if
$$\liminf_\alpha x_\alpha = \limsup_\alpha x_\alpha = x.$$

If $\mathbb{S}$ is compact then by Theorem A.9, the space of continuous functions $C(\mathbb{S}) = C_b(\mathbb{S})$ is separable and we can metrize weak convergence immediately; however, in the general case $C_b(\mathbb{S})$ is not separable. Is it possible to find a smaller space of functions which still guarantee weak convergence but which is separable? The first thought might be the functions $C_k(\mathbb{S})$ with compact support; however, these functions generate a different topology called the *vague* topology which is weaker than the weak topology. To see this, consider $\mathbb{S} = \mathbb{R}$ and $\mu_n = \delta_n$ the measure with an atom at $n \in \mathbb{N}$; clearly this sequence does not converge in the weak topology, but in the vague topology it converges to the zero measure. (Although this is not an element of $\mathcal{P}(\mathbb{S})$; it is an element of $\mathcal{M}(\mathbb{S})$.)

The Portmanteau theorem provides a crucial characterization of weak convergence; while an important part of the theory of weak convergences its main importance to us is a step in the metrization of the weak topology.

Theorem 2.17. *Let $\mathbb{S}$ be a metric space with metric d. Then the following are equivalent.*

1. *$\mu_\alpha \Rightarrow \mu$.*
2. *$\lim_\alpha \mu_\alpha g = \mu g$ for all uniformly continuous functions g, with respect to the metric d.*
3. *$\lim_\alpha \mu_\alpha g = \mu g$ for all Lipschitz functions g, with respect to the metric d.*
4. *$\limsup_\alpha \mu_\alpha(F) \leq \mu(F)$ for all F closed in $\mathbb{S}$.*
5. *$\liminf_\alpha \mu_\alpha(G) \geq \mu(G)$ for all G open in $\mathbb{S}$.*

Proof. The equivalence of (4) and (5) is immediate since the complement of an open set G is closed. That $(1)\Rightarrow(2)\Rightarrow(3)$ is immediate. So it is sufficient to prove that $(3)\Rightarrow(4)\Rightarrow(1)$. Start with $(3)\Rightarrow(4)$ and suppose that $\mu_\alpha f \to \mu f$ for all Lipschitz continuous $f \in C_b(\mathbb{S})$. Let F be a closed set in $\mathbb{S}$. We construct a sequence $f_n \downarrow 1_F$ viz for $n \geq 1$,

$$f_n(x) = (1 - nd(x, F))^+. \tag{2.10}$$

Clearly $f_n \in C_b(\mathbb{S})$ and f_n is Lipschitz continuous with Lipschitz constant n. But $0 \leq f_n \leq 1$ and for $x \in F$, $f_n(x) = 1$, so it follows that $f_n \geq 1_F$, and it is also immediate that this is a decreasing sequence. Thus by the monotone convergence theorem

$$\lim_{n\to\infty} \mu f_n = \mu(F). \tag{2.11}$$

Consider n fixed; since $1_F \leq f_n$ it follows that for $\alpha \in D$ $\mu_\alpha(F) \leq \mu_\alpha f_n$, and thus

$$\limsup_{\alpha \in D} \mu_\alpha(F) \leq \limsup_{\alpha \in D} \mu_\alpha f_n.$$

But by (3)

$$\limsup_{\alpha \in D} \mu_\alpha f_n = \lim_{\alpha \in D} \mu_\alpha f_n = \mu f_n;$$

it follows that for all $n \in \mathbb{N}$, $\limsup_{\alpha \in D} \mu_\alpha(F) \leq \mu f_n$ and by (2.11) it follows that $\limsup_{\alpha \in D} \mu_\alpha(F) \leq \mu(F)$, which is (4). The harder part is the proof is that $(4)\Rightarrow(1)$. Given $f \in C_b(\mathbb{S})$ we split it up horizontally as in the definition of the Lebesgue integral. Let

$$-\|f\|_\infty = a_0 < a_1 < \cdots < a_n = \|f\|_\infty + \varepsilon/2$$

be constructed with n sufficiently large to ensure that $a_{i+1} - a_i < \varepsilon$. Define

$$F_i \triangleq \{x : a_i \leq f(x)\},$$

which by continuity of f is clearly a closed set. It is clear that $\mu(F_0) = 1$ and $\mu(F_n) = 0$. Therefore

$$\sum_{i=1}^n a_{i-1}\left[\mu(F_{i-1}) - \mu(F_i)\right] \leq \mu f < \sum_{i=1}^n a_i\left[\mu(F_{i-1}) - \mu(F_i)\right].$$

By telescoping the sums on the left and right and using the fact that $a_0 = -\|f\|_\infty$, we obtain

$$-\|f\|_\infty + \varepsilon \sum_{i=1}^{n-1} \mu(F_i) \leq \mu f < -\|f\|_\infty + \varepsilon + \varepsilon \sum_{i=1}^{n-1} \mu(F_i). \tag{2.12}$$

By the assumption that (4) holds, $\limsup_\alpha \mu_\alpha(F_i) \leq \mu(F_i)$ for $i = 0, \ldots, n$ hence we obtain from the right-hand inequality in (2.12) that

$$\mu_\alpha f \leq -\|f\|_\infty + \varepsilon + \varepsilon \sum_{i=1}^{n-1} \mu_\alpha(F_i)$$

thus

$$\limsup_\alpha \mu_\alpha f \leq -\|f\|_\infty + \varepsilon + \varepsilon \sum_{i=1}^{n-1} \limsup_\alpha \mu_\alpha(F_i)$$

$$\leq -\|f\|_\infty + \varepsilon + \varepsilon \sum_{i=1}^{n-1} \mu(F_i)$$

and from the left-hand inequality in (2.12) this yields

$$\limsup_\alpha \mu_\alpha f \leq \varepsilon + \mu f.$$

As ε was arbitrary we obtain $\limsup_\alpha \mu_n f \leq \mu f$, and application to $-f$ yields $\liminf \mu_n f \geq \mu f$ which establishes (1). □

While it is clearly true that a convergence determining set of functions is separating, the converse is not true in general and in the case when $\mathbb{S}$ is not compact, there may exist separating sets which are not sufficiently large to be convergence determining. For further details see Ethier and Kurtz [95, Chapter 3, Theorem 4.5].

Theorem 2.18. *If $\mathbb{S}$ is a separable metric space then there exists a countable convergence determining class $\varphi_1, \varphi_2, \ldots$ where $\varphi_i \in C_b(\mathbb{S})$.*

Proof. By Lemma A.6 a separable metric space is homeomorphic to a subset of $[0,1]^\mathbb{N}$; let the homeomorphism be denoted α. As the space $[0,1]^\mathbb{N}$ is compact, the closure $\overline{\alpha(\mathbb{S})}$ is also compact. Thus by Theorem A.9 the space $C(\overline{\alpha(\mathbb{S})})$ is separable. Let $\psi_1, \psi_2, \ldots$ be a countable dense family, where $\psi_i \in C(\overline{\alpha(\mathbb{S})})$.

It is therefore immediate that we can approximate any function $\psi \in C(\overline{\alpha(\mathbb{S})})$ arbitrarily closely in the uniform metric by suitable choice of ψ_i provided that ψ is the restriction to $\alpha(\mathbb{S})$ of a function in $C(\overline{\alpha(\mathbb{S})})$.

Now define $\varphi_i = \psi_i \circ \alpha$ for each i. By the same reasoning, we can approximate $f \in C(\mathbb{S})$ arbitrarily closely in the uniform metric by some f_i provided that $f = g \circ \alpha$ where g is the restriction to $\alpha(\mathbb{S})$ of a function in $C(\overline{\alpha(\mathbb{S})})$.

Define a metric on $\mathbb{S}$, $\hat{\rho}(x,y) = d(\alpha(x), \alpha(y))$, where d is a metric induced by the topology of co-ordinatewise convergence on $[0,1]^\mathbb{N}$. As α is a homeomorphism, this is a metric on $\mathbb{S}$. For F closed in $\mathbb{S}$, define the function

$$f_n^F(x) \triangleq (1 - n\hat{\rho}(x,F))^+ = (1 - nd(\alpha(x), \alpha(F)))^+ = (g_n^F \circ \alpha)(x), \quad (2.13)$$

where

$$g_n^F(x) \triangleq (1 - nd(x, \alpha(F)))^+.$$

This function g_n^F is an element of $C([0,1]^\mathbb{N})$, and hence is an element of $C(\overline{\alpha(\mathbb{S})})$; thus by the foregoing argument, we can approximate f_n^F arbitrarily closely by one of the functions φ_i. But we have seen from the proof that (3)⇒(4) in Theorem 2.17 that f_n^F of the form (2.13) for all F closed, $n \in \mathbb{N}$ form a convergence determining class. Suppose that for all i, we have that $\lim_\alpha \mu_\alpha \varphi_i = \mu \varphi_i$; then for each i

$$|\mu_\alpha f_n^F - \mu f_n^F| \leq 2\|f_n^F - \varphi_i\|_\infty + |\mu_\alpha \varphi_i - \mu \varphi_i|,$$

by the postulated convergence for all i of $\mu_\alpha \varphi_i$; it follows that the second term vanishes and thus for all i,

$$\limsup_\alpha |\mu_\alpha f_n^F - \mu f_n^F| \leq 2\|f_n^F - \varphi_i\|_\infty.$$

As i was arbitrary, it is immediate that

$$\limsup_\alpha |\mu_\alpha f_n^F - \mu f_n^F| \leq 2\liminf_i \|f_n^F - \varphi_i\|_\infty,$$

and since f_n^F can be arbitrarily approximated in the uniform norm by a φ_i, it follows $\lim_\alpha \mu_\alpha f_n^F = \mu f_n^F$, and since this holds for all n, and F is closed, it follows that $\mu_\alpha \Rightarrow \mu$. □

Theorem 2.19. *If $\mathbb{S}$ is a separable metric space, then $\mathcal{P}(\mathbb{S})$ with the weak topology is separable. We can then find a countable subset $\varphi_1, \varphi_2, \ldots$ of $C_b(\mathbb{S})$, with $\|\varphi_i\|_\infty = 1$ for all i, such that*

$$d: \mathcal{P}(\mathbb{S}) \times \mathcal{P}(\mathbb{S}) \to [0, \infty), \qquad d(\mu, \nu) = \sum_{i=1}^\infty \frac{|\mu \varphi_i - \nu \varphi_i|}{2^i} \qquad (2.14)$$

defines a metric on $\mathcal{P}(\mathbb{S})$ which generates the weak topology; i.e., a net μ_α converges to μ weakly if and only if $\lim_\alpha d(\mu_\alpha, \mu) = 0$.

Proof. By Theorem 2.18 there exists a countable set $f_1, f_2, \ldots$ of elements of $C_b(\mathbb{S})$ which is convergence determining for weak convergence. Define $\varphi_i \triangleq f_i/\|f_i\|_\infty$; clearly $\|\varphi_i\|_\infty = 1$, and the φ_is also form a convergence determining set. Define the map

$$\beta: \mathcal{P}(\mathbb{S}) \to [0,1]^\mathbb{N} \quad \beta: \mu \mapsto (\mu \varphi_1, \mu \varphi_2, \ldots).$$

Since the φ_is are convergence determining; they must also be separating and thus the map β is one to one. It is clear that if $\mu_\alpha \Rightarrow \mu$ then from the definition of weak convergence, $\lim_\alpha \beta(\mu_\alpha) = \beta(\mu)$. Conversely, since the φ_is are convergence determining, if $\lim_\alpha \mu_\alpha \varphi_i = \mu \varphi_i$ for all i then $\mu_\alpha \Rightarrow \mu$. Thus β is a homeomorphism from $\mathcal{P}(\mathbb{S})$ with the topology of weak convergence to

$[0,1]^{\mathbb{N}}$ with the topology of co-ordinatewise convergence. Thus since $[0,1]^{\mathbb{N}}$ is separable, this implies that $\mathcal{P}(\mathbb{S})$ is separable.

The space $[0,1]^{\mathbb{N}}$ admits a metric which generates the topology of co-ordinatewise convergence, given for $x, y \in [0,1]^{\mathbb{N}}$ by

$$D(x,y) = \sum_{n=1}^{\infty} \frac{|x_i - y_i|}{2^i}. \tag{2.15}$$

Therefore it follows that $d(x,y) = D(\beta(x), \beta(y))$ is a metric on $\mathcal{P}(\mathbb{S})$ which generates the weak topology. $\square$

As a consequence of this theorem, when $\mathbb{S}$ is a complete separable metric space the weak topology on $\mathcal{P}(\mathbb{S})$ is metrizable, so it is possible to consider convergence in terms of convergent sequences instead of using nets.

Exercise 2.20. Exhibit a countable dense subset of the space $\mathcal{P}(\mathbb{R})$ endowed with the weak topology. (Such a set must exist since $\mathbb{R}$ is a complete separable metric space, which implies that $\mathcal{P}(\mathbb{R})$ is separable.) Show further that $\mathcal{P}(\mathbb{R})$ is not complete under the metric d defined by (2.14).

Separability is a topological property of the space (i.e. it is independent of both existence and choice of metric), whereas completeness is a property of the metric. The topology of weak convergence on a complete separable space $\mathbb{S}$ can be metrized by a different metric called the Prohorov metric, under which it is complete (see, e.g. Theorem 1.7 of Chapter 3 of Ethier and Kurtz [95]).

Exercise 2.21. Let $(\Omega, \mathcal{F})$ be a probability space and $\mathbb{S}$ be a separable metric space. Let $\zeta : \Omega \to \mathcal{P}(\mathbb{S})$ be a function. Write $\mathcal{B}(\mathcal{P}(\mathbb{S}))$ for the Borel σ-algebra on $\mathcal{P}(\mathbb{S})$ generated by the open sets in the weak topology. Let $\{\varphi_i\}_{i>0}$ be a countable convergence determining set of functions in $C_b(\mathbb{S})$, whose existence is guaranteed by Theorem 2.18. Prove that ζ is $\mathcal{F}/\mathcal{B}(\mathcal{P}(\mathbb{S}))$-measurable (and thus a random variable) if and only if $\zeta\varphi_i : \Omega \to \mathbb{R}$ is $\mathcal{F}/\mathcal{B}(\mathbb{R})$-measurable for all $i > 0$. [Hint: Consider the case where $\mathbb{S}$ is compact for a simpler argument.]

Let us now turn our attention to the case of a finite state space I. The situation is much easier in this case since both $\mathcal{M}(I)$ and $\mathcal{P}(I)$ can be viewed as subsets of the Euclidean space $\mathbb{R}^{|I|}$ with the product topology (which is separable), and equipped with a suitable complete metric.

$$\mathcal{M}(I) = \left\{ (x_i)_{i \in I} \in \mathbb{R}^{|I|} : \sum_{i \in I} x_i < \infty, \ x_i \geq 0 \, \forall i \in I \right\}$$

$$\mathcal{P}(I) = \left\{ (x_i)_{i \in I} \in \mathcal{M}(I) : \sum_{i \in I} x_i = 1 \right\}.$$

The Borel sets in $\mathcal{M}(I)$, viz $\mathcal{B}(\mathcal{M}(I))$, are generated by the cylinder sets $\{R_{i,a,b}\}_{i \in I; a, b \geq 0}$, where $R_{i,a,b} = \{(x_j)_{j \in I} \in \mathcal{M}(I) : a \leq x_i \leq b\}$ and $\mathcal{B}(\mathcal{P}(I))$ is similarly described in terms of cylinders.

Exercise 2.22. Let $d(x,y)$ be the Euclidean metric on $\mathbb{R}^{|I|}$. Prove that d metrizes the topology of weak convergence on $\mathcal{P}(I)$ and that $(\mathcal{P}(I), d)$ is a complete separable metric space.

2.4 The Stochastic Process π

The aim of this section is to construct a $\mathcal{P}(\mathbb{S})$-valued stochastic process π which is progressively measurable. In order to guarantee the existence of such a stochastic process some topological restrictions must be imposed on the state space $\mathbb{S}$. In this chapter we assume that $\mathbb{S}$ is a complete separable metric space.[†] While this topological restriction is not the most general possible, it includes all the cases which are of interest to us; extensions to more general spaces are possible at the expense of additional technical complications (for details of these extensions, see Getoor [105]).

If we only wished to construct for a fixed $t \in [0, \infty)$ a $\mathcal{P}(\mathbb{S})$-valued random variable π_t then we could use the theory of regular conditional probabilities. If the index set (in which t takes values) were countable then we could construct a suitable conditional distribution $\mathbb{Q}_t$ for each t. However, in the theory of continuous time processes the index set is $[0, \infty)$. If suitable conditions are satisfied, then by making a specific choice of Ω (usually the canonical path space), it is possible to regularize the sequence of regular conditional distributions $\{\mathbb{Q}_t : t \in \mathbb{Q}_+\}$ to obtain a càdlàg $\mathcal{P}(\Omega)$-valued stochastic process, $(\mathbb{Q}_t)_{t \geq 0}$ which is called a *kernel* for the optional projection. Such a kernel is independent of the signal process X and depends only on the probability space $(\Omega, \mathcal{F})$ and the filtration $\mathcal{Y}_t$.

Performing the construction in this way (see Meyer [206] for details) is somewhat involved and imposes unnecessary conditions on Ω, which are irrelevant since we are only interested in the distribution of the signal process X_t. Thus we do not follow this approach and instead choose to construct π_t by piecing together optional projections. The existence and uniqueness theorem for optional projections requires that we work with a filtration which satisfies the usual conditions, since the proof makes use of Doob's martingale regularisation theorem. Therefore since we have do not assume right continuity of $\mathcal{Y}_t$, in the following theorem the right continuous enlargement $\mathcal{Y}_{t+}$ is used as this satisfies the usual conditions.

Lemma 2.23. *Assume that $\mathbb{S}$ is a compact metric space and $\mathcal{S} = \mathcal{B}(\mathbb{S})$ is the corresponding Borel σ-algebra. Then there exists a $\mathcal{P}(\mathbb{S})$-valued stochastic process π_t which satisfies the following conditions.*

1. π_t *is a $\mathcal{Y}_{t+}$-optional process.*

[†] A complete separable metric space is sometimes called a Polish space following Bourbaki in recognition of the work of Kuratowksi.

2. For any $f \in B(\mathbb{S})$, the process $\pi_t f$ is indistinguishable from the $\mathcal{Y}_{t+}$-optional projection of $f(X_t)$.

Proof. The proof of this lemma is based upon the proofs of Proposition 1 in Yor [279], Theorem 4.1 in Getoor [105] and Theorem 5.1.15 in Stroock [262].

Let $\{f_i\}_{i=1}^{\infty}$ be a set of continuous bounded functions $f_i : \mathbb{S} \to \mathbb{R}$ whose linear span is dense in $C_b(\mathbb{S})$. The compactness of $\mathbb{S}$ implies by Corollary A.10 that such a set must exist. Set $f_0 = \mathbf{1}$. We may choose such a set so that $\{f_0, \ldots, f_n\}$ is linearly independent for each n. Set $g_0 = \mathbf{1}$, and for $n \geq 1$ set the process g_n equal to a $\mathcal{Y}_{t+}$-optional projection of $f_n(X)$. The existence of such an optional projection is guaranteed by Theorem 2.7.

Let $\mathcal{U}$ be the (countable) vector space generated by finite linear combinations of these f_is with rational coefficients. If for some $N \in \mathbb{N}$, $f = \sum_{i=1}^{N} \alpha_i f_i$ with $\alpha_i \in \mathbb{Q}$ then define the process $\Lambda^\omega = \sum_{i=1}^{N} \alpha_i g_i$. By the linear independence property, it is clear that any such representation is unique and therefore this is well defined.

Define a subspace, $\mathcal{U}^+ \triangleq \{v \in \mathcal{U}, \; v \geq 0\}$. For $v \in \mathcal{U}^+$ define

$$\mathcal{N}(v) = \{\omega \in \Omega : \Lambda_t^\omega(v) < 0 \text{ for some } t \geq 0\}.$$

It is immediate from Lemma A.26 that for each $v \in \mathcal{U}_+$, the process $\Lambda^\omega(v)$ has non-negative paths a.s., thus $\mathcal{N}(v)$ is a $\mathbb{P}$-null set. Define

$$\mathcal{N} = \bigcup_{f \in \mathcal{U}^+} \mathcal{N}_f,$$

which is also a $\mathbb{P}$-null set since this is a countable union. By construction Λ^ω is linear; $\Lambda(\mathbf{1}) = \mathbf{1}$.

Define a modified version of the process Λ^ω which is a functional on $\mathcal{U} \subset C_b(\mathbb{S})$ and retains the properties of non-negativity and linearity for all $\omega \in \Omega$,

$$\bar{\Lambda}^\omega(f) = \begin{cases} \Lambda^\omega(f) & \omega \notin \mathcal{N}, \\ 0 & \omega \in \mathcal{N}. \end{cases}$$

It only remains to check that $\bar{\Lambda}^\omega$ is a bounded operator. Let $f \in \mathcal{U} \subset C_b(\mathbb{S})$; then trivially $|f| \leq \|f\|_\infty \mathbf{1}$, so it follows that $\|f\|_\infty \mathbf{1} \pm f \geq 0$ and hence for all $t \geq 0$ $\bar{\Lambda}_t^\omega(\|f\|_\infty \mathbf{1} \pm f) \geq 0$ by the non-negativity property. But by linearity since $\Lambda^\omega(\mathbf{1}) = \mathbf{1}$, it follows that for all $t \geq 0$, $\|f\|_\infty \mathbf{1} \pm \Lambda_t^\omega(f) \geq 0$, from which we deduce $\sup_{t \in [0,\infty)} \|\Lambda_t^\omega(f)\|_\infty < \|f\|_\infty$.

Since $\bar{\Lambda}^\omega$ is bounded, and $\mathcal{U}$ is dense in $C_b(\mathbb{S})$ we can extend[†] the definition of $\bar{\Lambda}^\omega(f)$ for f outside of $\mathcal{U}$ as follows. Let $f \in C_b(\mathbb{S})$, since $\mathcal{U}$ is dense in $C_b(\mathbb{S})$, we can find a sequence $f_k \in \mathcal{U}$ such that $f_k \to f$ pointwise. Define

---

[†] Functional analysts will realise that we can use the Hahn–Banach theorem to construct a norm preserving extension. Since this is a metric space we can use the constructive proof given here instead.

$$\bar{\Lambda}^\omega(f) \triangleq \lim \bar{\Lambda}^\omega(f_k)$$

which is clearly well defined since if $f'_k \in \mathcal{U}$ is another sequence such that $f'_k \to f$, then by the boundedness of $\bar{\Lambda}$ and using the triangle inequality

$$\sup_{t \in [0,\infty)} \|\bar{\Lambda}^\omega_t(f_k) - \bar{\Lambda}^\omega_t(f'_n)\|_\infty \leq \|f_k - f'_n\|_\infty \leq \|f_k - f\|_\infty + \|f - f'_n\|_\infty.$$

Since $\mathbb{S}$ is compact and the convergence $f_k \to f$ and $f'_n \to f$ is uniform on $\mathbb{S}$, then given $\varepsilon > 0$, there exists k_0 such that $k \geq k_0$ implies $\|f_k - f\|_\infty < \varepsilon/2$ and similarly n_0 such that $n \geq n_0$ implies $\|f'_n - f\|_\infty < \varepsilon/2$ whence it follows that the limit as $n \to \infty$ of $\bar{\Lambda}^\omega(f'_n)$ is $\bar{\Lambda}^\omega(f)$.

We must check that for $f \in C_b(\mathbb{S})$, that $\bar{\Lambda}^\omega_t(f)$ is the $\mathcal{Y}_{t+}$-optional projection of $f(X_t)$. By the foregoing it is $\mathcal{Y}_{t+}$-optional. Let T be a $\mathcal{Y}_{t+}$-stopping time

$$\begin{aligned}
\mathbb{E}[\Lambda_T(f)1_{T<\infty}] &= \lim_{k \to \infty} \mathbb{E}\left[\bar{\Lambda}_T(f_k)1_{T<\infty}\right] \\
&= \lim_{k \to \infty} \mathbb{E}\left[f_k(X_T)1_{T<\infty}\right] \\
&= \mathbb{E}[f(X_T)1_{T<\infty}],
\end{aligned}$$

where the second equality follows since $\bar{\Lambda}(f_n)$ is a $\mathcal{Y}_{t+}$-optional projection of $f_n(X)$ and the other two inequalities follow by the dominated convergence theorem.

By the Riesz representation theorem, which applies since $\mathbb{S}$ is compact,[†] we can find a kernel $\pi^\omega(\cdot)$ such that for $\omega \in \Omega$,

$$\bar{\Lambda}^\omega_t(f) = \int_\mathbb{S} f(x) \pi^\omega_t(\mathrm{d}x) = \pi^\omega_t f, \quad \text{for all } t \geq 0. \tag{2.16}$$

To establish the first and second parts of the theorem, we need to check that for $f \in B(\mathbb{S})$ that $(\pi^\omega f)_t$ is the $\mathcal{Y}_{t+}$-optional projection of $f(X_t)$. We do this via the monotone class framework (see Theorem A.1 in the appendix) since on a metric space the σ-algebra generated by $C_b(\mathbb{S})$ is $\mathcal{B}(\mathbb{S})$. It is clear that for $f \in C_b(\mathbb{S})$ from (2.16) and the preceding argument that $(\pi^\omega f)_t$ is the $\mathcal{Y}_{t+}$-optional projection of $f(X_t)$.

Let $\mathcal{H}$ be the subset of $B(\mathbb{S})$ for which $(\pi^\omega f)_t$ is the $\mathcal{Y}_{t+}$-optional projection of $f(X_t)$. Clearly $\mathbf{1} \in \mathcal{H}$, $\mathcal{H}$ is a vector space and $C_b(\mathbb{S}) \subseteq \mathcal{H}$. The monotone convergence theorem for integration implies that $\mathcal{H}$ is a monotone class. Therefore by the monotone class theorem $\mathcal{H}$ contains $B(\mathbb{S})$. □

Theorem 2.24. *Let $\mathbb{S}$ be a complete separable metric space. Then there exists a $\mathcal{P}(\mathbb{S})$-valued stochastic process π_t which satisfies the following conditions.*

1. π_t is a $\mathcal{Y}_{t+}$ optional process.

[†] Without the compactness property, we can not guarantee that the kernel be σ-additive.

2. For any $f \in B(\mathbb{S})$, the process $\pi_t f$ is indistinguishable from the $\mathcal{Y}_{t+}$ optional projection of $f(X_t)$.

Proof. Since $\mathbb{S}$ is a complete separable metric space, by Theorem A.7 of the appendix, it is homeomorphic to a Borel subset of a compact metric space $\hat{\mathbb{S}}$; we denote the homeomorphism by α.

Define a process $\hat{X}_t = \alpha(X_t)$ taking values in $\hat{\mathbb{S}}$. Since $\hat{\mathbb{S}}$ is a compact separable metric space, by Lemma 2.23 there exists a $\mathcal{P}(\hat{\mathbb{S}})$-valued stochastic process $\hat{\pi}$, such that for each $\hat{f} \in B(\hat{\mathbb{S}})$, $\hat{\pi}_t \hat{f}$ is a $\mathcal{Y}_{t+}$-optional projection of $\hat{f}(\hat{X}_t)$.

Since the process $\hat{X}$ takes values in $\alpha(\mathbb{S}) \subset \hat{\mathbb{S}}$, it is immediate that

$$^o 1_{\hat{\mathbb{S}} \setminus \alpha(\mathbb{S})}(\hat{X}_t) = {^o 1}_{\hat{\mathbb{S}} \setminus \alpha(\mathbb{S})}(\alpha(X_t)) = {^o 0} = 0.$$

As the optional projection is only defined up to indistinguishability, it follows that

$$\hat{\pi}_t^\omega \left(\hat{\mathbb{S}} \setminus \alpha(\mathbb{S})\right) = \hat{\pi}_t^\omega \left(1_{\hat{\mathbb{S}} \setminus \alpha(\mathbb{S})}\right) = {^o 1}_{\hat{\mathbb{S}} \setminus \alpha(\mathbb{S})}(\hat{X}_t) = 0 \; \forall t \in [0, \infty) \quad \mathbb{P}\text{-a.s.}$$

Define

$$\bar{\mathcal{N}} \triangleq \left\{\omega \in \Omega : \hat{\pi}_t^\omega \left(\hat{\mathbb{S}} \setminus \alpha(\mathbb{S})\right) = 0 \; \forall t \in [0, \infty)\right\}^c,$$

which we have just shown to be a $\mathbb{P}$-null set. We define a $\mathcal{P}(\mathbb{S})$-valued random process π as follows; let $A \in \mathcal{B}(\mathbb{S})$,

$$\pi_t^\omega(A) \triangleq \begin{cases} \hat{\pi}_t^\omega(\alpha(A)) & \omega \notin \bar{\mathcal{N}}, \\ \mathbb{P} Y_t^{-1}(A) & \omega \in \bar{\mathcal{N}}. \end{cases}$$

Here the choice of π on $\bar{\mathcal{N}}$ is arbitrary; we cannot choose 0, because π_t^ω must be a probability measure on $\mathbb{S}$ for all $\omega \in \Omega$. Thus it is immediate that $\pi_t^\omega \in \mathcal{P}(\mathbb{S})$.

If $f \in B(\mathbb{S})$ then we can extend f to a function $\hat{f}$ in $B(\hat{\mathbb{S}})$ by defining

$$\hat{f}(x) = \begin{cases} f(\alpha^{-1}(x)) & \text{if } x \in \alpha(\mathbb{S}) \\ 0 & \text{otherwise.} \end{cases}$$

Clearly

$$\pi(f) = \pi\left(\hat{f} \circ \alpha\right) = \hat{\pi}\left(\hat{f} 1_{\alpha(\mathbb{S})}\right) = \hat{\pi}(\hat{f}) \quad \mathbb{P}\text{-a.s.,}$$

but $\hat{\pi}_t \hat{f}$ is the $\mathcal{Y}_{t+}$-optional projection of $\hat{f}(\hat{X}_t) = f(X_t)$, hence as required $\pi_t(f)$ is the $\mathcal{Y}_{t+}$-optional projection of $f(X_t)$. □

Exercise 2.25. Let π_t be defined as above. Show that for any $f \in B(\mathcal{S})$, then for any $t \in [0, \infty)$,

$$\pi_t f = \mathbb{E}\left[f(X_t) \mid \mathcal{Y}_{t+}\right]$$

holds $\mathbb{P}$-a.s.

2.4 The Stochastic Process π

Corollary 2.26. *If the sample paths of X are càdlàg then there is a version of π_t with càdlàg paths (where $\mathcal{P}(\mathbb{S})$ is endowed with the topology of weak convergence) and a countable set $Q \subset [0, \infty)$, such that for $t \in [0, \infty) \setminus Q$, for any $f \in B(\mathbb{S})$,*

$$\pi_t f = \mathbb{E}[f(X_t) \mid \mathcal{Y}_t].$$

Proof. For any $f \in C_b(\mathbb{S})$, the $\mathcal{Y}_{t+}$-optional projection of $f(X_t)$ is indistinguishable from a càdlàg process by Theorem 2.9. Since by Theorem 2.24 $\pi_t f$ is indistinguishable from the $\mathcal{Y}_{t+}$-optional projection of $f(X_t)$, it follows that $\pi_t f$ is indistinguishable from a càdlàg process.

By Theorem 2.18, there is a countable convergence determining class $\{\varphi_i\}_{i \geq 0}$, which is therefore also a separating class. We can therefore choose a modification of π_t such that $\pi_t \varphi_i$ is càdlàg for all i. Therefore π_t is càdlàg.

Since $\mathcal{P}(\mathbb{S})$ with the weak topology is metrizable it then follows by Lemma A.14 that

$$I \triangleq \{t > 0 : \mathbb{P}(\pi_{t-} \neq \pi_t) > 0\}$$

is countable. But for $t \notin I$, $\pi_t = \pi_{t-}$ a.s. thus $\pi_t = \lim_{s \uparrow\uparrow t} \pi_s$ (where the notation $s \uparrow\uparrow t$ is defined in Section A.7.1). Clearly π_s for $s < t$ is $\mathcal{Y}_t$-measurable and therefore so is the limit π_t. As $\mathcal{Y}_t \subset \mathcal{Y}_{t+}$ it follows from the definition of Kolmogorov conditional expectation that

$$\pi_t f = \mathbb{E}[\pi_t f \mid \mathcal{Y}_t] = \mathbb{E}[f(X_t) \mid \mathcal{Y}_t].$$

□

Remark 2.27. The theorem as stated above only guarantees $\pi_t f$ is a $\mathcal{Y}_{t+}$-optional projection of $f(X_t)$ for f a bounded measurable function. Examining the proof shows that this restriction to bounded f is the usual one arising from the use of the monotone class theorem A.1.

It is useful to consider πf when f is not bounded. Consider f non-negative and define $f^n \triangleq f \wedge n$, which is bounded, so by the above theorem $\pi(f^n)$ is $\mathcal{Y}_{t+}$-optional. Clearly $f_n \to f$ in a monotone fashion as $n \to \infty$, and since $\pi(f^n)$ is the expectation of f^n under the measure π_t, by the monotone convergence theorem $\pi(f^n) \to \pi(f)$. Since $\pi(f)$ is the limit of a sequence of $\mathcal{Y}_{t+}$-optional processes, it is $\mathcal{Y}_{t+}$-optional. By application of the monotone convergence theorem to the defining equation of optional projection (2.8), it follows that $\pi(f)$ is a $\mathcal{Y}_{t+}$-optional projection of $f(X_t)$.

In the general case where f is unbounded, but not necessarily non-negative, if $\pi_t |f| < \infty$ for all $t \in [0, \infty)$, $\mathbb{P}$-a.s., then writing $f_+ = f \wedge 0$ and $f_- = (-f) \wedge 0$, it follows that $|f| = f_+ + f_-$ and hence $\pi_t f_+ < \infty$ and $\pi_t f_- < \infty$ for all $t \in [0, \infty)$ $\mathbb{P}$-a.s. Thus $\pi_t f = \pi_t f_+ - \pi_t f_-$ is well defined (i.e. it cannot be $\infty - \infty$) and a similar argument verifies that it satisfies the conditions for the $\mathcal{Y}_{t+}$-optional projection of $f(X_t)$.

The pathwise regularity of $\pi_t f$ (i.e. showing that the trajectories of πf are càdlàg if X is càdlàg) requires stronger conditions in the unbounded case irrespective of whether f is non-negative. In particular we need to be able to exchange a limit and an expectation; a suitable condition for this to be valid is that the family of random variables $\{f(X_t) : t \in [0,\infty)\}$ is uniformly integrable. For example, this is true if this family is dominated by an integrable random variable, in other words if $\sup_{s \in [0,\infty)} |f(X_s)|$ is integrable.

2.4.1 Regular Conditional Probabilities

This section is not essential reading in order to understand the subsequent chapters. It describes the construction of a regular conditional probability. The ideas involved are important in many areas of probability theory and most of the work in establishing them has been done in the previous section, hence their inclusion here.

For many purposes a stronger notion of conditional probability is required than that provided by Kolmogorov conditional expectation (see Appendix A.2). The most useful form is that of *regular conditional probability*.

Definition 2.28. *Let $(\Omega, \mathcal{F}, \mathbb{P})$ be a probability space and $\mathcal{G}$ a sub-σ-algebra of $\mathcal{F}$. A function $\mathbb{Q}(\omega, B)$ defined for all $\omega \in \Omega$ and $B \in \mathcal{E}$ is called a regular conditional probability of $\mathbb{P}$ with respect to $\mathcal{G}$ if*

(a) For each $B \in \mathcal{F}$,
$$\mathbb{Q}(\omega, B) = \mathbb{E}[I_B \mid \mathcal{G}] \quad \mathbb{P}\text{-a.s.}$$

(b) For each $\omega \in \Omega$, $\mathbb{Q}(\omega, \cdot)$ is a probability measure on $(\Omega, \mathcal{F})$.
(c) For each $B \in \mathcal{E}$, the map $\mathbb{Q}(\cdot, B)$ is $\mathcal{G}$-measurable.
(d) If the σ-algebra $\mathcal{G}$ is countably generated then for all $G \in \mathcal{G}$,
$$\mathbb{Q}(\omega, G) = 1_G(\omega) \quad \mathbb{P}\text{-a.s.}$$

Regular conditional probabilities as described in Definition 2.28 do not always exist. For an example of non-existence of regular conditional probabilities due to Halmos, Dieudonné, Andersen and Jessen see Rogers and Williams [248, Section II.43].

Exercise 2.29. Prove by similar methods to those used in the proof of Theorem 2.24 that if Ω is a compact metric space then there exists a regular conditional probability distribution with respect to the σ-algebra $\mathcal{G} \subset \mathcal{F}$. Furthermore, show that in the case where $\mathcal{G}$ is finitely generated that if $A^{\mathcal{G}}(\omega)$ is the atom of $\mathcal{G}$ containing ω (i.e. $\bigcap\{G \in \mathcal{G} : \omega \in G\}$) then $\mathbb{Q}(\omega, A(\omega)) = 1$. This argument can be extended to complete separable metric spaces using Theorem A.7 using an argument similar to that in the proof of Theorem 2.24.

2.5 Right Continuity of Observation Filtration

The results in this section are proved under more restrictive conditions than those of Section 2.1. The observation process Y is assumed to satisfy the evolution equation (2.2). That is,

$$Y_t = Y_0 + \int_0^t h(X_s)\,ds + W_t, \quad t \geq 0,$$

where $W = \{W_t, t \geq 0\}$ is a standard $\mathcal{F}_t$-adapted m-dimensional Brownian motion and $h = (h_i)_{i=1}^m : \mathbb{S} \to \mathbb{R}^m$ is a measurable function. Assume that conditions (2.3) and (2.4) are satisfied; that is

$$\mathbb{E}\left[\int_0^t \|h(X_s)\|\,ds\right] < \infty,$$

and

$$\mathbb{P}\left(\int_0^t \|\pi_s(h)\|^2\,ds < \infty\right) = 1.$$

Let $I = \{I_t, t \geq 0\}$ be the following process, called the *innovation* process,

$$\begin{aligned} I_t &= Y_t - \int_0^t \pi_s(h)\,ds \\ &= W_t + \int_0^t h(X_s)\,ds - \int_0^t \pi_s(h)\,ds, \quad t \geq 0. \end{aligned} \quad (2.17)$$

For this innovation process to be well defined it is necessary that

$$\int_0^t \pi_s(\|h\|)\,ds < \infty \quad \mathbb{P}\text{-a.s.}, \quad (2.18)$$

which is clearly implied by the stronger condition (2.3). The condition (2.18) is not strong enough for the proof of the following theorem; consequently only condition (2.3) is referenced subsequently.

Proposition 2.30. *If condition (2.3) is satisfied then I_t is a $\mathcal{Y}_t$-adapted Brownian motion under the measure $\mathbb{P}$.*

Proof. Obviously I_t is $\mathcal{Y}_t$-adapted as both Y_t and $\int_0^t \pi_s(h)\,ds$ are. First it is shown that I_t is a continuous martingale. As a consequence of (2.3) I_t is integrable, hence taking conditional expectation

$$\mathbb{E}\left[I_t \mid \mathcal{Y}_s\right] - I_s = \mathbb{E}\left[W_t + \int_0^t h(X_r)\,\mathrm{d}r \,\Big|\, \mathcal{Y}_s\right] - \left(W_s + \int_0^s h(X_r)\,\mathrm{d}r\right)$$
$$- \mathbb{E}\left[\int_0^t \pi_r(h)\,\mathrm{d}r \,\Big|\, \mathcal{Y}_s\right] + \int_0^s \pi_r(h)\,\mathrm{d}r$$
$$= \mathbb{E}\left[Y_s + W_t - W_s + \int_s^t h(X_r)\,\mathrm{d}r \,\Big|\, \mathcal{Y}_s\right] - Y_s$$
$$- \mathbb{E}\left[\int_0^t \pi_r(h)\,\mathrm{d}r \,\Big|\, \mathcal{Y}_s\right] + \int_0^s \pi_r(h)\,\mathrm{d}r.$$

Since Y_t and $\int_0^t \pi_r\,\mathrm{d}r$ are $\mathcal{Y}_t$-measurable,

$$\mathbb{E}\left[I_t \mid \mathcal{Y}_s\right] - I_s = \mathbb{E}\left[W_t - W_s \mid \mathcal{Y}_s\right] + \int_s^t \mathbb{E}\left[h(X_r) - \pi_r(h) \mid \mathcal{Y}_s\right]\mathrm{d}r = 0,$$

where we have used the fact that for $r \geq s$, $\mathbb{E}\left[\pi_r(h) \mid \mathcal{Y}_s\right] = \mathbb{E}\left[h(X_r) \mid \mathcal{Y}_s\right]$ and $\mathbb{E}\left[W_t - W_s \mid \mathcal{Y}_s\right] = \mathbb{E}\left[\mathbb{E}\left[W_t - W_s \mid \mathcal{F}_s\right] \mid \mathcal{Y}_s\right] = 0$. The cross-variation of I is the same as the cross-variation of W as the other two terms in (2.17) give zero cross-variation. So I is a continuous martingale and its cross-variation is given by

$$\langle I^i, I^j \rangle_t = \langle W^i, W^j \rangle_t = t\delta_{ij}. \qquad (2.19)$$

Hence I is a Brownian motion by Lévy's characterisation of a Brownian motion (Theorem B.27). $\square$

From the first part of (2.17), for small δ,

$$Y_{t+\delta} - Y_t \simeq \pi_s(h)\delta + I_{t+\delta} - I_t.$$

Heuristically the incoming observation $Y_{t+\delta} - Y_t$ has a part which could be predicted from the knowledge of the system state $\pi_s(h)\delta$ and an additional component $I_{t+\delta} - I_t$, containing new information which is independent of the current knowledge. This is why I is called the *innovation process*.

Proposition 2.31 (Fujisaki, Kallianpur and Kunita [104]). *Assume the conditions (2.3) and (2.4) are satisfied. Then every square integrable random variable η which is $\mathcal{Y}_\infty$-measurable has a representation of the form*

$$\eta = \mathbb{E}[\eta] + \int_0^\infty \nu_s^\top \,\mathrm{d}I_s, \qquad (2.20)$$

where $\nu = \{\nu_t,\ t \geq 0\}$ is a progressively measurable $\mathcal{Y}_t$-adapted process such that

$$\mathbb{E}\left[\int_0^\infty \|\nu_s\|^2\,\mathrm{d}s\right] < \infty.$$

This theorem is often proved under the stronger condition

$$\mathbb{E}\left[\int_0^t \|h(X_s)\|^2 \, \mathrm{d}s\right] < \infty, \quad \forall t \geq 0,$$

which implies both conditions (2.4) and (2.3). The innovation process I_t is clearly $\mathcal{Y}_t$-adapted. If the converse result, that $\mathcal{Y}_t = \sigma(I_s : 0 \leq s \leq t) \vee \mathcal{N}$, were known[†] then this proposition would be a trivial application of the martingale representation theorem B.32. However, the representation provided by the proposition is the closest to a satisfactory converse which is known to hold.[‡]

The main element of the proof of Proposition 2.31 is an application of Girsanov's theorem followed by use of the martingale representation theorem. In Section 2.1 it was necessary to augment the filtration $\mathcal{Y}_{t+}$ with the null sets in order to construct the process π. This will cause some difficulties, because the process to be used as a change of measure is not necessarily a martingale. In order to construct a uniformly integrable martingale, a stopping argument must be used and this cannot be done directly working with the augmented filtration. This has the unfortunate effect of obscuring a simple and elegant proof. The proof for a simpler case, where the process is a martingale is discussed in Exercise 2.33 and the reader who is uninterested in measurability aspects would be well advised to consult the solution to this exercise instead of reading the proof. To be clear in notation, we denote by $\mathcal{Y}_t^o$ the unaugmented σ-algebra (i.e. without the addition of the null sets) corresponding to $\mathcal{Y}_t$.

The following technical lemma, whose conclusion might well seem to be 'obvious' is required. The proof of the lemma is not important for understanding the proof of the representation result, therefore it can be found in the appendix proved as Lemma A.24.

Lemma 2.32. *Let $\mathcal{X}_t^o$ be the unaugmented σ-algebra generated by a process X_t. Then for T a $\mathcal{X}^o$-stopping time, for all $t \geq 0$,*

$$\mathcal{X}_{t \wedge T}^o = \sigma\{X_{s \wedge T} : 0 \leq s \leq t\}.$$

[†] An example of Tsirel'son which is presented in a filtering context in Beneš [11] demonstrates that in general $\mathcal{Y}_t$ is not equal to $\sigma(I_s : 0 \leq s \leq t) \vee \mathcal{N}$.

[‡] In special cases the observation and innovation filtrations can be shown to be equal. Allinger and Mitter [3] extend an earlier result of Clark [55] (see also Theorem 11.4.1 in Kallianpur [145] and Meyer [205] pp 244–246) to show that if the observation and signal noise are uncorrelated and for some T, $\mathbb{E}[\int_0^T \|h(X_s)\|^2 \, \mathrm{d}s] < \infty$, then for $t \leq T$, $\sigma(I_s : 0 \leq s \leq t) \vee \mathcal{N} = \mathcal{Y}_t$. Their proof consists of an analysis of the Kallianpur–Striebel functional which leads to a pathwise uniqueness result for weak solutions of the equation $I_t = Y_t - \int_0^t \pi_s(h) \, \mathrm{d}s$. That is, if two valid weak solutions (Y, I) and $(\tilde{Y}, \tilde{I})$ of this equation have a common Brownian motion I (but not necessarily a common filtration) then Y and $\tilde{Y}$ are indistinguishable. From a result of Yamada and Watanabe (Remark 2, Corollary 1 of [275]; see also Chapter 8 of Stroock and Varadhan [261]) this establishes the result.

We are now in a position to prove the representation result, Proposition 2.31.

Proof. Since the integral in (2.4) is non-decreasing in t, this condition implies that
$$\mathbb{P}\left(\int_0^t \|\pi_r(h)\|^2 \, \mathrm{d}r < \infty, \ \forall t \in [0, \infty)\right) = 1. \tag{2.21}$$

Define
$$\bar{Z}_t \triangleq \exp\left(-\int_0^t \pi_r(h^\top) \, \mathrm{d}I_r - \frac{1}{2}\int_0^t \|\pi_r(h)\|^2 \, \mathrm{d}r\right), \tag{2.22}$$

and for $n > 0$ define
$$\bar{T}^n \triangleq \inf\left\{t \geq 0 : \left|\int_0^t \|\pi_r(h)\|^2 \, \mathrm{d}r\right| \geq n \text{ or } |\bar{Z}_t| \geq n\right\}, \tag{2.23}$$

which by Lemma A.19 is a $\mathcal{Y}_t$-stopping time, since the processes $t \mapsto \int_0^t \|\pi_r(h)\|^2 \, \mathrm{d}r$ and $\bar{Z}$ are both continuous and $\mathcal{Y}_t$-adapted. By Lemma A.21 the $\mathcal{Y}_t$-stopping time $\bar{T}^n$ is a.s. equal to a $\mathcal{Y}^o_{t+}$-stopping time. However, this is not strong enough; a $\mathcal{Y}^o_t$-stopping time is required.

The process $\pi_t(h)$ gives rise to a sequence of càdlàg step function approximations
$$\pi^n(h)(\omega) \triangleq \sum_{i=0}^{\infty} 1_{[2^{-n}i, 2^{-n}(i+1))}(t) \pi_{2^{-n}i}(h)(\omega).$$

Each $\pi_{2^{-n}i}(h)$ is a $\mathcal{Y}_{2^{-n}i}$-measurable random variable. From the definition of augmentation, by modification on a $\mathbb{P}$-null set a $\mathcal{Y}^o_{2^{-n}i}$-measurable random variable P^n_i can be defined such that $\pi_{2^{-n}i}(h) = P^n_i$ holds $\mathbb{P}$-a.s. Then define
$$\bar{\pi}^n(h)(\omega) \triangleq \sum_{i=0}^{\infty} 1_{[2^{-n}i, 2^{-n}(i+1))}(t) P^n_i(\omega),$$

and as a countable family of random variables has been modified on null sets, it follows that the processes $\bar{\pi}^n(h)$ and $\pi^n(h)$ are indistinguishable. The process $\bar{\pi}^n(h)$ is càdlàg and $\mathcal{Y}^o_t$-adapted, therefore it must be $\mathcal{Y}^o_t$-optional.

As the process $\pi(h)$ has by Lemma A.13 at most a countable number of discontinuities, it follows that $\pi^n(h)$ converges λ-a.s. to $\pi(h)$. Therefore the sequence $\bar{\pi}^n(h)$ converges $\lambda \otimes \mathbb{P}$-a.s. to $\pi(h)$. The limit $\bar{\pi}(h) \triangleq \liminf_{n \to \infty} \bar{\pi}^n(h)$ is a limit of $\mathcal{Y}^o_t$-optional processes and is therefore $\mathcal{Y}^o$-optional. Using this $\bar{\pi}$ process in place of $\pi(h)$ in the definition of $\bar{Z}$ we may define $\hat{Z}$. This process $\hat{Z}$ as constructed need not be continuous as it can explode from a finite value to infinity, because (2.21) only holds outside a null set. This process cannot simply be modified on a null set, as this might destroy the property of $\mathcal{Y}^o_t$-adaptedness. Instead define Z to be a modification of $\hat{Z}$, which is zero on the set
$$\left\{\omega \in \Omega : \int_0^r \|\bar{\pi}_s(h)\|^2 \, \mathrm{d}s = \infty \text{ for } r < t, r \in \mathbb{Q}\right\}.$$

This set is clearly $\mathcal{Y}_t^o$-measurable, hence this modified process Z is $\mathcal{Y}_t^o$-adapted and continuous.

As the processes Z and $\int_0^\cdot \|\bar{\pi}_s(h)\|^2 \, ds$ are continuous and $\mathcal{Y}_t^o$-adapted by Lemma A.19 $\inf\{t \geq 0 : Z_t \geq n\}$ and $\inf\{t \geq 0 : \int_0^t \|\bar{\pi}_s(h)\|^2 \, ds\}$ are both $\mathcal{Y}_t^o$-stopping times. The process $\bar{\pi}(h)$ is indistinguishable from $\pi(h)$, therefore define a second sequence of stopping times

$$T^n \triangleq \inf\left\{t \geq 0 : \left|\int_0^t \|\bar{\pi}_r(h)\|^2 \, dr\right| \geq n \text{ or } |Z_t| \geq n\right\} \quad (2.24)$$

and it follows that T_n is be a.s. equal to $\bar{T}_n$ and $Z_n \triangleq Z_{T^n}$ is $\mathbb{P}$-a.s. equal to $Z_{\bar{T}_n}$.

Clearly Z is a local martingale; but in general it is not a martingale. The next argument shows that by stopping at T^n, the process Z^{T^n} is a uniformly integrable martingale and therefore suitable for use as a measure change.

From (2.22), using Itô's formula

$$Z_t^{T_n} = Z_0 - \int_0^{t \wedge T_n} Z_s^{T_n} \pi_s(h^\top) \, dI_s,$$

and since by Proposition 2.30, I is a $\mathbb{P}$-Brownian motion adapted to $\mathcal{Y}_t$; it follows that the stochastic integral is a $\mathcal{Y}_t$-adapted martingale provided that

$$\mathbb{E}\left[\int_0^{t \wedge T_n} \|\pi_s(h)\|^2 \left(Z_s^{T_n}\right)^2 \, ds\right] < \infty, \quad \text{for all } t \geq 0.$$

It is clear that

$$\int_0^{t \wedge T_n} \|\pi_s(h)\|^2 \left(Z_s^{T_n}\right)^2 \, ds \leq n^2 \int_0^{t \wedge T_n} \|\pi_s(h)\|^2 \, ds \leq n^4 < \infty. \quad (2.25)$$

It follows that Z^{T^n} is a martingale which by (2.25) is uniformly bounded in L^2 and hence uniformly integrable. Define a change of measure by

$$\frac{d\tilde{\mathbb{P}}^n}{d\mathbb{P}} = Z_{T_n}.$$

As $\mathbb{P}$ and $\tilde{\mathbb{P}}^n$ are by construction equivalent probability measures, it follows that statements which hold $\mathbb{P}$-a.s. also hold $\tilde{\mathbb{P}}^n$-a.s. As a consequence of Girsanov's theorem (see Theorem B.28 of the appendix), since Z^{T_n} is a uniformly integrable martingale, under the measure $\tilde{\mathbb{P}}^n$, the process

$$Y_t^n \triangleq I_t + \int_0^{T^n \wedge t} \pi_r(h) \, dr,$$

is a Brownian motion with respect to the filtration $\mathcal{Y}_t$.

We are forced to use this Brownian motion Y^n in place of Y when applying the martingale representation theorem. Were Z itself a uniformly integrable martingale we could use this directly to construct a representation of $Z_\infty^{-1}\eta$ which is $\mathcal{Y}_\infty$-measurable and square integrable as an integral over Y. Using Y^n instead of Y as our Brownian motion is not itself a problem. However, the martingale representation theorem only allows representations to be constructed of random variables which are measurable with respect to the augmentation of the filtration generated by the Brownian motion. In this case this means measurable with respect to the augmentation of the filtration

$$\mathcal{Y}_t^{n,o} \triangleq \sigma\{Y_s^n : 0 \leq s \leq t\}.$$

Clearly this filtration $\mathcal{Y}_t^{n,o}$ is not the same as $\mathcal{Y}_t^o$.

From the definition of the innovation process

$$Y_t = I_t + \int_0^t \pi_r(h)\,\mathrm{d}r.$$

Thus Y^n and Y agree on the time interval $[0, T^n]$. It must now be shown that the σ-algebras generated by these processes stopped at T^n agree. From Lemma A.24 it follows that

$$\mathcal{Y}_{t \wedge T^n}^{n,o} = \sigma\{Y_{s \wedge T^n}^n : 0 \leq s \leq t\} = \sigma\{Y_{s \wedge T^n} : 0 \leq s \leq t\} = \mathcal{Y}_{t \wedge T^n}^o,$$

where the second equality follows from the fact that Y^n and Y agree on the interval $[0, T_n]$.

Suppose that η is an element of $L^2(\mathcal{Y}_{T_n}^o, \mathbb{P})$, that is η is $\mathcal{Y}_{T_n}^o$-measurable, and $\mathbb{E}[\eta^2] < \infty$. As the process Z_t is continuous, it is progressively measurable therefore Z_n is $\mathcal{Y}_{T_n}^o$-measurable. Thus $Z_n^{-1}\eta$ is also $\mathcal{Y}_{T^n}^o$-measurable. One of the conditions defining the stopping time T_n ensures that $|(Z_n)^{-1}| < \exp(2n)$, thus

$$\mathbb{E}_{\tilde{\mathbb{P}}^n}(Z_n^{-1}\eta)^2 = \mathbb{E}[Z_n Z_n^{-2}\eta^2] \leq \exp(2n)\mathbb{E}[\eta^2] < \infty,$$

and hence $Z_n^{-1}\eta$ is an element of $L^2(\mathcal{Y}_{T_n}^o, \tilde{\mathbb{P}}^n)$.

We can now apply the classical martingale representation theorem B.32 (together with Remark B.33) to construct a representation with respect to the Brownian motion Y^n of $\tilde{\eta}$ to establish the existence of a previsible process Φ^n adapted to the filtration $\mathcal{Y}_t^n$ (the representation theorem requires the use of the augmented filtration) such that

$$Z_n^{-1}\eta = \mathbb{E}_{\tilde{\mathbb{P}}^n}(Z_n^{-1}\eta) + \int_0^\infty (\Phi_s^n)^\top \,\mathrm{d}Y_s^n$$
$$= \mathbb{E}[\eta] + \int_0^\infty (\Phi_s^n)^\top \,\mathrm{d}I_s + \int_0^\infty (\Phi_s^n)^\top \pi_s(h)\,\mathrm{d}s.$$

As Φ_s^n is $\mathcal{Y}_s^n$-adapted, it follows that for $s > T^n$, $\Phi_s^n = 0$ and since Y and Y^n agree on $[0, T^n]$ it follows that Φ_s^n is adapted to $\mathcal{Y}_s$. We now construct a $\tilde{\mathbb{P}}^n$ martingale from ηZ_n^{-1} via

2.5 Right Continuity of Observation Filtration

$$\tilde{\eta}_t = \mathbb{E}_{\tilde{\mathbb{P}}^n}\left[\eta Z_n^{-1} \mid \mathcal{Y}_t^n\right].$$

Applying Itô's formula to the product $\tilde{\eta}_t Z_t^{T^n}$,

$$\mathrm{d}(\tilde{\eta}_t Z_{T_n \wedge t}) = 1_{t \leq T^n}\Big(-Z_t \tilde{\eta}_t \pi_t(h^\top)\,\mathrm{d}I_t + Z_t \left(\Phi_t^n\right)^\top \mathrm{d}I_t$$
$$+ Z_t \left(\Phi_t^n\right)^\top \pi_t(h)\,\mathrm{d}t - Z_t \left(\Phi_t^n\right)^\top \pi_t(h)\,\mathrm{d}t\Big).$$

The finite variation terms in this expression cancel and thus integrating from 0 to t,

$$\tilde{\eta}_t Z_t^{T_n} = \mathbb{E}[\eta] + \int_0^{t \wedge T^n} \left(Z_s \left(\Phi_s^n\right)^\top - \tilde{\eta}_s Z_s \pi_s(h^\top)\right) \mathrm{d}I_s.$$

Writing $\nu_t^n \triangleq Z_t \Phi_t^n - \tilde{\eta}_t Z_t \pi_t(h)$,

$$\tilde{\eta}_t Z_t^{T_n} = \mathbb{E}[\eta] + \int_0^{t \wedge T_n} \nu_s^\top \,\mathrm{d}I_s,$$

taking the limit as $t \to \infty$, yields a representation

$$Z_n^{-1} \eta Z_{T_n} = \mathbb{E}[\eta] + \int_0^{T_n} \nu_t^\top \,\mathrm{d}I_t.$$

By choice of Z_n, the left-hand side is a.s. equal to η and since Φ^n and Z are $\mathcal{Y}_t$-adapted, it follows that ν^n is also $\mathcal{Y}_t$-adapted. The fact that Φ^n is previsible implies that it is progressively measurable and hence since $\pi(h)$ is progressively measurable the progressive measurability of ν^n follows and we have established that for $\eta \in L^2(\mathcal{Y}_{T^n}^o, \mathbb{P})$ there is a representation

$$\eta = \mathbb{E}[\eta] + \int_0^{T_n} \nu_t^\top \,\mathrm{d}I_t, \qquad \mathbb{P}\text{-a.s.}, \tag{2.26}$$

where ν is progressively measurable.

Taking expectation of the square of (2.26) it follows that

$$\mathbb{E}[\eta - \mathbb{E}[\eta]]^2 = \mathbb{E}\left[\int_0^{T^n} (\nu_s^n)^2 \,\mathrm{d}s\right].$$

Since η is a priori a square integrable random variable, the left-hand side is finite and hence

$$\mathbb{E}\left[\int_0^{T^n} (\nu_s^n)^2 \,\mathrm{d}s\right] < \infty.$$

The representation of the form (2.26) must be unique, thus it follows that there exists ν_t such that $\nu_t = \nu_t^n$ on $t \leq T^n$ for all $n \in \mathbb{N}$.

To complete the proof let $\mathcal{H}$ be the set of all elements of $L^2(\mathcal{Y}_\infty, \mathbb{P})$ which have a representation of the form (2.20). By the foregoing argument, for any n, $L^2(\mathcal{Y}^o_{T^n}, \mathbb{P}) \subseteq \mathcal{H}$. Clearly the set $\mathcal{H}$ is closed and since

$$\bigcup_{n \in \mathbb{N}} L^2(\mathcal{Y}^o_{T^n}; \mathbb{P})$$

is dense in $L^2(\mathcal{Y}_\infty; \mathbb{P})$, hence $\mathcal{H} = L^2(\mathcal{Y}_\infty, \mathbb{P})$. □

Exercise 2.33. To ensure you understand the above proof, simplify the proof of Proposition 2.31 in the case where for ω not in some null set

$$\int_0^t \|\pi_r(h)\|^2 \, \mathrm{d}r < K(t) < \infty, \tag{2.27}$$

where $K(t)$ is independent of ω, a condition which is satisfied if h is bounded. In this case the condition (2.3) holds trivially.

Proposition 2.31 offers an easy route to showing that the filtration $\mathcal{Y}_t$ is right continuous.

Lemma 2.34. *Let $M = \{M_t, t \geq 0\}$ be a right continuous $\mathcal{Y}_{t+}$-adapted martingale that is bounded in $L^2(\Omega)$; that is M satisfies $\sup_{t \geq 0} \mathbb{E}[M_t^2] < \infty$. Then M is $\mathcal{Y}_t$-adapted and continuous.*

Proof. By the martingale convergence theorem (Theorem B.1) $M_t = \mathbb{E}[M_\infty \mid \mathcal{Y}_{t+}]$, and by Proposition 2.31

$$M_\infty = \mathbb{E}[M_\infty] + \int_0^\infty \nu_s^\top \, \mathrm{d}I_s,$$

so using the fact that I_t is $\mathcal{Y}_t$-adapted

$$M_t = \mathbb{E}[M_\infty] + \mathbb{E}\left[\int_0^\infty \nu_s^\top \, \mathrm{d}I_s \,\bigg|\, \mathcal{Y}_{t+}\right]$$
$$= \mathbb{E}[M_\infty] + \int_0^t \nu_s^\top \, \mathrm{d}I_s,$$

from which it follows both that M_t is $\mathcal{Y}_t$-measurable and that M is continuous. □

Theorem 2.35. *The observation σ-algebra is right continuous that is $\mathcal{Y}_{t+} = \mathcal{Y}_t$.*

Proof. For a given $t \geq 0$ let $A \in \mathcal{Y}_{t+}$. Then the process $M = \{M_s, s \geq 0\}$ defined by

$$M_s \triangleq \begin{cases} 1_A - \mathbb{E}[I_A \mid \mathcal{Y}_t] & \text{for } s \geq t \\ 0 & \text{for } s < t \end{cases}$$

is a $\mathcal{Y}_{s+}$-adapted right continuous martingale bounded in $L^2(\Omega)$. Hence, by Lemma 2.34, M is also a continuous $\mathcal{Y}_s$-adapted martingale. In particular $1_A - \mathbb{E}[I_A \mid \mathcal{Y}_t]$ is $\mathcal{Y}_t$-measurable so $A \in \mathcal{Y}_t$. Hence $\mathcal{Y}_{t+} \subseteq \mathcal{Y}_t$ and the conclusion follows since t was arbitrarily chosen. $\square$

Exercise 2.36. Let $\pi = \{\pi_t, \ t \geq 0\}$ be the $\mathcal{Y}_t$-adapted process defined in Theorem 2.24. Prove that for any $t \geq 0$, π_t has a $\sigma(Y_s, 0 \leq s \leq t)$-measurable modification.

2.6 Solutions to Exercises

2.4

i. Let $\mathcal{H}_t$ be set on the right-hand side of (2.7). Since, for any $G \in \mathcal{Y}_t^o$ and $N_1, N_2 \in \mathcal{N}$, $(G \backslash N_1) \cup N_2 \in \mathcal{Y}_t$ it follows that $\mathcal{H}_t \subseteq \mathcal{Y}_t$. Since $\mathcal{Y}_t^o$ and $\mathcal{N}$ are subsets of $\mathcal{H}_t$ and $\mathcal{H}$ is a σ-algebra $\mathcal{Y}_t = \mathcal{Y}_t^o \vee \mathcal{N} \subseteq \mathcal{H}_t$.
ii. From Part (i) the result is true for ξ, the indicator function of an arbitrary set in $\mathcal{Y}_t$. By linearity the result holds for simple random variables, that is, for linear combinations of indicator functions of sets in $\mathcal{Y}_t$. Finally let ξ be an arbitrary $\mathcal{Y}_t$-measurable function. Then there exists a sequence $(\xi_n)_{n>1}$ of simple random variables such that $\lim_{n \to \infty} \xi_n(\omega) = \xi(\omega)$ for any $\omega \in \Omega$. Let $(\eta_n)_{n \geq 1}$ be the corresponding sequence of $\mathcal{Y}_t^o$-measurable simple random variables such that, for any $n \geq 1$, $\xi_n(\omega) = \eta_n(\omega)$ for any $\omega \in \Omega \backslash N_n$ where $N_n \in \mathcal{N}$. Define $\eta = \limsup_{n \to \infty} \eta_n$. Hence η is $\mathcal{Y}_t^o$-measurable and $\xi(\omega) = \eta(\omega)$ for any $\omega \in \Omega \backslash (\cup_{n \geq 1} N_n)$ which establishes the result.

2.20 The rational numbers $\mathbb{Q}$ are a dense subset of $\mathbb{R}$. We show that the set $\mathcal{G} \subset \mathcal{P}(\mathbb{R})$ of measures $\sum_{k=1}^n \alpha_k \delta_{x_k}$, for $\alpha_k \in \mathbb{Q}^+$, and $x_k \in \mathbb{Q}$ for all k with $\sum_{k=1}^n \alpha_k = 1$, is dense in $\mathcal{P}(\mathbb{R})$ with the weak topology. Given $\mu \in \mathcal{P}(\mathbb{R})$ we must find a sequence $\mu_n \in \mathcal{G}$ such that $\mu_n \Rightarrow \mu$.

It is sufficient to show that we can find an approximating sequence μ_n in the space $\mathcal{H}$ of measures of the form $\sum_{i=1}^\infty \alpha_i \delta_{x_i}$ where $\alpha_i \in \mathbb{R}^+$, $x_i \in \mathbb{Q}$ and $\sum_{i=1}^\infty \alpha_i = 1$. It is clear that each such measure in $\mathcal{H}$ is the weak limit of a sequence of measures in $\mathcal{G}$.

We can cover $\mathbb{R}$ by the countable collection of disjoint sets of the form $[k/n, (k+1)/n)$ for $k \in \mathbb{Z}$. Define

$$\mu_n \triangleq \sum_{k=-\infty}^{\infty} \mu([k/n, (k+1)/n)) \delta_{k/n};$$

then $\mu_n \in \mathcal{H}$. Let $g \in C_b(\mathbb{R})$ be a Lipschitz continuous function. Define

$$a_k^n \triangleq \inf_{x \in [k/n, (k+1)/n)} g(x), \quad b_i^n \triangleq \sup_{x \in [k/n, (k+1)/n)} g(x).$$

As $n \to \infty$, since g is uniformly continuous it is clear that $\sup_i |a_i^n - b_i^n| \to 0$. Thus as

$$\mu_n g = \sum_{k=-\infty}^{\infty} g(k/n) \mu([k/n, (k+1)/n)),$$

and

$$\sum_{k=-\infty}^{\infty} a_k^n \mu([k/n, (k+1)/n)) \le \mu g \le \sum_{k=-\infty}^{\infty} b_k^n \mu([k/n, (k+1)/n)),$$

it follows that

$$|\mu_n g - \mu g| \le \sum_{k=-\infty}^{\infty} |b_k^n - a_k^n| \to 0.$$

As this holds for all uniformly continuous g, we have established (2) of Theorem 2.17 and thus $\mu_n \Rightarrow \mu$.

For the second part, define $\mu_n \triangleq \delta_n$ for $n \in \mathbb{N}$. This sequence does not converge weakly to any element of $\mathcal{P}(\mathbb{R})$ but the sequence is Cauchy in d, hence the space $(\mathcal{P}(\mathbb{R}), d)$ is not complete.

2.21 Suppose that $\zeta \varphi_i$ is $\mathcal{F}/\mathcal{B}(\mathbb{R})$-measurable for all i. To show that ζ is $\mathcal{F}/\mathcal{B}(\mathcal{P}(\mathbb{S}))$-measurable, it is sufficient to show that for all elements A of the neighbourhood basis of μ, the set $\zeta^{-1}(A) \in \mathcal{F}$. But the sets of the neighbourhood basis have the form given by (2.9). We show that the weak topology is also generated by the local neighbourhoods of μ of the form

$$B = \bigcap_{i=1}^{m} \{\nu \in \mathcal{P}(\mathbb{S}) : |\nu \varphi_{j_i} - \mu \varphi_{j_i}| < \varepsilon\}, \qquad (2.28)$$

where $\varepsilon > 0$, and $j_1, \ldots, j_m$ are elements of $\mathbb{N}$. Clearly the topology with this basis must be weaker than the weak topology. We establish the equivalence of the topologies if we also show that the weak topology is weaker than the topology with neighbourhoods of the form (2.28). To this end, consider an element A in the neighbourhood basis μ of the weak topology

$$A = \bigcap_{i=1}^{m} \{\nu \in \mathcal{P}(\mathbb{S}) : |\nu f_i - \mu f_i| < \varepsilon\};$$

we show that there is an element of the neighbourhood (2.28) which is a subset of A. Suppose no such subset exists; in this case we can find a sequence μ_n in $\mathcal{P}(\mathbb{S})$ such that $\mu_n \varphi_i \to \mu \varphi_i$ for all i, yet $\mu_n \notin A$ for all n. But since $\{\varphi_i\}_{i=1}^{\infty}$ is a convergence determining set, this implies that $\mu_n \Rightarrow \mu$ and hence $\mu_n f \to \mu f$ for all $f \in C_b(\mathbb{S})$, in which case for n sufficiently large μ_n must be in A, which is a contradiction. Thus we need only consider

$$\zeta^{-1}(B) = \bigcap_{i=1}^{m} \{\omega : |\zeta(\omega) \varphi_{j_i} - \mu \varphi_{j_i}| < \varepsilon\},$$

where $\varepsilon > 0$ and $j_1, \ldots, j_m$ in $\mathbb{N}$. Since $\zeta\varphi_i$ is $\mathcal{F}/\mathcal{B}(\mathbb{R})$-measurable, it follows that each element of the intersection is $\mathcal{F}$-measurable and thus $\zeta^{-1}(B) \in \mathcal{F}$. Thus we have established that ζ is $\mathcal{F}/\mathcal{B}(\mathcal{P}(\mathbb{S}))$-measurable.

For the converse implication suppose that ζ is $\mathcal{B}(\mathcal{P}(\mathbb{S}))$-measurable. We must show that ζf is $\mathcal{B}(\mathbb{R})$-measurable for any $f \in C_b(\mathbb{R})$. For any $x \in \mathbb{R}$, $\varepsilon > 0$ the set $\{\mu \in \mathcal{P}(\mathbb{S}) : |\mu f - x| < \varepsilon\}$ is open in the weak topology on $\mathcal{P}(\mathbb{S})$, hence $\{\omega : |\zeta f - x| < \varepsilon\}$ is $\mathcal{F}$-measurable; thus we have shown that $(\zeta f)^{-1}(x - \varepsilon, x + \varepsilon) \in \mathcal{F}$. The open intervals $(x - \varepsilon, x + \varepsilon)$ for all $x \in \mathbb{R}$, $\varepsilon > 0$ generate the open sets in $\mathbb{R}$, hence ζf is $\mathcal{F}/\mathcal{B}(\mathbb{R})$ measurable.

2.22 Considering $\mu \in \mathcal{P}(I)$ as a subset of $\mathbb{R}^{|I|}$, then a continuous bounded function φ on a finite set I may be thought of as elements of $\mathbb{R}^{|I|}$ and $\mu\varphi$ is the dot product $\mu \cdot \varphi$.

If $\mu_n, \mu \in \mathcal{P}(I)$ and $\mu_n \Rightarrow \mu$, then by choosing the functions to be the basis vectors of $\mathbb{R}^{|I|}$ we see that $\mu_n\{i\} \to \mu\{i\}$ as $n \to \infty$ for all $i \in I$. Thus weak convergence in $\mathcal{P}(I)$ is equivalent to co-ordinatewise convergence in $\mathbb{R}^{|I|}$.

It is then clear that $\mathcal{P}(I)$ is separable since the set $\mathbb{Q}^{|I|}$ is a countable dense subset of $\mathbb{R}^{|I|}$.

Since $(\mathbb{R}^{|I|}, d)$ is complete and since d is a metric for co-ordinatewise convergence in $\mathbb{R}^{|I|}$, it also metrizes weak convergence on $\mathcal{P}(I)$.

2.25 We know from Theorem 2.24 that πf is indistinguishable from the $\mathcal{Y}_{t+}$ optional projection of $f(X)$. As t is a bounded stopping time, for any $t \in [0, \infty)$,
$$\mathbb{E}[f(X_t) \mid \mathcal{Y}_{t+}] = {}^o(f(X_t)) \quad \mathbb{P}\text{-a.s.},$$
hence the result.

2.29 Parts (a) and (b) are similar to the argument given for the existence of the process π, but in this case taking $f_i \in C_b(\Omega, \mathbb{R})$ and $g_i = \mathbb{E}[f_i \mid \mathcal{G}]$ choosing some version of the conditional expectation. For (c) let G_i be a countable family generating $\mathcal{G}$. Define $\mathcal{K}$ to be the (countable) π system generated by these G_is. Clearly $\mathcal{G} = \sigma(\mathcal{K})$. Define
$$\Omega' \triangleq \{\omega \in \Omega : \mathbb{Q}(\omega, K) = 1_K(\omega), \forall K \in \mathcal{K}\}.$$
Since
$$\mathbb{E}[1_K \mid \mathcal{G}] = 1_K, \quad \mathbb{P}\text{-a.s.},$$
it follows that $\mathbb{P}(\Omega') = 1$. For $\omega \in \Omega'$ the set of $G \in \mathcal{G}$ on which $\mathbb{Q}(\omega, G) = 1_G(\omega)$ is a d-system; so by Dynkin's lemma (see A1.3 of Williams [272]) since this d-system includes the π-system $\mathcal{K}$ it must include $\sigma(\mathcal{K}) = \mathcal{G}$. Thus for $\omega \in \Omega'$ it follows that
$$\mathbb{Q}(\omega, G) = 1_G(\omega), \quad \forall G \in \mathcal{G}.$$

To show that $\mathbb{Q}(\omega, A^{\mathcal{G}}(\omega)) = 1$, observe that this would follow immediately from the above if $A^{\mathcal{G}}(\omega) \in \mathcal{G}$, but since it is defined in terms of an uncountable intersection we must use the countable generating system to write

$$A^{\mathcal{G}}(\omega) = \left(\bigcup_{G_i : \omega \in G_i} G_i\right) \cap \left(\bigcup_{G_i : \omega \notin G_i} G_i^c\right)$$

and since the expression on the right-hand side is in terms of a countable intersection of elements of $\mathcal{G}$, the result follows.

2.33 To keep the solution concise, consider the even simpler case where the process Z defined in (2.22) is itself a uniformly integrable martingale (the general case can be handled by defining the change of measure on each $\mathcal{F}_t$ to be given by Z_t as in Section 3.3). Thus we define a change of measure via

$$\frac{\mathrm{d}\tilde{\mathbb{P}}}{\mathrm{d}\mathbb{P}} = Z_\infty,$$

and consequently under $\tilde{\mathbb{P}}$ by Girsanov's theorem Y_t is a Brownian motion.

Let $\eta \in L^2(\mathcal{Y}_\infty, \mathbb{P})$, and apply the martingale representation theorem to $Z^{-1}\eta$, to obtain a previsible process ν_t such that

$$Z_\infty^{-1}\eta = \mathbb{E}_{\tilde{\mathbb{P}}}(Z^{-1}\eta) + \int_0^\infty \Phi_t^\top \, \mathrm{d}Y_s.$$

If we define a $\tilde{\mathbb{P}}$-martingale via $\tilde{\eta}_t = \mathbb{E}_{\tilde{\mathbb{P}}}[Z_\infty^{-1}\eta \mid \mathcal{Y}_t]$ and by stochastic integration by parts

$$\mathrm{d}(Z_t \tilde{\eta}_t) = \left(Z_t \Phi_t^\top - \tilde{\eta}_t Z_t \pi_t(h^\top)\right) \mathrm{d}I_t,$$

consequently we may define $\nu_t = Z_t \Phi_t^\top - \tilde{\eta}_t Z_t \pi_t(h^\top)$. We may integrate this to obtain

$$Z_t \tilde{\eta}_t = \mathbb{E}[\eta] + \int_0^t \nu_t^\top \, \mathrm{d}I_t,$$

and passing to the $t \to \infty$ limit

$$\eta = Z_\infty \tilde{\eta}_\infty = \mathbb{E}[\eta] + \int_0^\infty \nu_t^\top \, \mathrm{d}I_t.$$

2.36 Follow the same steps as in Lemma 2.23 for arbitrary fixed $t \geq 0$ only consider the random variables g_i to be given by the (Kolmogorov) conditional expectations $\mathbb{E}[f_i(X_t) \mid \sigma(Y_s, \, 0 \leq s \leq t)]$ instead of the $\mathcal{Y}_t$-optional projection. Then use Exercise 2.4 part (ii) to show that the two constructions give rise to the same (random) probability measure almost surely.

Alternatively, let $\bar{\pi}_t$ be the regular conditional distribution (in the sense of Definition A.2) of X_t given $\sigma(Y_s, \, 0 \leq s \leq t)$. Then for any $f \in B(\mathbb{S})$,

$$\bar{\pi}_t f = \mathbb{E}\left[f(X_t) \mid \sigma(Y_s, \, 0 \leq s \leq t)\right]$$

holds $\mathbb{P}$-a.s. Following Exercise 2.25 using the right continuity of the filtration $(\mathcal{Y}_t)_{t \geq 0}$ and Exercise 2.4, for any $f \in B(\mathbb{S})$,

$$\pi_t f = \mathbb{E}\left[f(X_t) \mid \mathcal{Y}_t\right] = \mathbb{E}\left[f(X_t) \mid \sigma(Y_s,\ 0 \leq s \leq t)\right] \qquad \mathbb{P}\text{-a.s.}$$

Since $\mathbb{S}$ is a complete separable metric space there exists a countable separating set $\mathcal{A} \subset C_b(\mathbb{S})$. Therefore, there exists a null set $N(\mathcal{A})$ such that for any $\omega \in \Omega \backslash N(\mathcal{A})$ we have

$$\bar{\pi}_t f(\omega) = \pi_t f(\omega)$$

for any $f \in \mathcal{A}$. Therefore $\bar{\pi}_t(\omega) = \pi_t(\omega)$ for any $\omega \in \Omega \backslash N(\mathcal{A})$.

2.7 Bibliographical Notes

The majority of the results about weak convergence and probability measures on metric spaces can be found in Prokhorov [246] and are part of the standard theory of probability measures.

The innovations argument originates in the work of Fujisaki, Kallianpur and Kunita [104], however, there are some technical difficulties whose resolution is not clear from this paper but which are discussed in detail in Meyer [205].

3
The Filtering Equations

3.1 The Filtering Framework

Let $(\Omega, \mathcal{F}, \mathbb{P})$ be a probability space together with a filtration $(\mathcal{F}_t)_{t \geq 0}$ which satisfies the usual conditions. (See Section 2.1 for a definition of the usual conditions.) On $(\Omega, \mathcal{F}, \mathbb{P})$ we consider an $\mathcal{F}_t$-adapted process $X = \{X_t, t \geq 0\}$ which takes values in a complete separable metric space $\mathbb{S}$ (the state space). Let $\mathcal{S}$ be the associated Borel σ-algebra $\mathcal{B}(\mathbb{S})$. The process X is assumed to have paths which are càdlàg. (See appendix A.5 for details.) In the following X is called the *signal* process. Let $\{\mathcal{X}_t, t \geq 0\}$ be the usual augmentation with null sets of the filtration associated with the process X. In other words define

$$\mathcal{X}_t = \sigma(X_s,\ s \in [0, t]) \vee \mathcal{N}, \tag{3.1}$$

where $\mathcal{N}$ is the collection of all $\mathbb{P}$-null sets of $(\Omega, \mathcal{F})$ and define

$$\mathcal{X} \triangleq \bigvee_{t \in \mathbb{R}_+} \mathcal{X}_t, \tag{3.2}$$

where the $\vee$ notation denotes taking the σ-algebra generated by the union $\cup_t \mathcal{X}_t$. That is,

$$\mathcal{X} = \sigma\left(\bigcup_{t \in \mathbb{R}_+} \mathcal{X}_t\right).$$

Recall that $B(\mathbb{S})$ is the space of bounded $\mathcal{B}(\mathbb{S})$-measurable functions. Let $A: B(\mathbb{S}) \to B(\mathbb{S})$ and write $\mathcal{D}(A)$ for the domain of A which is a subset of $B(\mathbb{S})$. We assume that $\mathbf{1} \in \mathcal{D}(A)$ and $A\mathbf{1} = 0$. This definition implies that if $f \in \mathcal{D}(A)$ then Af is bounded. This is a very important observation which is crucial for many of the bounds in this chapter.

Let $\pi_0 \in \mathcal{P}(\mathbb{S})$. Assume that X is a solution of the martingale problem for (A, π_0). In other words, assume that the distribution of X_0 is π_0 and that the process $M^\varphi = \{M_t^\varphi, t \geq 0\}$ defined as

A. Bain, D. Crisan, *Fundamentals of Stochastic Filtering*,
DOI 10.1007/978-0-387-76896-0_3, © Springer Science+Business Media, LLC 2009

$$M_t^\varphi = \varphi(X_t) - \varphi(X_0) - \int_0^t A\varphi(X_s)\,ds, \quad t \geq 0, \qquad (3.3)$$

is an $\mathcal{F}_t$-adapted martingale for any $\varphi \in \mathcal{D}(A)$. The operator A is called the *generator* of the process X.

Let $h = (h_i)_{i=1}^m : \mathbb{S} \to \mathbb{R}^m$ be a measurable function such that

$$\mathbb{P}\left(\int_0^t \|h(X_s)\|\,ds < \infty\right) = 1 \qquad (3.4)$$

for all $t \geq 0$. Let W be a standard $\mathcal{F}_t$-adapted m-dimensional Brownian motion on $(\Omega, \mathcal{F}, \mathbb{P})$ independent of X, and Y be the process satisfying the following evolution equation

$$Y_t = Y_0 + \int_0^t h(X_s)\,ds + W_t, \qquad (3.5)$$

where $h = (h_i)_{i=1}^m : \mathbb{S} \to \mathbb{R}^m$ is a measurable function. The condition (3.4) ensures that the Riemann integral in the definition of Y_t exists a.s. This process $\{Y_t,\ t \geq 0\}$ is the *observation* process. Let $\{\mathcal{Y}_t,\ t \geq 0\}$ be the usual augmentation of the filtration associated with the process Y, viz

$$\mathcal{Y}_t = \sigma(Y_s,\ s \in [0,t]) \vee \mathcal{N}, \qquad (3.6)$$

$$\mathcal{Y} = \bigvee_{t \in \mathbb{R}_+} \mathcal{Y}_t. \qquad (3.7)$$

Then note that since by the measurability of h, Y_t is $\mathcal{F}_t$-adapted, it follows that $\mathcal{Y}_t \subset \mathcal{F}_t$.

Remark 3.1. To simplify notation we have considered A and h as having no explicit time dependence. By addition of t as a component of the state vector X, most results immediately extend to the case when A and h are time dependent. The reason for adopting this approach is that it keeps the notation simple.

Definition 3.2. *The filtering problem consists in determining the conditional distribution π_t of the signal X at time t given the information accumulated from observing Y in the interval $[0,t]$; that is, for $\varphi \in B(\mathbb{S})$, computing*

$$\pi_t \varphi = \mathbb{E}[\varphi(X_t) \mid \mathcal{Y}_t]. \qquad (3.8)$$

As discussed in the previous chapter, we must choose a suitable regularisation of the process $\pi = \{\pi_t,\ t \geq 0\}$, and by Theorem 2.24 we can do this so that π_t is an optional (and hence progressively measurable), $\mathcal{Y}_t$-adapted probability measure-valued process for which (3.8) holds almost surely. While (3.8) was established for φ bounded, π_t as constructed is a probability measure-valued process, so it is quite legitimate to compute $\pi_t \varphi$ when φ is unbounded provided that the expectation in question is well defined, in other words when $\pi_t|\varphi| < \infty$. In the following, Y_0 is considered to be identically zero (there is no information available initially). Hence π_0, the initial distribution of X, is identical with the conditional distribution of X_0 given $\mathcal{Y}_0$ and we use the same notation for both

$$\pi_0 \varphi = \int_{\mathbb{S}} \varphi(x) \mathbb{P} X_0^{-1}(\mathrm{d}x).$$

In the following we deduce the evolution equation for π. We consider two possible approaches.

- *The change of measure method.* A new measure is constructed under which Y becomes a Brownian motion and π has a representation in terms of an associated unnormalised version ρ. This ρ is then shown to satisfy a linear evolution equation which leads to the evolution equation for π by an application of Itô's formula.
- *The innovation process method.* The second approach isolates the Brownian motion driving the evolution equation for π (called the innovation process) and then identifies the corresponding terms in the Doob–Meyer decomposition of π.

Before we proceed, we first present two important examples of the above framework.

3.2 Two Particular Cases

We consider here two particular cases. One is a diffusion process and the second is a Markov chain with a finite state space.

The results in the chapter are stated in as general a form as possible and the various exercises show how the results can be applied in these two particular cases. The exercises establish suitable conditions on the processes, under which the general results of the chapter are valid. The process of verifying these conditions is sequential and the exercises build upon the results of earlier exercises, thus they are best attempted in order. As usual, the solutions may be found at the end of the chapter.

3.2.1 X a Diffusion Process

Let $X = (X^i)_{i=1}^d$ be the solution of a d-dimensional stochastic differential equation driven by a p-dimensional Brownian motion $V = (V^j)_{j=1}^p$:

$$X_t^i = X_0^i + \int_0^t f^i(X_s)\,\mathrm{d}s + \sum_{j=1}^p \int_0^t \sigma^{ij}(X_s)\,\mathrm{d}V_s^j, \quad i = 1,\ldots,d. \quad (3.9)$$

We assume that both $f = (f^i)_{i=1}^d : \mathbb{R}^d \to \mathbb{R}^d$ and $\sigma = (\sigma^{ij})_{i=1,\ldots,d, j=1,\ldots,p} : \mathbb{R}^d \to \mathbb{R}^{d \times p}$ are globally Lipschitz: that is, there exists a positive constant K such that for all $x, y \in \mathbb{R}^d$ we have

$$\begin{aligned} \|f(x) - f(y)\| &\leq K\|x - y\| \\ \|\sigma(x) - \sigma(y)\| &\leq K\|x - y\|, \end{aligned} \tag{3.10}$$

where the Euclidean norm $\|\cdot\|$ is defined in the usual fashion for vectors, and extended to $d \times p$-matrices by considering them as $d \times p$-dimensional vectors, viz:

$$\|\sigma\| = \sqrt{\sum_{i=1}^d \sum_{j=1}^p \sigma_{ij}^2}.$$

Under the globally Lipschitz condition, (3.9) has a unique solution by Theorem B.38. The generator A associated with the process X is the second-order differential operator

$$A = \sum_{i=1}^d f^i \frac{\partial}{\partial x_i} + \sum_{i,j=1}^d a^{ij} \frac{\partial^2}{\partial x_i \partial x_j}, \tag{3.11}$$

where $a = (a^{ij})_{i,j=1,\ldots,d} : \mathbb{R}^d \to \mathbb{R}^{d \times d}$ is the matrix-valued function defined as

$$a^{ij} = \tfrac{1}{2} \sum_{k=1}^p \sigma^{ik} \sigma^{jk} = \tfrac{1}{2} \left(\sigma \sigma^\top\right)^{ij}. \tag{3.12}$$

for all $i, j = 1, \ldots, d$. Recall from the definition that Af must be bounded for $f \in \mathcal{D}(A)$. There are various possible choices of the domain. For example, we can choose $\mathcal{D}(A) = C_k^2(\mathbb{R}^d)$, the space of twice differentiable, compactly supported, continuous functions on $\mathbb{R}^d$, since $A\varphi \in B(\mathbb{R}^d)$ for all $\varphi \in C_k^2(\mathbb{R}^d)$ and the process $M^\varphi = \{M_t^\varphi, \, t \geq 0\}$ defined as in (3.3) is a martingale for any $\varphi \in C_k^2(\mathbb{R}^d)$.

Exercise 3.3. If the global Lipschitz condition (3.10) holds, show that there exists $\kappa > 0$ such that for $x \in \mathbb{R}^d$,

$$\begin{aligned} \|\sigma(x)\|^2 &\leq \kappa(1 + \|x\|)^2 & (3.13) \\ \|f(x)\| &\leq \kappa(1 + \|x\|). & (3.14) \end{aligned}$$

Consequently show that there exists $\kappa' > 0$ such that

$$\|\sigma(x)\sigma^\top(x)\| \leq \kappa'(1 + \|x\|^2). \tag{3.15}$$

Exercise 3.4. Let $\mathrm{SL}^2(\mathbb{R}^d)$ be the subset of all twice continuously differentiable real-valued functions on $\mathbb{R}^d$ for which there exists a constant C such that for all $i, j = 1, \ldots, d$ and $x \in \mathbb{R}^d$ we have

$$|\partial_i\varphi(x)| \leq \frac{C}{1+\|x\|}, \qquad |\partial_i\partial_j\varphi(x)| \leq \frac{C}{1+\|x\|^2}.$$

Prove that $A\varphi \in B(\mathbb{R}^d)$ for all $\varphi \in \mathrm{SL}^2(\mathbb{R}^d)$ and the process M^φ defined as in (3.3) is a martingale for any $\varphi \in \mathrm{SL}^2(\mathbb{R}^d)$.

We can also choose $\mathcal{D}(A)$ to be the *maximal* domain of A. That is, $\mathcal{D}(A)$ is the set of all $\varphi \in B(\mathbb{R}^d)$ for which $A\varphi \in B(\mathbb{R}^d)$ and M^φ is a martingale. In the following, unless otherwise stated, we assume that $\mathcal{D}(A)$ is the maximal domain of A.

Remark 3.5. The following question is interesting to answer. Under what conditions is the solution of a martingale problem associated with the second-order differential operator defined in (3.11) the solution of the SDE (3.9)? The answer is surprisingly complicated. If $\mathcal{D}(A)$ contains the sequences $(\varphi_k^i)_{k>0}$, $(\varphi_k^{i,j})_{k>0}$ of functions in $\mathcal{C}_k^2(\mathbb{R}^d)$ such that $\varphi_k^i = x^i$ and $\varphi_k^{i,j} = x^i x^j$ for $\|x\| \leq k$ then there exists a p-dimensional Brownian motion V defined on an extension $(\tilde{\Omega}, \tilde{\mathcal{F}}, \tilde{\mathbb{P}})$ of $(\Omega, \mathcal{F}, \mathbb{P})$ such that X is a weak solution of (3.9). For details see Proposition 4.6, page 315 together with Remark 4.12, page 318 in Karatzas and Shreve [149].

3.2.2 X a Markov Process with a Finite Number of States

Let X be an $\mathcal{F}_t$-adapted Markov process with values in a finite state space I. Then $B(\mathbb{S})$ is isomorphic to $\mathbb{R}^I$ and the rôle of A is taken by the Q-matrix $Q = \{q_{ij}(t),\ i, j \in I, t \geq 0\}$ associated with the process. The Q-matrix is defined so that for all $t, h \geq 0$ as $h \to 0$, uniformly in t, for any $i, j \in I$,

$$\mathbb{P}(X_{t+h} = j \mid X_t = i) = J_i(j) + q_{ij}(t)h + o(h). \tag{3.16}$$

In (3.16) J_i is the indicator function of the atom i. In other words, $q_{ij}(t)$ is the rate at which the process jumps from site i to site j and $-q_{ii}(t)$ is the rate at which the process leaves site i. Assume that Q has the properties:

a. $q_{ii}(t) \leq 0$ for all $i \in I$, $q_{ij}(t) \geq 0$ for all $i, j \in I$, $i \neq j$.
b. $\sum_{j \in I} q_{ij}(t) = 0$ for all $i \in I$.
c. $\sup_{t \geq 0} |q_{ij}(t)| < \infty$ for all $i, j \in I$.

Exercise 3.6. Prove that for all $\varphi \in B(\mathbb{S})$, the process $M^\varphi = \{M_t^\varphi,\ t \geq 0\}$ defined as

$$M_t^\varphi = \varphi(X_t) - \varphi(X_0) - \int_0^t Q\varphi(s, X_s)\,ds, \quad t \geq 0, \tag{3.17}$$

is an $\mathcal{F}_t$-adapted right-continuous martingale. In (3.17), $Q\varphi : [0, \infty) \times I \to \mathbb{R}$ is defined in a natural way as

$$(Q\varphi)(s, i) = \sum_{j \in I} q_{ij}(s)\varphi(j), \quad \text{for all } (s, i) \in [0, \infty) \times I.$$

Exercise 3.7. The following is a simple example with real-world applications which fits within the above framework. Let $X = \{X_t, \, t \geq 0\}$ be the process

$$X_t = I_{[T,\infty)}(t), \quad t \geq 0,$$

where T is a positive random variable with probability density p and tail probability

$$g_t = \mathbb{P}(T \geq t), \quad t > 0.$$

Prove that the Q-matrix associated with X has entries $q_{01}(t) = -q_{00}(t) = p_t/g_t$, $q_{11}(t) = q_{10}(t) = 0$. See Exercise 3.32 for more on how the associated filtering problem is solved.

Remark 3.8. We can think of T as the time of a certain event occurring, for example, the failure of a piece of signal processing equipment, or the onset of a medical condition, which we would like to detect based on the information given by observing Y. This is the so-called *change-detection* filtering problem.

3.3 The Change of Probability Measure Method

This method consists in modifying the probability measure on Ω, in order to transform the process Y into a Brownian motion by means of Girsanov's theorem. Let $Z = \{Z_t, t > 0\}$ be the process defined by

$$Z_t = \exp\left(-\sum_{i=1}^{m} \int_0^t h^i(X_s) \, dW_s^i - \frac{1}{2} \sum_{i=1}^{m} \int_0^t h^i(X_s)^2 \, ds\right), \quad t \geq 0. \quad (3.18)$$

We need to introduce conditions under which the process Z is a martingale. The classical condition is Novikov's condition (see Theorem B.34). If

$$\mathbb{E}\left[\exp\left(\frac{1}{2} \sum_{i=1}^{m} \int_0^t h^i(X_s)^2 \, ds\right)\right] < \infty \quad (3.19)$$

for all $t > 0$, then Z is a martingale. Since (3.19) is quite difficult to verify directly, we use an alternative condition provided by the following lemma.

Lemma 3.9. *Let $\xi = \{\xi_t, \, t \geq 0\}$ be a càdlàg m-dimensional process such that*

$$\mathbb{E}\left[\sum_{i=1}^{m} \int_0^t (\xi_s^i)^2 \, ds\right] < \infty \quad (3.20)$$

and $z = \{z_t, \, t > 0\}$ be the process defined as

$$z_t = \exp\left(\sum_{i=1}^{m} \int_0^t \xi_s^i \, dW_s^i - \frac{1}{2} \sum_{i=1}^{m} \int_0^t (\xi_s^i)^2 \, ds\right), \quad t \geq 0. \quad (3.21)$$

3.3 The Change of Probability Measure Method

If the pair (ξ, z) satisfies for all $t \geq 0$

$$\mathbb{E}\left[\sum_{i=1}^{m}\int_0^t z_s\left(\xi_s^i\right)^2 ds\right] < \infty, \tag{3.22}$$

then z is a martingale.

Proof. From (3.20), we see that the process

$$t \mapsto \sum_{i=1}^{m}\int_0^t \xi_s^i\, dW_s^i$$

is a continuous (square-integrable) martingale with quadratic variation process

$$t \mapsto \sum_{i=1}^{m}\int_0^t \left(\xi_s^i\right)^2 ds.$$

By Itô's formula, the process z satisfies the equation

$$z_t = 1 + \sum_{i=1}^{m}\int_0^t z_s \xi_s^i\, dW_s^i.$$

Hence z is a non-negative, continuous, local martingale and therefore by Fatou's lemma a continuous supermartingale. To prove that z is a (genuine) martingale it is enough to show that it has constant expectation. Using the supermartingale property we note that

$$\mathbb{E}[z_t] \leq \mathbb{E}[z_0] = 1.$$

By Itô's formula, for $\varepsilon > 0$,

$$\frac{z_t}{1+\varepsilon z_t} = \frac{1}{\varepsilon} - \frac{1}{\varepsilon(1+\varepsilon z_t)}$$

$$= \frac{1}{1+\varepsilon} + \sum_{i=1}^{m}\int_0^t \frac{z_s}{(1+\varepsilon z_s)^2}\xi_s^i\, dW_s^i$$

$$- \sum_{i=1}^{m}\int_0^t \frac{\varepsilon z_s^2}{(1+\varepsilon z_s)^3}\left(\xi_s^i\right)^2 ds. \tag{3.23}$$

From (3.20) it follows that

$$\mathbb{E}\left[\sum_{i=1}^{m}\int_0^t \left(\frac{z_s}{(1+\varepsilon z_s)^2}\right)^2 \left(\xi_s^i\right)^2 ds\right]$$

$$= \mathbb{E}\left[\sum_{i=1}^{m}\int_0^t \frac{1}{\varepsilon^2}\left(\frac{\varepsilon z_s}{1+\varepsilon z_s}\right)^2 \frac{1}{(1+\varepsilon z_s)^2}\left(\xi_s^i\right)^2 ds\right]$$

$$\leq \frac{1}{\varepsilon^2}\mathbb{E}\left[\sum_{i=1}^{m}\int_0^t \left(\xi_s^i\right)^2 ds\right] < \infty,$$

hence the second term in (3.23) is a martingale with zero expectation. By taking expectation in (3.23),

$$\mathbb{E}\left[\frac{z_t}{1+\varepsilon z_t}\right] = \frac{1}{1+\varepsilon} - \mathbb{E}\left[\sum_{i=1}^{m}\int_0^t \frac{1}{(1+\varepsilon z_s)^2}\frac{\varepsilon z_s}{1+\varepsilon z_s}z_s\left(\xi_s^i\right)^2 ds\right]. \quad (3.24)$$

We now take the limit in (3.24) as ε tends to 0. From (3.22) we obtain our claim by means of the dominated convergence theorem. $\square$

As we require Z to be a martingale in order to construct the change of measure, the preceding lemma suggests the following as a suitable condition to impose upon h,

$$\mathbb{E}\left[\int_0^t \|h(X_s)\|^2 ds\right] < \infty, \quad \mathbb{E}\left[\int_0^t Z_s\|h(X_s)\|^2 ds\right] < \infty, \quad \forall t > 0. \quad (3.25)$$

Note that, since X has càdlàg paths, the process $s \mapsto h(X_s)$ is progressively measurable. Condition (3.25) implies conditions (2.3) and (2.4) and hence $\mathcal{Y}_t$ is right continuous and π_t has a $\mathcal{Y}_t$-adapted progressively measurable version.

Exercise 3.10. Let X be the solution of (3.9). Prove that if (3.10) is satisfied and X_0 has finite second moment, then the second moment of $\|X_t\|$ is bounded on any finite time interval $[0,T]$. That is, there exists G_T such that for all $0 \le t \le T$,

$$\mathbb{E}[\|X_t\|^2] < G_T. \quad (3.26)$$

Further show that under the same conditions, if X_0 has finite third moment that for any time interval $[0,T]$, there exists H_T such that for $0 \le t \le T$,

$$\mathbb{E}[\|X_t\|^3] < H_T. \quad (3.27)$$

[Hint: Use Gronwall's lemma, in the form of Corollary A.40 in the appendix.]

Exercise 3.11. i. (Difficult) Let X be the solution of (3.9). Prove that if condition (3.10) is satisfied and X_0 has finite second moment and h has linear growth, that is, there exists C such that

$$\|h(x)\|^2 \le C(1+\|x\|^2) \quad \forall x \in \mathbb{R}^d, \quad (3.28)$$

then (3.25) is satisfied.
ii. Let X be the Markov process with values in the finite state space I as described in Section 3.2. Then show that (3.25) is satisfied.

Proposition 3.12. *If (3.25) holds then the process $Z = \{Z_t,\ t \ge 0\}$ is an $\mathcal{F}_t$-adapted martingale.*

Proof. Condition (3.25) implies condition (3.22) of Lemma 3.9, which implies the result. $\square$

3.3 The Change of Probability Measure Method

For fixed $t \geq 0$, since $Z_t > 0$ introduce a probability measure $\tilde{\mathbb{P}}^t$ on $\mathcal{F}_t$ by specifying its Radon–Nikodym derivative with respect to $\mathbb{P}$ to be given by Z_t, viz

$$\left.\frac{\mathrm{d}\tilde{\mathbb{P}}^t}{\mathrm{d}\mathbb{P}}\right|_{\mathcal{F}_t} = Z_t.$$

It is immediate from the martingale property of Z that the measures $\tilde{\mathbb{P}}^t$ form a consistent family. That is, if $A \in \mathcal{F}_t$ and $T \geq t$ then

$$\tilde{\mathbb{P}}^T(A) = \mathbb{E}[Z_T 1_A] = \mathbb{E}\left[\mathbb{E}[Z_T 1_A \mid \mathcal{F}_t]\right] = \mathbb{E}\left[1_A \mathbb{E}[Z_T \mid \mathcal{F}_t]\right] = \mathbb{E}[1_A Z_t] = \tilde{\mathbb{P}}^t(A),$$

where $\mathbb{E}$ denotes expectation with respect to the probability measure $\mathbb{P}$, a convention which we adhere to throughout this chapter. Therefore we can define a probability measure $\tilde{\mathbb{P}}$ which is equivalent to $\mathbb{P}$ on $\bigcup_{0 \leq t < \infty} \mathcal{F}_t$ and we are able to suppress the superscript t in subsequent calculations.

It is important to realise that we have not defined a measure on $\mathcal{F}_\infty$, where

$$\mathcal{F}_\infty = \bigvee_{t=0}^{\infty} \mathcal{F}_t = \sigma\left(\bigcup_{0 \leq t < \infty} \mathcal{F}_t\right).$$

We cannot in general use the Daniel–Kolmogorov theorem here to extend the definition of $\tilde{\mathbb{P}}$ to $\mathcal{F}_\infty$. Indeed there may not exist a measure defined on $\mathcal{F}_\infty$ which agrees with $\tilde{\mathbb{P}}^t$ on $\mathcal{F}_t$ for all $0 \leq t < \infty$. For a more detailed discussion of why this extension may not be possible, see the discussion in Section B.3.1 of the appendix.

Proposition 3.13. *If condition (3.25) is satisfied then under $\tilde{\mathbb{P}}$, the observation process Y is a Brownian motion independent of X; additionally the law of the signal process X under $\tilde{\mathbb{P}}$ is the same as its law under $\mathbb{P}$.*

Proof. By Corollary B.31 to Girsanov's theorem, the process

$$Y_t = W_t + \int_0^t h(X_s)\,\mathrm{d}s$$

is a Brownian motion with respect to $\tilde{\mathbb{P}}$. Also, the law of the pair process (X, Y) can be written as

$$(X, Y) = (X, W) + \left(0, \int_0^t h(X_s)\,\mathrm{d}s\right),$$

thus on the interval $[0, t]$ where t is arbitrary, the law of (X, W) is absolutely continuous with respect to the law of the process (X, Y), and its Radon–Nikodym derivative is Z_t (see Exercise 3.14). That is, for any bounded measurable function f defined on the product of the corresponding path spaces for the pair (X, Y),

56   3 The Filtering Equations

$$\mathbb{E}[f(X,Y)Z_t] = \mathbb{E}[f(X,W)], \quad (3.29)$$

where in (3.29) both processes are regarded up to time t. Hence

$$\tilde{\mathbb{E}}[f(X,Y)] = \mathbb{E}[f(X,Y)Z_t] = \mathbb{E}[f(X,W)]$$

and therefore X and Y are independent under $\tilde{\mathbb{P}}$ since (X,W) has the same joint distribution under $\mathbb{P}$ as (X,Y) has under $\tilde{\mathbb{P}}$ and a priori X and W are independent. □

Exercise 3.14. i. Show that the process $P = \{P_t,\ t \geq 0\}$ defined with $\beta \in \mathbb{R}^m$ as

$$P_t = \exp\left(i\beta^\top Y_t - \frac{1}{2}\|\beta\|^2 t\right) Z_t$$

is a $\mathcal{X} \vee \mathcal{F}_t$-martingale.

ii. Deduce from (i) that for any $n \geq 1$ and $0 \leq t_1 \leq t_2 \leq \cdots \leq t_n < \infty$ and any $\beta_1,\ldots,\beta_n \in \mathbb{R}^m$, we have

$$\mathbb{E}\left[\exp\left(\sum_{j=1}^n i\beta_j^\top Y_{t_j}\right) Z_{t_n}\,\bigg|\,\mathcal{X}\right] = \mathbb{E}\left[\exp\left(\sum_{j=1}^n i\beta_j^\top W_{t_j}\right)\,\bigg|\,\mathcal{X}\right].$$

iii. Deduce from (ii) that (3.29) holds true for any bounded measurable function f defined on the product of the corresponding path spaces for the pair (X,Y).

Let $\tilde{Z} = \{\tilde{Z}_t,\ t \geq 0\}$ be the process defined as $\tilde{Z}_t = Z_t^{-1}$ for $t \geq 0$. Under $\tilde{\mathbb{P}}$, $\tilde{Z}_t$ satisfies the following stochastic differential equation,

$$d\tilde{Z}_t = \sum_{i=1}^m \tilde{Z}_t h^i(X_t)\,dY_t^i \quad (3.30)$$

and since $\tilde{Z}_0 = 1$,

$$\tilde{Z}_t = \exp\left(\sum_{i=1}^m \int_0^t h^i(X_s)\,dY_s^i - \frac{1}{2}\sum_{i=1}^m \int_0^t h^i(X_s)^2\,ds\right), \quad (3.31)$$

then $\tilde{\mathbb{E}}[\tilde{Z}_t] = \mathbb{E}[\tilde{Z}_t Z_t] = 1$, so $\tilde{Z}_t$ is an $\mathcal{F}_t$-adapted martingale under $\tilde{\mathbb{P}}$ and we have

$$\left.\frac{d\mathbb{P}}{d\tilde{\mathbb{P}}}\right|_{\mathcal{F}_t} = \tilde{Z}_t \quad \text{for } t \geq 0.$$

Proposition 3.13 implies that under $\tilde{\mathbb{P}}$ the observation process Y is a $\mathcal{Y}_t$-adapted Brownian motion; we can make use of the fact that Brownian motion is a Markov process to derive the following proposition.

Proposition 3.15. Let U be an integrable $\mathcal{F}_t$-measurable random variable. Then we have

$$\tilde{\mathbb{E}}[U \mid \mathcal{Y}_t] = \tilde{\mathbb{E}}[U \mid \mathcal{Y}]. \quad (3.32)$$

Proof. Let us denote by

$$\mathcal{Y}'_t = \sigma(Y_{t+u} - Y_t;\ u \geq 0);$$

then $\mathcal{Y} = \sigma(\mathcal{Y}_t, \mathcal{Y}'_t)$. Under the probability measure $\tilde{\mathbb{P}}$ the σ-algebra $\mathcal{Y}'_t \subset \mathcal{Y}$ is independent of $\mathcal{F}_t$ because Y is an $\mathcal{F}_t$-adapted Brownian motion. Hence since U is $\mathcal{F}_t$-adapted using property (f) of conditional expectation

$$\tilde{\mathbb{E}}[U \mid \mathcal{Y}_t] = \tilde{\mathbb{E}}[U \mid \sigma(\mathcal{Y}_t, \mathcal{Y}'_t)] = \tilde{\mathbb{E}}[U \mid \mathcal{Y}].$$

□

This proposition is an important step in the change of measure route to deriving the equations of non-linear filtering. It allows us to replace the time-dependent family of σ-algebras $\mathcal{Y}_t$ in the conditional expectations with the fixed σ-algebra $\mathcal{Y}$. This enables us to use techniques based on results from Kolmogorov conditional expectation which would not be applicable if the conditioning set were time dependent (as in the case of $\mathcal{Y}_t$). The proposition also has an interesting physical interpretation: the solution of the filtering problem for an $\mathcal{F}_t$-adapted random variable U given all observations (future, present and past) is equal to $\tilde{\mathbb{E}}[U \mid \mathcal{Y}_t]$; that is, future observations will not influence the estimator.

3.4 Unnormalised Conditional Distribution

In this section we first prove the Kallianpur–Striebel formula and use this to define the unnormalized conditional distribution process.

The notation $\tilde{\mathbb{P}}(\mathbb{P})$-a.s. in Proposition 3.16 means that the result holds both $\tilde{\mathbb{P}}$-a.s. and $\tilde{\mathbb{P}}$-a.s. We only need to show that it holds true in the first sense since $\tilde{\mathbb{P}}$ and $\mathbb{P}$ are equivalent probability measures.

Proposition 3.16 (Kallianpur–Striebel). *Assume that condition (3.25) holds. For every $\varphi \in B(\mathbb{S})$, for fixed $t \in [0, \infty)$,*

$$\pi_t(\varphi) = \frac{\tilde{\mathbb{E}}[\tilde{Z}_t \varphi(X_t) \mid \mathcal{Y}]}{\tilde{\mathbb{E}}[\tilde{Z}_t \mid \mathcal{Y}]} \qquad \tilde{\mathbb{P}}(\mathbb{P})\text{-}a.s. \qquad (3.33)$$

Proof. It is clear from the definition that $\tilde{Z}_t \geq 0$; furthermore it is readily observed that

$$0 = \tilde{\mathbb{E}}\left[1_{\{\tilde{Z}_t = 0\}} \tilde{Z}_t\right] = \mathbb{E}\left[1_{\{\tilde{Z}_t = 0\}}\right] = \mathbb{P}(\tilde{Z}_t = 0),$$

whence it follows that $\tilde{Z}_t > 0$ $\mathbb{P}$-a.s. as a consequence of which $\tilde{\mathbb{E}}[\tilde{Z}_t \mid \mathcal{Y}] > 0$ $\mathbb{P}$-a.s. and the right-hand side of (3.33) is well defined. Hence using Proposition 3.15 it suffices to show that

$$\pi_t(\varphi)\tilde{\mathbb{E}}[\tilde{Z}_t \mid \mathcal{Y}_t] = \tilde{\mathbb{E}}[\tilde{Z}_t\varphi(X_t) \mid \mathcal{Y}_t] \qquad \tilde{\mathbb{P}}\text{-a.s.}$$

As both the left- and right-hand sides of this equation are $\mathcal{Y}_t$-measurable, this is equivalent to showing that for any bounded $\mathcal{Y}_t$-measurable random variable b,

$$\tilde{\mathbb{E}}[\pi_t(\varphi)\tilde{\mathbb{E}}[\tilde{Z}_t \mid \mathcal{Y}_t]b] = \tilde{\mathbb{E}}[\tilde{\mathbb{E}}[\tilde{Z}_t\varphi(X_t) \mid \mathcal{Y}_t]b].$$

A consequence of the definition of the process π_t is that $\pi_t\varphi = \mathbb{E}[\varphi(X_t) \mid \mathcal{Y}_t]$ $\mathbb{P}$-a.s., so from the definition of Kolmogorov conditional expectation

$$\mathbb{E}\left[\pi_t(\varphi)b\right] = \mathbb{E}\left[\varphi(X_t)b\right].$$

Writing this under the measure $\tilde{\mathbb{P}}$,

$$\tilde{\mathbb{E}}\left[\pi_t(\varphi)b\tilde{Z}_t\right] = \tilde{\mathbb{E}}\left[\varphi(X_t)b\tilde{Z}_t\right].$$

By the tower property of the conditional expectation, since by assumption the function b is $\mathcal{Y}_t$-measurable

$$\tilde{\mathbb{E}}\left[\pi_t(\varphi)\tilde{\mathbb{E}}[\tilde{Z}_t \mid \mathcal{Y}_t]b\right] = \tilde{\mathbb{E}}\left[\tilde{\mathbb{E}}[\varphi(X_t)\tilde{Z}_t \mid \mathcal{Y}_t]b\right]$$

which proves that the result holds $\tilde{\mathbb{P}}$-a.s. $\square$

Let $\zeta = \{\zeta_t,\ t \geq 0\}$ be the process defined by

$$\zeta_t = \tilde{\mathbb{E}}[\tilde{Z}_t \mid \mathcal{Y}_t], \tag{3.34}$$

then as $\tilde{Z}_t$ is an $\mathcal{F}_t$-martingale under $\tilde{\mathbb{P}}$ and $\mathcal{Y}_s \subseteq \mathcal{F}_s$, it follows that for $0 \leq s < t$,

$$\tilde{\mathbb{E}}[\zeta_t \mid \mathcal{Y}_s] = \tilde{\mathbb{E}}[\tilde{Z}_t \mid \mathcal{Y}_s] = \tilde{\mathbb{E}}\left[\tilde{\mathbb{E}}[\tilde{Z}_t \mid \mathcal{F}_s] \mid \mathcal{Y}_s\right] = \tilde{\mathbb{E}}[\tilde{Z}_s \mid \mathcal{Y}_s] = \zeta_s.$$

Therefore by Doob's regularization theorem (see Rogers and Williams [248, Theorem II.67.7]) since the filtration $\mathcal{Y}_t$ satisfies the usual conditions we can choose a càdlàg version of ζ_t which is a $\mathcal{Y}_t$-martingale. In what follows, assume that $\{\zeta_t, t \geq 0\}$ has been chosen to be such a version. Given such a ζ, Proposition 3.16 suggests the following definition.

Definition 3.17. *Define the* unnormalised conditional distribution *of X to be the measure-valued process $\rho = \{\rho_t,\ t \geq 0\}$ which is determined (see Theorem 2.13) by the values of $\rho_t(\varphi)$ for $\varphi \in B(\mathbb{S})$ which are given for $t \geq 0$ by*

$$\rho_t(\varphi) \triangleq \pi_t(\varphi)\zeta_t.$$

Lemma 3.18. *The process $\{\rho_t,\ t \geq 0\}$ is càdlàg and $\mathcal{Y}_t$-adapted. Furthermore, for any $t \geq 0$,*

$$\rho_t(\varphi) = \tilde{\mathbb{E}}\left[\tilde{Z}_t\varphi(X_t) \mid \mathcal{Y}_t\right] \qquad \tilde{\mathbb{P}}(\mathbb{P})\text{-a.s.} \tag{3.35}$$

3.4 Unnormalised Conditional Distribution

Proof. Both $\pi_t(\varphi)$ and ζ_t are $\mathcal{Y}_t$-adapted. By construction $\{\zeta, t \geq 0\}$ is also càdlàg.[†] By Theorem 2.24 and Corollary 2.26 $\{\pi_t, t \geq 0\}$ is càdlàg and $\mathcal{Y}_t$-adapted, therefore the process $\{\rho_t, t \geq 0\}$ is also càdlàg and $\mathcal{Y}_t$-adapted.

For the second part, from Proposition 3.15 and Proposition 3.16 it follows that
$$\pi_t(\varphi)\tilde{\mathbb{E}}[\tilde{Z}_t \mid \mathcal{Y}_t] = \tilde{\mathbb{E}}[\tilde{Z}_t \varphi(X_t) \mid \mathcal{Y}_t] \quad \tilde{\mathbb{P}}\text{-a.s.},$$
From (3.34), $\tilde{\mathbb{E}}[\tilde{Z}_t \mid \mathcal{Y}_t] = \zeta_t$ a.s. from which the result follows. □

It may be useful to point out that for general φ, the process $\rho_t(\varphi)$ is not a $\mathcal{Y}_t$-martingale but a semimartingale. This misconception arising from (2.8) is due to confusion with the well-known result that taking conditional expectation of an integrable random variable Z with respect to the family $\mathcal{Y}_t$ gives rise to a (uniformly integrable) martingale $\mathbb{E}[Z \mid \mathcal{Y}_t]$. But this is only true for a fixed random variable Z which does not depend upon t.

Corollary 3.19. *Assume that condition (3.25) holds. For every $\varphi \in B(\mathbb{S})$,*
$$\pi_t(\varphi) = \frac{\rho_t(\varphi)}{\rho_t(\mathbf{1})} \quad \forall t \in [0, \infty) \quad \tilde{\mathbb{P}}(\mathbb{P})\text{-a.s.} \quad (3.36)$$

Proof. It is clear from Definition 3.17 that $\zeta_t = \rho_t(\mathbf{1})$. The result then follows immediately. □

The Kallianpur–Striebel formula explains the usage of the term *unnormalised* in the definition of ρ_t as the denominator $\rho_t(\mathbf{1})$ can be viewed as the normalising factor. The result can also be viewed as the abstract version of Bayes' identity in this filtering framework. In theory at least the Kallianpur–Striebel formula provides a method for solving the filtering problem.

Remark 3.20. The Kallianpur–Striebel formula (3.33) holds true for any Borel-measurable φ, not necessarily bounded, such that $\mathbb{E}[|\varphi(X_t)|] < \infty$; see Exercise 5.1 for details.

Lemma 3.21. *i. Let $\{u_t, t \geq 0\}$ be an $\mathcal{F}_t$-progressively measurable process such that for all $t \geq 0$, we have*
$$\tilde{\mathbb{E}}\left[\int_0^t u_s^2 \, ds\right] < \infty; \quad (3.37)$$
then, for all $t \geq 0$, and $j = 1, \ldots, m$, we have
$$\tilde{\mathbb{E}}\left[\int_0^t u_s \, dY_s^j \,\bigg|\, \mathcal{Y}\right] = \int_0^t \tilde{\mathbb{E}}[u_s \mid \mathcal{Y}] \, dY_s^j. \quad (3.38)$$

[†] It is in fact the case that $\zeta_t = \exp\left(\int_0^t \pi_s(h^\top) \, dY_s - \frac{1}{2}\int_0^t \|\pi_s(h)\|^2 \, ds\right)$; see Lemma 3.29.

ii. Now let $\{u_t,\ t \geq 0\}$ be an $\mathcal{F}_t$-progressively measurable process such that for all $t \geq 0$, we have

$$\tilde{\mathbb{E}}\left[\int_0^t u_s^2 \, \mathrm{d}\langle M^\varphi \rangle_s\right] < \infty; \tag{3.39}$$

then

$$\tilde{\mathbb{E}}\left[\int_0^t u_s \, \mathrm{d}M_s^\varphi \,\bigg|\, \mathcal{Y}\right] = 0. \tag{3.40}$$

Proof. i. Every ε_t from the total set S_t as defined in Lemma B.39 satisfies the following stochastic differential equation

$$\varepsilon_t = 1 + \int_0^t i\varepsilon_s r_s^\top \, \mathrm{d}Y_s.$$

We observe the following sequence of identities

$$\tilde{\mathbb{E}}\left[\varepsilon_t \tilde{\mathbb{E}}\left[\int_0^t u_s \, \mathrm{d}Y_s^j \,\bigg|\, \mathcal{Y}\right]\right] = \tilde{\mathbb{E}}\left[\varepsilon_t \int_0^t u_s \, \mathrm{d}Y_s^j\right]$$

$$= \tilde{\mathbb{E}}\left[\int_0^t u_s \, \mathrm{d}Y_s^j\right] + \tilde{\mathbb{E}}\left[\int_0^t i\varepsilon_s r_s^j u_s \, \mathrm{d}s\right]$$

$$= \tilde{\mathbb{E}}\left[\tilde{\mathbb{E}}\left[\int_0^t i\varepsilon_s r_s^j u_s \, \mathrm{d}s \,\bigg|\, \mathcal{Y}\right]\right]$$

$$= \tilde{\mathbb{E}}\left[\int_0^t i\varepsilon_s r_s^j \tilde{\mathbb{E}}[u_s \mid \mathcal{Y}] \, \mathrm{d}s\right]$$

$$= \tilde{\mathbb{E}}\left[\varepsilon_t \int_0^t \tilde{\mathbb{E}}[u_s \mid \mathcal{Y}] \, \mathrm{d}Y_s^j\right],$$

which completes the proof of (3.38).

ii. Since for all $\varphi \in \mathcal{D}(A)$, $\{M_t^\varphi, \mathcal{F}_t\}$ is a square integrable martingale, we can define the Itô integral with respect to it. The proof of (3.40) is similar to that of (3.38). We once again choose ε_t from the set S_t and obtain the following sequence of identities (we use the fact that the quadratic covariation between M_t^φ and Y is 0).

$$\tilde{\mathbb{E}}\left[\varepsilon_t \tilde{\mathbb{E}}\left[\int_0^t u_s \, dM_s^\varphi \,\Big|\, \mathcal{Y}\right]\right] = \tilde{\mathbb{E}}\left[\varepsilon_t \int_0^t u_s \, dM_s^\varphi\right]$$

$$= \tilde{\mathbb{E}}\left[\int_0^t u_s \, dM_s^\varphi\right]$$

$$+ \sum_{i=1}^m \tilde{\mathbb{E}} \left\langle \int_0^\cdot i\varepsilon_s r_s^j \, dY_s^j, \int_0^\cdot u_s \, dM_s^\varphi \right\rangle_t$$

$$= \tilde{\mathbb{E}}\left[\int_0^t u_s \, dM_s^\varphi\right] + \sum_{i=1}^m \tilde{\mathbb{E}} \int_0^t i\varepsilon_s r_s^j u_s \, d\langle M_\cdot^\varphi, Y_\cdot^j\rangle_s$$

$$= \tilde{\mathbb{E}}\left[\int_0^t u_s \, dM_s^\varphi\right]$$

$$= 0,$$

where the final equality follows from the fact that the condition (3.39) ensures that the stochastic integral is a martingale. □

Exercise 3.22. Prove that if $\varphi, \varphi^2 \in \mathcal{D}(A)$ then

$$\langle M^\varphi \rangle_t = \int_0^t \left(A\varphi^2 - 2\varphi A\varphi\right)(X_s) \, ds. \quad (3.41)$$

Hence, show in this case that condition (3.37) implies condition (3.39) of Lemma 3.21.

3.5 The Zakai Equation

In the following, we further assume that for all $t \geq 0$,

$$\tilde{\mathbb{P}}\left[\int_0^t [\rho_s(\|h\|)]^2 \, ds < \infty\right] = 1. \quad (3.42)$$

Exercise 3.25 gives some convenient conditions under which (3.42) holds for the two example classes of signal processes considered in this chapter.

Exercise 3.23. Show that the stochastic integral $\int_0^t \rho_s(\varphi h^\top) \, dY_s$ is well defined for any $\varphi \in B(\mathbb{S})$ under condition (3.42). Hence the process

$$t \mapsto \int_0^t \rho_s(\varphi h^\top) \, dY_s,$$

is a local martingale with paths which are almost surely continuous, since it is $\mathcal{Y}_t$-adapted and $(\mathcal{Y}_t)_{t \geq 0}$ is a Brownian filtration.

Theorem 3.24. *If conditions (3.25) and (3.42) are satisfied then the process ρ_t satisfies the following evolution equation, called the Zakai equation,*

$$\rho_t(\varphi) = \pi_0(\varphi) + \int_0^t \rho_s(A\varphi)\,\mathrm{d}s + \int_0^t \rho_s(\varphi h^\top)\,\mathrm{d}Y_s, \quad \tilde{\mathbb{P}}\text{-a.s. } \forall t \geq 0 \quad (3.43)$$

for any $\varphi \in \mathcal{D}(A)$.

Proof. We first approximate $\tilde{Z}_t$ with $\tilde{Z}_t^\varepsilon$ given by

$$\tilde{Z}_t^\varepsilon = \frac{\tilde{Z}_t}{1 + \varepsilon \tilde{Z}_t}.$$

Using Itô's rule and integration by parts, we find

$$\mathrm{d}\left(\tilde{Z}_t^\varepsilon \varphi(X_t)\right) = \tilde{Z}_t^\varepsilon A\varphi(X_t)\,\mathrm{d}t + \tilde{Z}_t^\varepsilon\,\mathrm{d}M_t^\varphi$$
$$- \varepsilon\varphi(X_t)(1+\varepsilon\tilde{Z}_t)^{-3}\tilde{Z}_t^2 \|h(X_t)\|^2\,\mathrm{d}t$$
$$+ \varphi(X_t)(1+\varepsilon\tilde{Z}_t)^{-2}\tilde{Z}_t h^\top(X_t)\,\mathrm{d}Y_t.$$

Since $\tilde{Z}_t^\varepsilon$ is bounded, (3.39) is satisfied; hence by Lemma 3.21

$$\tilde{\mathbb{E}}\left[\int_0^t \tilde{Z}_s^\varepsilon\,\mathrm{d}M_s^\varphi \,\bigg|\, \mathcal{Y}\right] = 0.$$

Also since

$$\tilde{\mathbb{E}}\left[\int_0^t \varphi^2(X_s)\frac{1}{(1+\varepsilon\tilde{Z}_s)^2}\frac{1}{\varepsilon^2}\left(\frac{\varepsilon\tilde{Z}_s}{1+\varepsilon\tilde{Z}_s}\right)^2\|h(X_s)\|^2\,\mathrm{d}s\right]$$
$$\leq \frac{\|\varphi\|_\infty^2}{\varepsilon^2}\tilde{\mathbb{E}}\left[\int_0^t \|h(X_s)\|^2\,\mathrm{d}s\right]$$
$$= \frac{\|\varphi\|_\infty^2}{\varepsilon^2}\mathbb{E}\left[\int_0^t Z_s\|h(X_s)\|^2\,\mathrm{d}s\right] < \infty,$$

where the final inequality is a consequence of (3.25). Therefore condition (3.37) is satisfied. Hence, by taking conditional expectation with respect to $\mathcal{Y}$ and applying (3.38) and (3.40), we obtain

$$\tilde{\mathbb{E}}[\tilde{Z}_t^\varepsilon \varphi(X_t) \mid \mathcal{Y}] = \frac{\pi_0(\varphi)}{1+\varepsilon} + \int_0^t \tilde{\mathbb{E}}[\tilde{Z}_s^\varepsilon A\varphi(X_s) \mid \mathcal{Y}]\,\mathrm{d}s$$
$$- \int_0^t \tilde{\mathbb{E}}\left[\varepsilon\varphi(X_s)(\tilde{Z}_t^\varepsilon)^2 \frac{1}{(1+\varepsilon\tilde{Z}_s)}\|h(X_s)\|^2 \,\bigg|\, \mathcal{Y}\right]\,\mathrm{d}s$$
$$+ \int_0^t \tilde{\mathbb{E}}\left[\tilde{Z}_t^\varepsilon \frac{1}{1+\varepsilon\tilde{Z}_s}\varphi(X_s)h^\top(X_s) \,\bigg|\, \mathcal{Y}\right]\,\mathrm{d}Y_s. \quad (3.44)$$

Now let ε tend to 0. We have, writing λ for Lebesgue measure on $[0,\infty)$,

$$\lim_{\varepsilon \to 0} \tilde{Z}_t^\varepsilon = \tilde{Z}_t$$

$$\lim_{\varepsilon \to 0} \tilde{\mathbb{E}}[\tilde{Z}_t^\varepsilon \varphi(X_t) \mid \mathcal{Y}] = \rho_t(\varphi), \quad \tilde{\mathbb{P}}\text{-a.s.}$$

$$\lim_{\varepsilon \to 0} \tilde{\mathbb{E}}[\tilde{Z}_t^\varepsilon A\varphi(X_t) \mid \mathcal{Y}] = \rho_t(A\varphi), \quad \lambda \otimes \tilde{\mathbb{P}}\text{-a.e.}$$

This last sequence remains bounded by the random variable $\|A\varphi\|_\infty \tilde{\mathbb{E}}[\tilde{Z}_t \mid \mathcal{Y}]$, which can be seen to be in $L^1([0,t] \times \Omega; \lambda \otimes \tilde{\mathbb{P}})$ since

$$\tilde{\mathbb{E}}\left[\int_0^t \|A\varphi\|_\infty \tilde{\mathbb{E}}[\tilde{Z}_s \mid \mathcal{Y}] \, ds\right] \leq \|A\varphi\|_\infty \int_0^t \tilde{\mathbb{E}}[\tilde{Z}_s] \, ds \leq \|A\varphi\|_\infty t < \infty.$$

Consequently by the conditional form of the dominated convergence theorem as $\varepsilon \to 0$,

$$\tilde{\mathbb{E}}\left[\int_0^t \tilde{\mathbb{E}}[\tilde{Z}_s^\varepsilon A\varphi(X_s) \mid \mathcal{Y}] \, ds \,\bigg|\, \mathcal{Y}\right] \to \tilde{\mathbb{E}}\left[\int_0^t \rho_s(A\varphi) \, ds \,\bigg|\, \mathcal{Y}\right], \quad \tilde{\mathbb{P}}\text{-a.s.}$$

Using the definition of ρ_t, we see that by Fubini's theorem

$$\int_0^t \tilde{\mathbb{E}}[\tilde{Z}_s^\varepsilon A\varphi(X_s) \mid \mathcal{Y}] \, ds \to \int_0^t \rho_s(A\varphi) \, ds, \quad \tilde{\mathbb{P}}\text{-a.s.}$$

Next we have that for almost every t,

$$\lim_{\varepsilon \to 0} \varepsilon \varphi(X_s)(\tilde{Z}_s^\varepsilon)^2 (1 + \varepsilon \tilde{Z}_s)^{-1} \|h(X_s)\|^2 = 0, \quad \tilde{\mathbb{P}}\text{-a.s.},$$

and

$$\left|\varepsilon \varphi(X_s)(\tilde{Z}_s^\varepsilon)^2 (1 + \varepsilon \tilde{Z}_s)^{-1} \|h(X_s)\|^2\right|$$

$$= \left|\varphi(X_s) \tilde{Z}_s \|h(X_s)\|^2 \frac{\varepsilon \tilde{Z}_s}{1 + \varepsilon \tilde{Z}_s} \left(1 + \varepsilon \tilde{Z}_s\right)^{-2}\right|$$

$$\leq \|\varphi\|_\infty \tilde{Z}_s \|h(X_s)\|^2. \quad (3.45)$$

The right-hand side of (3.45) is integrable over $[0,t] \times \Omega$ with respect to $\lambda \otimes \tilde{\mathbb{P}}$ using (3.25):

$$\tilde{\mathbb{E}}\left[\int_0^t \tilde{Z}_s \|h(X_s)\|^2 \, ds\right] = \mathbb{E}\left[\int_0^t \|h(X_s)\|^2 \, ds\right] < \infty.$$

Thus we can use the conditional form of the dominated convergence theorem to obtain that

$$\lim_{\varepsilon \to 0} \int_0^t \varepsilon \tilde{\mathbb{E}}\left[\varphi(X_s) \left(\tilde{Z}_s^\varepsilon\right)^2 (1 + \varepsilon \tilde{Z}_s)^{-1} \|h(X_s)\|^2 \,\bigg|\, \mathcal{Y}\right] ds = 0.$$

To complete the proof it only remains to show that as $\varepsilon \to 0$,

$$\int_0^t \tilde{\mathbb{E}}\left[\tilde{Z}_s^\varepsilon \frac{1}{1+\varepsilon \tilde{Z}_s} \varphi(X_s) h^\top(X_s) \mid \mathcal{Y}\right] dY_s \to \int_0^t \rho_s(\varphi h^\top) dY_s. \tag{3.46}$$

Consider the process

$$t \mapsto \int_0^t \tilde{\mathbb{E}}\left[\tilde{Z}_t^\varepsilon \frac{1}{1+\varepsilon \tilde{Z}_s} \varphi(X_s) h^\top(X_s) \mid \mathcal{Y}\right] dY_s; \tag{3.47}$$

we show that this is a martingale. By Jensen's inequality, Fubini's theorem and (3.25),

$$\tilde{\mathbb{E}}\left[\int_0^t \tilde{\mathbb{E}}\left[\left(\tilde{Z}_t^\varepsilon \frac{1}{1+\varepsilon \tilde{Z}_s} \varphi(X_s) h^\top(X_s)\right)^2 \mid \mathcal{Y}\right] ds\right]$$

$$\leq \frac{\|\varphi\|_\infty^2}{\varepsilon^2} \tilde{\mathbb{E}}\left[\int_0^t \tilde{\mathbb{E}}[\|h(X_s)\|^2 \mid \mathcal{Y}] ds\right]$$

$$= \varepsilon^2 \|\varphi\|_\infty^2 \int_0^t \tilde{\mathbb{E}}[\|h(X_s)\|^2] ds$$

$$= \varepsilon^2 \|\varphi\|_\infty^2 \mathbb{E}\left[\int_0^t Z_s \|h(X_s)\|^2 ds\right]$$

$$< \infty.$$

Thus the process defined in (3.47) is an $\mathcal{F}_t$-martingale. From condition (3.42) and Exercise 3.23 the postulated limit process as $\varepsilon \to 0$,

$$t \mapsto \int_0^t \rho_s(\varphi h^\top) dY_s, \tag{3.48}$$

is a well defined local martingale. Thus the difference of (3.47) and (3.48) is a well defined local martingale,

$$t \mapsto \int_0^t \tilde{\mathbb{E}}\left[\frac{\varepsilon \tilde{Z}_s^2(2+\varepsilon \tilde{Z}_s)}{(1+\varepsilon \tilde{Z}_s)^2} \varphi(X_s) h^\top(X_s) \mid \mathcal{Y}\right] dY_s. \tag{3.49}$$

We use Proposition B.41 to prove that the integral in (3.49) converges to 0, $\tilde{\mathbb{P}}$-almost surely. Since, for all $i = 1, \ldots, m$,

$$\lim_{\varepsilon \to 0} \frac{\varepsilon \tilde{Z}_s^2(2+\varepsilon \tilde{Z}_s)}{(1+\varepsilon \tilde{Z}_s)^2} \varphi(X_s) h^i(X_s) = 0, \qquad \tilde{\mathbb{P}}\text{-a.s.}$$

and

$$\left|\tilde{Z}_s \frac{\varepsilon \tilde{Z}_s}{(1+\varepsilon \tilde{Z}_s)} \frac{(2+\varepsilon \tilde{Z}_s)}{(1+\varepsilon \tilde{Z}_s)} \varphi(X_s) h^i(X_s)\right| \leq 2\|\varphi\|_\infty \tilde{Z}_s \left|h^i(X_s)\right|, \tag{3.50}$$

using (3.25) it follows that for Lebesgue a.e. $s \geq 0$, the right-hand side is $\tilde{\mathbb{P}}$-integrable, and hence it follows by the dominated convergence theorem that for almost every $s \geq 0$,

$$\lim_{\varepsilon \to 0} \tilde{\mathbb{E}}\left[\frac{\varepsilon \tilde{Z}_s^2(2+\varepsilon \tilde{Z}_s)}{(1+\varepsilon \tilde{Z}_s)^2}\varphi(X_s)h^i(X_s) \mid \mathcal{Y}\right] = 0, \qquad \tilde{\mathbb{P}}\text{-a.s.}$$

As a consequence of (3.50),

$$\tilde{\mathbb{E}}\left[\frac{\varepsilon \tilde{Z}_s^2(2+\varepsilon \tilde{Z}_s)}{(1+\varepsilon \tilde{Z}_s)^2}\varphi(X_s)h^i(X_s) \mid \mathcal{Y}\right] \leq 2\|\varphi\|_\infty \rho_s(\|h\|),$$

and using the assumed condition (3.42), it follows that $\tilde{\mathbb{P}}$-a.s.

$$\int_0^t \left(\tilde{\mathbb{E}}\left[\frac{\varepsilon \tilde{Z}_s^2(2+\varepsilon \tilde{Z}_s)}{(1+\varepsilon \tilde{Z}_s)^2}\varphi(X_s)h^i(X_s) \mid \mathcal{Y}\right]\right)^2 ds$$

$$\leq 4\|\varphi\|_\infty^2 \int_0^t [\rho_s(\|h\|)]^2\, ds < \infty.$$

Thus using the dominated convergence theorem for $L^2([0,t])$, we obtain that

$$\int_0^t \sum_{i=1}^m \left(\tilde{\mathbb{E}}\left[\frac{\varepsilon \tilde{Z}_s^2(2+\varepsilon \tilde{Z}_s)}{(1+\varepsilon \tilde{Z}_s)^2}\varphi(X_s)h^i(X_s) \mid \mathcal{Y}\right]\right)^2 ds \to 0 \qquad \tilde{\mathbb{P}}\text{-a.s.} \quad (3.51)$$

Because this convergence only holds almost surely we cannot apply the Itô isometry to conclude that the stochastic integrals in (3.46) converge. However, Proposition B.41 of the appendix is applicable as a consequence of (3.51), which establishes the convergence in (3.46).[†] □

Exercise 3.25. i. (Difficult) Let X be the solution of (3.9). Prove that if (3.10) is satisfied, X_0 has finite third moment and h has linear growth (3.28), then (3.42) is satisfied. [Hint: Use the result of Exercise 3.10.]
ii. Let X be the Markov process with values in the finite state space I as described in Section 3.2. Then (3.42) is satisfied.

Remark 3.26. If X is a Markov process with finite state space I, then the Zakai equation is, in fact, a (finite-dimensional) linear stochastic differential equation. To see this, let us define by ρ_t^i the mass that ρ_t puts on site $\{i\}$ for any $i \in I$. In particular,

$$\rho_t^i = \rho_t(\{i\})$$
$$= \tilde{\mathbb{E}}[J_i(X_t)\tilde{Z}_t \mid \mathcal{Y}_t], \qquad i \in I,$$

[†] The convergence established in Proposition B.41 is in probability only. Therefore the convergence in (3.46) follows for a suitably chosen sequence (ε_n) such that $\varepsilon_n \to 0$. The theorem follows by taking the limit in (3.44) as $\varepsilon_n \to 0$.

where J_i is the indicator function of the singleton set $\{i\}$ and for an arbitrary function $\varphi : I \to \mathbb{R}$, we have

$$\rho_t(\varphi) = \sum_{i \in I} \varphi(i) \rho_t^i.$$

Hence the measure ρ_t and the $|I|$-dimensional vector $(\rho_t^i)_{i \in I}$ can be identified as one and the same object and from (3.43) we get that

$$\rho_t(\varphi) = \sum_{i \in I} \varphi(i) \rho_t^i$$
$$= \sum_{i \in I} \varphi(i) \left(\pi_0^i + \int_0^t \sum_{j \in I} Q_{ji} \rho_s^j \, \mathrm{d}s + \sum_{j=1}^m \int_0^t \rho_s^i h^j(i) \, \mathrm{d}Y_s^j \right).$$

Hence $\rho_t = (\rho_t^i)_{i \in I}$ satisfies the $|I|$-dimensional linear stochastic differential equation

$$\rho_t = \pi_0 + \int_0^t Q^\top \rho_s \, \mathrm{d}s + \sum_{j=1}^m \int_0^t H^j \rho_s \, \mathrm{d}Y_s^j, \tag{3.52}$$

where, for $j = 1, \ldots, m$, $H^j = \mathrm{diag}(h^j)$ is the $|I| \times |I|$ diagonal matrix with entries $H_{ii} = h_i^j$, and π_0 is the $|I|$-dimensional vector with entries

$$\pi_0^i = \pi_0(\{i\}) = \mathbb{P}(X_0 = i).$$

The use of the same notation for the vector and the corresponding measure is warranted for the same reasons as above. Evidently, due to its linearity, (3.52) has a unique solution.

Exercise 3.27. Let X be a Markov process with finite state space I with associated Q-matrix Q and $\pi = \{(\pi_t^i)_{i \in I}, t \geq 0\}$ be the conditional distribution of X given the σ-algebra $\mathcal{Y}_t$ viewed as a process with values in $\mathbb{R}^I$.

i. Deduce from (3.52) that the $|I|$-dimensional process π solves the following (non-linear) stochastic differential equation,

$$\pi_t = \pi_0 + \int_0^t Q^\top \pi_s \, \mathrm{d}s$$
$$+ \sum_{j=1}^m \int_0^t \left(H^j - \pi_s(h^j) \mathbb{I}_{|I|} \right) \pi_s (\mathrm{d}Y_s^j - \pi_s(h^j) \, \mathrm{d}s), \tag{3.53}$$

where $\mathbb{I}_{|I|}$ is the identity matrix of size $|I|$.

ii. Prove that (3.53) has a unique solution in the space of continuous $\mathcal{Y}_t$-adapted $|I|$-dimensional processes.

Remark 3.28. There is a corresponding treatment of the Zakai equation for the case $\mathbb{S} = \mathbb{R}^d$ and X is the solution of the stochastic differential (3.9). This be done in Chapter 7. In this case, ρ_t can no longer be associated with a finite-dimensional object (a vector). Under additional assumptions, it can be associated with functions defined on $\mathbb{R}^d$ which represent the density of the measure ρ_t with respect to the Lebesgue measure. The analysis goes in two steps. First one needs to make sense of the stochastic partial differential equation satisfied by the density of ρ_t (the equivalent of (3.52)). That is, one shows the existence and uniqueness of its solution in a suitably chosen space of functions. Next one shows that the measure with that given density solves the Zakai equation which we establish beforehand that it has a unique solution. This implies that ρ_t has the solution of the stochastic partial differential equation as its density with respect to the Lebesgue measure.

3.6 The Kushner–Stratonovich Equation

An equation has been derived for the unnormalised conditional distribution ρ. In order to solve the filtering problem the normalised conditional distribution π is required. In this section an equation is derived which π satisfies. The condition (2.4) viz:

$$\mathbb{P}\left(\int_0^t \|\pi_s(h)\|^2 \, \mathrm{d}s < \infty\right) = 1, \quad \text{for all } t \geq 0, \tag{3.54}$$

turns out to be fundamental to the derivation of the Kushner–Stratonovich equation by various methods This technical condition (3.54) is unfortunate since it depends on the process π which we are trying to find, rather than being a direct condition on the system. It is, however, a consequence of the stronger condition which was required for the change of measure approach to the derivation of the Zakai equation, which is the first part of (3.25), since π_t is a probability measure for all $t \in [0, \infty)$.

Lemma 3.29. *If conditions (3.25) and (3.42) are satisfied then the process $t \mapsto \rho_t(\mathbf{1})$ has the following explicit representation,*

$$\rho_t(\mathbf{1}) = \exp\left(\int_0^t \pi_s(h^\top) \, \mathrm{d}Y_s - \frac{1}{2}\int_0^t \pi_s(h^\top)\pi_s(h) \, \mathrm{d}s\right). \tag{3.55}$$

Proof. Because h is not bounded, it is not automatic that $\pi_t(h)$ is defined (h might not be integrable with respect to π_t). However (3.25) ensures that it is defined $\lambda \otimes \mathbb{P}$-a.s. which suffices. From the Zakai equation (3.43), since $A\mathbf{1} = 0$, one obtains that $\rho_t(\mathbf{1})$ satisfies the following equation,

$$\rho_t(\mathbf{1}) = 1 + \int_0^t \rho_s(h^\top) \, \mathrm{d}Y_s,$$

which gives
$$\rho_t(\mathbf{1}) = 1 + \int_0^t \rho_s(\mathbf{1}) \pi_s(h^\top) \, dY_s.$$

We cannot simply apply Itô's formula to $\log \rho_t(\mathbf{1})$ to conclude that $\rho_t(\mathbf{1})$ has the explicit form (3.55), because the function $x \mapsto \log x$ is not continuous at $x = 0$ (it is not even defined at 0) and we do not know a priori that $\rho_t(\mathbf{1}) > 0$.

Using the fact that $\rho_t(\mathbf{1})$ is non-negative, we use Itô's formula to compute for $\varepsilon > 0$

$$\begin{aligned}
d\left(\log \sqrt{\varepsilon + \rho_t(\mathbf{1})^2}\right) &= \frac{\rho_t(\mathbf{1})^2}{\varepsilon + \rho_t(\mathbf{1})^2} \pi_t(h^\top) \, dY_t \\
&\quad + \frac{1}{2} \frac{\varepsilon - \rho_t(\mathbf{1})^2}{(\varepsilon + \rho_t(\mathbf{1})^2)^2} \pi_s(h^\top) \pi_s(h) \, dt \\
&= \frac{\rho_t(\mathbf{1})^2}{\varepsilon + \rho_t(\mathbf{1})^2} \pi_t(h^\top) h(X_t) dt + \frac{\rho_t(\mathbf{1})^2}{\varepsilon + \rho_t(\mathbf{1})^2} \pi_t(h^\top) \, dW_t \\
&\quad + \frac{1}{2} \frac{\varepsilon - \rho_t(\mathbf{1})^2}{(\varepsilon + \rho_t(\mathbf{1})^2)^2} \pi_s(h^\top) \pi_s(h) \, dt. \quad (3.56)
\end{aligned}$$

From (3.25) the condition (2.4) is satisfied; thus

$$\int_0^t \left(\frac{\rho_s(\mathbf{1})^2}{\varepsilon + \rho_s(\mathbf{1})^2}\right)^2 \|\pi_s(h)\|^2 \, ds \le \int_0^t \|\pi_s(h)\|^2 \, ds < \infty \qquad \mathbb{P}\text{-a.s.}$$

and from (3.25) and (2.4)

$$\int_0^t \pi_s(h^\top) h(X_s) \, ds \le \sqrt{\int_0^t \|\pi_s(h)\|^2 \, ds \int_0^t \|h(X_s)\|^2 \, ds} \qquad \mathbb{P}\text{-a.s.}$$

Thus $s \mapsto \pi_s(h^\top) h(X_s)$ is integrable, so by dominated convergence the limit as $\varepsilon \to 0$ in (3.56) yields

$$\begin{aligned}
d(\log \rho_t(\mathbf{1})) &= \pi_t(h^\top) \left(h(X_t) dt + dW_t\right) - \tfrac{1}{2} \pi_t(h^\top) \pi_t(h) \, dt \\
&= \pi_t(h^\top) \, dY_t - \tfrac{1}{2} \pi_t(h^\top) \pi_t(h) \, dt.
\end{aligned}$$

Integrating this SDE, followed by exponentiation yields the desired result. □

Theorem 3.30. *If conditions (3.25) and (3.42) are satisfied then the conditional distribution of the signal π_t satisfies the following evolution equation, called the* Kushner–Stratonovich equation,

$$\begin{aligned}
\pi_t(\varphi) = \pi_0(\varphi) &+ \int_0^t \pi_s(A\varphi) \, ds \\
&+ \int_0^t \left(\pi_s(\varphi h^\top) - \pi_s(h^\top) \pi_s(\varphi)\right) (dY_s - \pi_s(h) \, ds), \quad (3.57)
\end{aligned}$$

for any $\varphi \in \mathcal{D}(A)$.

3.6 The Kushner–Stratonovich Equation

Proof. From Lemma 3.29 we obtain

$$\frac{1}{\rho_t(\mathbf{1})} = \exp\left(-\int_0^t \pi_s(h^\top)\,\mathrm{d}Y_s + \frac{1}{2}\int_0^t \pi_s(h^\top)\pi_s(h)\,\mathrm{d}s\right)$$

$$\mathrm{d}\left(\frac{1}{\rho_t(\mathbf{1})}\right) = \frac{1}{\rho_t(\mathbf{1})}\left[-\pi_t(h^\top)\mathrm{d}Y_t + \pi_t(h^\top)\pi_t(h)\mathrm{d}t\right]. \tag{3.58}$$

By using (stochastic) integration by parts, (3.58), the Zakai equation for $\rho_t(\varphi)$ and the Kallianpur–Striebel formula, we obtain the stochastic differential equation satisfied by π_t,

$$\pi_t(\varphi) = \rho_t(\varphi) \cdot \frac{1}{\rho_t(\mathbf{1})}$$

$$\mathrm{d}\pi_t(\varphi) = \pi_t(A\varphi)\mathrm{d}t + \pi_t(\varphi h^\top)\mathrm{d}Y_t - \pi_t(\varphi)\pi_t(h^\top)\mathrm{d}Y_t$$
$$+ \pi_t(\varphi)\pi_t(h^\top)\pi_t(h)\mathrm{d}t - \pi_t(\varphi h^\top)\pi_t(h)\mathrm{d}t$$

which gives us the result. □

Remark 3.31. The Zakai and Kushner–Stratonovich equations can be extended for time inhomogeneous test functions. Let $\varphi : [0,\infty) \times \mathbb{S} \to \mathbb{R}$ be a bounded measurable function and let $\varphi_t(\cdot) = \varphi(t,\cdot)$ for any $t \geq 0$. Then

$$\rho_t(\varphi_t) = \pi_0(\varphi_0) + \int_0^t \rho_s(\partial_s\varphi_s + A\varphi_s)\,\mathrm{d}s + \int_0^t \rho_s(\varphi_s h^\top)\,\mathrm{d}Y_s \tag{3.59}$$

$$\pi_t(\varphi_t) = \pi_0(\varphi_0) + \int_0^t \pi_s(\partial_s\varphi_s + A\varphi_s)\,\mathrm{d}s$$
$$+ \int_0^t (\pi_s(\varphi_s h^\top) - \pi_s(h^\top)\pi_s(\varphi_s))(\mathrm{d}Y_s - \pi_s(h)\,\mathrm{d}s) \tag{3.60}$$

for any $\varphi \in \mathcal{D}(A)$. This extension is carried out in Lemma 4.8.

Exercise 3.32. Consider once again the change detection filter introduced in Exercise 3.7. Starting from the result of this exercise define an observation process

$$Y_t = \int_0^t X_s\,\mathrm{d}s + W_t.$$

Show that the Kushner–Stratonovich equation for the process X takes the form

$$\mathrm{d}\pi_t(J_1) = \pi_t(J_1)(1 - \pi_t(J_1))\,(\mathrm{d}Y_t - \pi_t(J_1)\mathrm{d}t) + (1 - \pi_t(J_1))p_t/g_t\,\mathrm{d}t. \tag{3.61}$$

where J_1 is the indicator function of the singleton set $\{1\}$.

3.7 The Innovation Process Approach

Here we use the representation implied by Proposition 2.31 to derive the Kushner–Stratonovich equation. The following corollary gives us a representation for $\mathcal{Y}_t$-adapted martingales.

Corollary 3.33. *Under the conditions of Proposition 2.31 every right continuous square integrable martingale which is $\mathcal{Y}_t$-adapted has a representation*

$$\eta_t = \eta_0 + \int_0^t \nu_s^\top \, dI_s \qquad t \geq 0. \tag{3.62}$$

Proof. Following Proposition 2.31, for any $n \geq 0$, the $\mathcal{Y}_\infty$-measurable (square integrable) random variable $\eta_n - \eta_0$ has a representation of the form

$$\eta_n - \eta_0 = \int_0^\infty (\nu_s^n)^\top \, dI_s.$$

By conditioning with respect to $\mathcal{Y}_t$, for arbitrary $t \in [0, n]$, we get that

$$\eta_t = \eta_0 + \int_0^t (\nu_s^n)^\top \, dI_s, \qquad t \in [0, n].$$

The result follows by observing that the processes ν^n, $n = 1, 2, \ldots$ must be compatible. That is, for any $n, m > 0$, ν^n and ν^m are equal on the set $[0, \min(n, m)]$. □

We therefore identify a square integrable martingale to which the corollary 3.33 may be applied.

Lemma 3.34. *Define $N_t \triangleq \pi_t\varphi - \int_0^t \pi_s(A\varphi) \, ds$, then N is a $\mathcal{Y}_t$-adapted square integrable martingale under the probability measure $\mathbb{P}$.*

Proof. Recall that $\pi_t\varphi$ is indistinguishable from the $\mathcal{Y}_t$-optional projection of $\varphi(X_t)$, hence let T be a bounded $\mathcal{Y}_t$-stopping time such that $T(\omega) \leq K$ for all $\omega \in \Omega$. Then since $A\varphi$ is bounded it follows that we can apply Fubini's theorem combined with the definition of optional projection to obtain,

$$\mathbb{E} N_T = \mathbb{E}\left[\pi_T\varphi - \int_0^T \pi_s(A\varphi) \, ds\right]$$

$$= \mathbb{E}[\pi_T\varphi] - \mathbb{E}\left[\int_0^K 1_{[0,T]}(s)\pi_s(A\varphi) \, ds\right]$$

$$= \mathbb{E}[\varphi(X_T)] - \int_0^K \mathbb{E}\left[1_{[0,T]}(s)\pi_s(A\varphi)\right] \, ds$$

$$= \mathbb{E}[\varphi(X_T)] - \int_0^K \mathbb{E}\left[1_{[0,T]}(s)A\varphi(X_s)\right] \, ds$$

$$= \mathbb{E}[\varphi(X_T)] - \mathbb{E}\left[\int_0^T A\varphi(X_s) \, ds\right].$$

Then using the definition of the generator A in the form of (3.3), we can find M_t^φ an $\mathcal{F}_t$-adapted martingale such that

$$\mathbb{E}N_T = \mathbb{E}[\varphi(X_T)] - \mathbb{E}\left[\varphi(X_T) - \varphi(X_0) - M_T^\varphi\right]$$
$$= \mathbb{E}[\varphi(X_0)].$$

Thus since N_t is $\mathcal{Y}_t$-adapted, and this holds for all bounded $\mathcal{Y}_t$-stopping times, it follows by Lemma B.2 that N is a $\mathcal{Y}_t$-adapted martingale. Furthermore since $A\varphi$ is bounded for $\varphi \in \mathcal{D}(A)$, it follows that N_t is bounded and hence square integrable. $\square$

An alternative proof of Proposition 3.30 can now be given using the innovation process approach. The proposition is restated because the conditions under which it is proved via the innovations method differ slightly from those in Proposition 3.30.

Theorem 3.35. *If the conditions (2.3) and (2.4) are satisfied then the conditional distribution of the signal π satisfies the following evolution equation,*

$$\pi_t(\varphi) = \pi_0(\varphi) + \int_0^t \pi_s(A\varphi)\,\mathrm{d}s$$
$$+ \int_0^t \left(\pi_s(\varphi h^\top) - \pi_s(h^\top)\pi_s(\varphi)\right)(\mathrm{d}Y_s - \pi_s(h)\,\mathrm{d}s), \qquad (3.63)$$

for any $\varphi \in \mathcal{D}(A)$.

Proof. Let φ be an element of $\mathcal{D}(A)$. The process $N_t = \pi_t\varphi - \int_0^t \pi_s(A\varphi)\,\mathrm{d}s$ is by Lemma 3.34 a square integrable $\mathcal{Y}_t$-martingale. By assumption, condition (2.21) is satisfied, thus Corollary 3.33 allows us to find an integral representation for N_t. This means that there exists a progressively measurable process ν such that

$$N_t = \mathbb{E}N_0 + \int_0^t \nu_s^\top\,\mathrm{d}I_s = \pi_0(\varphi) + \int_0^t \nu_s^\top\,\mathrm{d}I_s; \qquad (3.64)$$

thus using the definition of N_t, we obtain the following evolution equation for the conditional distribution process π,

$$\pi_t(\varphi) = \pi_0(\varphi) + \int_0^t \pi_s(A\varphi)\,\mathrm{d}s + \int_0^t \nu_s^\top\,\mathrm{d}I_s. \qquad (3.65)$$

To complete the proof, it only remains to identify explicitly the process ν_t. Let $\varepsilon = (\varepsilon_t)_{t \geq 0}$ be the process as defined in (B.19), Lemma B.39. Thus

$$\mathrm{d}\varepsilon_t = i\varepsilon_t r_t^\top\,\mathrm{d}Y_t,$$

hence, by stochastic integration by parts (i.e. by applying Itô's formula to the products $\pi_t(\varphi)\varepsilon_t$ and $\varphi(X_t)\varepsilon_t$)

$$\pi_t(\varphi)\varepsilon_t = \pi_0(\varphi)\varepsilon_0 + \int_0^t \pi_s(A\varphi)\varepsilon_s\,ds + \int_0^t \nu_s^\top \varepsilon_s\,dI_s$$
$$+ \int_0^t \pi_s(\varphi)i\varepsilon_s r_s^\top (dI_s + \pi_s(h)ds) + \int_0^t i\varepsilon_s r_s^\top \nu_s\,ds \quad (3.66)$$

$$\varphi(X_t)\varepsilon_t = \varphi(X_0)\varepsilon_0 + \int_0^t A\varphi(X_s)\varepsilon_s\,ds + \int_0^t \varepsilon_s\,dM_s^\varphi + \int_0^t i\varepsilon_s r_s^\top\,d\langle M^\varphi, W\rangle_s$$
$$+ \int_0^t \varphi(X_s)i\varepsilon_s r_s^\top (h(X_s)ds + dW_s). \quad (3.67)$$

Since we have assumed that the signal process and the observation process noise are uncorrelated, $\langle M^\varphi, Y\rangle_t = \langle M^\varphi, W\rangle_t = 0$ consequently subtracting (3.67) from (3.66) and taking the expectation, all of the martingale terms vanish and we obtain

$$\int_0^t ir_s^\top \mathbb{E}\left[\varepsilon_s\left(\nu_s - \varphi(X_s)h(X_s) + \pi_s(h)\pi_s(\varphi)\right)\right]ds$$
$$= \mathbb{E}\left[\varepsilon_t\left(\pi_t(\varphi) - \varphi(X_t)\right)\right] + \mathbb{E}\left[\varepsilon_0\left(\pi_0(\varphi) - \varphi(X_0)\right)\right]$$
$$+ \mathbb{E}\left[\int_0^t \varepsilon_s\left(A\varphi(X_s) - \pi_s(A\varphi)\right)ds\right]$$
$$= \mathbb{E}\left[\varepsilon_t\left(\mathbb{E}\left[\varphi(X_t)\mid \mathcal{Y}_t\right] - \varphi(X_t)\right)\right]$$
$$= 0.$$

Hence, for almost all $t \geq 0$,
$$\mathbb{E}\left[\varepsilon_t\left(\nu_t - \varphi(X_t)h(X_t) + \pi_t(\varphi)\pi_t(h)\right)\right] = 0,$$
so since ε_t belongs to a total set it follows that
$$\nu_t = \pi_t(\varphi h) - \pi_t(\varphi)\pi_t(h), \qquad \text{P-a.s.} \quad (3.68)$$

Using the expression for $\pi_t(\varphi)$ given by (3.65) expressing the final term using the representation (3.64) with ν_t given by (3.68)

$$\pi_t(\varphi) = \pi_0(\varphi) + \int_0^t \pi_s(A\varphi)\,ds + \int_0^t \left(\pi_s\left(\varphi h^\top\right) - \pi_s(\varphi)\pi_s\left(h^\top\right)\right)dI_s, \quad (3.69)$$

which is the Kushner–Stratonovich equation as desired. □

The following exercise shows how the filtering equations can be derived in a situation which on first inspection does not appear to have an interpretation as a filtering problem, but which can be approached via the innovation process method.

Exercise 3.36. Define the $\mathcal{F}_t$-adapted semimartingale α via
$$\alpha_t = \alpha_0 + \int_0^t \beta_s\,ds + V_t, \qquad t \geq 0$$

and
$$\delta_t = \delta_0 + \int_0^t \gamma_s \, ds + W_t, \quad t \geq 0,$$

where β_t and γ_t are bounded progressively measurable processes and where W is an $\mathcal{F}_t$-adapted Brownian motion which is independent of β and γ. Define $\mathcal{D}_t = \sigma(\delta_s; \ 0 \leq s \leq t) \vee \mathcal{N}$. Find the equivalent of the Kushner–Stratonovich equation for $\pi_t(\varphi) = \mathbb{E}\left[\varphi(\alpha_t) \mid \mathcal{D}_t\right]$.

The following exercise shows how one can deduce the Zakai equation from the Kushner–Stratonovich equation. For this introduce the exponential martingale $\hat{Z} = \{\hat{Z}_t, \ t > 0\}$ defined by

$$\hat{Z}_t \triangleq \exp\left(\int_0^t \pi_s\left(h^\top\right) \, dY_s - \frac{1}{2}\int_0^t \|\pi_s(h)\|^2 \, ds\right), \quad t \geq 0.$$

Exercise 3.37. i. Show that
$$d\left(\frac{1}{\hat{Z}_t}\right) = -\frac{1}{\hat{Z}_t}\pi_t\left(h^\top\right) \, dI_t.$$

ii. Show that for any ε_t from the total set S_t as defined in Lemma B.39,
$$\mathbb{E}\left[\frac{\varepsilon_t}{\hat{Z}_t}\right] = \mathbb{E}\left[\varepsilon_t Z_t\right].$$

iii. Show that $\hat{Z}_t = \tilde{\mathbb{E}}\left[\tilde{Z}_t \mid \mathcal{Y}_t\right] = \rho_t(\mathbf{1})$.

iv. Use the Kallianpur–Striebel formula to deduce the Zakai equation.

3.8 The Correlated Noise Framework

Hitherto the noise in the observations W has been assumed to be independent of the signal process X. In this section we extend the results to the case when this noise W is correlated to the signal.

As in the previous section, the signal process $\{X_t, \ t \geq 0\}$ is the solution of a martingale problem associated with the generator A. That is, for $\varphi \in \mathcal{D}(A)$,
$$M_t^\varphi \triangleq \varphi(X_t) - \varphi(X_0) - \int_0^t A\varphi(X_s) \quad t \geq 0$$

is a martingale.

We assume that there exists a vector of operators $B = (B_1, \ldots, B_m)^\top$ such that $B_i \colon B(\mathbb{S}) \to B(\mathbb{S})$ for $i = 1, \ldots, m$. Let $\mathcal{D}(B_i) \subseteq B(\mathbb{S})$ denote the domain of the operator B_i. We require for each $i = 1, \ldots, m$ that $B_i \mathbf{1} = 0$ and for $\varphi \in \mathcal{D}(B_i)$,

$$\langle M^\varphi, W^i \rangle_t = \int_0^t B_i\varphi(X_s) \, ds. \tag{3.70}$$

Define
$$\mathcal{D}(B) \triangleq \bigcap_{i=1}^{n} \mathcal{D}(B_i).$$

Corollary 3.38. *In the correlated noise case, the Kushner–Stratonovich equation is*
$$d\pi_t(\varphi) = \pi_t(A\varphi)dt + (\pi_t(h^\top \varphi) - \pi_t(h^\top)\pi_t(\varphi) + \pi_t(B^\top \varphi))$$
$$\times (dY_t - \pi_t(h)dt), \quad \text{for all } \varphi \in \mathcal{D}(A) \cap \mathcal{D}(B). \quad (3.71)$$

Proof. We now follow the innovations proof of the Kushner–Stratonovich equation. However, using (3.70) the term
$$\int_0^t i\varepsilon_s r_s^\top \, d\langle M^\varphi, W\rangle_s = \int_0^t i\varepsilon_s r_s^\top B\varphi(X_s) \, ds.$$
Inserting this term, we obtain instead of (3.68),
$$\nu_t = \pi_t(\varphi h) - \pi_t(\varphi)\pi_t(h) + \pi_t(B\varphi), \quad \mathbb{P}\text{-a.s.}$$
and using this in (3.65) yields the result. □

Corollary 3.39. *In the correlated noise case, for $\varphi \in B(\mathbb{S})$, the Zakai equation is*
$$\rho_t(\varphi) = \rho_0(\varphi) + \int_0^t \rho_s(A\varphi) \, ds + \int_0^t \rho_s((h^\top + B^\top)\varphi) \, dY_s. \quad (3.72)$$

Consider the obvious extension of the diffusion process example studied earlier to the case where the signal process is a diffusion given by
$$dX_t = b(X_t) \, dt + \sigma(X_t) \, dV_t + \bar{\sigma}(X_t) \, dW_t; \quad (3.73)$$
thus $\bar{\sigma}$ is a $d \times m$ matrix-valued process. If $\bar{\sigma} \equiv 0$ this case reduces to the uncorrelated case which was studied previously.

Corollary 3.40. *When the signal process is given by (3.73), the operator $B = (B_i)_{i=1}^m$ defined by (3.70) is given for $k = 1, \ldots, m$ by*
$$B_k = \sum_{i=1}^d \bar{\sigma}_{ik} \frac{\partial}{\partial x_i}. \quad (3.74)$$

Proof. Denoting by $\mathcal{A}$ the generator of X,
$$M_t^\varphi = \varphi(X_t) - \varphi(X_0) - \int_0^t \mathcal{A}\varphi(X_s) \, ds$$
$$= \sum_{i=1}^d \int_0^t \frac{\partial \varphi}{\partial x_i}(\sigma dV_s)^i + \sum_{i=1}^d \int_0^t \frac{\partial \varphi}{\partial x_i}(\bar{\sigma} dW_s)^i.$$

Thus

$$\langle M^\varphi, W^k\rangle_t = \sum_{i=1}^d \sum_{j=1}^m \int_0^t \frac{\partial \varphi}{\partial x_i} \bar\sigma_{ij}\, \mathrm{d}\langle W^j, W^k\rangle_s$$

$$= \sum_{i=1}^d \int_0^t \frac{\partial \varphi}{\partial x_i} \bar\sigma_{ik}\, \mathrm{d}s$$

and the result follows from (3.70). □

3.9 Solutions to Exercises

3.3 From (3.10) with $y = 0$,

$$\|\sigma(x) - \sigma(0)\| \le K\|x\|,$$

by the triangle inequality

$$\|\sigma(x)\| \le \|\sigma(x) - \sigma(0)\| + \|\sigma(0)\|$$
$$\le \|\sigma(0)\| + K\|x\|.$$

Thus since $(a+b)^2 \le 2a^2 + 2b^2$,

$$\|\sigma(x)\|^2 \le 2\|\sigma(0)\|^2 + 2K^2\|x\|^2;$$

thus setting $\kappa_1 = \max(2\|\sigma(0)\|^2, 2K^2)$, we see that

$$\|\sigma(x)\|^2 \le \kappa_1(1 + \|x\|^2).$$

Similarly from (3.10) with $y = 0$, and the triangle inequality, it follows that

$$\|f(x)\| \le \|f(0)\| + K\|x\|,$$

so setting $\kappa_2 = \max(\|f(0)\|, K)$,

$$\|f(x)\| \le \kappa_2(1 + \|x\|).$$

The result follows if we take $\kappa = \max(\kappa_1, \kappa_2)$.

For the final part, note that

$$(\sigma\sigma^\top)_{ij} = \sum_{k=1}^p \sigma_{ik}\sigma_{jk},$$

hence $|(\sigma\sigma^\top)_{ij}(x)| \le p\|\sigma\|^2$, consequently

$$\|\sigma(x)\sigma^\top(x)\| \le pd^2\kappa(1 + \|x\|^2);$$

thus we set $\kappa' = pd^2\kappa$ to get the required result.

3.4 First we must check that $A\varphi$ is bounded for $\varphi \in \mathrm{SL}^2(\mathbb{R}^d)$. By the result of Exercise 3.3, with $\kappa' = \kappa pd^2/2$,

$$\|a\| = \tfrac{1}{2}\|\sigma(x)\sigma^\top(x)\| \leq \kappa'(1+\|x\|^2).$$

Hence

$$|A\varphi(x)| \leq \sum_{i=1}^d |f_i(x)||\partial_i\varphi(x)| + \sum_{i,j=1}^d |a^{ij}(x)||\partial_i\partial_j\varphi(x)|$$

$$\leq \sum_{i=1}^d |f_i(x)|\frac{C}{1+\|x\|} + \sum_{i,j=1}^d |a^{ij}(x)|\frac{C}{1+\|x\|^2}$$

$$\leq Cd\kappa + Cpd^2\kappa' < \infty,$$

so $A\varphi \in B(\mathbb{R}^d)$. By Itô's formula since $\varphi \in C^2(\mathbb{R}^d)$,

$$\varphi(X_t) = \varphi(X_0) + \int_0^t \sum_{i=1}^d \partial_i\varphi(X_s)\left(f^i(X_s)\,\mathrm{d}s + \sum_{j=1}^p \sigma^{ij}\,\mathrm{d}V_s^j\right)$$

$$+ \frac{1}{2}\int_0^t \sum_{i,j=1}^d \partial_i\partial_j\varphi(X_s) \sum_{k=1}^p \sigma^{ik}(X_s)\sigma^{jk}(X_s)\,\mathrm{d}s.$$

Hence

$$M_t^\varphi = \sum_{i=1}^d \int_0^t \partial_i\varphi(X_s) \sum_{j=1}^p \sigma^{ij}(X_s)\,\mathrm{d}V_s^j,$$

which is clearly a local martingale. Consider

$$\sum_{i=1}^d \int_0^t |\partial_i\varphi(X_s)|^2 \left|\sum_{j=1}^p \sigma^{ij}(X_s)\right|^2 \mathrm{d}s \leq p\int_0^t \frac{C^2\|\sigma(X_s)\|^2}{(1+\|X_s\|)^2}\,\mathrm{d}s$$

$$\leq C^2 p \int_0^t \frac{pd^2\kappa(1+\|X_s\|^2)}{(1+\|X_s\|)^2}\,\mathrm{d}s$$

$$\leq C^2 p^2 d^2 \kappa t < \infty.$$

Hence M^φ is a martingale.

3.6 It is sufficient to show that for all $i \in I$, the process $M^i = \{M_t^i,\ t \geq 0\}$ defined as

$$M_t^i = J_i(X_t) - J_i(X_0) - \int_0^t q_{X_s,i}(s)\,\mathrm{d}s, \quad t \geq 0,$$

where J_i is the indicator function of the singleton set $\{i\}$, is an $\mathcal{F}_t$-adapted right-continuous martingale. This is sufficient since

$$M^\varphi = \sum_{i \in I} \varphi(i) M^i, \quad \text{for all } \varphi \in B(\mathbb{S}).$$

Thus if M^i is a martingale for $i \in I$ then so is M^φ which establishes the result.

The adaptedness, integrability and right continuity of M_t^i are straightforward. From (3.16) and using the Markov property for $0 \le s \le t$,

$$\begin{aligned}
\mathbb{P}(X_t = i \mid \mathcal{F}_s) &= \mathbb{E}\left(\mathbb{E}\left(1_{\{X_t = i\}} \mid \mathcal{F}_{t-h}\right) \mid \mathcal{F}_s\right) \\
&= \mathbb{E}\left[\mathbb{P}(X_t = i \mid X_{t-h}) \mid \mathcal{F}_s\right] \\
&= \mathbb{E}[J_i(X_{t-h}) \mid \mathcal{F}_s] + \mathbb{E}\left[q_{X_{t-h}, i}(t-h) \mid \mathcal{F}_s\right] h + o(h) \\
&= \mathbb{P}(X_{t-h} = i \mid \mathcal{F}_s) + \mathbb{E}\left[q_{X_{t-h}, i}(t-h) \mid \mathcal{F}_s\right] h + o(h).
\end{aligned}$$

It is clear that we may apply this iteratively; the error term is $o(h)/h$ which by definition tends to zero as $h \to 0$. Doing this and passing to the limit as $h \to 0$ we obtain

$$\mathbb{P}(X_t = i \mid X_s) = J_i(X_s) + \mathbb{E}\left[\int_s^t q_{X_r, i}(r) \, \mathrm{d}r \,\bigg|\, \mathcal{F}_s\right].$$

Now

$$\begin{aligned}
\mathbb{E}[M_t^i \mid \mathcal{F}_s] &= \mathbb{P}(X_t = i \mid \mathcal{F}_s) - J_i(X_0) - \mathbb{E}\left[\int_0^t q_{X_r, i}(r) \, \mathrm{d}r \,\bigg|\, \mathcal{F}_s\right] \\
&= J_i(X_s) - J_i(X_0) - \int_0^s q_{X_r, i}(r) \, \mathrm{d}r \\
&= M_s^i.
\end{aligned}$$

It follows that M_t^i is a martingale.

3.7 Clearly the state space of X is $\{0, 1\}$. Once in state 1 the process never leaves the state 1 hence $q_{10}(t) = q_{11}(t) = 0$. Consider the transition from state 0 to 1,

$$\begin{aligned}
\mathbb{P}(X_{t+h} = 1 \mid X_t = 0) &= \mathbb{P}(T \le t + h \mid T > t) = \frac{\mathbb{P}(t < T \le t + h)}{\mathbb{P}(T > t)} \\
&= \frac{p_t}{g_t} h + o(h).
\end{aligned}$$

Thus $q_{01}(t) = p_t/g_t$ and hence $q_{00}(t) = -q_{01}(t) = -p_t/g_t$.

3.10 By Itô's formula

$$\mathrm{d}\left(\|X_t\|^2\right) = 2 X_t^\top \left(f(X_t) \mathrm{d}t + \sigma(X_t) \mathrm{d}V_t\right) + \mathrm{tr}\left(\sigma(X_t) \sigma^\top(X_t)\right) \mathrm{d}t, \quad (3.75)$$

Thus if we define

$$M_t \triangleq \int_0^t 2 X_s^\top \sigma(X_s) \, \mathrm{d}V_s,$$

this is clearly a local martingale. Take T_n a reducing sequence (see Definition B.4) such that $M_t^{T_n}$ is a martingale for all n and $T_n \to \infty$. Integrating between 0 and $t \wedge T_n$ and taking expectation, $\mathbb{E}M_{t \wedge T_n} = 0$, hence

$$\mathbb{E}\|X_{t \wedge T_n}\|^2 = \mathbb{E}\|X_0\|^2 + \mathbb{E}\int_0^{t \wedge T_n} 2X_s^T f(X_s) + \operatorname{tr}(\sigma(X_s)\sigma^\top(X_s))\,\mathrm{d}s.$$

By the results of Exercise 3.3,

$$\mathbb{E}\|X_{t \wedge T_n}\|^2 \leq \mathbb{E}\|X_0\|^2 + \mathbb{E}\int_0^{t \wedge T_n} 2d\kappa\|X_s\|(1+\|X_s\|) + \kappa'(1+\|X_s\|^2)\,\mathrm{d}s$$

so setting $c = \max(2d\kappa, 2d\kappa + \kappa', \kappa') > 0$,

$$\mathbb{E}\|X_{t \wedge T_n}\|^2 \leq \mathbb{E}\|X_0\|^2 + c\mathbb{E}\int_0^{t \wedge T_n}(1+\|X\|_s + \|X_s\|^2)\,\mathrm{d}s.$$

But by Jensen's inequality for $p > 1$, it follows that for Y a non-negative random variable

$$\mathbb{E}[Y] \leq (\mathbb{E}[Y^p])^{1/p} \leq 1 + \mathbb{E}[Y^p].$$

Thus

$$1 + \mathbb{E}\|X_{t \wedge T_n}\|^2 \leq 1 + \mathbb{E}\|X_0\|^2 + 2c\int_0^{t \wedge T_n} \mathbb{E}[1+\|X_s\|^2]\,\mathrm{d}s,$$

and by Corollary A.40 to Gronwall's lemma

$$1 + \mathbb{E}\|X_{t \wedge T_n}\|^2 \leq (1 + \mathbb{E}\|X_0\|^2)e^{2c(t \wedge T_n)}.$$

We may take the limit as $n \to \infty$ by Fatou's lemma to obtain

$$\mathbb{E}\|X_t\|^2 \leq (1 + \mathbb{E}\|X_0\|^2)e^{2ct} - 1, \tag{3.76}$$

which establishes the result for the second moment.

In the case of the third moment, applying Itô's formula to $f(x) = x^{3/2}$ and the process $\|X_t\|^2$ yields

$$\mathrm{d}\left(\|X_t\|^3\right) = 3\|X_t\|\left(2X_t^\top(f(X_t)\mathrm{d}t + \sigma(X_t)\mathrm{d}V_t) + \operatorname{tr}(\sigma(X_t)\sigma^\top(X_t))\mathrm{d}t\right)$$
$$+ \frac{3}{2\|X_t\|}X_t^\top \sigma(X_t)\sigma^\top(X_t)X_t\mathrm{d}t.$$

Define

$$N_t \triangleq 6\int_0^t \|X_s\|X_s^\top \sigma(X_s)\,\mathrm{d}V_s,$$

and let T_n be a reducing sequence for the local martingale N_t. Integrating between 0 and $t \wedge T_n$ and taking expectation, we obtain for some constant $c > 0$ (independent of n, t) that

$$\mathbb{E}[\|X_{t\wedge T_n}\|^3] \leq \mathbb{E}[\|X_0\|^3] + c\int_0^{t\wedge T_n} \mathbb{E}[\|X_s\| + \|X_s\|^2 + \|X_s\|^3]\,\mathrm{d}s,$$

using Jensen's inequality as before,

$$\mathbb{E}[\|X_{t\wedge T_n}\|^3] \leq \mathbb{E}[\|X_0\|^3] + 3c\int_0^{t\wedge T_n} 1 + \mathbb{E}[\|X_s\|^3]\,\mathrm{d}s,$$

thus by Corollary A.40 to Gronwall's lemma

$$\mathbb{E}[\|X_{t\wedge T_n}\|^3 + 1] \leq \mathbb{E}[\|X_0\|^3] + (1+\mathbb{E}\|X_0\|^3)\mathrm{e}^{3c(t\wedge T_n)},$$

passing to the limit as $n \to \infty$ using Fatou's lemma

$$\mathbb{E}[\|X_t\|^3] \leq (1 + \mathbb{E}[\|X_0\|^3])\mathrm{e}^{3ct} - 1, \tag{3.77}$$

and since $\mathbb{E}[\|X_0\|^3] < \infty$ (X_0 has finite third moment) this yields the result.

3.11

i. As a consequence of the linear growth bound on h,

$$\mathbb{E}\left(\int_0^t \|h(X_s)\|^2\,\mathrm{d}s\right) \leq C\mathbb{E}\left(\int_0^t (1+\|X_s\|^2)\,\mathrm{d}s\right) \leq Ct + C\mathbb{E}\int_0^t \|X_s\|^2\,\mathrm{d}s.$$

It follows by Jensen's inequality that

$$\mathbb{E}[\|X_t\|^2] \leq \left[\mathbb{E}\|X_t\|^3\right]^{2/3}.$$

Since the conditions (3.10) are satisfied and the second moment of X_0 is finite, we can use the bound derived in Exercise 3.7 as (3.76); viz

$$\mathbb{E}[\|X_t\|^2] \leq (\mathbb{E}\|X_0\|^2 + 1)\mathrm{e}^{2ct}.$$

Consequently for $t \geq 0$,

$$\mathbb{E}\left(\int_0^t \|h(X_s)\|^2\,\mathrm{d}s\right) \leq Ct + C\left(\mathbb{E}[\|X_0\|^2] + 1\right)\frac{\mathrm{e}^{2ct}-1}{2c}\right) < \infty. \tag{3.78}$$

This establishes the first of the conditions (3.25). For the second condition, using the result of (3.75), Itô's formula yields

$$\mathrm{d}\left(Z_t\|X_t\|^2\right) = Z_t\left(2X_t^\top(f(X_t)\mathrm{d}t + \sigma(X_t)\mathrm{d}V_t) + \mathrm{tr}\left(\sigma(X_t)\sigma^\top(X_t)\right)\mathrm{d}t\right)$$
$$- Z_t\|X_t\|^2 h^\top(X_t)\mathrm{d}Y_t.$$

Thus applying Itô's formula to the function $f(x) = x/(1+\varepsilon x)$ and the process $Z_t\|X_t\|^2$ yields

3 The Filtering Equations

$$d\left(\frac{Z_t\|X_t\|^2}{1+\varepsilon Z_t\|X_t\|^2}\right) = \frac{1}{(1+\varepsilon Z_t\|X_t\|^2)^2} d\left(Z_t\|X_t\|^2\right)$$
$$- \frac{\varepsilon}{(1+\varepsilon Z_t\|X_t\|^2)^3}\left(Z_t^2\|X_t\|^4 h^\top(X_t)h(X_t)\right.$$
$$\left. + 4Z_t^2 X_t^\top \sigma(X_t)\sigma^\top(X_t)X_t\right)dt. \qquad (3.79)$$

Integrating between 0 and t and taking expectation, the stochastic integrals are local martingales; we must show that they are martingales. Consider first the term

$$\int_0^t \frac{Z_s 2 X_s^\top \sigma(X_s)}{(1+\epsilon Z_s\|X_s\|^2)^2} dV_s;$$

to show that this is a martingale we must therefore establish that

$$\mathbb{E}\left[\int_0^t \left\|\frac{Z_s 2 X_s^\top \sigma}{(1+\epsilon Z_s\|X_s\|^2)^2}\right\|^2 ds\right] = 4\mathbb{E}\left[\int_0^t \frac{Z_s^2 X_s^\top \sigma\sigma^\top X_s}{(1+\epsilon Z_s\|X_s\|^2)^4} ds\right] < \infty.$$

In order to establish this inequality notice that

$$|X_t^\top \sigma(X_t)\sigma^\top(X_t)X_t| \le d^2 \|X_t\|^2 \|\sigma(X_t)\sigma^\top(X_t)\|,$$

and from Exercise 3.3

$$\|\sigma\sigma^\top\| \le \kappa'(1+\|X\|^2),$$

hence

$$|X_t\sigma(X_t)\sigma(X_t)X_t| \le d^2\kappa'\|X_t\|^2\left(1+\|X_t\|^2\right),$$

so the integral may be bounded by

$$\int_0^t \frac{Z_s^2 X_s^\top \sigma\sigma^\top X_s}{(1+\varepsilon Z_t\|X_t\|^2)^4} ds \le \kappa' d^2 \int_0^t \frac{Z_s^2 \|X_s\|^2 \left(1+\|X_s\|^2\right)}{(1+\varepsilon Z_t\|X_s\|^2)^4} ds$$
$$= \kappa' d^2 \int_0^t \frac{Z_s^2 \|X_s\|^2}{(1+\varepsilon Z_s\|X_s\|^2)^4} + \frac{Z_s^2\|X_s\|^4}{(1+\varepsilon Z_s\|X_s\|^2)^4} ds.$$

Considering each term of the integral separately, the first satisfies

$$\int_0^t \frac{Z_s^2\|X_s\|^2}{(1+\varepsilon Z_s\|X_s\|^2)^4} ds \le \int_0^t Z_s \times \frac{Z_s\|X_s\|^2}{(1+\varepsilon Z_t\|X_t\|^2)} \times \frac{1}{(1+\varepsilon Z_t\|X_t\|^2)^3} ds$$
$$\le \int_0^t \frac{Z_s}{\varepsilon} ds \le \frac{1}{\varepsilon}\int_0^t Z_s\, ds.$$

Thus the expectation of this integral is bounded by t/ε, because $\mathbb{E}[Z_s] \le 1$. Similarly for the second term,

$$\int_0^t \left[\frac{Z_s\|X_s\|^2}{(1+\varepsilon Z_s\|X_s\|^2)^2}\right]^2 ds \le \int_0^t \frac{Z_s^2\|X_s\|^4}{(1+\varepsilon Z_s\|X_s\|^2)^2} \times \frac{1}{(1+\epsilon Z_s\|X_s\|^2)^2} ds$$
$$\le \frac{1}{\varepsilon^2} t < \infty.$$

For the second stochastic integral term,
$$-\int_0^t \frac{Z_s\|X_s\|^2 h^\top(X_s)}{(1+\varepsilon Z_s\|X_s\|^2)^2} \, dV_s,$$
to show that this is a martingale, we must show that
$$\mathbb{E}\left[\int_0^t \frac{Z_s^2\|X_s\|^4 \|h(X_s)\|^2}{(1+\varepsilon Z_s\|X_s\|^2)^4} \, ds\right] < \infty.$$

Thus bounding this integral
$$\int_0^t \frac{Z_s^2\|X_s\|^4 \|h(X_s)\|^2}{(1+\varepsilon Z_s\|X_s\|^2)^4} ds \le \int_0^t \left(\frac{Z_s\|X_s\|^2}{(1+\varepsilon Z_s\|X_s\|^2)}\right)^2 \frac{\|h(X_s)\|^2}{(1+\varepsilon Z_t\|X\|^2)^2} ds$$
$$\le \frac{C}{\varepsilon^2}\int_0^t \|h(X_s)\|^2 \, ds.$$

Taking expectation, and using the result (3.78),
$$\mathbb{E}\left[\int_0^t \frac{Z_s^2\|X_s\|^4 \|h(X_s)\|^2}{(1+\varepsilon Z_s\|X_s\|^2)^4} ds\right] \le \frac{C}{\varepsilon^2}\mathbb{E}\left(\int_0^t \|h(X_s)\|^2 \, ds\right) < \infty.$$

Therefore we have established that the stochastic integrals in (3.79) are martingales and have zero expectation. Consider now the remaining terms; by an application of Fubini's theorem, we see that
$$\frac{d}{dt}\mathbb{E}\left[\frac{Z_t\|X_t\|^2}{1+\varepsilon Z_t\|X_t\|^2}\right] \le \mathbb{E}\left[\frac{Z_t\left(2X_t^\top f(X_t) + \operatorname{tr}\left(\sigma(X_t)\sigma^\top(X_t)\right)\right)}{1+\varepsilon Z_t\|X_t\|^2}\right]$$
$$\le K\left(\mathbb{E}\left[\frac{Z_t\|X_t\|^2}{1+\varepsilon Z_t\|X_t\|^2}\right] + 1\right),$$

where we used the fact that $\mathbb{E}[Z_t] \le 1$. Hence, by Corollary A.40 to Gronwall's inequality there exists K_t such that for $0 \le s \le t$,
$$\mathbb{E}\left[\frac{Z_s\|X_s\|^2}{1+\varepsilon Z_s\|X_s\|^2}\right] \le K_t < \infty,$$

by Fatou's lemma as $\varepsilon \to 0$,
$$\mathbb{E}\left[Z_s\|X_s\|^2\right] \le K_t < \infty.$$

Then by Fubini's theorem

$$\mathbb{E}\left[\int_0^t Z_s\|h(X_s)\|^2\,ds\right] = \mathbb{E}\left[\int_0^t Z_s \sum_{i=1}^m h^i(X_s)^2\,ds\right]$$
$$= \int_0^t \mathbb{E}\left[Z_s\,\|h(X_s)\|^2\right]ds$$
$$\leq C\int_0^t \mathbb{E}\left[Z_s\left(1+\|X_s\|^2\right)\right]ds \leq Ct(1+K_t) < \infty,$$

which establishes the second condition in (3.25).

ii. Let $H = \max_{i\in I}|h(\{i\})|$, as the state space I is finite, it is clear that $H < \infty$. Therefore

$$\mathbb{E}\left[\int_0^t \|h(X_s)\|^2\,ds\right] \leq \mathbb{E}[Ht] = Ht < \infty,$$

which establishes the first condition of (3.25). For the second condition by Fubini's theorem and the fact that $Z_t \geq 0$,

$$\mathbb{E}\left[\int_0^t Z_s\|h(X_s)\|^2\,ds\right] \leq H\int_0^t \mathbb{E}[Z_s]\,ds \leq Ht < \infty.$$

Thus both conditions in (3.25) are satisfied ($\mathbb{E}[Z_s] \leq 1$ for any $s \in [0,\infty)$).

3.14

i. It is clear that P_t is $\mathcal{X} \vee \mathcal{F}_t$-measurable and that it is integrable. Now for $0 \leq s \leq t$,

$$\mathbb{E}\left[P_t \mid \mathcal{X} \vee \mathcal{F}_s\right] = \mathbb{E}\left[\exp\left(i\beta^\top Y_t - \tfrac{1}{2}\|\beta\|^2 t\right) Z_t \,\big|\, \mathcal{X} \vee \mathcal{F}_s\right]$$
$$= \frac{\tilde{\mathbb{E}}\left[\exp\left(i\beta^\top Y_t - \tfrac{1}{2}\|\beta\|^2 t\right) \mid \mathcal{X} \vee \mathcal{F}_s\right]}{\tilde{\mathbb{E}}\left[\tilde{Z}_t \mid \mathcal{X} \vee \mathcal{F}_s\right]}$$
$$= \tilde{Z}_s^{-1}\exp\left(i\beta^\top Y_s - \tfrac{1}{2}\|\beta\|^2 s\right) = P_s.$$

Hence P_t is a $\mathcal{X} \vee \mathcal{F}_t$ martingale under $\mathbb{P}$.

ii. For notational convenience let us fix $t_0 = 0$ and define

$$l_i = \sum_{j=i}^n \beta_j.$$

Since W is independent of $\mathcal{X}$ it follows that

$$\mathbb{E}\left[\exp\left(\sum_{j=1}^{n} i\beta_j^\top W_{t_j}\right)\bigg|\mathcal{X}\right] = \mathbb{E}\left[\exp\left(\sum_{j=1}^{n} i\beta_j^\top W_{t_j}\right)\right]$$

$$= \mathbb{E}\left[\exp\left(\sum_{j=1}^{n} il_j^\top (W_{t_j} - W_{t_{j-1}})\right)\right]$$

$$= \exp\left(\sum_{j=1}^{n} \tfrac{1}{2}\|l_j\|^2 (t_j - t_{j-1})\right).$$

For the left-hand side we write

$$\mathbb{E}\left[\exp\left(\sum_{j=1}^{n} i\beta_j^\top Y_{t_j}\right) Z_{t_n}\bigg|\mathcal{X}\right] = \mathbb{E}\left[\exp\left(i\sum_{j=1}^{n} l_j^\top (Y_{t_j} - Y_{t_{j-1}})\right) Z_{t_n}\bigg|\mathcal{X}\right]$$

$$= \mathbb{E}\bigg[Z_{t_1}\exp(il_1^\top Y_{t_1})\frac{Z_{t_2}\exp(il_2^\top Y_{t_2})}{Z_{t_1}\exp(il_2^\top Y_{t_1})}$$

$$\times \cdots \times \frac{Z_{t_n}\exp(il_n^\top Y_{t_n})}{Z_{t_{n-1}}\exp(il_n^\top Y_{t_{n-1}})}\bigg|\mathcal{X}\bigg].$$

Write $P_t(l) = \exp\big(il^\top Y_t - \tfrac{1}{2}\|l\|^2 t\big) Z_t$; then

$$\mathbb{E}\left[\exp\left(\sum_{j=1}^{n} i\beta_j^\top Y_{t_j}\right) Z_{t_n}\bigg|\mathcal{X}\right]$$

$$= \mathbb{E}\left[P_{t_1}(l_1)\frac{P_{t_2}(l_2)}{P_{t_1}(l_2)}\cdots\frac{P_{t_{n-1}}(l_{n-1})}{P_{t_{n-2}}(l_{n-1})}\frac{P_{t_n}(l_n)}{P_{t_{n-1}}(l_n)}\bigg|\mathcal{X}\right]$$

$$\times \left[\exp\left(\sum_{j=1}^{n} \tfrac{1}{2}\|l_j\|^2 (t_j - t_{j-1})\right)\right].$$

From part (i) we know that $P_t(l)$ is a $\mathcal{X} \vee \mathcal{F}_t$ martingale for each $l \in \mathbb{R}^m$; thus conditioning on $\mathcal{X} \vee \mathcal{F}_{t_{n-1}}$,

$$\mathbb{E}\left[P_{t_1}(l_1)\frac{P_{t_2}(l_2)}{P_{t_1}(l_2)}\cdots\frac{P_{t_{n-1}}(l_{n-1})}{P_{t_{n-2}}(l_{n-1})}\frac{P_{t_n}(l_n)}{P_{t_{n-1}}(l_n)}\bigg|\mathcal{X}\right]$$

$$= \mathbb{E}\left[\frac{P_{t_1}(l_1)}{P_{t_1}(l_2)}\frac{P_{t_2}(l_2)}{P_{t_2}(l_3)}\cdots\frac{P_{t_{n-1}}(l_{n-1})}{P_{t_{n-1}}(l_n)}P_{t_n}(l_n)\bigg|\mathcal{X}\right]$$

$$= \mathbb{E}\left[\frac{P_{t_1}(l_1)}{P_{t_1}(l_2)}\frac{P_{t_2}(l_2)}{P_{t_2}(l_3)}\cdots\frac{P_{t_{n-1}}(l_{n-1})}{P_{t_{n-1}}(l_n)}\mathbb{E}\left[P_{t_n}(l_n)\mid\mathcal{X}\vee\mathcal{F}_{t_{n-1}}\right]\bigg|\mathcal{X}\right]$$

$$= \mathbb{E}\left[\frac{P_{t_1}(l_1)}{P_{t_1}(l_2)}\frac{P_{t_2}(l_2)}{P_{t_2}(l_3)}\cdots\frac{P_{t_{n-2}}(l_{n-2})}{P_{t_{n-2}}(l_{n-1})}P_{t_{n-1}}(l_{n-1})\bigg|\mathcal{X}\right].$$

Repeating this conditioning we obtain

$$\mathbb{E}\left[P_{t_1}(l_1)\frac{P_{t_2}(l_2)}{P_{t_1}(l_2)}\cdots\frac{P_{t_{n-1}}(l_{n-1})}{P_{t_{n-2}}(l_{n-1})}\frac{P_{t_n}(l_n)}{P_{t_{n-1}}(l_n)}\bigg|\mathcal{X}\right]$$
$$=\mathbb{E}\left[P_{t_1}(l_1)\mid\mathcal{X}\right]$$
$$=\mathbb{E}\left[\mathbb{E}\left[P_{t_1}(l_1)\mid\mathcal{X}\vee\mathcal{F}_{t_0}\right]\mid\mathcal{X}\right]$$
$$=\mathbb{E}\left[P_{t_0}(l_1)\mid\mathcal{X}\right]=1.$$

Hence

$$\mathbb{E}\left[\exp\left(\sum_{j=1}^{n}i\beta_j^\top Y_{t_j}\right)Z_{t_n}\bigg|\mathcal{X}\right]=\exp\left(\sum_{j=1}^{n}\tfrac{1}{2}\|l_j\|^2(t_j-t_{j-1})\right),$$

which is the same as the result computed earlier for the right-hand side.

iii. By Weierstrass' approximation theorem any bounded continuous complex valued function $g(Y_{t_1},\ldots,Y_{t_p})$ can be approximated by a sequence as $r\to\infty$,

$$g^{(r)}(Y_{t_1},\ldots,Y_{t_p})\triangleq\sum_{k=1}^{m_r}a_k^r\exp\left(i\sum_{j=1}^{p}(\beta_{k,j}^r)^\top Y_{t_j}\right).$$

Thus as a consequence of (ii) it follows that for such a function g,

$$\mathbb{E}[g(Y_{t_1},\ldots,Y_{t_p})Z_t\mid\mathcal{X}]=\mathbb{E}[g(Y_{t_1},\ldots,Y_{t_p})\mid\mathcal{X}],$$

which since p was arbitrary by a further standard approximation argument extends to any bounded Borel measurable function g,

$$\mathbb{E}[g(Y)Z_t\mid\mathcal{X}]=\mathbb{E}[g(Y)\mid\mathcal{X}].$$

Thus given $f(X,Y)$ bounded and measurable on the path spaces of X and Y it follows that

$$\mathbb{E}[f(X,Y)Z_t]=\mathbb{E}\left[\mathbb{E}[f(X,Y)Z_t\mid\mathcal{X}]\right].$$

Conditional on $\mathcal{X}$, $f(X,Y)$ may be considered as a function $g^X(Y)$ on the path space of Y and hence

$$\mathbb{E}[f(X,Y)Z_t]=\mathbb{E}\left[\mathbb{E}[g^X(Y)Z_t\mid\mathcal{X}]\right]$$
$$=\mathbb{E}\left[\mathbb{E}[g^X(W)\mid\mathcal{X}]\right]=\mathbb{E}[f(X,W)].$$

3.22 The result (3.41) is immediate from the following identities,

$$\varphi(X_t)=\varphi(X_0)+M_t^\varphi+\int_0^t A\varphi(X_s)\,\mathrm{d}s,$$

$$\varphi^2(X_t)=\varphi^2(X_0)+2\int_0^t\varphi(X_s)\,\mathrm{d}M_s^\varphi+\int_0^t 2\varphi A\varphi(X_s)\,\mathrm{d}s+\langle M^\varphi\rangle_t,$$

$$\varphi^2(X_t)=\varphi^2(X_0)+M_t^{\varphi^2}+\int_0^t A\varphi^2(X_s)\,\mathrm{d}s;$$

thus
$$\langle M^\varphi\rangle_t = \int_0^t (A\varphi^2 - 2\varphi A\varphi)(X_s)\,\mathrm{d}s.$$

Hence (3.39) becomes
$$\int_0^t u_s^2(A\varphi^2 - 2\varphi A\varphi)\,\mathrm{d}s \leq (\|A\varphi^2\|_\infty + 2\|\varphi\|_\infty\|A\varphi\|_\infty)\int_0^t u_s^2\,\mathrm{d}s < \infty.$$

3.23 Since under $\tilde{\mathbb{P}}$ the process Y is a Brownian motion a sufficient condition for the stochastic integral to be well defined is given by (B.9) which in this case takes the form, for all $t \geq 0$, that
$$\tilde{\mathbb{P}}\left[\int_0^t \sum_{i=1}^d (\rho_s(\varphi h_i))^2\,\mathrm{d}s < \infty\right] = 1.$$

But since $\varphi \in B(\mathbb{R}^d)$ it follows that
$$\int_0^t \sum_{i=1}^d \rho_s(\varphi h_i)^2\,\mathrm{d}s \leq \|\varphi\|_\infty^2 \int_0^t \sum_{i=1}^d \rho_s(h_i)^2\,\mathrm{d}s$$
$$\leq d\|\varphi\|_\infty^2 \int_0^t \rho_s(\|h\|)^2\,\mathrm{d}s.$$

Thus under (3.42) for all $t \geq 0$
$$\tilde{\mathbb{P}}\left[\int_0^t \rho_s(\|h\|)^2\,\mathrm{d}s < \infty\right] = 1,$$

and the result follows.

3.25

i. As a consequence of the linear growth condition (3.28) we have that
$$\rho_t(\|h\|) \leq C\rho_t\left(\sqrt{1 + \|X_t\|^2}\right),$$

and we prove that
$$t \mapsto \rho_t\left(\sqrt{1 + \|X_t\|^2}\right) \tag{3.80}$$

is uniformly bounded on compact intervals. The derivation of (3.44) did not require condition (3.42). We should like to apply this to $\psi(x) = \sqrt{1 + \|x\|^2}$, but while continuous this is not bounded. Thus choosing an approximating test function
$$\varphi_\lambda(x) = \sqrt{\frac{1 + \|x\|^2}{1 + \lambda\|x\|^2}}$$

in (3.44), we wish to take the limit as λ tends to 0 as φ_λ converges pointwise to ψ. Note that

$$\left\| \frac{\varphi_\lambda(X_s)\tilde{Z}_s}{(1+\varepsilon\tilde{Z}_s)^2} h(X_s) \right\| = \frac{1}{\varepsilon} \left\| \varphi_\lambda(X_s) \frac{\varepsilon\tilde{Z}_s}{1+\varepsilon\tilde{Z}_s} \frac{1}{1+\varepsilon\tilde{Z}_s} h(X_s) \right\|$$

$$\leq \frac{1}{\varepsilon} \varphi_\lambda(X_s) \|h(X_s)\|$$

$$\leq \frac{\sqrt{C}}{\varepsilon} \left(\sqrt{\frac{1+\|X_s\|^2}{1+\lambda\|X_s\|^2}} \sqrt{1+\|X_s\|^2} \right)$$

$$\leq \frac{\sqrt{C}}{\varepsilon} \left(1+\|X_s\|^2\right).$$

Therefore we have the bound,

$$\left\| \tilde{\mathbb{E}}\left[\frac{(\varphi_\lambda(X_s) - \psi(X_s))\tilde{Z}_s}{(1+\varepsilon\tilde{Z}_s)^2} h(X_s) \mid \mathcal{Y} \right] \right\| \leq \frac{\sqrt{C}}{\varepsilon} \left(1+\tilde{\mathbb{E}}[\|X_s\|^2 \mid \mathcal{Y}]\right).$$

But by Proposition 3.13 since under $\tilde{\mathbb{P}}$ the process X is independent of Y, and since the law of X is the same under $\mathbb{P}$ as it is under $\tilde{\mathbb{P}}$, it follows that

$$\left\| \tilde{\mathbb{E}}\left[\frac{(\varphi_\lambda(X_s) - \psi(X_s))\tilde{Z}_s}{(1+\varepsilon\tilde{Z}_s)^2} h(X_s) \mid \mathcal{Y} \right] \right\| \leq \frac{\sqrt{C}}{\varepsilon} \left(1+\mathbb{E}[\|X_s\|^2]\right). \quad (3.81)$$

Using the result (3.76) of Exercise 3.10 conclude that

$$\int_0^t \left(\frac{\sqrt{C}}{\varepsilon} \left(1+\mathbb{E}\|X_s\|^2\right) \right)^2 ds \leq \frac{C}{\varepsilon^2} \left(1+\mathbb{E}\|X_0\|^2\right)^2 \int_0^t e^{4cs} ds < \infty.$$

Thus by the dominated convergence theorem using the right-hand side of (3.81) as a dominating function, $\lambda \to 0$,

$$\int_0^t \left(\tilde{\mathbb{E}}\left[\varphi_\lambda(X_s)\tilde{Z}_s(1+\varepsilon\tilde{Z}_s)^{-2} h(X_s) \mid \mathcal{Y} \right] \right.$$
$$\left. - \tilde{\mathbb{E}}\left[\psi(X_s)\tilde{Z}_s(1+\varepsilon\tilde{Z}_s)^{-2} h(X_s) \mid \mathcal{Y} \right] \right)^2 ds \to 0;$$

thus using Itô's isometry it follows that as $\lambda \to 0$,

$$\int_0^t \tilde{\mathbb{E}}\left[\varphi_\lambda(X_s)\tilde{Z}_s(1+\varepsilon\tilde{Z}_s)^{-2} h(X_s) \mid \mathcal{Y} \right] dY_s$$
$$\to \int_0^t \tilde{\mathbb{E}}\left[\varphi(X_s)\tilde{Z}_s(1+\varepsilon\tilde{Z}_s)^{-2} h(X_s) \mid \mathcal{Y} \right] dY_s \to 0,$$

whence we see that (3.44) holds for the unbounded test function ψ. This ψ is not contained in $\mathcal{D}(A)$ since it is not bounded; however, computing using (3.11) directly

$$A\psi = \frac{1}{\psi}\left(f^\top x + \frac{1}{2}\operatorname{tr}(\sigma\sigma^\top) - \frac{1}{2\psi^2}(X^\top\sigma\sigma^\top X)\right).$$

Thus using the bounds in (3.14) and (3.15) which follow from (3.10),

$$|A\psi|/\psi \leq \frac{1}{\psi^2}\left(\kappa d(1+\|X\|)\|X\| + \tfrac{1}{2}\kappa'(1+\|X\|^2) + \tfrac{1}{2}\kappa'd^2\|X\|^2\right)$$

$$\leq \tfrac{1}{2}\kappa' + \kappa d + \tfrac{1}{2}d^2\kappa'.$$

For future reference we define

$$k_A \triangleq \tfrac{1}{2}\kappa' + \kappa d + \tfrac{1}{2}d^2\kappa'. \tag{3.82}$$

We also need a simple bound which follows from (3.26) and Jensen's inequality

$$\tilde{\mathbb{E}}[\tilde{Z}_t\psi(X_t)] = \mathbb{E}[\psi(X_t)] \leq \sqrt{1+\mathbb{E}[\|X_t\|^2]} \leq \sqrt{1+G_t}. \tag{3.83}$$

In the argument following (3.47) the stochastic integral in (3.44) was shown to be a $\mathcal{Y}_t$-adapted martingale under the measure $\tilde{\mathbb{P}}$. Therefore for $0 \leq r \leq t$,

$$\tilde{\mathbb{E}}\left[\tilde{\mathbb{E}}[\tilde{Z}_t^\varepsilon\psi(X_t)\mid\mathcal{Y}] - \frac{\pi_0(\psi)}{1+\varepsilon} + \int_0^t \tilde{\mathbb{E}}[\tilde{Z}_s^\varepsilon A\psi(X_s)\mid\mathcal{Y}]\,ds\right.$$
$$\left. - \int_0^t \tilde{\mathbb{E}}\left[\varepsilon\psi(X_s)\left(\tilde{Z}_s\right)^2\left(1+\varepsilon\tilde{Z}_s\right)^{-3}\|h(X_s)\|^2\mid\mathcal{Y}\right]ds\,\bigg|\,\mathcal{Y}_r\right]$$
$$= \tilde{\mathbb{E}}[\tilde{Z}_r^\varepsilon\psi(X_r)\mid\mathcal{Y}] - \frac{\pi_0(\psi)}{1+\varepsilon} + \int_0^r \tilde{\mathbb{E}}\left[\tilde{Z}_s^\varepsilon A\psi(X_s)\mid\mathcal{Y}\right]ds$$
$$- \int_0^r \tilde{\mathbb{E}}\left[\varepsilon\psi(X_s)\left(\tilde{Z}_s\right)^2\left(1+\varepsilon\tilde{Z}_s\right)^{-3}\|h(X_s)\|^2\mid\mathcal{Y}\right]ds.$$

Then we can take the limit on both sides of this equality as $\varepsilon \to 0$. For the term $\tilde{\mathbb{E}}[\tilde{Z}_t^\varepsilon\psi(X_t)\mid\mathcal{Y}]$ the limit follows by monotone convergence. For the term involving $\pi_0(\psi)$, since X_0 has finite third moment,

$$\tilde{\mathbb{E}}(\pi_0(\psi)) = \mathbb{E}[\psi(X_0)] < \sqrt{1+\mathbb{E}\|X_0\|^2} < \infty, \tag{3.84}$$

the limit follows by the dominated convergence theorem. For the integral involving the generator A we use the bound (3.82) to construct a dominating function since using (3.83) it follows that $\tilde{\mathbb{E}}[\tilde{Z}_t k_A \psi(X_t)] < \infty$, the limit then follows by the dominated convergence theorem. This only leaves the integral term which does not involve A; as this is not monotone in ε we must construct a dominating function. As a consequence of (3.28) and the definition of $\psi(x)$,

$$\left|\frac{\varepsilon\psi(X_s)\tilde{Z}_s^2}{(1+\varepsilon\tilde{Z}_s)^3}\|h(X_s)\|^2\right| = \left|\psi(X_s)\tilde{Z}_s\|h(X_s)\|^2\frac{\varepsilon\tilde{Z}_s}{1+\varepsilon\tilde{Z}_s}\left(1+\varepsilon\tilde{Z}_s\right)^{-2}\right|$$
$$\leq \psi(X_s)\tilde{Z}_s\|h(X_s)\|^2$$
$$\leq C\tilde{Z}_s(1+\|X_s\|^2)^{1/2}(1+\|X_s\|^2)$$
$$\leq C\tilde{Z}_s(1+\|X_s\|^2)^{3/2}.$$

and use the fact that the third moment of $\|X_t\|$ is bounded (3.27) to see that this is a suitable dominating function. Hence as $\varepsilon \to 0$,

$$\int_0^t \tilde{\mathbb{E}}\left[\varepsilon\varphi(X_s)\left(\tilde{Z}_s\right)^2\left(1+\varepsilon\tilde{Z}_s\right)^{-3}\|h(X_s)\|^2 \mid \mathcal{Y}\right]ds \to 0,$$

and thus passing to the $\varepsilon \to 0$ limit we obtain that

$$M_t \triangleq \rho_t(\psi) - \pi_0(\psi) + \int_0^t \rho_s(A\psi)\,ds \qquad (3.85)$$

satisfies $\tilde{\mathbb{E}}[M_t \mid \mathcal{F}_r] = M_r$ for $0 \leq r \leq t$, and M_t is $\mathcal{Y}_t$-adapted. To show that M_t is a martingale, it only remains to show that $\tilde{\mathbb{E}}|M_t| < \infty$, but this follows from the fact that for $s \in [0,t]$ using (3.83),

$$\tilde{\mathbb{E}}[\rho_t(\psi)] = \tilde{\mathbb{E}}\left[\tilde{\mathbb{E}}[\tilde{Z}_t\psi(X_t) \mid \mathcal{Y}]\right] = \tilde{\mathbb{E}}(\tilde{Z}_t\psi(X_t)) < \infty,$$

together with the bounds (3.82) and (3.84) this implies

$$\tilde{\mathbb{E}}[|M_t|] \leq \tilde{\mathbb{E}}(\rho_t(\psi))) + \tilde{\mathbb{E}}[\pi_0(\psi)] + k_A\int_0^t \tilde{\mathbb{E}}[\rho_s(\psi)]\,ds < \infty$$
$$\leq \sqrt{1+G_t}(1+k_at) + \sqrt{1+\mathbb{E}\|X_0\|^2} < \infty.$$

But since $\rho_t(\psi)$ is càdlàg (from the properties of ρ_t) it follows that M_t is a càdlàg $\mathcal{Y}_t$-adapted martingale under $\tilde{\mathbb{P}}$. Finally we use the fact that a càdlàg martingale has paths which are bounded on compact intervals in time (a consequence of Doob's submartingale inequality, see Theorem 3.8 page 13 of Karatzas and Shreve [149] for a proof) to see that $\tilde{\mathbb{P}}(\sup_{s\in[0,t]}|M_t| < \infty) = 1$. Then for ω fixed we have from (3.82) that

$$|\rho_t(\psi)| \leq \sup_{s\in[0,t]}|M_t| + |\pi_0(\psi)| + k_A\int_0^t |\rho_s(\psi)|\,ds,$$

so Gronwall's inequality implies that

$$|\rho_t(\psi)(\omega)| \leq \left(\sup_{s\in[0,t]}|M_t| + |\pi_0(\psi)|\right)e^{k_At},$$

whence for ω not in a null set $\rho_s(\psi)$ is bounded for $s \in [0,t]$. Hence the result.

ii. Setting $H = \max_{i \in I} \|h(\{i\})\|$, since I is finite, $H < \infty$, thus using the fact that ρ_s is a probability measure

$$\int_0^t \rho_s(\|h\|)^2 \, \mathrm{d}s \leq H^2 \int_0^t \rho_s(\mathbf{1})^2 \, \mathrm{d}s.$$

From (3.44) with $\varphi = \mathbf{1}$, since $A\mathbf{1} = 0$,

$$\tilde{\mathbb{E}}[\tilde{Z}_t^\varepsilon \mid \mathcal{Y}] = \frac{\pi_0(\mathbf{1})}{1+\varepsilon} - \int_0^t \tilde{\mathbb{E}}\left[\varepsilon(\tilde{Z}_t^\varepsilon)^2 \frac{1}{(1+\varepsilon\tilde{Z}_s)} \|h(X_s)\|^2 \mid \mathcal{Y}\right] \mathrm{d}s$$
$$+ \int_0^t \tilde{\mathbb{E}}\left[\tilde{Z}_t^\varepsilon \frac{1}{1+\varepsilon\tilde{Z}_s} h^\top(X_s) \mid \mathcal{Y}\right] \mathrm{d}Y_s.$$

Taking conditional expectation with respect to $\mathcal{Y}_r$ for $0 \leq r \leq t$,

$$\tilde{\mathbb{E}}\left(\tilde{\mathbb{E}}[\tilde{Z}_t^\varepsilon] + \int_0^t \tilde{\mathbb{E}}\left[\varepsilon(\tilde{Z}_t^\varepsilon)^2 \frac{1}{(1+\varepsilon\tilde{Z}_s)} \|h(X_s)\|^2 \mid \mathcal{Y}\right] \mathrm{d}s \,\bigg|\, \mathcal{Y}_r\right)$$
$$= \tilde{\mathbb{E}}[\tilde{Z}_t^\varepsilon \mid \mathcal{Y}] + \int_0^r \tilde{\mathbb{E}}\left[\varepsilon(\tilde{Z}_t^\varepsilon)^2 \frac{1}{(1+\varepsilon\tilde{Z}_s)} \|h(X_s)\|^2 \mid \mathcal{Y}\right] \mathrm{d}s.$$

Since $\|h\| \leq H$, it is straightforward to pass to the limit as $\varepsilon \to 0$ which yields $\rho_t(\mathbf{1})$ is a $\mathcal{Y}_t$-martingale. As in case (i) above then this has a càdlàg version which is a.s. bounded on finite intervals. Thus

$$\int_0^t \rho_s(\mathbf{1}) \, \mathrm{d}s < \infty \qquad \tilde{\mathbb{P}}\text{-a.s.,}$$

which establishes (3.42) since the measures $\tilde{\mathbb{P}}$ and $\mathbb{P}$ are equivalent on $\mathcal{F}_t$ and thus have the same null sets.

3.27

i. Observe first that (using the properties of the matrix Q):

$$\rho_t(\mathbf{1}) = \sum_{i \in I} \rho_t^i = 1 + \sum_{j=1}^m \int_0^t \rho_s\left(h^j\right) \mathrm{d}Y_s^j.$$

Next apply Itô's formula and integration by parts to obtain the evolution equation of

$$\pi_t^i = \frac{\rho_t^i}{\sum_{i \in I} \rho_t^i}.$$

ii. Assume that there are two continuous $\mathcal{Y}_t$-adapted $|I|$-dimensional processes, π and $\bar{\pi}$, solutions of the equation (3.53). Show that the processes continuous $\mathcal{Y}_t$-adapted $|I|$-dimensional processes ρ and $\bar{\rho}$ defined as

90   3 The Filtering Equations

$$\rho_t = \exp\left(\sum_{j=1}^{m} \int_0^t \pi_s(h^j)\,dY_s^j - \tfrac{1}{2}\int_0^t \pi_s(h^j)^2\,ds\right)\pi_t, \quad t \geq 0$$

$$\bar{\rho}_t = \exp\left(\sum_{j=1}^{m} \int_0^t \bar{\pi}_s(h^j)\,dY_s^j - \tfrac{1}{2}\int_0^t \bar{\pi}_s(h^j)^2\,ds\right)\bar{\pi}_t, \quad t \geq 0$$

satisfy equation (3.52) hence must coincide. Hence their normalised version must do so, too. Note that the continuity and the adaptedness of the processes are used to ensure that the stochastic integrals appearing in (3.52) and, respectively, (3.53) are well defined.

3.32 It is easiest to start from the finite-dimensional form of the Kushner–Stratonovich equation which was derived as (3.53). The Markov chain has two states, 0 and 1 depending upon whether the event is yet to happen, or has happened. Since it is clear that $\pi_t^0 + \pi_t^1 = 1$, then it suffices to write the equation for the component corresponding to the state 1 as this is $\pi_t^1 = \pi_t(J_1)$. Then h is given by $1_{\{T \leq t\}}$ and hence $h = J_1$. Writing the equation for state $\{1\}$,

$$\pi_t^1 = \pi_0^1 + \int_0^t (q_{01}\pi_s^0 + q_{11}\pi_s^1)\,ds + \int_0^t (h(1) - \pi_s^1(h))\pi_s^1(dY_s - \pi_s^1 ds)$$

$$= \pi_0^1 + \int_0^t (1 - \pi_s^1)p_t/g_t\,ds + \int_0^t (1 - \pi_s^1)\pi_s^1(dY_s - \pi_s^1 ds).$$

3.36 Since β is bounded for $\varphi \in C_b^2(\mathbb{R})$ by Itô's formula

$$\varphi(\alpha_t) - \varphi(\alpha_0) = \int_0^t A_s\varphi(\alpha_s)\,ds + M_t^\varphi,$$

where

$$A_s = \beta_s \frac{\partial}{\partial x} + \frac{1}{2}\frac{\partial^2}{\partial x^2},$$

and $M_t^\varphi = \int_0^t \varphi'(X_s)\,dV_s$ is an $\mathcal{F}_t$-adapted martingale.

Analogously to Theorem 2.24, we can define a probability measure-valued process π_t, such that for f_t a bounded $\mathcal{F}_t$-adapted process, $\pi(f_t)$ is a version of the $\mathcal{D}_t$-optional projection of f_t. The equivalent of the innovations process I_t for this problem is

$$I_t \triangleq \delta_t - \int_0^t \pi_s(\gamma_s)\,ds,$$

which is a $\mathcal{D}_t$-adapted Brownian motion under $\mathbb{P}$. By the representation result, Proposition 2.31, we can find a progressively measurable process ν_t such that

$$\pi_t(\varphi(\alpha_t)) - \int_0^t \pi_s(A_s\varphi(\alpha_s))\,ds = \pi_0(\varphi(\alpha_0)) + \int_0^t \nu_s\,dI_s,$$

therefore it follows that
$$\pi_t(\varphi(\alpha_t)) = \pi_0(\varphi(\alpha_0)) + \int_0^t \pi_s(A_s\varphi(\alpha_s))\,ds + \int_0^t \nu_s\,dI_s.$$

As in the innovations proof of the Kushner–Stratonovich equation, to identify ν, we can compute $d(\pi_t(\varphi(\alpha_t))\varepsilon_t)$ and $d(\varepsilon_t\varphi(\alpha_t))$ whence subtracting and taking expectations and using the independence of W and V we obtain that
$$\nu_t = \pi_t(\gamma_t\varphi(\alpha_t)) - \pi(\gamma_t)\pi(\varphi(\alpha_t)),$$
whence
$$\pi_t(\varphi(\alpha_s)) = \pi_0(\varphi(\alpha_0)) + \int_0^t \pi_s\left(\beta_s\varphi'(\alpha_s) + \tfrac{1}{2}\varphi''(\alpha_s)\right)ds$$
$$+ \int_0^t (\pi_s(\gamma_s\varphi(\alpha_s)) - \pi_s(\gamma_s)\pi_s(\varphi(\alpha_s)))\,(d\delta_s - \pi_s(\gamma_s)ds).$$

3.37

i. By Itô's formula
$$\begin{aligned}d(\hat{Z}_t^{-1}) &= \hat{Z}_t^{-1}(-\pi_t(h^\top)dY_t + \tfrac{1}{2}\|\pi_t(h)\|^2 dt) + \tfrac{1}{2}\hat{Z}_t^{-1}\|\pi_t(h)\|^2 dt\\ &= -\hat{Z}_t^{-1}\pi_t(h^\top)(dY_t - \pi_t(h)dt)\\ &= -\hat{Z}_t^{-1}\pi_t(h^\top)dI_t.\end{aligned}$$

ii. Let $\varepsilon_t \in S_t$ be such that $d\varepsilon_t = i\varepsilon_t r^\top dY_t$ and apply Itô's formula to the product $d(\varepsilon_t \hat{Z}_t^{-1})$ which yields
$$\begin{aligned}d(\varepsilon_t\hat{Z}_t^{-1}) &= -\varepsilon_t\hat{Z}_t^{-1}\pi_t(h^\top)dI_t + i\hat{Z}_t^{-1}\varepsilon_t r_t^\top dY_t - i\varepsilon_t\hat{Z}_t^{-1}\langle r_t^\top dY_t, \pi_t(h^\top)dI_t\rangle\\ &= \varepsilon_t\hat{Z}_t^{-1}\left(-\pi_t(h^\top)dI_t + ir_t^\top dY_t - ir_t^\top \pi_t(h)ds\right)\\ &= \varepsilon_t\hat{Z}_t^{-1}\left(-\pi_t(h^\top) + ir_t^\top\right)dI_t.\end{aligned}$$

Since by Proposition 2.30 the innovation process I_t is a $\mathcal{Y}_t$-adapted Brownian motion under the measure $\mathbb{P}$ it follows that taking expectation
$$\mathbb{E}[\varepsilon_t\hat{Z}_t^{-1}] = \mathbb{E}[\varepsilon_0\hat{Z}_0^{-1}] = 1.$$

Now consider
$$\mathbb{E}[Z_t\varepsilon_t] = \tilde{\mathbb{E}}[\varepsilon_t] = \tilde{\mathbb{E}}\left[1 + \int_0^t i\varepsilon_s r_s^\top dY_s\right] = 1,$$
since Y_t is a Brownian motion under $\tilde{\mathbb{P}}$. Thus
$$\mathbb{E}[\hat{Z}_t^{-1}\varepsilon_t] = \mathbb{E}[Z_t\varepsilon_t].$$

iii. It follows from the result of the previous part that

$$\tilde{\mathbb{E}}\left[\tilde{Z}_t \varepsilon_t / \hat{Z}_t\right] = \tilde{\mathbb{E}}[\tilde{Z}_t Z_t \varepsilon_t].$$

Hence

$$\tilde{\mathbb{E}}\left[\varepsilon_t \left(\hat{Z}_t^{-1}\tilde{Z}_t - 1\right)\right] = 0.$$

Clearly $\hat{Z}_t$ and ε_t are $\mathcal{Y}_t$-measurable

$$\tilde{\mathbb{E}}\left[\varepsilon_t \left(\hat{Z}_t^{-1}\tilde{\mathbb{E}}[\tilde{Z}_t \mid \mathcal{Y}_t] - 1\right)\right] = 0.$$

Since $\hat{Z}_t^{-1}\tilde{\mathbb{E}}[\tilde{Z}_t \mid \mathcal{Y}_t] - 1$ is $\mathcal{Y}_t$-measurable, it follows from the total set property of S_t that

$$\hat{Z}_t^{-1}\tilde{\mathbb{E}}[\tilde{Z}_t \mid \mathcal{Y}_t] = 1, \qquad \mathbb{P}\text{-a.s.}$$

Since $\hat{Z}_t > 0$ it follows that

$$\hat{Z}_t = \tilde{\mathbb{E}}[Z_t \mid \mathcal{Y}_t].$$

We may drop the a.s. qualification since it is implicit from the fact that conditional expectations are only defined almost surely.

iv. By the Kallianpur–Striebel formula $\mathbb{P}$-a.s. using the result of part (iii)

$$\pi_t(\varphi) = \frac{\rho_t(\varphi)}{\rho_t(\mathbf{1})} = \hat{Z}_t^{-1}\rho_t(\varphi).$$

Hence $\rho_t(\varphi) = \hat{Z}_t \pi_t$, and note that by a simple application of Itô's formula $\mathrm{d}\hat{Z}_t = \hat{Z}_t \pi_t(h^\top)\mathrm{d}Y_t$. Starting from the Kushner–Stratonovich equation

$$\mathrm{d}\pi_t(\varphi) = \pi_t(A\varphi)\mathrm{d}t + \pi_t(\varphi h^\top)\mathrm{d}I_t - \pi_t(\varphi)\pi_t(h^\top)\mathrm{d}I_t.$$

Applying Itô's formula to the product $\hat{Z}_t \pi_t$ we find

$$\begin{aligned}\mathrm{d}\rho_t(\varphi) &= \mathrm{d}\pi_t(\varphi)\hat{Z}_t + \pi_t \hat{Z}_t \pi_t(h^\top)\mathrm{d}Y_t + \mathrm{d}\langle \hat{Z}_t, \pi_t(\varphi)\rangle \\ &= \left(\pi_t(A\varphi)\mathrm{d}t + \pi_t(\varphi h^\top)\mathrm{d}I_t - \pi_t(\varphi)\pi_t(h^\top)\mathrm{d}I_t\right)\hat{Z}_t + \pi_t(\varphi)\hat{Z}_t \pi_t(h^\top)\mathrm{d}Y_t \\ &\quad + \hat{Z}_t \pi_t(h)(\pi_t(\varphi h^\top) - \pi_t(\varphi)\pi_t(h^\top))\mathrm{d}t \\ &= \hat{Z}_t \left(\pi_t(A\varphi)\mathrm{d}t + \pi_t(\varphi h^\top)\mathrm{d}Y_t\right) \\ &= \rho_t(A\varphi)\mathrm{d}t + \rho_t(\varphi h^\top)\mathrm{d}Y_t.\end{aligned}$$

But this is the Zakai equation as required.

3.10 Bibliographical Notes

In [160], Krylov and Rozovskii develop the theory of strong solutions of Itô equations in Banach spaces and use this theory to deduce the filtering equations in a different manner from the two methods presented here.

In [163], Krylov and Zatezalo deduce the filtering equations using a PDE, rather than probabilistic, approach. They use extensively the elaborate theoretical framework for analyzing SPDEs developed by Krylov in [157] and [158]. The approach requires boundedness of the coefficients and strict ellipticity of the signal's diffusion matrix.

4
Uniqueness of the Solution to the Zakai and the Kushner–Stratonovich Equations

The conditional distribution of the signal $\pi = \{\pi_t,\ t \geq 0\}$ is a solution of the Kushner–Stratonovich equation, whilst its unnormalised version $\rho = \{\rho_t,\ t \geq 0\}$ solves the Zakai equation. It then becomes natural to ask whether the Zakai equation uniquely characterizes ρ, and the Kushner–Stratonovich equation uniquely characterizes π. In other words, we should like to know under what assumptions on the coefficients of the signal and observation processes the two equations have a unique solution. The question of uniqueness of the solutions of the two equations is central when attempting to approximate numerically π or ρ as most of the analysis of existing numerical algorithms relies on the SPDE characterization of the two processes.

To answer the uniqueness question one has to identify suitable spaces of possible solutions to the equations (3.43) and (3.57). These spaces must be large enough to allow for the existence of solutions of the corresponding SPDE. Thus π should naturally belong to the space of possible solutions for the Kushner–Stratonovich equation, and ρ to the space of possible solutions to the Zakai equation. However, if we choose a space of possible solutions which is too large this may make the analysis more difficult, and even allow multiple solutions.

In the following we present two approaches to prove the uniqueness of the solutions to the two equations: the first one is a PDE approach, inspired by Bensoussan [13]; the second one is a more recent functional analytic approach introduced by Lucic and Heunis [200]. For both approaches the following result is useful.

Exercise 4.1. Let $\mu^1 = \{\mu_t^1,\ t \geq 0\}$ and $\mu^2 = \{\mu_t^2,\ t \geq 0\}$ be two $\mathcal{M}(\mathbb{S})$-valued stochastic processes with càdlàg paths and $(\varphi_i)_{i \geq 0}$ be a separating set of bounded measurable functions (in the sense of Definition 2.12). If for each $t \geq 0$ and $i \geq 0$, the identity $\mu_t^1(\varphi_i) = \mu_t^2(\varphi_i)$ holds almost surely, then μ^1 and μ^2 are indistinguishable.

4.1 The PDE Approach to Uniqueness

In this section we assume that the state space of the signal is $\mathbb{S} = \mathbb{R}^d$ and that the signal process is a diffusion process as described in Section 3.2.1.

First we define the space of measure-valued stochastic processes within which we prove uniqueness of the solution. This space has to be chosen so that it contains only measures with respect to which the integral of any function with linear growth is finite. The reason for this is that we want to allow the coefficients of the signal and observation processes to be unbounded. Define first the class of integrands for these measures. Let $\psi : \mathbb{R}^d \to \mathbb{R}$ be the function

$$\psi(x) = 1 + \|x\|, \tag{4.1}$$

for any $x \in \mathbb{R}^d$ and define $C^l(\mathbb{R}^d)$ to be the space of continuous functions φ such that $\varphi/\psi \in C_b(\mathbb{R}^d)$. Endow the space $C^l(\mathbb{R}^d)$ with the norm

$$\|\varphi\|_\infty^l = \sup_{x \in \mathbb{R}^d} \frac{|\varphi(x)|}{\psi(x)}.$$

Also let $\mathcal{E}$ be the space of continuous functions $\varphi : [0, \infty) \times \mathbb{R}^d \to \mathbb{R}$ such that for all $t \geq 0$, we have

$$\sup_{s \in [0,t]} \|\varphi_s\|_\infty^l < \infty, \tag{4.2}$$

where $\varphi_s(x) = \varphi(s, x)$ for any $(s, x) \in [0, \infty) \times \mathbb{R}^d$.

Let $\mathcal{M}^l(\mathbb{R}^d) \subset \mathcal{M}(\mathbb{R}^d)$ be the space of finite measures μ over $\mathcal{B}(\mathbb{R}^d)$ such that $\mu(\psi) < \infty$. In particular, this implies that $\mu(\varphi) < \infty$ for all $\varphi \in C^l(\mathbb{R}^d)$. We endow $\mathcal{M}^l(\mathbb{R}^d)$ with the corresponding weak topology. That is, a sequence (μ_n) of measures in $\mathcal{M}^l(\mathbb{R}^d)$ converges to $\mu \in \mathcal{M}^l(\mathbb{R}^d)$ if and only if

$$\lim_{n \to \infty} \mu_n(\varphi) = \mu(\varphi), \tag{4.3}$$

for all $\varphi \in C^l(\mathbb{R}^d)$. Obviously this topology is finer than the usual weak topology (i.e. the topology under which (4.3) holds true only for $\varphi \in C_b(\mathbb{R}^d)$).

Exercise 4.2. For any $\mu \in \mathcal{M}^l(\mathbb{R}^d)$ define $\nu_\mu \in \mathcal{M}(\mathbb{R}^d)$ to be the measure whose Radon–Nikodym derivative with respect to μ is ψ (defined in (4.1)). Let $\mu, \mu_n, n \geq 1$ be measures in $\mathcal{M}^l(\mathbb{R}^d)$. Then μ_n converges to μ in $\mathcal{M}^l(\mathbb{R}^d)$ if and only if (ν_{μ_n}) converges weakly to ν_μ in $\mathcal{M}(\mathbb{R}^d)$.

Definition 4.3. *The class $\mathcal{U}$ is the space of all $\mathcal{Y}_t$-adapted $\mathcal{M}^l(\mathbb{R}^d)$-valued stochastic processes $\mu = \{\mu_t, t \geq 0\}$ with càdlàg paths such that, for all $t \geq 0$, we have*

$$\tilde{\mathbb{E}}\left[\int_0^t (\mu_s(\psi))^2 \, ds\right] < \infty. \tag{4.4}$$

Exercise 4.4. (Difficult) Let X be the solution of (3.9). Prove that if (3.10) is satisfied, X_0 has finite second moment, and h is bounded then ρ belongs to the class $\mathcal{U}$. [Hint: You will need to use the Kallianpur–Striebel formula and the normalised conditional distribution π_t.]

We prove that the Zakai equation (3.43) has a unique solution in the class $\mathcal{U}$ subject to the following conditions on the processes.

Condition 4.5 (U). The functions $f = (f^i)_{i=1}^d : \mathbb{R}^d \to \mathbb{R}^d$ appearing in the signal equation (3.9), $a = (a^{ij})_{i,j=1,\ldots,d} : \mathbb{R}^d \to \mathbb{R}^{d \times d}$ as defined in (3.12) and $h = (h_i)_{i=1}^m : \mathbb{R}^d \to \mathbb{R}^m$ appearing in the observation equation (3.5) have twice continuously differentiable components and all their derivatives of first- and second-order are bounded.

Remark 4.6. Under condition U all components of the functions a, f and h are in $C^l(\mathbb{R}^d)$, but need not be bounded. However, condition U does imply that a, f and h satisfy the linear growth condition (see Exercise 4.11 for details).

Exercise 4.7. i. Show that if the process μ belongs to the class $\mathcal{U}$ then $t \mapsto \mu_t(\varphi_t)$ is a $\mathcal{Y}_t$-adapted process for all $\varphi \in \mathcal{E}$ (where $\mathcal{E}$ is defined in (4.2)).
ii. Let φ be a function in $C_b^{1,2}([0,t] \times \mathbb{R}^d)$ and μ be a process in the class $\mathcal{U}$. Assume that h satisfied the bounded growth condition (3.28). Then the processes

$$t \mapsto \int_0^t \mu_s \left(\frac{\partial \varphi_s}{\partial s} + A\varphi_s \right) ds, \quad t \geq 0$$

$$t \mapsto \int_0^t \mu_s(\varphi_s h^\top) \, dY_s, \quad t \geq 0$$

are well defined $\mathcal{Y}_t$-adapted processes. In particular, the second process is a square integrable continuous martingale under the measure $\tilde{\mathbb{P}}$.

When establishing uniqueness of the solution of the Zakai equation, we need to make use of a time-inhomogeneous version of (3.43).

Lemma 4.8. *Assume that the coefficients a, f and g satisfy condition U. Let μ be a process belonging to the class $\mathcal{U}$ which satisfies (3.43) for any $\varphi \in \mathcal{D}(A)$. Then, $\tilde{\mathbb{P}}$-almost surely,*

$$\mu_t(\varphi_t) = \pi_0(\varphi_0) + \int_0^t \mu_s \left(\frac{\partial \varphi_s}{\partial s} + A\varphi_s \right) ds + \int_0^t \mu_s(\varphi_s h^\top) \, dY_s, \quad (4.5)$$

for any $\varphi \in C_b^{1,2}([0,t] \times \mathbb{R}^d)$.

Proof. Let us first prove that under condition U, μ satisfies equation (3.43) for any function $\varphi \in C_b^2(\mathbb{R}^d)$ not just for φ in the domain of the infinitesimal generator $\varphi \in \mathcal{D}(A) \subset C_b^2(\mathbb{R}^d)$. We do this via an approximation argument.

Choose a sequence (φ_n) such that $\varphi_n \in \mathcal{D}(A)$ (e.g. $\varphi_n \in C_k^2(\mathbb{R}^d)$) such that, φ_n, $\partial_\alpha \varphi_n$, $\alpha = 1, \ldots, d$ and $\partial_\alpha \partial_\beta \varphi_n$, $\alpha, \beta = 1, \ldots, d$ converge boundedly pointwise to φ, $\partial_\alpha \varphi$, $\alpha = 1, \ldots, d$ and $\partial_\alpha \partial_\beta \varphi$, $\alpha, \beta = 1, \ldots, d$. In other words the sequence (φ_n) is uniformly bounded and for all $x \in \mathbb{R}^d$, $\lim_{n \to \infty} \varphi_n(x) = \varphi(x)$, with a similar convergence assumed for the first and second partial derivatives of φ_n. Then, $\tilde{\mathbb{P}}$-almost surely

$$\mu_t(\varphi_n) = \pi_0(\varphi_n) + \int_0^t \mu_s(A\varphi_n)\, ds + \int_0^t \mu_s(\varphi_n h^\top)\, dY_s. \tag{4.6}$$

Since (φ_n) is uniformly bounded and pointwise convergent, by the dominated convergence theorem, we get that

$$\lim_{n \to \infty} \mu_t(\varphi_n) = \mu_t(\varphi), \tag{4.7}$$

and similarly

$$\lim_{n \to \infty} \pi_0(\varphi_n) = \pi_0(\varphi). \tag{4.8}$$

The use of bounded pointwise convergence and condition U implies that there exists a constant K such that

$$|A\varphi_n(x)| \le K\psi(x),$$

for any $x \in \mathbb{R}^d$ and $n > 0$. Since $\mu \in \mathcal{U}$ implies that $\mu_s(\psi) < \infty$, by the dominated convergence theorem $\lim_{n \to \infty} \mu_s(A\varphi^n) = \mu_s(A\varphi)$. Also, from (4.4) it follows that

$$\tilde{\mathbb{E}}\left[\int_0^t \mu_s(\psi)\, ds\right] \le \tilde{\mathbb{E}}\left[\int_0^t \tfrac{1}{2}\left(1 + \mu_s(\psi)^2\right) ds\right] < \infty. \tag{4.9}$$

Therefore, $\tilde{\mathbb{P}}$-almost surely

$$\int_0^t \mu_s(\psi)\, ds < \infty$$

and, again by the dominated convergence theorem, it follows that

$$\lim_{n \to \infty} \int_0^t \mu_s(A\varphi^n)\, ds = \int_0^t \mu_s(A\varphi)\, ds \quad \tilde{\mathbb{P}}\text{-a.s.} \tag{4.10}$$

Similarly, one uses the integrability condition (4.4) and again the dominated convergence theorem to show that for $i = 1, \ldots, m$,

$$\lim_{n \to \infty} \tilde{\mathbb{E}}\left[\int_0^t \left(\mu_s(\varphi^n h_i) - \mu_s(\varphi h_i)\right)^2 ds\right] = 0;$$

hence by Itô's isometry property, we get that

$$\lim_{n\to\infty} \int_0^t \mu_s(\varphi^n h^\top)\,\mathrm{d}Y_s = \int_0^t \mu_s(\varphi h^\top)\,\mathrm{d}Y_s. \qquad (4.11)$$

Finally, by taking the limit of both sides of the identity (4.6) and using the results (4.7), (4.8), (4.10) and (4.11) we obtain that μ satisfies equation (3.43) for any function $\varphi \in C_b^2(\mathbb{R}^d)$. The limiting processes $t \mapsto \int_0^t \mu_s(A\varphi)\,\mathrm{d}s$ and $t \mapsto \int_0^t \mu_s(\varphi_s h^\top)\,\mathrm{d}Y_s$, $t \geq 0$ are well defined as a consequence of Exercise 4.7.

Let us extend the result to the case of time-dependent test functions $\varphi \in C_b^{1,2}([0,t] \times \mathbb{R}^d)$. Once again by Exercise 4.7 all the integral terms in (4.5) are well defined and finite. Also from (3.43), for $i = 0, 1, \ldots, n-1$ we have

$$\mu_{(i+1)t/n}(\varphi_{it/n}) = \mu_{it/n}(\varphi_{it/n}) + \int_{it/n}^{(i+1)t/n} \mu_s(A\varphi_{it/n})\,\mathrm{d}s$$
$$+ \int_{it/n}^{(i+1)t/n} \mu_s(\varphi_{it/n} h^\top)\,\mathrm{d}Y_s$$

for $i = 0, 1, \ldots, n-1$. By Fubini's theorem we have that

$$\mu_{(i+1)t/n}(\varphi_{(i+1)t/n} - \varphi_{it/n}) = \int_{it/n}^{(i+1)t/n} \mu_{(i+1)t/n}\left(\frac{\partial \varphi_s}{\partial s}\right)\,\mathrm{d}s.$$

Hence

$$\mu_{(i+1)t/n}(\varphi_{(i+1)t/n}) = \mu_{(i+1)t/n}(\varphi_{(i+1)t/n} - \varphi_{it/n}) + \mu_{(i+1)t/n}(\varphi_{it/n})$$
$$= \mu_{it/n}(\varphi_{it/n}) + \int_{it/n}^{(i+1)t/n} \mu_{(i+1)t/n}\left(\frac{\partial \varphi_s}{\partial s}\right)\,\mathrm{d}s$$
$$+ \int_{it/n}^{(i+1)t/n} \mu_s\left(A\varphi_{it/n}\right)\,\mathrm{d}s$$
$$+ \int_{it/n}^{(i+1)t/n} \mu_s(\varphi_{it/n} h^\top)\,\mathrm{d}Y_s.$$

Summing over the intervals $[it/n, (i+1)t/n]$ from $i = 0$ to $n-1$,

$$\mu_t(\varphi_t) = \pi_0(\varphi_0) + \int_0^t \mu_{([ns/t]+1)t/n}\left(\frac{\partial \varphi_s}{\partial s}\right)\,\mathrm{d}s + \int_0^t \mu_s\left(A\varphi_{[ns/t]t/n}\right)\,\mathrm{d}s$$
$$+ \int_0^t \mu_s\left(\varphi_{[ns/t]t/n} h^\top\right)\,\mathrm{d}Y_s. \qquad (4.12)$$

The claim follows by taking the limit as n tends to infinity of both sides of the identity (4.12) and using repeatedly the dominated convergence theorem. Note that we use the càdlàg property of the paths of μ to find the upper bound for the second term. □

Exercise 4.9. Assume that the coefficients a, f and g satisfy condition U. Let μ be a process belonging to the class $\mathcal{U}$ which satisfies the Zakai equation (3.43) and φ be a function in $C_b^{1,2}([0,t] \times \mathbb{R}^d)$. Let $\varepsilon_t \in S_t$, where S_t is the set defined in Corollary B.40, that is,

$$\varepsilon_t = \exp\left(i \int_0^t r_s^\top \, \mathrm{d}Y_s + \frac{1}{2} \int_0^t \|r_s\|^2 \, \mathrm{d}s\right),$$

where $r \in C_b^m([0,t], \mathbb{R}^m)$. Then

$$\tilde{\mathbb{E}}[\varepsilon_t \mu_t(\varphi_t)] = \pi_0(\varphi_0) + \tilde{\mathbb{E}}\left[\int_0^t \varepsilon_s \mu_s \left(\frac{\partial \varphi_s}{\partial s} + A\varphi_s + i\varphi_s h^\top r_s\right) \mathrm{d}s\right] \quad (4.13)$$

for any $\varphi \in C_b^{1,2}([0,t] \times \mathbb{R}^d)$.

In the following we establish the existence of a function $\varphi \in C_b^{1,2}([0,t] \times \mathbb{R}^d)$ which plays the rôle of a (partial) function dual of the process μ; in other words we seek φ such that for $s \in [0,t]$, $\mu_s(\varphi_s) = 0$. In particular as a consequence of (4.13) and the fact that the set S_t is total, such a function could arise as a solution $\varphi \in C_b^{1,2}([0,t] \times \mathbb{R}^d)$ of the second-order parabolic partial differential equation

$$\frac{\partial \varphi_s(s,x)}{\partial s} + A\varphi_s(s,x) + i\varphi_s(s,x) h^\top(x) r_s = 0, \quad (4.14)$$

where the operator A is given by

$$A\varphi = \sum_{i,j=1}^d a^{ij} \frac{\partial^2}{\partial x_i \partial x_j} \varphi + \sum_{i=1}^d f^i \frac{\partial}{\partial x_i} \varphi.$$

This leads to a unique characterisation of μ. The partial differential equation (4.14) turns out to be very hard to analyse for two reasons. Firstly, the coefficients $a^{ij}(x)$ for $i,j = 1, \ldots, d$, $f^i(x)$, and $h^i(x)$ for $i = 1, \ldots, d$ are not in general bounded as functions of x. Secondly, the matrix $a(x)$ may be degenerate at some points $x \in \mathbb{R}^d$. A few remarks on this degeneracy may be helpful. Since $a(x) = \frac{1}{2}\sigma^\top(x)\sigma(x)$ it is clear that $y^\top a(x) y = \frac{1}{2} y^\top \sigma^\top(x) \sigma(x) y = \frac{1}{2}(\sigma(x)y)^\top (\sigma(x)y) \geq 0$, thus for all $x \in \mathbb{R}^d$, $a(x)$ is positive semidefinite. However, $a(x)$ is not guaranteed to be positive definite for all $x \in \mathbb{R}^d$; in other words there may exist $x \in \mathbb{R}^d$ such that there is a non-zero y such that $y^\top a(x) y = 0$, for example, if for some x, $a(x) = 0$ and this is not positive definite. Such a situation is not physically unrealistic since it has the interpretation of an absence of noise in the signal process at the point x.

A typical existence and uniqueness result for parabolic PDEs is the following

Theorem 4.10. *If the PDE*

$$\frac{\partial \varphi_t}{\partial t} = \sum_{i,j=1}^d a^{ij} \frac{\partial^2 \varphi_t}{\partial x_i \partial x_j} + \sum_{i=1}^d f^i \frac{\partial \varphi_t}{\partial x_i} \quad (4.15)$$

is uniformly parabolic, that is, if there exists $\lambda > 0$ such that $x^\top a x \geq \lambda \|x\|^2$ for every $x \neq 0$, the functions f and a bounded and Hölder continuous with exponent α and Φ is a $C^{2+\alpha}$ function, then there exists a unique solution to the initial condition problem given by (4.15) and the condition $\varphi_0(x) = \Phi(x)$. Furthermore if the coefficients a, f and the initial condition Φ are infinitely differentiable then the solution φ is infinitely differentiable in the spatial variable x.

The proof of the existence of solutions to the parabolic PDE is fairly difficult and its length precludes its inclusion here. These details can be found in Friedman [102] as Theorem 7 of Chapter 3 and the continuity result follows from Corollary 2 in Chapter 3. Recall that the Hölder continuity condition is satisfied with $\alpha = 1$ for Lipschitz functions.

As these conditions are not satisfied by the PDE (4.14), we use a sequence of functions (v^n) which solves uniformly parabolic PDEs with smooth bounded coefficients. For this, we approximate a, f and h by bounded continuous functions. More precisely let $(a_n)_{n \geq 1}$ be a sequence of functions $a_n : \mathbb{R}^d \to \mathbb{R}^{d \times d}$, $(f_n)_{n \geq 1}$ a sequence of functions $f_n : \mathbb{R}^d \to \mathbb{R}^d$ and $(h_n)_{n \geq 1}$ a sequence of functions $h_n : \mathbb{R}^d \to \mathbb{R}^m$. We denote components as usual by superscript indices. We require that these sequences of functions have the following properties. All the component functions have bounded continuous derivatives of all orders; in other words each component is an element of $C_b^\infty(\mathbb{R}^d)$. There exists a constant K_0 such that the bounds on the first- and second-order derivatives (but not necessarily on the function values) hold uniformly in n,

$$\sup_n \max_{i,j,\alpha} \left\| \partial_\alpha a_n^{ij} \right\|_\infty \leq K_0, \qquad \sup_n \max_{i,j,\alpha,\beta} \left\| \partial_\alpha \partial_\beta a_n^{ij} \right\|_\infty \leq K_0, \qquad (4.16)$$

and the same inequality holds true for the partial derivatives of the components of f_n and h_n. We also require that these sequences converge to the original functions a, f and h; i.e. $\lim_{n \to \infty} a_n(x) = a(x)$, $\lim_{n \to \infty} f_n(x) = f(x)$ and $\lim_{n \to \infty} h_n(x) = h(x)$ for any $x \in \mathbb{R}^d$. Finally we require that the matrix a_n is uniformly elliptic; in other words for each n, there exists λ_n such that $x^\top a_n x \geq \lambda_n \|x\|^2$ for all $x \in \mathbb{R}^d$. We write

$$A_n \triangleq \sum_{i,j=1}^d a_n^{ij} \frac{\partial^2}{\partial x_i \partial x_j} + \sum_{i=1}^d f_n^i \frac{\partial}{\partial x_i},$$

for the associated generator of the nth approximating system.[†]

[†] To obtain a_n, we use first the procedure detailed in section 6.2.1. That is, we consider first the function $\psi^n a$, where ψ^n is the function defined in (6.23) (see also the limits (6.24), (6.25) and (6.26)). Then we regularize $\psi^n a$ by using the convolution operator $T_{1/n}$ as defined in (7.4), to obtain the function $T_{1/n}(\psi^n a)$. More precisely, $T_{1/n}(\psi^n a)$ is a matrix-valued function with components $T_{1/n}(\psi^n a^{ij})$, $1 \leq i, j \leq d$. Finally, we define the function a_n to be equal to $T_{1/n}(\psi^n a) + \frac{1}{n} \mathbb{I}_d$, where $\mathbb{I}_d$ is the $d \times d$ identity matrix. The functions f_n and h_n are constructed in the same manner (without using the last step).

Exercise 4.11. If condition U holds, show that the entries of the sequences $(a_n)_{n\geq 1}$, $(f_n)_{n\geq 1}$ and $(h_n)_{n\geq 1}$ belong to $C^l(\mathbb{R}^d)$. Moreover show that there exists a constant K_1 such that

$$\sup_n \left(\max_{i,j}\|a_n^{ij}\|_\infty^l, \max_i\|f_n^i\|_\infty^l, \max_i\|h_n^i\|_\infty^l\right) \leq K_1.$$

Next we use a result from the theory of systems of parabolic partial differential equations. Consider the following partial differential equation

$$\frac{\partial v_s^n}{\partial s} = -A_n v_s^n - i v_s^n h_n^\top r_s, \qquad s \in [0,t] \qquad (4.17)$$

with final condition

$$v_t^n(x) = \Phi(x), \qquad (4.18)$$

where $r \in C_b^m([0,t], \mathbb{R}^m)$ and Φ is a complex-valued C^∞ function. In other words, if $v_s^n = v_s^{n,1} + i v_s^{n,2}$, $s \in [0,t]$, $\Phi = \Phi^1 + i\Phi^2$ then we have the equivalent system of real-valued PDEs

$$\begin{aligned}\frac{\partial v_s^{n,1}}{\partial s} &= -A_n v_s^{n,1} + v_s^{n,2} h_n^\top r_s & v_t^{n,1}(x) &= \Phi^1(x),\\ \frac{\partial v_s^{n,2}}{\partial s} &= -A_n v_s^{n,2} - v_s^{n,1} h_n^\top r_s & v_t^{n,2}(x) &= \Phi^2(x).\end{aligned} \qquad (4.19)$$

We need to make use of the maximum principle for parabolic PDEs in the domain $[0,T] \times \mathbb{R}^d$.

Lemma 4.12. *Let*

$$A = \sum_{i,j=1}^d a_{ij}(x)\frac{\partial^2}{\partial x_i \partial x_j} + f_i(x)\frac{\partial}{\partial x_i}$$

be an elliptic operator; that is, for all $x \in \mathbb{R}^d$, it holds that $\sum_{i,j=1}^d y_i a_{ij}(x) y_j > 0$ for all $y \in \mathbb{R}^d \setminus \{0\}$. Let the coefficients $a_{ij}(x)$ and $f_i(x)$ be continuous in x. If $u \in C^{1,2}([0,\infty) \times \mathbb{R}^d)$ is such that

$$Au - \frac{\partial u}{\partial t} \geq 0 \qquad (4.20)$$

in $(0,\infty) \times \mathbb{R}^d$ with $u(0,x) = \Phi(x)$ and u is bounded above, then for all $t \in [0,\infty)$,

$$\|u(t,x)\|_\infty \leq \|\Phi\|_\infty. \qquad (4.21)$$

Proof. Define $w(t,x) = u(t,x) - \|\Phi\|_\infty$. It is immediate that $Aw - \frac{\partial w}{\partial t} \geq 0$. Clearly $w(0,x) \leq 0$ for all $x \in \mathbb{R}^d$. Consider the region $(0,t] \times \mathbb{R}^d$ for t fixed. If (4.21) does not hold for $s \in [0,t]$ then $w(t,x) > 0$ for some $0 < s \leq t$, $x \in \mathbb{R}^d$. As we have assumed that u is bounded above, the same holds for w, which implies that w has a positive maximum in the region $(0, t] \times \mathbb{R}^d$ (including the boundary at t). Suppose this occurs at the point $P_0 = (x, t)$; then it follows by Theorem 4' of Chapter 2 of Friedman [102] that w assumes this positive constant value over the whole region $S(P_0) = [0, t] \times \mathbb{R}^d$ which is clearly a contradiction since $w(0, x) \leq 0$ and w is continuous in t. Thus $w(t, x) \leq 0$ for all $x \in \mathbb{R}^d$ which establishes the result. □

Exercise 4.13. Prove the above result in the case where the coefficients a_{ij} for $i, j = 1, \ldots, d$ and f_i for $i = 1, \ldots, d$ are bounded, without appealing to general results from the theory of parabolic PDEs. By modifying the above proof of Lemma 4.12 it is clear that it is sufficient to prove directly that if $u \in C^{1,2}([0, \infty) \times \mathbb{R}^d)$ is bounded above, satisfies (4.20), and $u(0, x) \leq 0$, then $u(t, x) \leq 0$ for $t \in [0, \infty)$ and $x \in \mathbb{R}^d$. This may be done in the following stages.

i. First, by considering derivatives prove that if (4.20) were replaced by

$$Au - \frac{\partial u}{\partial t} > 0 \qquad (4.22)$$

then $u(t, x)$ cannot have a maximum in $(0, t] \times \mathbb{R}^d$.

ii. Show that if u satisfies the original condition (4.20) then show that we can find δ and ε such that $w_{\delta,\varepsilon} \triangleq u(t, x) - \delta t - \varepsilon e^{-t} \|x\|^2$ satisfies the stronger condition (4.22).

iii. Show that if $u(t, x) \geq 0$ then $w_{\delta,\varepsilon}$ must have a maximum in $(0, t] \times \mathbb{R}^d$; hence use (i) to establish the result.

Proposition 4.14. *If $\Phi^1, \Phi^2 \in C_b^\infty(\mathbb{R}^d)$, then the system of PDEs (4.19) has a solution $(v^{n,1}, v^{n,2})$ where $v^{n,i} \in C_b^{1,2}([0, t] \times \mathbb{R}^d)$ for $i = 1, 2$, for which there exists a constant K_2 independent of n such that $\|v^{n,i}\|$, $\|\partial_\alpha v^{n,i}\|$, $\|\partial_\alpha \partial_\beta v^{n,i}\|$, for $i = 1, 2$, $\alpha, \beta = 1, \ldots, d$ are bounded by K_2 on $[0, t] \times \mathbb{R}^d$.*

Proof. We must rewrite our PDE as an initial value problem, by reversing time. That is, we define $\bar{v}_s^n \triangleq v_{t-s}^n$ for $s \in [0, t]$. Then we have the following system of real-valued partial differential equations and initial conditions

$$\begin{aligned} \frac{\partial \bar{v}_s^{n,1}}{\partial s} &= A_n \bar{v}_s^{n,1} - \bar{v}_s^{n,2} h_n^\top r_{t-s} & \bar{v}_0^{n,1}(x) &= \Phi^1(x), \\ \frac{\partial \bar{v}_s^{n,2}}{\partial s} &= A_n \bar{v}_s^{n,2} + \bar{v}_s^{n,1} h_n^\top r_{t-s} & \bar{v}_0^{n,2}(x) &= \Phi^2(x). \end{aligned} \qquad (4.23)$$

As the operator A_n is uniformly elliptic and has smooth bounded coefficients, the existence of the solution of (4.23) is justified by Theorem 4.10 (the coefficients have uniformly bounded first derivative and are therefore Lipschitz and thus satisfy the Hölder continuity condition). Furthermore since the initial condition and coefficients are also smooth, the solution $\bar{v}^n$ (and thus v^n) is also smooth (has continuous derivatives of all orders) in the spatial variable.

It only remains to prove the boundedness of the solution and of its first and second derivatives. Here we follow the argument in Proposition 4.2.1, page 90 from Bensoussan [13]. Define

$$z_t^n \triangleq \frac{1}{2}\left(\left(\bar{v}_t^{n,1}\right)^2 + \left(\bar{v}_t^{n,2}\right)^2\right). \tag{4.24}$$

Then

$$\frac{\partial z_s^n}{\partial s} - A_n z_s^n = -\sum_{\alpha,\beta=1}^d a_n^{\alpha\beta}\left(\partial_\alpha \bar{v}_s^{n,1} \partial_\beta \bar{v}_s^{n,1} + \partial_\alpha \bar{v}_s^{n,2} \partial_\beta \bar{v}_s^{n,2}\right) \le 0.$$

Therefore from our version of the positive maximum principle, Lemma 4.12, it follows that

$$\|\bar{v}_s^{n,1}\|_\infty^2 + \|\bar{v}_s^{n,2}\|_\infty^2 \le \|\Phi^1\|_\infty^2 + \|\Phi^2\|_\infty^2, \tag{4.25}$$

for any $s \in [0,t]$, which establishes the bound on $\|v^{n,i}\|$. Define

$$u_s^n \triangleq \frac{1}{2}\sum_{\alpha=1}^d \left(\left(\partial_\alpha \bar{v}_s^{n,1}\right)^2 + \left(\partial_\alpha \bar{v}_s^{n,2}\right)^2\right). \tag{4.26}$$

Then

$$\frac{\partial u_s^n}{\partial s} - A_n u_s^n =$$

$$-\sum_{\alpha,\beta,\gamma=1}^d a_n^{\alpha\beta}\left(\left(\partial_\alpha \partial_\gamma \bar{v}_s^{n,1}\right)\left(\partial_\beta \partial_\gamma \bar{v}_s^{n,1}\right) + \left(\partial_\alpha \partial_\gamma \bar{v}_s^{n,2}\right)\left(\partial_\beta \partial_\gamma \bar{v}_s^{n,2}\right)\right)$$

$$+\sum_{\alpha,\beta,\gamma=1}^d \partial_\gamma a_n^{\alpha\beta}\left(\left(\partial_\alpha \partial_\beta \bar{v}_s^{n,1}\right)\left(\partial_\gamma \bar{v}_s^{n,1}\right) + \left(\partial_\alpha \partial_\beta \bar{v}_s^{n,2}\right)\left(\partial_\gamma \bar{v}_s^{n,2}\right)\right)$$

$$+\sum_{\alpha,\beta=1}^d \partial_\beta f_n^\alpha \left(\partial_\alpha \bar{v}_s^{n,1} \partial_\beta \bar{v}_s^{n,1} + \partial_\alpha \bar{v}_s^{n,2} \partial_\beta \bar{v}_s^{n,2}\right)$$

$$+\sum_{\alpha=1}^d \partial_\alpha g_{n,s}(-\bar{v}_s^{n,2}\partial_\alpha \bar{v}_s^{n,1} + \bar{v}_s^{n,1}\partial_\alpha \bar{v}_s^{n,2}), \tag{4.27}$$

where $g_{n,s} = h_n^\top r_{t-s}$. The first term in (4.27) is non-positive as a consequence of the non-negative definiteness of a. Then by (4.16), since $|\partial_\beta f_n^\alpha|$ is uniformly bounded by K_0, using the inequality $(\sum_{i=1}^d a_i)^2 \le d\sum_{i=1}^d a_i^2$, the third term of (4.27) satisfies

$$\sum_{\alpha,\beta=1}^d \partial_\beta f_n^\alpha \left(\partial_\alpha \bar{v}^{n,1}\partial_\beta \bar{v}^{n,1} + \partial_\alpha \bar{v}^{n,2}\partial_\beta \bar{v}^{n,2}\right) \le 2K_0 d u_s^n. \tag{4.28}$$

Similarly, from (4.16) and (4.25) we see that the fourth term of (4.27) satisfies

$$\sum_{\alpha=1}^{d} \partial_\alpha g_{n,s}(-\bar{v}_s^{n,2}\partial_\alpha \bar{v}_s^{n,1} + \bar{v}_s^{n,1}\partial_\alpha \bar{v}_s^{n,2}) \leq K_0 \sum_{\alpha=1}^{d} \left(|\bar{v}_s^{n,2}|\left|\partial_\alpha \bar{v}_s^{n,1}\right| + |\bar{v}_s^{n,1}|\left|\partial_\alpha \bar{v}_s^{n,2}\right|\right)$$

$$\leq K_0 \left(\|\Phi^1\|_\infty + \|\Phi^2\|_\infty\right) \sum_{\alpha=1}^{d} \left(\left|\partial_\alpha \bar{v}_s^{n,1}\right| + \left|\partial_\alpha \bar{v}_s^{n,2}\right|\right)$$

$$\leq K_0 \left(\|\Phi^1\|_\infty + \|\Phi^2\|_\infty\right)(u_s^n + d)$$

$$\leq C_4(u_s^n + d), \qquad (4.29)$$

where the constant $C_4 \triangleq K_0(\|\Phi^1\|_\infty + \|\Phi^2\|_\infty)$. It only remains to find a suitable bound for the second term in (4.27). This is done using the following lemma, which is due to Oleinik–Radkevic (see [234, page 64]). Recall that a $d \times d$-matrix a is said to be non-negative definite if $\theta^\top a \theta \geq 0$ for all $\theta \in \mathbb{R}^d$.

Lemma 4.15. *Let $a : \mathbb{R} \to \mathbb{R}^{d \times d}$, be a symmetric non-negative definite matrix-valued function which is twice continuously differentiable and denote its components $a_{ij}(x)$ for $1 \leq i, j \leq d$. Let u be any symmetric $d \times d$-matrix; then*

$$(\operatorname{tr}(a'(x)u))^2 \leq 2d^2 \lambda \operatorname{tr}(u a(x) u) \qquad \forall x \in \mathbb{R},$$

where primes denote differentiation with respect to x, and

$$\lambda = \sup\left\{\frac{|\theta^\top a''(x)\theta|}{\|\theta\|^2} : x \in \mathbb{R},\ \theta \in \mathbb{R}^d \backslash \{0\}\right\}.$$

Proof. We start by showing that

$$|a'_{ij}(x)| \leq \sqrt{\lambda(a_{ii}(x) + a_{jj}(x))} \qquad \forall x \in \mathbb{R}. \qquad (4.30)$$

Let $\varphi \in C^2(\mathbb{R})$ be a non-negative function with uniformly bounded second derivative; let $\alpha = \sup_{x \in \mathbb{R}} |\varphi''(x)|$. Then Taylor's theorem implies that

$$0 \leq \varphi(x+y) \leq \varphi(x) + y\varphi'(x) + \alpha y^2/2;$$

thus the quadratic in y must have no real roots, which implies that the discriminant is non-positive thus

$$|\varphi'(x)| \leq \sqrt{2\alpha \varphi(x)}.$$

Let e_i denote the standard basis of $\mathbb{R}^d$; define the functions

$$\varphi_\pm^{ij}(x) = (e_i \pm e_j)^\top a(x)(e_i \pm e_j) = a_{ii}(x) \pm 2a_{ij}(x) + a_{jj}(x).$$

From the fact that a is non-negative definite, it follows that $\varphi_\pm^{ij}(x) \geq 0$. From the definition of λ, since $\|e_i \pm e_j\| = \sqrt{2}$, it follows that $|\varphi''_\pm(x)| < 2\lambda$; thus applying the above result

4 Uniqueness of the Solution

$$|\varphi'_\pm(x)| \le \sqrt{4\lambda \varphi_\pm(x)}.$$

From the definition $a_{ij}(x) = (\varphi_+ - \varphi_-)/4$, using (4.30)

$$\begin{aligned} |a'_{ij}(x)| &\le (|\varphi'_+(x)| + |\varphi'_-(x)|)/4 \\ &\le \tfrac{1}{2}\sqrt{\lambda\varphi_+(x)} + \sqrt{\lambda\varphi_-(x)} \\ &\le \sqrt{\lambda(\varphi_+(x) + \varphi_-(x))}/\sqrt{2} \\ &\le \sqrt{\lambda(a_{ii}(x) + a_{jj}(x))}. \end{aligned}$$

To establish the main result, by Cauchy–Schwartz

$$\begin{aligned} (\mathrm{tr}(a'(x)u))^2 &= \left(\sum_{i,j=1}^d a'_{ij}(x)u_{ji}\right)^2 \\ &\le d^2 \sum_{i,j=1}^d \left(a'_{ij}(x)u_{ji}\right)^2 \\ &\le 2\lambda d^2 \sum_{i,j=1}^d (a_{ii}(x) + a_{jj}(x))(u_{ji})^2 \\ &\le 2d^2\lambda \sum_{i,j=1}^d u_{ij} a_{jj}(x) u_{ji}. \end{aligned}$$

In general since a is real-valued and symmetric, at any x we can find an orthogonal matrix q such that $q^\top a(x) q$ is diagonal. We fix this matrix q and then since $\mathrm{tr}(q^\top u q) = \mathrm{tr}(q q^\top u) = \mathrm{tr}\, u$, it follows that

$$\begin{aligned} (\mathrm{tr}(a'(x)u))^2 &= \left(\mathrm{tr}(q^\top a'(x) q q^\top u q)\right)^2 \\ &\le 2d^2 \lambda \sum_{i,j=1}^d (q^\top u q)_{ij} (q^\top a(x) q)_{jj} (q^\top u q)_{ji} \\ &\le 2d^2 \lambda \, \mathrm{tr}\big((q^\top u q)(q^\top a(x) q)(q^\top u q)\big) \\ &\le 2\lambda d^2\, \mathrm{tr}(u a(x) u). \end{aligned}$$

$\square$

Taking $u^{\alpha,\beta} = \partial_\alpha \partial_\beta \bar{v}_s^{n,i}$, Lemma 4.15 implies that

$$\sum_{\alpha,\beta,\gamma=1}^d \left(\partial_\gamma a_n^{\alpha\beta} \partial_\alpha \partial_\beta \bar{v}_s^{n,i}\right)^2 \le C_2 \sum_{\alpha,\beta,\gamma=1}^d a_n^{\alpha\beta} \left(\partial_\alpha \partial_\gamma \bar{v}_s^{n,i}\right)\left(\partial_\beta \partial_\gamma \bar{v}_s^{n,i}\right), \quad i=1,2,$$

where C_2 only depends upon the dimension of the space and K_0 (in particular, it depends on the bound on the second partial derivatives of the entries of a_n). Hence, by using the elementary inequality, for $C > 0$,

$$\tau\zeta \leq \frac{1}{2C}\tau^2 + \frac{1}{2}C\zeta^2, \qquad (4.31)$$

on each term in the summation in the second term of (4.27) one can find an upper bound for the second sum of the form

$$\tfrac{1}{2}\Theta_s^n + C_2 u_s^n,$$

where Θ_s^n is given by

$$\Theta_s^n \triangleq \sum_{\alpha,\beta,\gamma=1}^{d} a_n^{\alpha\beta}\left(\left(\partial_\alpha \partial_\gamma \bar{v}_s^{n,1}\right)\left(\partial_\beta \partial_\gamma \bar{v}_s^{n,1}\right) + \left(\partial_\alpha \partial_\gamma \bar{v}_s^{n,2}\right)\left(\partial_\beta \partial_\gamma \bar{v}_s^{n,2}\right)\right),$$

and as a is non-negative definite $\Theta_s^n \geq 0$. By substituting the bounds (4.28), (4.29) and (4.31) into (4.27) we obtain the bound

$$\begin{aligned}\frac{\partial u_s^n}{\partial s} - A_n u_s^n &\leq -\Theta_s^n + \tfrac{1}{2}\Theta_s^n + C_2 u_s^n + 2K_0 d u_s^n + C_4(u_s^n + d) \\ &\leq C_2 u_s^n + 2K_0 d u_s^n + C_4(u_s^n + d) \\ &\leq C_0 u_s^n + C_1,\end{aligned}$$

where the constants C_0 and C_1 only depend upon the dimension of the space and K_0 (and not upon s or x). Thus

$$\hat{u}_s^n = \frac{C_1}{C_0}\mathrm{e}^{-C_0 s} + u_s^n \mathrm{e}^{-C_0 s}$$

satisfies

$$\frac{\partial \hat{u}_s^n}{\partial t} - A_n \hat{u}_s^n \leq 0;$$

thus from the maximum principle in the form of Lemma 4.12 we have that $\|\hat{u}_s^n\|_\infty \leq \|\hat{u}_0^n\|_\infty$, but $\hat{u}_0 = C_1/C_0 + u_0^n$,

$$\|u_s^n\|_\infty \leq \mathrm{e}^{C_0 T}\left(\frac{1}{2}\sum_{\alpha=1}^{d}\left(\|\partial_\alpha \Phi^1\|_\infty^2 + \|\partial_\alpha \Phi^2\|_\infty^2\right) + \frac{C_1}{C_0}\right),$$

which establishes the uniform bound on the first derivatives. The bound on the second-order partial derivatives of $\bar{v}$ is obtained by performing a similar, but more tedious, analysis of the function

$$w_t^n \triangleq \frac{1}{2}\sum_{\alpha,\beta=1}^{d}\left(\left(\partial_\alpha \partial_\beta \bar{v}_t^{n,1}\right)^2 + \left(\partial_\alpha \partial_\beta \bar{v}_t^{n,2}\right)^2\right).$$

Similar bounds will not hold for higher-order partial derivatives. □

Theorem 4.16. *Assuming condition U on the coefficients a, f and g, the equation (4.5) has a unique solution in the class $\mathcal{U}$, up to indistinguishability.*

Proof. Let v^n be the solution to the PDE (4.17). Applying Exercise 4.9 to v^n yields that for any solution μ of (3.43) in the class $\mathcal{U}$ we have

$$\tilde{\mathbb{E}}[\varepsilon_t \mu_t(v^n_t)] = \pi_0(v^n_0) + \tilde{\mathbb{E}}\left[\int_0^t \varepsilon_s \mu_s \left(\frac{\partial v^n_s}{\partial t} + A v^n_s + i h^\top v^n_s r_s\right) ds\right]$$

and using the fact that v^n_s satisfies (4.17) we see that

$$\tilde{\mathbb{E}}[\varepsilon_t \mu_t(v^n_t)] = \pi_0(v^n_0)$$
$$+ \tilde{\mathbb{E}}\left[\int_0^t \varepsilon_s \mu_s \left((A - A_n) v^n_s + i v^n_s (h - h_n)^\top r_s\right) ds\right]. \quad (4.32)$$

As a consequence of Proposition 4.14, v^n and its first- and second-order partial derivatives are uniformly bounded and consequently,

$$\lim_{n \to \infty} (A - A_n) v^n_s(x) = 0, \quad \lim_{n \to \infty} v^n_s(x)(h^\top(x) - h_n^\top(x)) r_s(x) = 0$$

for any $x \in \mathbb{R}^{d \times d}$. Also there exists a constant C_t independent of n such that

$$|(A - A_n) v^n_s(x)|, |v^n_s(x)(h(x) - h_n(x))^\top r_s| \le C_t \psi(x)$$

for any $x \in \mathbb{R}^{d \times d}$ and $s \in [0, t]$. Hence, as $\mu_s \in \mathcal{U}$ it follows that $\mu_s(\psi) < \infty$ and thus by the dominated convergence theorem we have that

$$\lim_{n \to \infty} \mu_s \left((A - A_n) v^n_s + i v^n_s (h - h_n)^\top r_s\right) = 0.$$

Next let us observe that $\sup_{s \in [0,t]} |\varepsilon_s| < \exp(\sup_{s \in [0,t]} \|r_s\| t/2) < \infty$, hence there exists a constant C'_t such that for $s \in [0, t]$,

$$\left|\varepsilon_s \mu_s \left((A - A_n) v^n_s + i v^n_s (h - h_n)^\top r_s\right)\right| \le C'_t \mu_s(\psi)$$

and since as a consequence of (4.4), it follows that (4.9) holds; thus

$$\tilde{\mathbb{E}}\left[\int_0^t \mu_s(\psi) ds\right] < \infty.$$

It follows that $C'_t \mu_s(\psi)$ is a dominating function, thus by the dominated convergence theorem it follows that

$$\lim_{n \to \infty} \tilde{\mathbb{E}}\left[\int_0^t \varepsilon_s \mu_s \left((A - A_n) v^n_s + i v^n_s (h - h_n)^\top r_s\right) ds\right] = 0. \quad (4.33)$$

Finally, let μ^1 and μ^2 be two solutions of the Zakai equation (3.43) in the class $\mathcal{U}$. Then from (4.32),

$$\tilde{\mathbb{E}}[\varepsilon_t \mu^1_t(v^n_t)] - \tilde{\mathbb{E}}[\varepsilon_t \mu^2_t(v^n_t)]$$
$$= \tilde{\mathbb{E}}\left[\int_0^t \varepsilon_s \left(\mu^1_s - \mu^2_s\right) \left((A - A_n) v^n_s + i v^n_s (h - h_n)^\top r_s\right) ds\right].$$

The final condition of the partial differential equation (4.18) implies that $v_t^n(x) = \Phi(x)$ for all $x \in \mathbb{R}^d$; thus

$$\tilde{\mathbb{E}}[\varepsilon_t \mu_t^1(\Phi)] - \tilde{\mathbb{E}}[\varepsilon_t \mu_t^2(\Phi)]$$
$$= \tilde{\mathbb{E}}\left[\int_0^t \varepsilon_s \left(\mu_s^1 - \mu_s^2\right) \left((A - A_n)v_s^n + iv_s^n(h - h_n)^\top r_s\right) \mathrm{d}s\right]$$

and we may then pass to the limit as $n \to \infty$ using (4.33) to obtain

$$\tilde{\mathbb{E}}(\varepsilon_t \mu_t^1(\Phi)) = \tilde{\mathbb{E}}(\varepsilon_t \mu_t^2(\Phi)). \tag{4.34}$$

The function Φ was an arbitrary C_b^∞ function, therefore using the fact that the set S_t is total, for φ any smooth bounded function, $\tilde{\mathbb{P}}$-almost surely $\mu_t^1(\varphi) = \mu_t^2(\varphi)$. From the bounds we know that $\|v_0^n\|_\infty \leq \|\Phi\|_\infty$, thus by the dominated convergence theorem since π_0 is a probability measure

$$\lim_{n \to \infty} \pi_0(v_0^n) = \pi_0\left(\lim_{n \to \infty} v_0^n\right);$$

passing to $n \to \infty$ we get

$$\tilde{\mathbb{E}}(\varepsilon_t \mu_t(\Phi)) = \pi_0\left(\lim_{n \to \infty} v_0^n\right)$$

whence

$$\left|\tilde{\mathbb{E}}(\varepsilon_t \mu_t(\Phi))\right| \leq \|\Phi\|_\infty.$$

By the dominated convergence theorem, we can extend (4.34) to any φ which is a continuous bounded function. Hence by Exercise 4.1 μ_t^1 and μ_t^2 are indistinguishable. □

Exercise 4.17. (Difficult) Extend Theorem 4.16 to the correlated noise framework.

Now let $\mu = \{\mu_t, t \geq 0\}$ be a $\mathcal{Y}_t$-adapted $\mathcal{M}^l(\mathbb{R}^d)$-valued stochastic process with càdlàg paths and $m^\mu = \{m_t^\mu, t \geq 0\}$ be the $\mathcal{Y}_t$-adapted real-valued process

$$m_t^\mu = \exp\left(\int_0^t \mu_s(h^\top) \mathrm{d}Y_s - \frac{1}{2}\int_0^t \mu_s(h^\top)\mu_s(h) \mathrm{d}s\right), \qquad t \geq 0.$$

We prove uniqueness for the Kushner–Stratonovich equation (3.57) in the class $\bar{\mathcal{U}}$ of all $\mathcal{Y}_t$-adapted $\mathcal{M}^l(\mathbb{R}^d)$-valued stochastic processes $\mu = \{\mu_t, t \geq 0\}$ with càdlàg paths such that the process $m^\mu \mu$ belongs to the class $\mathcal{U}$.

Exercise 4.18. Let X be the solution of the SDE (3.9). Prove that if (3.10) is satisfied, π_0 has finite third moment and h satisfies the linear growth condition (3.28) then the process π belongs to the class $\bar{\mathcal{U}}$.

Theorem 4.19. *Assuming condition U on the coefficients a, f and g the equation (3.57) has a unique solution in the class $\bar{\mathcal{U}}$, up to indistinguishability.*

Proof. Let π^1 and π^2 be two solutions of the equation (3.57) belonging to the class $\bar{\mathcal{U}}$. Then by a straightforward integration by parts, one shows that $\rho^i = m^{\pi^i}\pi^i$, $i = 1, 2$ are solutions of the Zakai equation (3.43). However, by Theorem 4.16, equation (3.43) has a unique solution in the class $\mathcal{U}$ (where both ρ^1 and ρ^2 reside). Hence, ρ^1 and ρ^2 coincide. In particular, $\mathbb{P}$-almost surely

$$m_t^{\pi^1} = \rho_t^1(\mathbf{1}) = \rho_t^2(\mathbf{1}) = m_t^{\pi^2}$$

for all $t \geq 0$. and hence

$$\pi_t^1 = \frac{1}{\rho_t^1(\mathbf{1})}\rho_t^1 = \frac{1}{\rho_t^2(\mathbf{1})}\rho_t^2 = \pi_t^2$$

for all $t \geq 0$, $\mathbb{P}$-almost surely. □

4.2 The Functional Analytic Approach

In this section, uniqueness is proved directly for the case when the signal and observation noise are correlated. However, in contrast to all of the arguments which have preceded this we assume that the function h is bounded. We recall that $A, B_i : B(\mathbb{S}) \to B(\mathbb{S}), i = 1, \ldots, m$ are operators with domains, respectively, $\mathcal{D}(A), \mathcal{D}(B_i) \subseteq B(\mathbb{S})$, $i = 1, \ldots, m$ with

$$\mathbf{1} \in \mathcal{D} \triangleq \mathcal{D}(A) \cap \bigcap_{i=1}^{m} \mathcal{D}(B_i) \quad \text{and} \quad A\mathbf{1} = B_1\mathbf{1} = \cdots = B_n\mathbf{1} = 0. \quad (4.35)$$

As in the previous section we need to define the space of measure-valued stochastic processes within which we prove uniqueness of the solution. We recall that $(\Omega, \mathcal{F}, \tilde{\mathbb{P}})$ is a complete probability space and that the filtration $(\mathcal{F}_t)_{t\geq 0}$ satisfies the usual conditions. Also recall that, under $\tilde{\mathbb{P}}$, the process Y is an $\mathcal{F}_t$-adapted Brownian motion. The conditions (4.35) imply that for all $t \geq 0$ and $\varphi \in \mathcal{D}$ since $B\varphi$ is bounded,

$$\int_0^t (\mu_s(\|B\varphi\|))^2 \, ds < \|B\varphi\|_\infty^2 \int_0^t (\mu_s(\mathbf{1}))^2 \, ds, \quad (4.36)$$

for any $\mu = \{\mu_t, t \geq 0\}$ which is an $\mathcal{F}_t$-adapted $\mathcal{M}(\mathbb{S})$-valued stochastic process.

Definition 4.20. *Let $\mathcal{U}'$ be the class of $\mathcal{F}_t$-adapted $\mathcal{M}(\mathbb{S})$-valued stochastic processes $\mu = \{\mu_t, t \geq 0\}$ with càdlàg paths that satisfy conditions (4.36) and (3.42); that is, for all $t \geq 0$, $\varphi \in \mathcal{D}$,*

$$\tilde{\mathbb{P}}\left[\int_0^t \sum_{i=1}^m [\mu_s(|(h_i + B_i)\varphi|)]^2 \, ds < \infty\right] = 1. \quad (4.37)$$

Let $\rho = \{\rho_s,\ s \geq 0\}$ be the $\mathcal{M}(\mathbb{S})$-valued process with càdlàg paths which is the unnormalised conditional distribution of the signal given the observation process as defined in Section 3.4. We have assumed that $h = (h_i)_{i=1}^m : \mathbb{S} \to \mathbb{R}$ for $i = 1, \ldots, m$ is a bounded measurable function hence it satisfies condition (3.25) which in turn ensures that the process $\tilde{Z} = \{\tilde{Z}_t,\ t \geq 0\}$ introduced in (3.30) and (3.31) is a (genuine) martingale under $\tilde{\mathbb{P}}$, where $\tilde{\mathbb{P}}$ is the probability measure defined in Section 3.3.

Exercise 4.21. Prove that the mass process $\rho(\mathbf{1}) = \{\rho_t(\mathbf{1}),\ t \geq 0\}$ is a $\mathcal{Y}_t$-adapted martingale under $\tilde{\mathbb{P}}$.

Since the mass process $\rho(\mathbf{1}) = \{\rho_t(\mathbf{1}),\ t \geq 0\}$ is a martingale under $\tilde{\mathbb{P}}$ which is càdlàg by Lemma 3.18, it is almost surely bounded on compact intervals.

Exercise 4.22. Prove that if (3.42) is satisfied, then the process ρ as defined by Definition 3.17 belongs to the class $\mathcal{U}'$.

Recall that, for any $t \geq 0$ and $\varphi \in \mathcal{D}$ we have, $\tilde{\mathbb{P}}$-almost surely that the unnormalised conditional distribution satisfies the Zakai equation, which in the correlated noise situation which we are considering here is

$$\rho_t(\varphi) = \pi_0(\varphi) + \int_0^t \rho_s(A\varphi)\,\mathrm{d}s + \int_0^t \rho_s((h^\top + B^\top)\varphi)\,\mathrm{d}Y_s, \qquad (4.38)$$

where condition (4.37) ensures that the stochastic integral in this equation is well defined.

Proposition 4.23. *If h is a bounded measurable function and $\rho = \{\rho_t,\ t \geq 0\}$ is an $\mathcal{F}_t$-adapted $\mathcal{M}(\mathbb{S})$-valued stochastic process belonging to the class $\mathcal{U}'$ which satisfies (4.38), then for any $\alpha > 0$, there exists a constant $k(\alpha)$ such that*

$$\tilde{\mathbb{E}}\left[\sup_{s \in [0,t]} (\rho_s(\mathbf{1}))^\alpha\right] < k(\alpha) < \infty. \qquad (4.39)$$

Proof. From condition (4.35) and equation (4.38) for $\varphi = \mathbf{1}$, we get that

$$\rho_t(\mathbf{1}) = 1 + \int_0^t \rho_s(h^\top)\,\mathrm{d}Y_s. \qquad (4.40)$$

In the following we make use of the normalised version of $\rho_t(h_i)$. Since we do not know that $\rho_t(\mathbf{1})$ is strictly positive this normalisation must be defined with some care. Let $\bar{\rho}_t(h_i)$ be defined as

$$\bar{\rho}_t(h_i) = \begin{cases} \dfrac{\rho_t(h_i)}{\rho_t(\mathbf{1})} & \text{if } \rho_t(\mathbf{1}) > 0 \\ 0 & \text{if } \rho_t(\mathbf{1}) = 0. \end{cases}$$

Since h is bounded it follows that $\rho_t(h_i) \leq \|h_i\|\rho_t(\mathbf{1})$; hence $\bar{\rho}_t(h_i) \leq \|h_i\|$. Hence $\rho_t(\mathbf{1})$ satisfies the equation

$$\rho_t(\mathbf{1}) = 1 + \int_0^t \bar{\rho}_t(h^\top)\rho_t(\mathbf{1})\,dY_s \tag{4.41}$$

and has the explicit representation (as in Lemma 3.29)

$$\rho_t(\mathbf{1}) = \exp\left(\sum_{i=1}^m \left(\int_0^t \bar{\rho}_s(h_i)\,dY_s^i - \frac{1}{2}\int_0^t (\bar{\rho}_s(h_i))^2\,ds\right)\right).$$

We apply Lemma 3.9 to the bounded m-dimensional process $\xi = \{\xi_t,\, t \geq 0\}$ defined as $\xi_t^i \triangleq \bar{\rho}_t(h_i)$, $i = 1,\ldots,m$, $t \geq 0$ and deduce from the boundedness of $\bar{\rho}_t$ that $\rho_t(\mathbf{1})$ is a (genuine) $\mathcal{Y}_t$-adapted martingale under $\tilde{\mathbb{P}}$. Also

$$\begin{aligned}(\rho_t(\mathbf{1}))^\alpha &= z_t^\alpha \exp\left(\sum_{i=1}^m \frac{\alpha^2 - \alpha}{2}\int_0^t (\bar{\rho}_s(h_i))^2\,ds\right) \\ &\leq z_t^\alpha \exp\left(\frac{m}{2}t\,|\alpha^2 - \alpha|\,\|h\|_\infty^2\right),\end{aligned} \tag{4.42}$$

where the process $z^\alpha = \{z_t^\alpha,\, t \geq 0\}$ is defined by

$$z_t^\alpha \triangleq \exp\left(\sum_{i=1}^m \left(\alpha\int_0^t \bar{\rho}_s(h_i)\,dY_s^i - \frac{\alpha^2}{2}\int_0^t (\bar{\rho}_s(h_i))^2\,ds\right)\right), \quad t \geq 0.$$

and is again a genuine $\tilde{\mathbb{P}}$ martingale by using Lemma 3.9. By Doob's maximal inequality we get from (4.42) that for $\alpha > 1$,

$$\begin{aligned}\tilde{\mathbb{E}}\left[\sup_{s \in [0,t]} (\rho_s(\mathbf{1}))^\alpha\right] &\leq \left(\frac{\alpha}{\alpha - 1}\right)^\alpha \tilde{\mathbb{E}}[(\rho_t(\mathbf{1}))^\alpha] \\ &\leq \left(\frac{\alpha}{\alpha - 1}\right)^\alpha \exp\left(\frac{m}{2}t\,(\alpha^2 - \alpha)\,\|h\|_\infty^2\right).\end{aligned}$$

Hence defining

$$k(\alpha) = \left(\frac{\alpha}{\alpha - 1}\right)^\alpha \exp\left(\frac{m}{2}t\,(\alpha^2 - \alpha)\,\|h\|_\infty^2\right),$$

we have established the required bound for $\alpha > 1$. The bound (4.39) for $0 < \alpha \leq 1$ follows by a straightforward application of Jensen's inequality. For example,

$$\tilde{\mathbb{E}}\left[\sup_{s \in [0,t]} (\rho_s(\mathbf{1}))^\alpha\right] \leq \left(\tilde{\mathbb{E}}\left[\sup_{s \in [0,t]} (\rho_s(\mathbf{1}))^2\right]\right)^{\alpha/2} \leq k(2)^{\alpha/2}.$$

$\square$

The class $\mathcal{U}'$ of measure-valued stochastic processes is larger than the class $\mathcal{U}$ defined in the Section 4.1. This is for two reasons; firstly because the constituent processes are no longer required to be adapted to the observation filtration $\mathcal{Y}_t$, but to the larger filtration $\mathcal{F}_t$. This relaxation is quite important as it leads to the uniqueness in distribution of the weak solutions of the Zakai equation (4.38) (see Lucic and Heunis [200] for details). The second relaxation is that condition (4.4) is no longer imposed. Unfortunately, this has to be done at the expense of the boundedness assumption on the function h.

Following Proposition 4.23, assumption (4.37) can be strengthened to

$$\tilde{\mathbb{E}}\left[\int_0^t \sum_{i=1}^m \rho_s(|(h_i+B_i)\varphi|)^2 \, ds\right]$$

$$\leq m\left(\|B\varphi\|_\infty + \|h\|_\infty\|\varphi\|_\infty\right)^2 \tilde{\mathbb{E}}\left[\int_0^t (\rho_s(\mathbf{1}))^2 \, ds\right]$$

$$\leq m\left(\|B\varphi\|_\infty + \|h\|_\infty\|\varphi\|_\infty\right)^2 tk(2) < \infty. \quad (4.43)$$

In particular, this implies that the stochastic integral in (4.38) is a (genuine) martingale. Let us define the operator $\Phi : B(\mathbb{S} \times \mathbb{S}) \to B(\mathbb{S} \times \mathbb{S})$ with domain

$$\mathcal{D}(\Phi) = \{\varphi \in B(\mathbb{S} \times \mathbb{S}) : \varphi(x_1, x_2) = \varphi_1(x_1)\varphi_2(x_2), \forall x_1, x_2 \in \mathbb{S}, \ \varphi_1, \varphi_2 \in \mathcal{D}\}$$

defined as follows. For $\varphi \in \mathcal{D}(\Phi)$ such that $\varphi(x_1, x_2) = \varphi_1(x_1)\varphi_2(x_2)$, for all $x_1, x_2 \in \mathbb{S}$ we have

$$\Phi\varphi(x_1, x_2) = \varphi_1(x_1)A\varphi_2(x_2) + \varphi_2(x_2)A\varphi_1(x_1)$$
$$+ \sum_{i=1}^m (h_i+B_i)\varphi_1(x_1)(h_i+B_i)\varphi_2(x_2). \quad (4.44)$$

We introduce next the following deterministic evolution equation

$$\nu_t\varphi = \nu_0(\varphi) + \int_0^t \nu_s(\Phi\varphi) \, ds, \quad (4.45)$$

where $\nu = \{\nu_t, \ t \geq 0\}$ is an $\mathcal{M}(\mathbb{S} \times \mathbb{S})$-valued stochastic process, with the property that the map $t \mapsto \nu_t\varphi : [0, \infty) \to [0, \infty)$ is Borel-measurable for any $\varphi \in B(\mathbb{S} \times \mathbb{S})$ and integrable for any φ in the range of Φ.

Condition 4.24 (U′). The function $h = (h_i)_{i=1}^m : \mathbb{S} \to \mathbb{R}^m$ appearing in the observation equation (3.5) is a bounded measurable function and the deterministic evolution equation (4.45) has a unique solution.

Of course, condition U′ is not as easy to verify as the corresponding condition U which is used in the PDE approach of Section 4.1. However Lucic and Heunis [200] prove that, in the case when the signal satisfies the stochastic differential equation,

$$dX_t^i = f^i(X_t)\,dt + \sum_{j=1}^n \sigma^{ij}(X_t)\,dV_t^j + \sum_{j=1}^m \bar{\sigma}^{ij}(X_t)\,dW_t^j, \tag{4.46}$$

then condition U' is implied by the following condition which is easier to verify.

Condition 4.25 (U''). The function $f = (f^i)_{i=1}^d : \mathbb{R}^d \to \mathbb{R}^d$ appearing in the signal equation (4.46) is Borel-measurable, whilst the functions $\sigma = (\sigma^{ij})_{i=1,\ldots,d, j=1,\ldots,n} : \mathbb{R}^d \to \mathbb{R}^{d\times n}$ and $\bar{\sigma} = (\bar{\sigma}^{ik})_{i=1,\ldots,d, k=1,\ldots,m} : \mathbb{R}^d \to \mathbb{R}^{d\times m}$ are continuous and there exists a constant K such that, for $x \in \mathbb{R}^d$, they satisfy the following linear growth condition

$$\max_{i,j,k} \left\{|f^i(x)|, |\sigma^{ij}(x)|, |\bar{\sigma}^{ik}(x)|\right\} \leq K(1+|x|).$$

Also $\bar{\sigma}\bar{\sigma}^\top$ is a strictly positive definite matrix for any $x \in \mathbb{R}^d$. Finally, the function $h = (h_i)_{i=1}^m : \mathbb{S} \to \mathbb{R}^m$ appearing in the observation equation (3.5) is a bounded measurable function.

The importance of Condition U' is that it ensures that there are enough functions in the domain of Φ so that $\nu = \{\nu_t,\ t \geq 0\}$ is uniquely characterized by (4.45). Lucic and Heunis [200] show that, under condition U'', the closure of the domain of Φ contains the set of bounded continuous functions which in turn implies the uniqueness of (4.45).

Theorem 4.26. *Assuming condition U', the equation (4.38) has a unique solution in the class $\mathcal{U}'$, up to indistinguishability.*

Proof. Let $\rho^1 = \{\rho_t^1,\ t \geq 0\}$ and $\rho^2 = \{\rho_t^2,\ t \geq 0\}$ be two processes belonging to the class $\mathcal{U}'$ and define the $\mathcal{M}(\mathbb{S} \times \mathbb{S})$-valued processes

$$\rho^{\alpha\beta} = \{\rho_t^{\alpha\beta},\ t \geq 0\}, \qquad \alpha, \beta = 1, 2$$

to be the unique processes for which

$$\rho_t^{\alpha\beta}(\Gamma_1 \times \Gamma_2) = \rho_t^\alpha(\Gamma_1)\rho_t^\beta(\Gamma_2), \quad \text{for any } \Gamma_1, \Gamma_2 \in \mathcal{B}(\mathbb{S}) \text{ and } t \geq 0.$$

Of course $\rho^{\alpha\beta}$ is an $\mathcal{F}_t$-adapted, progressively measurable process. Also define $\nu^{\alpha\beta} = \{\nu_t^{\alpha\beta},\ t \geq 0\}$ for $\alpha, \beta = 1, 2$ as follows

$$\nu_t^{\alpha\beta}(\Gamma) = \tilde{\mathbb{E}}\left[\rho_t^{\alpha\beta}(\Gamma)\right] \quad \text{for any } \Gamma \in \mathcal{B}(\mathbb{S} \times \mathbb{S}) \text{ and } t \geq 0.$$

It follows that $\nu_t^{\alpha\beta}$ is a positive measure on $(\mathbb{S} \times \mathbb{S}, \mathcal{B}(\mathbb{S} \times \mathbb{S}))$ and from Proposition 4.23 we get that, for any $t \geq 0$,

$$\sup_{s\in[0,t]} \nu_s^{\alpha\beta}(\mathbb{S}\times\mathbb{S}) = \sup_{s\in[0,t]} \tilde{\mathbb{E}}\left[\rho_t^\alpha(\mathbb{S})\rho_t^\beta(\mathbb{S})\right] \leq k(2);$$

hence $\nu^{\alpha\beta}$ is uniformly bounded with respect to s in any interval $[0,t]$ and by Fubini's theorem $t \mapsto \nu_t^{\alpha\beta}(\Gamma)$ is Borel-measurable for any $\Gamma \in \mathcal{B}(\mathbb{S}\times\mathbb{S})$. Let

$\varphi \in \mathcal{B}(\mathbb{S} \times \mathbb{S})$ such that $\varphi \in \mathcal{D}(\Phi)$. By definition, $\varphi(x_1, x_2) = \varphi_1(x_1)\varphi_2(x_2)$ and for all $x_1, x_2 \in \mathbb{S}$, $\varphi_1, \varphi_2 \in \mathcal{D}$ and

$$\begin{aligned}
\mathrm{d}\rho_t^{\alpha\beta}(\varphi) &= \mathrm{d}\left(\rho_t^\alpha(\varphi_1)\rho_t^\beta(\varphi_2)\right) \\
&= \rho_t^\alpha(\varphi_1)\,\mathrm{d}\rho_t^\beta(\varphi_2) + \rho_t^\alpha(\varphi_2)\,\mathrm{d}\rho_t^\beta(\varphi_2) + \mathrm{d}\left\langle \rho^\alpha(\varphi_1), \rho^\beta(\varphi_2)\right\rangle_t \\
&= \rho_t^\alpha(\varphi_a)\left(\rho_t^\beta(A\varphi_2)\,\mathrm{d}t + \rho_t^\beta((h^\top + B^\top)\varphi_2)\,\mathrm{d}Y_t\right) \\
&\quad + \rho_t^\beta(\varphi_2)\left(\rho_t^\alpha(A\varphi_1)\,\mathrm{d}t + \rho_t^\alpha((h^\top + B^\top)\varphi_1)\,\mathrm{d}Y_t\right) \\
&\quad + \sum_{i=1}^m \rho_t^\alpha((h_i + B_i)\varphi_1)\rho_t^\beta((h_i + B_i)\varphi_2)\,\mathrm{d}t.
\end{aligned}$$

In other words using Φ defined in (4.44) for $\varphi \in \mathcal{D}(\Phi)$,

$$\rho_t^{\alpha\beta}(\varphi) = \rho_0^{\alpha\beta}(\varphi) + \int_0^t \rho_s^{\alpha\beta}(\Phi\varphi)\,\mathrm{d}s + \int_0^t \Lambda_s^{\alpha\beta}(\varphi)\,\mathrm{d}Y_s, \qquad (4.47)$$

where $\Lambda_s^{\alpha\beta}(\varphi) \triangleq \rho_s^\alpha(\varphi_1)\rho_s^\beta((h^\top + B^\top)\varphi_2) + \rho_s^\beta(\varphi_2)\rho_s^\alpha((h^\top + B^\top)\varphi_1)$. By Proposition 4.23 and the Cauchy–Schwartz inequality we have that

$$\begin{aligned}
\mathbb{E}\left[\int_0^t \left(\Lambda_s^{\alpha\beta}(\varphi)\right)^2 \mathrm{d}s\right] &\leq M\mathbb{E}\left[\int_0^t \rho_s^\alpha(\mathbf{1})^2 \rho_s^\beta(\mathbf{1})^2\,\mathrm{d}s\right] \\
&\leq Mt\mathbb{E}\left[\sup_{s\in[0,T]} \rho_s^\alpha(\mathbf{1})^2 \sup_{s\in[0,T]} \rho_s^\beta(\mathbf{1})^2\right] \\
&\leq Mt\sqrt{\mathbb{E}\left[\sup_{s\in[0,T]} \rho_s^\alpha(\mathbf{1})^4\right]\mathbb{E}\left[\sup_{s\in[0,T]} \rho_s^\beta(\mathbf{1})^4\right]} \\
&\leq Mtk(4) < \infty,
\end{aligned}$$

where the constant M is given by

$$M = 4\max\left(\|\varphi_1\|_\infty^2, \|\varphi_2\|_\infty^2, \sum_{i=1}^m \|(h_i + B_i)\varphi_1\|_\infty^2, \sum_{i=1}^m \|(h_i + B_i)\varphi_2\|_\infty^2\right),$$

which is finite since $\varphi_1, \varphi_2 \in \mathcal{D}$ and consequently they belong to the domain of B_i, $i = 1, \ldots, m$. It follows that the stochastic integral in (4.47) is a martingale with zero expectation. In particular, from (4.47) and Fubini's theorem we get that for $\varphi \in \mathcal{D}(\Phi)$,

$$\begin{aligned}
\nu_t^{\alpha\beta}(\varphi) &= \tilde{\mathbb{E}}\left[\rho_t^{\alpha\beta}(\varphi)\right] \\
&= \tilde{\mathbb{E}}\left[\rho_0^{\alpha\beta}(\varphi) + \int_0^t \rho_s^{\alpha\beta}(\Phi\varphi)\,\mathrm{d}s\right] \\
&= \nu_0^{\alpha\beta}(\varphi) + \int_0^t \nu_s^{\alpha\beta}(\Phi\varphi)\,\mathrm{d}s. \qquad (4.48)
\end{aligned}$$

In (4.48), the use of the Fubini's theorem is justified as the mapping

$$(\omega, s) \in \Omega \times [0, t] \mapsto \rho_s^{\alpha\beta}(\Phi\varphi) \in \mathbb{R}$$

is $\mathcal{F} \times \mathcal{B}([0, t])$-measurable (it is a product of two $\mathcal{F} \times \mathcal{B}([0, t])$-measurable mappings) and integrable (following Proposition 4.23). From (4.48), we deduce that $\nu^{\alpha\beta}$ is a solution of the equation (4.45), hence by condition U' the deterministic evolution equation has a unique solution and since $\nu_0^{11} = \nu_0^{12} = \nu_0^{22}$, we have that for any $t \geq 0$,

$$\nu_t^{11} = \nu_t^{22} = \nu_t^{12}.$$

This implies that for any φ bounded Borel-measurable function we have

$$\mathbb{E}\left[\left(\rho_t^1(\varphi) - \rho_t^2(\varphi)\right)^2\right] = \nu_t^{11}(\varphi \times \varphi) + \nu_t^{11}(\varphi \times \varphi) - 2\nu_t^{12}(\varphi \times \varphi) = 0.$$

Hence $\rho_t^1(\varphi) = \rho_t^2(\varphi)$ holds $\tilde{\mathbb{P}}$-almost surely and by Exercise 4.1, the measure-valued processes ρ^1 and ρ^2 are indistinguishable. □

As in the previous section, now let $\mu = \{\mu_t, \, t \geq 0\}$ be an $\mathcal{F}_t$-adapted $\mathcal{M}(\mathbb{S})$-valued stochastic processes with càdlàg paths and $m^\mu = \{m_t^\mu, \, t \geq 0\}$ be the $\mathcal{F}_t$-adapted real-valued process

$$m_t^\mu = \exp\left(\int_0^t \mu_s(h^\top)\, dY_s - \frac{1}{2}\int_0^t \mu_s(h^\top)\mu_s(h)\, ds\right), \quad t \geq 0.$$

Define the class $\bar{\mathcal{U}}'$ of all $\mathcal{F}_t$-adapted $\mathcal{M}(\mathbb{S})$-valued stochastic processes with càdlàg paths such that the process $m^\mu \mu$ belongs to the class $\mathcal{U}'$.

Exercise 4.27. Let X be the solution of the SDE (4.46). Prove that if h is bounded then π belongs to the class $\bar{\mathcal{U}}'$.

Exercise 4.28. Assume that condition U' holds. Prove that the Kushner–Stratonovich equation has a unique solution (up to indistinguishability) in the class $\bar{\mathcal{U}}'$.

4.3 Solutions to Exercises

4.1 Since $\mu_t^1(\varphi_i) = \mu_t^2(\varphi_i)$ almost surely for any $i \geq 0$ one can find a set $\hat{\Omega}_t$ of measure one, independent of $i \geq 0$, such that for any $\omega \in \hat{\Omega}_t$, $\mu_t^1(\varphi)(\omega) = \mu_t^2(\varphi_i)(\omega)$ for all $i \geq 0$. Since $(\varphi_i)_{i\geq 0}$ is a separating sequence, it follows that for any $\omega \in \hat{\Omega}_t$, $\mu_t^1(\omega) = \mu_t^2(\omega)$. Hence one can find a set $\hat{\Omega}$ of measure one independent of t such that for any $\omega \in \hat{\Omega}$, $\mu_t^1(\omega) = \mu_t^2(\omega)$ for all $t \in \mathbb{Q}_+$ (the positive rational numbers). This together with the right continuity of the sample paths of μ^1 and μ^2 implies that for any $\omega \in \hat{\Omega}$, $\mu_t^1(\omega) = \mu_t^2(\omega)$ for all $t \geq 0$.

4.2 Suppose $\nu_{\mu_n} \Rightarrow \nu_\mu$; then from the definition of weak convergence, for any $\varphi \in C_b(\mathbb{R}^d)$ it follows that $\nu_{\mu_n}\varphi \to \nu_\mu\varphi$ as $n \to \infty$. Thus $\mu_n(\varphi\psi) \to \mu(\varphi\psi)$. Since any function in $C^l(\mathbb{R}^d)$ is of the form $\varphi\psi$ where $\varphi \in C_b(\mathbb{R}^d)$, it follows that μ_n converges to μ in $\mathcal{M}^l(\mathbb{R}^d)$.

Conversely suppose that μ_n converges to μ in $\mathcal{M}^l(\mathbb{R}^d)$; thus $\mu_n\varphi \to \mu\varphi$ for $\varphi \in C^l(\mathbb{R}^d)$. If we set $\varphi = \psi\theta$ for $\theta \in C_b(\mathbb{R}^d)$, then as $\varphi/\psi \in C_b(\mathbb{R}^d)$, it follows that $\varphi \in C^l(\mathbb{R}^d)$. Thus $\mu_n(\psi\theta) \to \mu(\psi\theta)$ for all $\theta \in C_b(\mathbb{R}^d)$, whence $\nu_{\mu_n} \Rightarrow \nu_\mu$.

4.4 We have by the Kallianpur–Striebel formula

$$\tilde{\mathbb{E}}\left[\int_0^t (\rho_s(\psi))^2\, ds\right] = \tilde{\mathbb{E}}\left[\int_0^t (\pi_s(\psi))^2\, \rho_s^2(\mathbf{1})\, ds\right]$$
$$= \int_0^t \tilde{\mathbb{E}}\left[(\pi_s(\psi))^2\, \rho_s^2(\mathbf{1})\right] ds.$$

Now

$$\tilde{\mathbb{E}}\left[(\pi_s(\psi))^2\, \rho_s^2(\mathbf{1})\right] \le \tilde{\mathbb{E}}\left[\pi_s(\psi^2)\, \rho_s^2(\mathbf{1})\right]$$
$$= \mathbb{E}\left[\pi_s(\psi^2)\, \rho_s^2(\mathbf{1}) Z_s\right]$$
$$= \mathbb{E}\left[\pi_s(\psi^2)\rho_s^2(\mathbf{1}) \mathbb{E}[Z_s|\mathcal{Y}_s]\right].$$

Since $\rho_s(\mathbf{1}) = 1/\mathbb{E}[Z_s \mid \mathcal{Y}_s]$ (see Exercise 3.37 part (iii)) we get that

$$\tilde{\mathbb{E}}\left[\pi_s(\psi^2)\, \rho_s^2(\mathbf{1})\right] = \mathbb{E}\left[\pi_s(\psi^2)\, \rho_s(\mathbf{1})\right]$$
$$= \mathbb{E}\left[\mathbb{E}\left[\psi^2(X_s)|\mathcal{Y}_s\right] \rho_s(\mathbf{1})\right]$$
$$= \mathbb{E}\left[\psi^2(X_s)\, \rho_s(\mathbf{1})\right].$$

Now since h is bounded,

$$\rho_s(\mathbf{1}) = \exp\left(\int_0^s \pi_r(h^\top)\, dY_r - \frac{1}{2}\int_0^s \|\pi_r(h)\|^2\, dr\right)$$
$$= \exp\left(\int_0^s \pi_r(h^\top)\, dW_r + \int_0^s \pi_r(h^\top) h(X_r)\, dr - \frac{1}{2}\int_0^s \|\pi_r(h)\|^2\, dr\right)$$
$$\le e^{s\|h\|_\infty^2} \exp\left(\int_0^s \pi_r(h^\top)\, dW_s - \frac{1}{2}\int_0^s \|\pi_r(h)\|^2\, dr\right).$$

Using the independence of W and X we see that

$$\mathbb{E}\left[\exp\left(\int_0^s \pi_r(h^\top)\, dW_s - \frac{1}{2}\int_0^s \pi_r(h^\top)\pi_r(h) dr\right)\bigg|\sigma(X_r,\ r \in [0,s])\right] = 1,$$

hence

$$\mathbb{E}[\rho_s(\mathbf{1})|\sigma(X_r,\ r \in [0,s])] \le e^{s\|h\|_\infty^2}.$$

It follows that

118   4 Uniqueness of the Solution

$$\tilde{\mathbb{E}}\left[\pi_s(\psi^2)\rho_s^2(1)\right] \le e^{s\|h\|_\infty^2}\mathbb{E}\left[(1+\|X_s\|)^2\right],$$

and therefore

$$\tilde{\mathbb{E}}\left[\int_0^t (\rho_s(\psi))^2\,ds\right] \le te^{t\|h\|_\infty^2}\sup_{s\in[0,t]}\mathbb{E}\left[(1+\|X_s\|)^2\right]$$

$$\le 2te^{t\|h\|_\infty^2}\left(1+\sup_{s\in[0,t]}\mathbb{E}\left[\|X_s\|^2\right]\right).$$

As a consequence of Exercise 3.10, the last term in this equation is finite if X_0 has finite second moment and (3.10) is satisfied. Thus ρ satisfies condition (4.4) and hence it belongs to the class $\mathcal{U}$.

4.7

i. We know that for t in $[0,\infty)$ the process μ_t is $\mathcal{Y}_t$-measurable. As $\varphi \in \mathcal{E}$, this implies that $\varphi_t \in C^l(\mathbb{R}^d)$ and thus $|\varphi_t(x)| \le \|\varphi_t\|_\infty^l \psi(x)$. Define the sequence

$$\varphi_t^n(x) \triangleq \varphi_t(x)1_{\{|\varphi_t(x)|\le n\}}.$$

By the argument used for Exercise 2.21 we know that $\mu_t(\varphi^n)$ is $\mathcal{Y}_t$-adapted since φ^n is bounded. But $\|\varphi_t\|_\infty^l \psi$ is a dominating function, and since $\mu \in \mathcal{U}$, it follows that $\mu_t(\psi) < \infty$ hence it is a μ_t-measurable dominating function. Thus $\mu_t(\varphi_t^n) \to \mu_t(\varphi_t)$ as $n \to \infty$, which implies that $\mu_t(\varphi_t)$ is $\mathcal{Y}_t$-measurable. As this holds for all $t \in [0,\infty)$ it follows that $\mu_t(\varphi_t)$ is $\mathcal{Y}_t$-adapted.

ii. From the solution to Exercise 3.23, a sufficient condition for the stochastic integral to be well defined is

$$\tilde{\mathbb{P}}\left[\int_0^t (\mu_s(\varphi\|h\|))^2\,ds < \infty\right] = 1.$$

We establish the stronger condition for the stochastic integral to be a martingale; viz for all $t \ge 0$,

$$\tilde{\mathbb{E}}\left[\int_0^t (\mu_s(\varphi\|h\|))^2\,ds\right] < \infty.$$

Using the boundedness of φ and the linear growth condition

$$\varphi(x)h(x) \le \sqrt{C}\|\varphi\|_\infty\sqrt{1+\|x\|^2} = \sqrt{C}\|\varphi\|_\infty\psi(x),$$

but since $\mu_s \in \mathcal{M}^l(\mathbb{R}^d)$, it follows that $\mu_s(\psi) < \infty$. Thus

$$\int_0^t (\mu_s(\varphi\|h\|))^2\,ds \le \|\varphi\|_\infty C\int_0^t (\mu_s(\psi))^2\,ds,$$

and by condition (4.4) it follows that

$$\tilde{\mathbb{E}}\left[\int_0^t (\mu_s(\psi))^2\right] < \infty$$

so the stochastic integral is both well defined and a martingale.

4.9 Starting from (4.5) we apply Itô's formula to the product $\varepsilon_t \mu_t(\varphi_t)$, obtaining

$$\varepsilon_t \mu_t(\varphi_t) = \varepsilon_0 \pi_0(\varphi_0) + \int_0^t \varepsilon_s \mu_s \left(\frac{\partial \varphi_t}{\partial t} + A\varphi_s\right) ds$$
$$+ \int_0^t \varepsilon_s \mu_s(\varphi_t h^\top) dY_s + \int_0^t i\varepsilon_s r_s^\top \mu_t(\varphi_t) dY_s + \int_0^t i\varepsilon_s r_s \mu_s(\varphi_s h^\top) ds.$$

Next we take expectation under $\tilde{\mathbb{P}}$. We now show that as a consequence of condition (4.4) both stochastic integrals are genuine martingales. Because ε_t is complex-valued we need to introduce the notation

$$\|\varepsilon(\omega)\|_\infty = \sup_{t \in [0,\infty)} |\varepsilon_t(\omega)|$$

where $|\cdot|$ denotes the modulus of the complex number. The following bound is elementary,

$$\|\varepsilon_t\|_\infty \leq \exp\left(\tfrac{1}{2} \max_{i=1,\dots,m} \|r_i\|_\infty^2 t\right) < \infty;$$

for notational conciseness write $R = \max_{i=1,\dots,m} \|r_i\|_\infty$. By assumption there is a uniform bound on $\|\varphi_s\|_\infty$ for $s \in [0,t]$; hence

$$\tilde{\mathbb{E}}\left[\int_0^t \varepsilon_s^2 \left(\mu_s(\varphi_s h^\top)\right)^2 ds\right] \leq e^{R^2 t} \sup_{[0,t]} \|\varphi_s\|_\infty \tilde{\mathbb{E}}\left[\int_0^t (\mu_s(\|h\|))^2 ds\right]$$

and the right-hand side is finite by (4.4). The second stochastic integral is treated in a similar manner

$$\tilde{\mathbb{E}}\left[\int_0^t \varepsilon_s^2 \|r_s\|^2 (\mu_s(\varphi_s))^2 ds\right] \leq R^2 e^{R^2 t} \sup_{[0,t]} \|\varphi_s\|_\infty^2 \tilde{\mathbb{E}}\left[\int_0^t (\mu_s(\mathbf{1}))^2 ds\right].$$

Therefore

$$\tilde{\mathbb{E}}(\varepsilon_t \mu_t(\varphi_t)) = \pi_0(\varphi_0) + \tilde{\mathbb{E}}\left[\int_0^t \varepsilon_s \mu_s \left(\frac{\partial \varphi_s}{\partial t} + A\varphi_s + ir_s \varphi_s h^\top\right) ds\right],$$

which is (4.13).

4.11 Since the components of a_n, f_n and h_n are bounded it is immediate that they belong to $C_b(\mathbb{R}^d)$ and consequently to the larger space $C^l(\mathbb{R}^d)$.

For the bound, as there are a finite number of components it is sufficient to establish the result for one of them. Clearly

$$a_n^{ij}(x) = a_n^{ij}(0) + \sum_{k=1}^{d} \int_0^1 \frac{\partial a_n^{ij}}{\partial x_k}(xs) x_k \, ds.$$

By (4.16), uniformly in x and i,

$$\left| \frac{\partial a_n^{ij}}{\partial x_i} \right| \leq K_0;$$

thus

$$|a_n^{ij}(x)| \leq |a_n^{ij}(0)| + dK_0 \|x\|.$$

Secondly, since $a_n^{ij} \to a^{ij}$ it follows that $a_n^{ij}(0) \to a^{ij}(0)$; thus given $\varepsilon > 0$, there exists n_0 such that for $n \geq n_0$, $|a_n^{ij}(0) - a^{ij}(0)| < \varepsilon$. Thus we obtain the bound

$$\|a_n^{ij}(x)\| \leq \max_{1 \leq i \leq n_0} \|a_i^{ij}(0)\| + \|a^{ij}(0)\| + \varepsilon + dK_0 \|x\|.$$

Hence, since

$$\|a_n^{ij}\|_\infty^l = \sup_{x \in \mathbb{R}^d} \frac{|a_n^{ij}(x)|}{1 + \|x\|}$$

setting $A = \max(\max_{1 \leq i \leq n_0} \|a_i^{ij}(0)\| + \|a^{ij}(0)\| + \varepsilon, dK_0)$, it follows that $\|a_n^{ij}\|_\infty^l \leq A$.

4.13

i. At such a maximum (t_0, x_0) in $(0, t] \times \mathbb{R}^d$,

$$\frac{\partial u}{\partial t}(t_0, x_0) \geq 0, \qquad \frac{\partial u}{\partial x_i}(t_0, x_0) = 0, \quad i = 1, \ldots d,$$

(we cannot assert that the time derivative is zero, since the maximum might occur on the boundary at t) and the Hessian matrix of u (i.e. $(\partial_i \partial_j u)$) is negative definite. Thus since a is positive definite, it follows that

$$\sum_{i,j=1}^{d} a^{ij}(x_0) \frac{\partial^2 u}{\partial x_i \partial x_j}(t_0, x_0) \leq 0, \qquad \sum_{i=1}^{d} f^i(x_0) \frac{\partial u}{\partial x_i}(t_0, x_0) = 0;$$

consequently

$$Au(x_0) - \frac{\partial u}{\partial t}(t_0, x_0) \leq 0$$

which is a contradiction since we had assumed that the left-hand side was strictly positive.

ii. It is easy to verify that

$$\frac{\partial w}{\partial t} = \frac{\partial u}{\partial t} - \delta + \varepsilon e^{-t} \|x\|^2,$$

and

$$Aw = Au - \varepsilon e^{-t}\left(2\operatorname{tr} a + 2b^\top x\right).$$

Thus
$$Aw - \frac{\partial w}{\partial t} \geq -\varepsilon e^{-t}\left(2\operatorname{tr} a + 2(b-x)^\top x\right) + \delta.$$

Thus given $\delta > 0$ using the fact that a and b are bounded, we can find $\varepsilon(\delta)$ so that this right-hand side is strictly positive.

iii. Choose δ, ε so that the condition in part (ii) is satisfied. It is clear that $w_{\delta,\varepsilon}(0,x) = u(0,x) - \varepsilon\|x\|^2$. Thus since $\varepsilon > 0$, if $u(0,x) \leq 0$, it follows that $w_{\delta,\varepsilon}(0,x) \leq 0$. Also since u is bounded above, it is clear that as $\|x\| \to \infty$, $w_{\delta,\varepsilon}(t,x) \to -\infty$. Therefore if $u(t,x) \geq 0$ at some point, it is clear that $w_{\delta,\varepsilon}$ has a maximum. But by part (i) $w_{\delta,\varepsilon}(t,x)$ cannot have such a maximum on $(0,t] \times \mathbb{R}^d$. Hence $u(t,x) \leq 0$ for all $t \in [0,\infty)$ and $x \in \mathbb{R}^d$.

4.17 Under the condition that
$$\widetilde{\mathbb{E}}\left(\int_0^t \left[\rho_s(\|(h^\top + B^\top)\varphi\|)\right]^2 ds\right) < \infty,$$

we deduce that the corresponding complex values PDE for a functional dual φ is
$$A\varphi_t + \frac{\partial \varphi_t}{\partial t} + ir_t^\top(h\varphi_t + B\varphi_t) = 0.$$

If we write $\varphi_t = v_1^t + iv_2^t$, then the time reversed equation is
$$\frac{\partial \bar{v}^1}{\partial t} = A\bar{v}^1 - \bar{v}^2 g_s - \bar{r}^\top B\bar{v}^2$$
$$\frac{\partial \bar{v}^2}{\partial t} = A\bar{v}^2 + \bar{v}^1 g_s + \bar{r}^\top B\bar{v}^1,$$

where $r_s = r_{t-s}$, and $g_s = h^\top \bar{r}$. As in the proof for the uncorrelated case an approximating sequence of uniformly parabolic PDEs is taken, with smooth bounded coefficients and so that (4.16) holds together with the analogue for f. Then with z_t^n defined by (4.24),

$$\frac{\partial z_s}{\partial s} - Az_s = -\sum_{\alpha,\beta=1}^d a^{\alpha\beta}\left(\partial_\alpha \bar{v}_s^{n,1}\partial_\beta \bar{v}_s^{n,1} + \partial_\alpha \bar{v}_s^{n,2}\partial_\beta \bar{v}_s^{n,2}\right)$$
$$- \bar{v}_s^{n,1}\bar{r}^\top B\bar{v}_s^{n,2} + \bar{v}_s^{n,2}\bar{r}^\top B\bar{v}_s^{n,1}.$$

If we consider the special case of Corollary 3.40, and write $c_t = \bar{\sigma}\bar{r}_t$, which we assume to be uniformly bounded, then

$$\frac{\partial z_s^n}{\partial s} - Az_s^n = -\sum_{\alpha,\beta=1}^d a^{\alpha\beta}\left(\partial_\alpha \bar{v}_s^{n,1}\partial_\beta \bar{v}_s^{n,1} + \partial_\alpha \bar{v}_s^{n,2}\partial_\beta \bar{v}_s^{n,2}\right)$$
$$+ \sum_{\gamma=1}^d c_t^\gamma\left(-\bar{v}_s^{n,1}\partial_\gamma \bar{v}_s^{n,2} + \bar{v}_s^{n,2}\partial_\gamma \bar{v}_s^{n,1}\right).$$

Using the inequality $ab \le \frac{1}{2}(a^2+b^2)$, it follows that for $\varepsilon > 0$,

$$\frac{\partial z_s^n}{\partial s} - A z_s^n \le -\sum_{\alpha,\beta=1}^{d} a^{\alpha\beta}\left(\partial_\alpha \bar{v}_s^{n,1} \partial_\beta \bar{v}_s^{n,1} + \partial_\alpha \bar{v}_s^{n,2} \partial_\beta \bar{v}_s^{n,2}\right)$$

$$+ \frac{1}{2\varepsilon} \sum_{\gamma=1}^{d} |c_t^\gamma|\left((\bar{v}_s^{n,1})^2 + (\bar{v}_s^{n,2})^2\right)$$

$$+ \frac{\varepsilon}{2} \sum_{\gamma=1}^{d} \left((\partial_\gamma \bar{v}_s^{n,1})^2 + (\partial_\gamma \bar{v}_s^{n,2})^2\right)$$

$$\le \frac{z_s^n d \|c\|_\infty}{\varepsilon}$$

$$- \sum_{\alpha,\beta=1}^{d} (a - \varepsilon/2\mathbb{I})^{\alpha\beta}\left(\partial_\alpha \bar{v}_s^{n,1} \partial_\beta \bar{v}_s^{n,1} + \partial_\alpha \bar{v}_s^{n,2} \partial_\beta \bar{v}_s^{n,2}\right).$$

As a is uniformly elliptic, $x^\top a x \ge \lambda \|x\|^2$, therefore, by choosing ε sufficiently small (i.e. $\varepsilon < 2\lambda$) then the matrix $a - \varepsilon/2\mathbb{I}$ is positive definite. Thus

$$\frac{\partial z_s^n}{\partial s} - A z_s^n \le \frac{z_s^n d \|c\|_\infty}{\varepsilon}.$$

Writing $\bar{C}_0 = d\|c\|_\infty/\varepsilon$ and $\hat{z}_t = e^{-\bar{C}_0 t} z_t$, then

$$\frac{\partial \hat{z}_s^n}{\partial s} - A \hat{z}_s^n \le 0,$$

from which the positive maximum principle (Lemma 4.12) implies that

$$\|\bar{v}_t^{n,1}\|_\infty^2 + \|\bar{v}_t^{n,2}\|_\infty^2 \le e^{\bar{C}_0 t}\left(\|\Phi_t^1\|_\infty^2 + \|\Phi_t^2\|_\infty^2\right)$$

and the boundedness of $\bar{v}^{n,1}$ and $\bar{v}^{n,2}$ follows. To show the boundedness of the first derivatives, define u_s^n as in (4.26); then

$$\frac{\partial u_s^n}{\partial s} - A_n u_s^n =$$

$$- \sum_{\alpha,\beta,\gamma=1}^{d} a_n^{\alpha\beta} \left((\partial_\alpha \partial_\gamma \bar{v}_s^{n,1})(\partial_\beta \partial_\gamma \bar{v}_s^{n,1}) + (\partial_\alpha \partial_\gamma \bar{v}_s^{n,2})(\partial_\beta \partial_\gamma \bar{v}_s^{n,2}) \right)$$

$$+ \sum_{\alpha,\beta,\gamma=1}^{d} \partial_\gamma a_n^{\alpha\beta} \left((\partial_\alpha \partial_\beta \bar{v}_s^{n,1})(\partial_\gamma \bar{v}_s^{n,1}) + (\partial_\alpha \partial_\beta \bar{v}_s^{n,2})(\partial_\gamma \bar{v}_s^{n,2}) \right)$$

$$+ \sum_{\alpha,\beta=1}^{d} \partial_\beta f_n^{\alpha} \left(\partial_\alpha \bar{v}_s^{n,1} \partial_\beta \bar{v}_s^{n,1} + \partial_\alpha \bar{v}_s^{n,2} \partial_\beta \bar{v}_s^{n,2} \right)$$

$$+ \sum_{\alpha=1}^{d} \partial_\alpha g_{n,s}(-\bar{v}_s^{n,2} \partial_\alpha \bar{v}_s^{n,1} + \bar{v}_s^{n,1} \partial_\alpha \bar{v}_s^{n,2})$$

$$+ \sum_{\alpha=1}^{d} \left(-(\partial_\alpha \bar{v}_s^{n,1})(\partial_\alpha(\bar{r}^\top B \bar{v}_s^{n,2})) + (\partial_\alpha \bar{v}_s^{n,2})(\partial_\alpha(\bar{r}^\top B \bar{v}_s^{n,1})) \right).$$

Bounds on the first four summations are identical to those used in the proof in the uncorrelated noise case, so

$$\frac{\partial u_s^n}{\partial s} - A_n u_s^n \leq -\Theta_s^n + \tfrac{1}{2}\Theta_s^n + C_2 u_s^n + 2K_0 d u_s^n + C_4(u_s^n + d)$$

$$+ \sum_{\alpha=1}^{d} \left(-(\partial_\alpha \bar{v}_s^{n,1})(\partial_\alpha(\bar{r}^\top B \bar{v}_s^{n,2})) + (\partial_\alpha \bar{v}_s^{n,2})(\partial_\alpha(\bar{r}^\top B \bar{v}_s^{n,1})) \right).$$

To bound the final summation again use the special form of Corollary 3.40,

$$\frac{\partial u_s^n}{\partial s} - A_n u_s^n \leq \tfrac{1}{2}\Theta_s^n + C_0 u_s^n + C_1$$

$$+ \sum_{\alpha,\gamma=1}^{d} c_s^\gamma \left(-(\partial_\alpha \bar{v}_s^{n,1})(\partial_\alpha \partial_\gamma \bar{v}_s^{n,2}) + (\partial_\alpha \bar{v}_s^{n,2})(\partial_\alpha \partial_\gamma \bar{v}_s^{n,1}) \right)$$

$$+ \sum_{\alpha,\gamma=1}^{d} (\partial_\alpha c_s^\gamma) \left(-(\partial_\alpha \bar{v}_s^{n,1})(\partial_\gamma \bar{v}_s^{n,2}) + (\partial_\alpha \bar{v}_s^{n,2})(\partial_\gamma \bar{v}_s^{n,1}) \right).$$

The first summation can be bounded using $ab \leq \tfrac{1}{2}(a^2 + b^2)$ for $\varepsilon > 0$,

$$\sum_{\alpha,\gamma=1}^{d} c_s^\gamma \left(-(\partial_\alpha \bar{v}_s^{n,1})(\partial_\alpha \partial_\gamma \bar{v}_s^{n,2}) + (\partial_\alpha \bar{v}_s^{n,2})(\partial_\alpha \partial_\gamma \bar{v}_s^{n,1}) \right)$$

$$\leq \frac{d \|c\|_\infty u_s^n}{\varepsilon}$$

$$+ \frac{\varepsilon}{2} \|c\|_\infty \sum_{\alpha,\gamma=1}^{d} \left((\partial_\alpha \partial_\gamma \bar{v}_s^{n,1})^2 + (\partial_\alpha \partial_\gamma \bar{v}_s^{n,2})^2 \right).$$

Again by choice of ε sufficiently small, the matrix $a - \varepsilon\|c\|_\infty I$ remains positive definite (for $\varepsilon < \lambda$), therefore

$$-\frac{1}{2}\Theta^n_s + \frac{\varepsilon}{2}\|c\|_\infty \sum_{\alpha,\gamma=1}^d \left((\partial_\alpha\partial_\gamma \bar{v}^{n,1}_s)^2 + (\partial_\alpha\partial_\gamma \bar{v}^{n,2}_s)^2\right) \leq 0.$$

Since $\partial_\alpha c^\gamma_t$ is uniformly bounded by C_5, it follows that

$$\sum_{\alpha,\gamma=1}^d (\partial_\alpha c^\gamma_s)\left(-(\partial_\alpha \bar{v}^{n,1}_s)(\partial_\gamma \bar{v}^{n,2}_s) + (\partial_\alpha \bar{v}^{n,2}_s)(\partial_\gamma \bar{v}^{n,1}_s)\right)$$

$$\leq C_5 \sum_{\alpha,\gamma=1}^d \left(|\partial_\alpha \bar{v}^{n,2}_s||\partial_\gamma \bar{v}^{n,1}_s| + |\partial_\alpha \bar{v}^{n,1}_s||\partial_\gamma \bar{v}^{n,2}_s|\right)$$

$$\leq \frac{C_5}{2} \sum_{\alpha,\gamma=1}^d \left(|\partial_\alpha \bar{v}^{n,1}_s|^2 + |\partial_\gamma \bar{v}^{n,2}_s|^2 + |\partial_\alpha \bar{v}^{n,2}_s|^2 + |\partial_\gamma \bar{v}^{n,1}_s|^2\right)$$

$$\leq dC_5 \sum_{\alpha=1}^d \left(|\partial_\alpha \bar{v}^{n,1}_s|^2 + |\partial_\alpha \bar{v}^{n,2}_s|^2\right)$$

$$\leq 2dC_5 u^n_s.$$

Using all these bounds

$$\frac{\partial u^n_s}{\partial s} - A_n u^n_s \leq \hat{C}_0 u^n_s + \hat{C}_1,$$

where $\hat{C}_0 \triangleq C_2 + 2K_0 d + C_4 + d\|c\|_\infty/\varepsilon + 2dC_5$ and $\hat{C}_1 \triangleq dC_4$; thus as in the correlated case

$$\|u^n_s\|_\infty \leq e^{\hat{C}_0 T}\left(\frac{1}{2}\sum_{\alpha=1}^d \left(\|\partial_\alpha \Phi^1\|^2_\infty + \|\partial_\alpha \Phi^2\|^2_\infty\right) + \frac{\hat{C}_1}{\hat{C}_0}\right),$$

from which the bound follows. The boundedness of the second derivatives is established by a similar but longer argument.

4.18 Using Exercises 3.11 and 3.25 the conditions (3.25) and (3.42) are satisfied. Lemma 3.29 then implies that $m^\pi_t = \rho_t(1)$. From the Kallianpur–Striebel formula (3.36), for any φ bounded Borel-measurable, $\rho_t(\varphi) = \pi_t(\varphi)\rho_t(1)$, and by Exercise 4.4 the process ρ_t belongs to $\mathcal{U}$.

4.21 Since $\rho_t(1) = \tilde{\mathbb{E}}[\tilde{Z}_t|\mathcal{Y}_t]$, we need to prove that $\tilde{\mathbb{E}}[\rho_t(1)\xi] = \tilde{\mathbb{E}}[\rho_s(1)\xi]$ for any $\mathcal{Y}_s$-measurable function. We have, using the martingale property of $\tilde{Z}$ that

$$\tilde{\mathbb{E}}\left[\tilde{\mathbb{E}}[\tilde{Z}_t|\mathcal{Y}_t]\xi\right] = \tilde{\mathbb{E}}\left[\tilde{Z}_t\xi\right] = \tilde{\mathbb{E}}\left[\tilde{\mathbb{E}}\left[\tilde{Z}_t\xi|\mathcal{Y}_s\right]\right] = \tilde{\mathbb{E}}\left[\tilde{Z}_s\xi\right] = \tilde{\mathbb{E}}\left[\tilde{\mathbb{E}}[\tilde{Z}_s|\mathcal{Y}_s]\xi\right],$$

which implies that $\rho_t(1)$ is a $\mathcal{Y}_t$-martingale.

4.22 From Lemma 3.18 it follows that ρ_t is càdlàg, and ρ_t is $\mathcal{Y}_t$-adapted which implies that it is $\mathcal{F}_t$-adapted since $\mathcal{Y}_t \subset \mathcal{F}_t$. To check the condition (4.37), note that

$$(\mu_t(|(h_i + B_i)\varphi|))^2 \leq 2\left(\mu_t(|h_i\varphi|)\right)^2 + 2\left(\mu_t(|B_i\varphi|)\right)^2$$
$$\leq 2\|\varphi\|_\infty^2 \left(\mu_t(\|h\|)\right)^2 + 2\|B\varphi\|_\infty^2 \left(\mu_t(\mathbf{1})\right)^2.$$

Thus

$$\int_0^t \sum_{i=1}^m [\mu_s(|(h_i + B_i)\varphi|)]^2 \, \mathrm{d}s$$
$$\leq 2m \left(\|\varphi\|_\infty^2 \int_0^t \left(\mu_s(\|h\|)\right)^2 \mathrm{d}s + \|B\varphi\|_\infty^2 \int_0^t \left(\mu_s(\mathbf{1})\right)^2 \mathrm{d}s \right)$$
$$\leq 2m \left(\|\varphi\|_\infty^2 \int_0^t \left(\mu_s(\|h\|)\right)^2 \mathrm{d}s + t\|B\varphi\|_\infty^2 \left(\sup_{s \in [0,t]} \mu_s(\mathbf{1}) \right)^2 \right).$$

Since (3.42) is satisfied, the first term is $\tilde{\mathbb{P}}$-a.s. finite. As $\mu_t(\mathbf{1})$ has càdlàg paths, it follows that the second term is $\tilde{\mathbb{P}}$-a.s. finite.

4.27 If h is bounded then conditions (3.25) and (3.42) are automatically satisfied. If π_t is the normalised conditional distribution, by Lemma 3.29, $m_t^\pi = \rho_t(\mathbf{1})$, hence from the Kallianpur–Striebel formula (3.36) $m_t^\pi \pi_t(\varphi) = \rho_t(\varphi)$, and from Exercise 4.22 it then follows that $m^\pi \pi$ is in $\mathcal{U}'$. As π_t is $\mathcal{Y}_t$-adapted, it is $\mathcal{F}_t$-adapted. Furthermore, from Corollary 2.26 the process π_t has càdlàg paths; thus π_t is in $\bar{\mathcal{U}}'$.

4.28 Suppose that there are two solutions π_1 and π_2 in $\bar{\mathcal{U}}'$. Then $\rho_i \triangleq m^{\pi^i} \pi_i$ are corresponding solutions of the Zakai equation, and from the definition of $\bar{\mathcal{U}}'$ must lie in $\mathcal{U}'$. As condition U' holds, by Theorem 4.26, it follows that ρ_1 and ρ_2 are indistinguishable. The remainder of the proof is identical to that of Theorem 4.19.

4.4 Bibliographical Notes

There are numerous other approaches to establish uniqueness of solution to the filtering equations. Several papers address the question of uniqueness without assuming that the solution of the two SPDEs (Zakai's equation or the Kushner–Stratonovich equation) is adapted with respect to the given observation σ-field $\mathcal{Y}_t$. A benefit of this approach is that it allows uniqueness in law of the solution to be established. In Szpirglas [264], the author shows that in the absence of correlation between the observation noise and the signal, the Zakai equation is equivalent to the equation

$$\rho_t(\varphi) = \pi_0(P_t \varphi) + \int_0^t \rho_s(P_{t-s}\varphi h^\top) \, \mathrm{d}Y_s, \qquad (4.49)$$

for all $\varphi \in B(\mathbb{S})$, where P_t is the semigroup associated with the generator A. This equivalence means that a solution of the Zakai equation is a solution of (4.49) and vice versa. The uniqueness of the solution of (4.49) is established by iterating a simple integral inequality (Section V2, [264]). However, this technique does not appear to extend to the case of correlated noise.

More recently, Lucic and Heunis [200] prove uniqueness for the correlated case, again without the assumption of adaptedness of the solution to the observation σ-algebra. There are no smoothness conditions imposed on the coefficients of the signal or observation equation. However h is assumed to be bounded and the signal non-degenerate (i.e. $\sigma^\top \sigma$ is required to be positive definite).

The problem of establishing uniqueness when ρ_t and π_t are required to be adapted to a specified σ-algebra $\mathcal{Y}_t$ is considered in Kurtz and Ocone [170] and further in Bhatt et al. [18]. This form of uniqueness can be established under much less restrictive conditions on the system.

5

The Robust Representation Formula

5.1 The Framework

Throughout this section we assume that the pair (X, Y) are as defined in Chapter 3. That is, X is a solution of the martingale problem for (A, π_0) and Y satisfies the evolution equation (3.5) with null initial condition; that is,

$$Y_s = \int_0^s h(X_r)\,\mathrm{d}r + W_s, \quad s \geq 0. \tag{5.1}$$

To start off with, we assume that the function $h = (h_i)_{i=1}^m : \mathbb{S} \to \mathbb{R}^m$ satisfies either Novikov's condition (3.19) or condition (3.25) so that the process $Z = \{Z_t,\ t > 0\}$ defined by

$$Z_t = \exp\left(-\int_0^t h(X_s)^\top \,\mathrm{d}W_s - \frac{1}{2}\int_0^t \|h(X_s)\|^2\,\mathrm{d}s\right), \quad t \geq 0, \tag{5.2}$$

is a genuine martingale and the probability measure $\tilde{\mathbb{P}}$ defined on $\mathcal{F}_t$ by taking its Radon–Nikodym derivative with respect to $\mathbb{P}$ to be given by Z_t, viz

$$\left.\frac{\mathrm{d}\tilde{\mathbb{P}}}{\mathrm{d}\mathbb{P}}\right|_{\mathcal{F}_t} = Z_t$$

is well defined (see Section 3.3 for details; see also Theorem B.34 and Corollary B.31). We remind the reader that, under $\tilde{\mathbb{P}}$ the process Y is a Brownian motion independent of X. The Kallianpur–Striebel formula (3.33) implies that for any φ a bounded Borel-measurable function

$$\pi_t(\varphi) = \frac{\rho_t(\varphi)}{\rho_t(\mathbf{1})} \quad \tilde{\mathbb{P}}(\mathbb{P})\text{-a.s.,}$$

where ρ_t is the unnormalised conditional distribution of X,

A. Bain, D. Crisan, *Fundamentals of Stochastic Filtering*,
DOI 10.1007/978-0-387-76896-0_5, © Springer Science+Business Media, LLC 2009

$$\rho_t(\varphi) = \tilde{\mathbb{E}}\left[\varphi(X_t)\tilde{Z}_t \Big| \mathcal{Y}_t\right],$$

and

$$\tilde{Z}_t = \exp\left(\int_0^t h(X_s)^\top \, dY_s - \frac{1}{2}\int_0^t \|h(X_s)\|^2 \, ds\right). \tag{5.3}$$

Exercise 5.1. Show that the Kallianpur–Striebel formula holds true for any Borel-measurable function φ such that $\mathbb{E}\left[|\varphi(X_t)|\right] < \infty$.

In the following, we require that $s \mapsto h(X_s)$ be a semimartingale. Let

$$h(X_s) = H_s^{\text{fv}} + H_s^{\text{m}}, \quad s \geq 0$$

be the Doob–Meyer decomposition of $h(X_s)$ with $H_\cdot^{\text{fv}} = (H_\cdot^{\text{fv},i})_{i=1}^m$ the finite variation part of $h(X)$, and $H_\cdot^{\text{m}} = (H_\cdot^{\text{m},i})_{i=1}^m$ the martingale part, which is assumed to be square integrable. We require that for all positive $k > 0$, the following conditions be satisfied,

$$c^{\text{fv},k} = \tilde{\mathbb{E}}\left[\exp\left(k\sum_{i=1}^m \int_0^t |dH_s^{\text{fv},i}|\right)\right] < \infty \tag{5.4}$$

$$c^{\text{m},k} = \tilde{\mathbb{E}}\left[\exp\left(k\sum_{i=1}^m \int_0^t d\left\langle H^{\text{m},i}\right\rangle_s\right)\right] < \infty, \tag{5.5}$$

where $s \mapsto \left\langle H^{\text{m},i}\right\rangle_s$ is the quadratic variation of $H^{\text{m},i}$, for $i = 1, \ldots, m$ and $\int_0^t |dH_s^{\text{fv},i}|$ is the total variation of $H^{\text{fv},i}$ on $[0,t]$ for $i = 1, \ldots, m$.

Exercise 5.2. Using the notation from Chapter 3, show that if X is a solution of the martingale problem for (A, π_0) and $h^i, (h^i)^2 \in \mathcal{D}(A)$, $i = 1, \ldots, m$, then conditions (5.4) and (5.5) are satisfied. [Hint: Use Exercise 3.22.]

5.2 The Importance of a Robust Representation

In the following we denote by $y_\cdot$ an arbitrary element of the set $C_{\mathbb{R}^m}[0,t]$, where $t \geq 0$ is arbitrary but fixed throughout the section. In other words $s \mapsto y_s$ is a continuous function $y_\cdot : [0,t] \to \mathbb{R}^m$. Also let $Y_\cdot$ be the path-valued random variable

$$Y_\cdot : \Omega \to C_{\mathbb{R}^m}[0,t], \quad Y_\cdot(\omega) = (Y_s(\omega), \ 0 \leq s \leq t).$$

Similar to Theorem 1.1, one can show that if φ is, for example, a bounded Borel-measurable function, then $\pi_t(\varphi)$ can be written as a function of the observation path. That is, there exists a bounded measurable function $f^\varphi : C_{\mathbb{R}^m}[0,t] \to \mathbb{R}$ such that

$$\pi_t(\varphi) = f^\varphi(Y_\cdot) \quad \text{P-a.s.} \tag{5.6}$$

Of course, f^φ is not unique. Any other function $\bar{f}^\varphi$ such that

$$\mathbb{P} \circ Y_{\cdot}^{-1}\left(\bar{f}^\varphi \neq f^\varphi\right) = 0,$$

where $\mathbb{P} \circ Y_{\cdot}^{-1}$ is the distribution of $Y_{\cdot}$ on the path space $C_{\mathbb{R}^m}[0,t]$ can replace f^φ in (5.6). In the following we obtain a robust representation of the conditional expectation $\pi_t(\varphi)$ (following Clark [56]). That is, we show that there exists a continuous function $\hat{f}^\varphi : C_{\mathbb{R}^m}[0,t] \to \mathbb{R}$ (with respect to the supremum norm on $C_{\mathbb{R}^m}[0,t]$) such that

$$\pi_t(\varphi) = \hat{f}^\varphi(Y_{\cdot}) \qquad \mathbb{P}\text{-a.s.} \qquad (5.7)$$

The following exercise shows that such a continuous $\hat{f}^\varphi$ has the virtue of uniqueness.

Exercise 5.3. Show that if $\mathbb{P} \circ Y_{\cdot}^{-1}$ positively charges all non-empty open sets in $C_{\mathbb{R}^m}[0,t]$, then there exists a unique continuous function $\hat{f}^\varphi : C_{\mathbb{R}^m}[0,t] \to \mathbb{R}$ for which (5.7) holds true. Finally show that if Y satisfies evolution equation (5.1) then it charges all non-empty open sets.

The need for this type of representation arises when the filtering framework is used to model and solve 'real-life' problems. As explained in a substantial number of papers (e.g. [56, 74, 73, 75, 76, 179, 180]) the model Y chosen for the "real-life" observation process $\bar{Y}$ may not be a perfect one. However, as long as the distribution of $\bar{Y}_{\cdot}$ is close in a weak sense to that of $Y_{\cdot}$ (and some integrability assumptions hold), the estimate $\hat{f}(\bar{Y}_{\cdot})$ computed on the actual observation will still be reasonable, as $\mathbb{E}[(\varphi(X_t) - \hat{f}^\varphi(\bar{Y}_{\cdot}))^2]$ is well approximated by the idealized error $\mathbb{E}[(\varphi(X_t) - \hat{f}^\varphi(Y_{\cdot}))^2]$.

Even when Y and $\bar{Y}$ coincide, one is never able to obtain and exploit a continuous stream of data as modelled by the continuous path $Y_{\cdot}(\omega)$. Instead the observation arrives and is processed at discrete moments in time

$$0 = t_0 < t_1 < t_2 < \cdots < t_n = t.$$

However the continuous path $\hat{Y}_{\cdot}(\omega)$ obtained from the discrete observations $(Y_{t_i}(\omega))_{i=1}^n$ by linear interpolation is close to $Y_{\cdot}(\omega)$ (with respect to the supremum norm on $C_{\mathbb{R}^m}[0,t]$); hence, by the same argument, $\hat{f}^\varphi(\hat{Y}_{\cdot})$ will be a sensible approximation to $\pi_t(\varphi)$.

5.3 Preliminary Bounds

Let $\Theta(y_{\cdot})$ be the following random variable

$$\Theta(y_{\cdot}) \triangleq \exp\left(h(X_t)^\top y_t - I(y_{\cdot}) - \tfrac{1}{2}\int_0^t \|h(X_s)\|^2\,\mathrm{d}s\right), \qquad (5.8)$$

where $I(y.)$, is a version of the stochastic integral $\int_0^t y_s^\top \, dh(X_s)$. The argument of the exponent in the definition of $\Theta(y.)$ will be recognized as a formal integration by parts of the argument of the exponential in (5.3). In the following, for any random variable ξ we denote by $\|\xi\|_{\Omega,p}$ the usual L_p norm of ξ,

$$\|\xi\|_{\Omega,p} = \tilde{\mathbb{E}}\left[|\xi|^p\right]^{1/p},$$

Lemma 5.4. *For any $R > 0$ and $p \geq 1$ there exists a positive constant $M_{R,p}^\Theta$ such that*

$$\sup_{\|y.\| \leq R} \|\Theta(y.)\|_{\Omega,p} \leq M_{R,p}^\Theta. \tag{5.9}$$

Proof. In the following, for notational conciseness, for arbitrary $y. \in C_{\mathbb{R}^m}[0,t]$, define $\bar{y}. \in C_{\mathbb{R}^m}[0,t]$ by

$$\bar{y}_s \triangleq y_t - y_s, \qquad s \in [0,t].$$

If $\|y.\| \leq R$, then it is clear that $\|\bar{y}.\| \leq 2R$. From (5.8) we get that

$$\Theta(y.) = \exp\left(\int_0^t \bar{y}_s^\top \, dh(X_s) - \tfrac{1}{2}\int_0^t \|h(X_s)\|^2 \, ds\right)$$

$$\leq \exp\left(\int_0^t \bar{y}_s^\top \, dH_s^{\mathrm{fv}} + \int_0^t \bar{y}_s^\top \, dH_s^{\mathrm{m}}\right).$$

Next observe that, from (5.4) we have

$$\tilde{\mathbb{E}}\left[\exp\left(2p\int_0^t \bar{y}_s^\top \, dH_s^{\mathrm{fv}}\right)\right] \leq \tilde{\mathbb{E}}\left[\exp\left(4pR\int_0^t |dH_s^{\mathrm{fv}}|\right)\right] = c^{\mathrm{fv},4pR},$$

and by using the Cauchy–Schwartz inequality

$$\tilde{\mathbb{E}}\left[\exp\left(2p\int_0^t \bar{y}_s^\top \, dH_s^{\mathrm{m}}\right)\right]$$

$$= \tilde{\mathbb{E}}\left[\exp\left(2p\int_0^t \bar{y}_s^\top \, dH_s^{\mathrm{m}} - 4p^2 \sum_{i,j=1}^m \int_0^t \bar{y}_s^i \bar{y}_s^j \, d\langle H^{\mathrm{m},i}, H^{\mathrm{m},j}\rangle_s \right. \right.$$

$$\left. \left. + 4p^2 \sum_{i,j=1}^m \int_0^t \bar{y}_s^i \bar{y}_s^j \, d\langle H^{\mathrm{m},i}, H^{\mathrm{m},j}\rangle_s \right)\right]$$

$$\leq \sqrt{\tilde{\mathbb{E}}\left[\Theta_r'(y.)\right]} \sqrt{\tilde{\mathbb{E}}\left[\exp\left(8p^2 \sum_{i,j=1}^m \int_0^t \bar{y}_s^i \bar{y}_s^j \, d\langle H^{\mathrm{m},i}, H^{\mathrm{m},j}\rangle_s\right)\right]}$$

$$\leq \sqrt{\tilde{\mathbb{E}}\left[\Theta_r'(y.)\right]} \sqrt{\tilde{\mathbb{E}}\left[\exp\left(32p^2 R^2 \sum_{i,j=1}^m \int_0^t |d\langle H^{\mathrm{m},i}, H^{\mathrm{m},j}\rangle|_s\right)\right]},$$

where

$$\Theta'_r(y_\cdot) \triangleq \exp\left(4p \int_0^r \bar{y}_s^\top \mathrm{d}H_s^m - \frac{(4p)^2}{2} \sum_{i,j=1}^m \int_0^r \bar{y}_s^i \bar{y}_s^j \, \mathrm{d}\langle H^{m,i}, H^{m,j}\rangle_s\right).$$

The process $r \mapsto \Theta'_r(y_\cdot)$ is clearly an exponential local martingale and by Novikov's condition and (5.5) it is a martingale, so

$$\tilde{\mathbb{E}}\left[\Theta'_r(y_\cdot)\right] = 1.$$

From this, the fact that

$$\int_0^t \left|\mathrm{d}\langle H^{m,i}, H^{m,j}\rangle\right|_s \leq \frac{1}{2}\int_0^t \mathrm{d}\langle H^{m,i}\rangle_s + \frac{1}{2}\int_0^t \mathrm{d}\langle H^{m,j}\rangle_s,$$

and (5.5) we get

$$\tilde{\mathbb{E}}\left[\exp\left(2p \int_0^t \bar{y}_s^\top \mathrm{d}H_s^m\right)\right] \leq \sqrt{c^{m,32p^2 R^2 m}}.$$

Hence, again by applying Cauchy–Schwarz's inequality, (5.9) follows with $M_{R,p}^\Theta = (c^{\mathrm{fv},4pR}\sqrt{c^{m,32p^2 R^2 m}})^{1/2p}$. □

Now let φ be a Borel-measurable function such that $\|\varphi(X_t)\|_{\Omega,p} < \infty$ for some $p > 1$. Note that $\|\varphi(X_t)\|_{\Omega,p}$ is the same whether we integrate with respect to $\mathbb{P}$ or $\tilde{\mathbb{P}}$. Let $\hat{g}^\varphi$, $\hat{g}^\mathbf{1}$, $\hat{f}^\varphi : C_{\mathbb{R}^m}[0,t] \to \mathbb{R}$ be the following functions,

$$\hat{g}^\varphi(y_\cdot) = \tilde{\mathbb{E}}\left[\varphi\Theta(y_\cdot)\right], \qquad \hat{g}^\mathbf{1}(y_\cdot) = \tilde{\mathbb{E}}\left[\Theta(y_\cdot)\right], \qquad \hat{f}(y_\cdot) = \frac{\hat{g}^\varphi(y_\cdot)}{\hat{g}^\mathbf{1}(y_\cdot)}. \tag{5.10}$$

Lemma 5.5. *For any $R > 0$ and $q \geq 1$ there exists a positive constant $M_{R,q}^\Theta$ such that*

$$\|\Theta(y_\cdot^1) - \Theta(y_\cdot^2)\|_{\Omega,q} \leq M_{R,q}^\Theta \|y_\cdot^1 - y_\cdot^2\| \tag{5.11}$$

for any two paths $y_\cdot^1, y_\cdot^2$ such that $|y_\cdot^1|, |y_\cdot^2| \leq R$. In particular, (5.11) implies that $\hat{g}^\mathbf{1}$ is locally Lipschitz; more precisely

$$\left|\hat{g}^\mathbf{1}(y_\cdot^1) - \hat{g}^\mathbf{1}(y_\cdot^2)\right| \leq M_R^\Theta \|y_\cdot^1 - y_\cdot^2\|$$

for any two paths $y_\cdot^1, y_\cdot^2$ such that $\|y_\cdot^1\|, \|y_\cdot^2\| \leq R$ and $M_R^\Theta = \inf_{q \geq 1} M_{R,q}^\Theta$.

Proof. For the two paths $y_\cdot^1, y_\cdot^2$ denote by $y_\cdot^{12}$ the difference path defined as $y_\cdot^{12} \triangleq y_\cdot^1 - y_\cdot^2$. Then

$$\left|\Theta(y_\cdot^1) - \Theta(y_\cdot^2)\right| \leq (\Theta(y_\cdot^1) + \Theta(y_\cdot^2))\left|\int_0^t (\bar{y}_s^{12})^\top \mathrm{d}h(X_s)\right|,$$

Using the Cauchy–Schwartz inequality

$$\left\|\Theta(y^1_{\cdot}) - \Theta(y^2_{\cdot})\right\|_{\Omega,q} \leq 2M^{\Theta}_{R,2q} \left\|\int_0^t (\bar{y}^{12}_s)^{\top} \, \mathrm{d}h(X_s)\right\|_{\Omega,2q}. \tag{5.12}$$

Finally, since $\|\bar{y}^{12}_{\cdot}\| \leq 2\|y^1_{\cdot} - y^2_{\cdot}\|$, a standard argument based on Burkholder–Davis–Gundy's inequality shows that the expectation on the right-hand side of (5.12) is bounded by

$$\left\|\int_0^t (\bar{y}^{12}_s)^{\top} \, \mathrm{d}h(X_s)\right\|_{\Omega,2q} \leq \left\|\int_0^t (\bar{y}^{12}_s)^{\top} \, \mathrm{d}H^{\mathrm{fv}}_s\right\|_{\Omega,2q} + \left\|\int_0^t (\bar{y}^{12}_s)^{\top} \, \mathrm{d}H^{\mathrm{m}}_s\right\|_{\Omega,2q}$$

$$\leq 2\|y^1_{\cdot} - y^2_{\cdot}\| \left\|\int_0^t |\mathrm{d}H^{\mathrm{fv}}_s|\right\|_{\Omega,2q}$$

$$+ 2c_q \|y^1_{\cdot} - y^2_{\cdot}\| \sum_{i=1}^m \left\|\int_0^t \mathrm{d}\langle H^{\mathrm{m},i}\rangle_s\right\|_{\Omega,q}^{1/2},$$

where c_q is the constant appearing in the Burkholder–Davis–Gundy inequality. Hence (5.11) holds true. □

Lemma 5.6. *The function $\hat{g}^{\varphi}$ is locally Lipschitz and locally bounded.*

Proof. Fix $R > 0$ and let $y^1_{\cdot}, y^2_{\cdot}$ be two paths such that $\|y^1_{\cdot}\|, \|y^2_{\cdot}\| \leq R$. By Hölder's inequality and (5.11), we see that

$$\tilde{\mathbb{E}}\left[|\varphi(X_t)|\left|\Theta(y^1_{\cdot}) - \Theta(y^2_{\cdot})\right|\right] \leq \|\varphi(X_t)\|_{\Omega,p} M^{\Theta}_{R,q} \|y^1_{\cdot} - y^2_{\cdot}\|. \tag{5.13}$$

where q is such that $p^{-1} + q^{-1} = 1$. Hence $\hat{g}^{\varphi}$ is locally Lipschitz, since

$$\hat{g}^{\varphi}(y^1_{\cdot}) - \hat{g}^{\varphi}(y^2_{\cdot}) = \tilde{\mathbb{E}}\left[\varphi(X_t)\left(\Theta(y^1_{\cdot}) - \Theta(y^2_{\cdot})\right)\right]$$

and $R > 0$ was arbitrarily chosen. Next let $y_{\cdot}$ be a path such that $\|y_{\cdot}\| \leq R$. Again, by Hölder's inequality and (5.9), we get that

$$\sup_{\|y_{\cdot}\| \leq R} |\hat{g}^{\varphi}(y_{\cdot})| = \sup_{\|y_{\cdot}\| \leq R} \left|\tilde{\mathbb{E}}\left[\varphi(X_t)\Theta(y^1_{\cdot})\right]\right| \leq \|\varphi(X_t)\|_p M^{\Theta}_{R,q} < \infty.$$

Hence $\hat{g}^{\varphi}$ is locally bounded. □

Theorem 5.7. *The function $\hat{f}^{\varphi}$ is locally Lipschitz.*

Proof. The ratio $\hat{g}^{\varphi}/\hat{g}^{\mathbf{1}}$ of the two locally Lipschitz functions $\hat{g}^{\varphi}$ and $\hat{g}^{\mathbf{1}}$ (Lemma 5.5 and Lemma 5.6) is locally Lipschitz provided both $\hat{g}^{\varphi}$ and $1/\hat{g}^{\mathbf{1}}_t$ are locally bounded. The local boundedness property of $\hat{g}^{\varphi}$ is shown in Lemma 5.6 and that of $1/\hat{g}^{\mathbf{1}}_t$ follows from the following simple argument. If $\|y_{\cdot}\| \leq R$ Jensen's inequality implies that

$$\tilde{\mathbb{E}}[\Theta(y_{\cdot})] \geq \exp\left(\mathbb{E}\left[\int_0^t \bar{y}^{\top}_s \, \mathrm{d}H^{\mathrm{m}}_s + \int_0^t \bar{y}^{\top}_s \, \mathrm{d}H^{\mathrm{fv}}_s - \frac{1}{2}\int_0^t \|h(X_s)\|^2 \, \mathrm{d}s\right]\right)$$

$$\geq \exp\left(-2R \sum_{i=1}^m \mathbb{E}\left[\int_0^t |\mathrm{d}H^{\mathrm{fv},i}_s|\right] - \frac{1}{2}\mathbb{E}\left[\int_0^t \|h(X_s)\|^2 \, \mathrm{d}s\right]\right). \tag{5.14}$$

Note that both expectations in (5.14) are finite, by virtue of condition (5.4). □

5.4 Clark's Robustness Result

We proceed next to show that $\hat{f}^{\varphi}(Y.)$ is a version of $\pi_t(\varphi)$. This fact is much more delicate than showing that $\hat{f}^{\varphi}$ is locally Lipschitz. The main difficulty is the fact that the mapping

$$(y., \omega) \in C_{\mathbb{R}^m}[0,t] \times \Omega \to I(y.) \in \mathbb{R}$$

is not $\mathcal{B}\left(C_{\mathbb{R}^m}[0,t]\right) \times \mathcal{F}$-measurable since the integral $I(y.)$ is constructed path by path (where $\mathcal{B}(C_{\mathbb{R}^m}[0,t])$ is the Borel σ-field on $C_{\mathbb{R}^m}[0,t]$). Let $\mathcal{H}_{1/3}$ be the following subset of $C_{\mathbb{R}^m}[0,t]$,

$$\mathcal{H}_{1/3} = \left\{ y. \in C_{\mathbb{R}^m}[0,t] : K(y.) \triangleq \sup_{s_1, s_2 \in [0,t]} \frac{\|y_{s_1} - y_{s_2}\|_{\infty}}{|s_1 - s_2|^{1/3}} < \infty \right\}.$$

Exercise 5.8. Show that almost all paths of Y belong to $\mathcal{H}_{1/3}$, in other words show that

$$\tilde{\mathbb{P}}\left(\omega \in \Omega : Y.(\omega) \in \mathcal{H}_{1/3}\right) = 1.$$

[Hint: Use the modulus of continuity for Brownian motion; see, for example, [149, page 114].]

Lemma 5.9. *There exists a version of the stochastic integral $I(y.)$ which has the property that the mapping $(y., \omega) \in C_{\mathbb{R}^m}[0,t] \times \Omega \to I(y.) \in \mathbb{R}$, whilst still non-measurable, is equal on $\mathcal{H}_{1/3} \times \Omega$ to a $\mathcal{B}\left(C_{\mathbb{R}^m}[0,t]\right) \times \Omega$-measurable mapping.*

Proof. Denote by $I^{\mathrm{fv}}(y.)$ the Stieltjes integral with respect to H^{fv}. $I^{\mathrm{fv}}(y.)$ is defined unambiguously pathwise. To avoid ambiguity, for arbitrary $y. \in C_{\mathbb{R}^m}[0,t]$ and all $\omega \in \Omega$, we have

$$I^{\mathrm{fv}}(y.)(\omega) = \lim_{n \to \infty} \sum_{i=0}^{n-1} y_{it/n}^{\top} \left(H^{\mathrm{fv}}_{(i+1)t/n}(\omega) - H^{\mathrm{fv}}_{it/n}(\omega) \right).$$

Hence defining $I(y.)$ only depends on selecting the version of $\int_0^t y_s^\top \, dH_s^{\mathrm{m}}$, the stochastic integral with respect to the martingale part of $h(X.)$, which we denote by $I^{\mathrm{m}}(y.)$. Recall that for integrators which have unbounded variation on locally compact intervals it is not possible to define a stochastic integral pathwise for general integrands. However, if we restrict to a suitable class of integrands (such as $\mathcal{H}_{1/3}$) then this is possible.

$$I^{\mathrm{m}}_n(y.)(\omega) \triangleq \sum_{i=0}^{n-1} y_{it/n}^{\top} \left(H^{\mathrm{m}}_{(i+1)t/n}(\omega) - H^{\mathrm{m}}_{it/n}(\omega) \right).$$

Since, for $y. \in \mathcal{H}_{1/3}$,

$$\tilde{\mathbb{E}}\left[\left(I^m_{2^k}(y.) - \int_0^t y_s^\top \, dH^m_s\right)^2\right]$$

$$= \tilde{\mathbb{E}}\left[\left(\sum_{i=1}^m \int_0^t \left(y^i_s - y^i_{[s2^k/t]t2^{-k}}\right) dH^{m,i}_s\right)^2\right]$$

$$\leq m \sum_{i=1}^m \tilde{\mathbb{E}}\left[\int_0^t \left(y^i_s - y^i_{[s2^k/t]t2^{-k}}\right)^2 d\langle H^{m,i}\rangle_s\right]$$

$$\leq \frac{mc_X K(y.)^2 t^{2/3}}{2^{2k/3}},$$

where

$$c_X = \sum_{i=1}^m \tilde{\mathbb{E}}\left[\left(H^{m,i}_t\right)^2\right] < \infty.$$

Hence by Chebychev's inequality

$$\tilde{\mathbb{P}}\left(\left|I^m_{2^k}(y.) - \int_0^t y_s^\top \, dH^m_s\right| > \varepsilon\right) \leq \frac{1}{\varepsilon^2} \frac{mc_X K(y.)^2 t^{2/3}}{2^{2k/3}}.$$

But since

$$\sum_{k=1}^\infty \frac{mc_X K(y.)^2 t^{2/3}}{2^{2k/3}} < \infty,$$

by the first Borel–Cantelli lemma it follows that

$$\tilde{\mathbb{P}}\left(\limsup_{k\to\infty} \left|I^m_{2^k}(y.) - \int_0^t y_s^\top \, dH^m_s\right| > \varepsilon\right) = 0;$$

hence for $y \in \mathcal{H}_{1/3}$, $I^m_{2^k}(y.)$ converges to $\int_0^t y_s^\top \, dH^m_s$, $\tilde{\mathbb{P}}$-almost surely. We define $I^m(y.)$ to be the limit

$$I^m(y.)(\omega) \triangleq \limsup_{k\to\infty} I^m_{2^k}(y.)(\omega)$$

for any $(\omega, y.) \in \Omega \times \mathcal{H}_{1/3}$ and any version of $\int_0^t y_s^\top \, dH^m_s$ on $(C_{\mathbb{R}^m}[0,t] \setminus \mathcal{H}_{1/3}) \times \Omega$. Although the resulting map is generally non-measurable with respect to $\mathcal{B}(C_{\mathbb{R}^m}[0,t]) \otimes \mathcal{F}$, it is equal on $\mathcal{H}_{1/3} \times \Omega$ to the following jointly measurable function

$$J^m(y.) \triangleq \limsup_{k\to\infty} I^m_{2^k}(y.) \tag{5.15}$$

defined on the whole of $C_{\mathbb{R}^m}[0,t] \times \Omega$. We emphasize that for $y \notin \mathcal{H}_{1/3}$ it is quite possible that $J^m(y)$ differs from the value of $\int_0^t y_s^\top \, dH^m_s$. □

In order to simplify the proof of the robustness result which follows, it is useful to decouple the two processes X and Y. Let $(\hat{\Omega}, \hat{\mathcal{F}}, \hat{\mathbb{P}})$ be an identical copy of $(\Omega, \mathcal{F}, \tilde{\mathbb{P}})$ and let $\hat{X}$ be the copy of X within the new space

$(\hat{\Omega}, \hat{\mathcal{F}}, \tilde{\mathbb{P}})$. Let $\hat{H}^{\mathrm{m}}$ and $\hat{H}^{\mathrm{fv}}$ be the processes within the new space $(\hat{\Omega}, \hat{\mathcal{F}}, \tilde{\mathbb{P}})$ corresponding to the original H^{m} and H^{fv}. Then the function $\hat{g}^{\varphi}$ has the following representation,

$$\hat{g}^{\varphi}(y.) = \hat{\mathbb{E}}\left[\varphi(\hat{X}_t)\hat{\Theta}(y.)\right] \tag{5.16}$$

$$\hat{\Theta}(y.) = \exp\left(h(\hat{X}_t)^\top y_t - \hat{I}(y.) - \tfrac{1}{2}\int_0^t \|h(\hat{X}_t)\|^2 \, \mathrm{d}s\right), \tag{5.17}$$

where $\hat{\mathbb{E}}$ denotes integration on $(\hat{\Omega}, \hat{\mathcal{F}}, \hat{\mathbb{P}})$, and $\hat{I}(y.)$ is the version of the stochastic integral $\int_0^t y_s^\top \, \mathrm{d}h(\hat{X}_s)$ corresponding to $I(y.)$ as constructed above. Denote by $\hat{I}^{\mathrm{m}}(y.)$ the respective version of the stochastic integral with respect to the martingale $\hat{H}^{\mathrm{m}}$ and by $\hat{I}^{\mathrm{fv}}(y.)$ the Stieltjes integral with respect to $\hat{H}^{\mathrm{fv}}$. Let $\hat{J}^{\mathrm{m}}(y.)$ be the function corresponding to $J^{\mathrm{m}}(y.)$ as defined in (5.15). Then, for $y. \in \mathcal{H}_{1/3}$, $\hat{\Theta}(y.)$ can be written as

$$\hat{\Theta}(y.) = \exp\left(h^\top(\hat{X}_t)y_t - \hat{I}^{\mathrm{fv}}(y.) - \hat{J}^{\mathrm{m}}(y.) - \tfrac{1}{2}\int_0^t \|h(\hat{X}_s)\|^2 \, \mathrm{d}s\right). \tag{5.18}$$

Finally, let $(\bar{\Omega}, \bar{\mathcal{F}}, \bar{\mathbb{P}})$ be the product space

$$(\bar{\Omega}, \bar{\mathcal{F}}, \bar{\mathbb{P}}) = (\Omega \times \hat{\Omega}, \mathcal{F} \otimes \hat{\mathcal{F}}, \tilde{\mathbb{P}} \otimes \hat{\mathbb{P}})$$

on which we 'lift' the processes $\hat{H}$ and Y from the component spaces. In other words, $Y(\omega, \hat{\omega}) = Y(\omega)$ and $\hat{H}(\omega, \hat{\omega}) = \hat{H}(\hat{\omega})$ for all $(\omega, \hat{\omega}) \in \Omega \times \hat{\Omega}$.

Lemma 5.10. *There exists a null set $\mathcal{N} \in \mathcal{F}$ such that the mapping $(\omega, \hat{\omega}) \in \bar{\Omega} \mapsto \hat{I}(Y(\omega))(\hat{\omega})$ coincides on $(\Omega \setminus \mathcal{N}) \times \hat{\Omega}$ with an $\bar{\mathcal{F}}$-measurable mapping.*

Proof. First let us remark that $(\omega, \hat{\omega}) \mapsto \hat{I}^{\mathrm{fv}}(Y(\omega))(\hat{\omega})$ is equal to

$$\hat{I}^{\mathrm{fv}}(Y.(\omega))(\hat{\omega}) = \lim_{n\to\infty} \sum_{i=0}^{n-1} Y_{it/n}^\top(\omega)\left(\hat{H}^{\mathrm{fv}}_{(i+1)t/n}(\hat{\omega}) - \hat{H}^{\mathrm{fv}}_{it/n}(\hat{\omega})\right) \tag{5.19}$$

and since

$$(\omega, \hat{\omega}) \in \bar{\Omega} \mapsto \sum_{i=0}^{n-1} Y_{it/n}^\top(\omega)\left(\hat{H}^{\mathrm{fv}}_{(i+1)t/n}(\hat{\omega}) - \hat{H}^{\mathrm{fv}}_{it/n}(\hat{\omega})\right)$$

is $\bar{\mathcal{F}}$-measurable then so is its limit. Define $\mathcal{N} \triangleq \{\omega \in \Omega : Y.(\omega) \notin \mathcal{H}_{1/3}\}$. Then $\mathcal{N} \in \mathcal{F}$ and $\tilde{\mathbb{P}}(\mathcal{N}) = 0$. Following the definition of $I^{\mathrm{m}}(y.)$, the mapping $(\omega, \hat{\omega}) \mapsto \hat{I}^{\mathrm{m}}(Y(\omega))(\hat{\omega})$ coincides with the mapping $(\omega, \hat{\omega}) \mapsto \hat{J}^{\mathrm{m}}(Y(\omega))(\hat{\omega})$ on $(\Omega \setminus \mathcal{N}) \times \hat{\Omega}$. Then $\hat{J}^{\mathrm{m}}$ is an $\bar{\mathcal{F}}$-measurable random variable, since

$$\hat{J}^{\mathrm{m}}(Y(\omega))(\hat{\omega}) = \limsup_{k\to\infty} \sum_{i=0}^{2^k-1} Y_{it/2^k}^\top(\omega)\left(\hat{H}^{\mathrm{m}}_{(i+1)t/2^k}(\hat{\omega}) - \hat{H}^{\mathrm{m}}_{it/2^k}(\hat{\omega})\right). \tag{5.20}$$

Combining this with the measurability of $\hat{I}^{\mathrm{fv}}(Y.)$ gives us the lemma. $\square$

Lemma 5.11. $\bar{\mathbb{P}}$-almost surely

$$\int_0^t Y_s^\top \, \mathrm{d}\hat{H}_s = \hat{I}^{\mathrm{fv}}(Y_\cdot) + \hat{J}^{\mathrm{m}}(Y_\cdot). \tag{5.21}$$

Proof. We have

$$\int_0^t Y_s^\top \, \mathrm{d}\hat{H}_s = \int_0^t Y_s^\top \, \mathrm{d}\hat{H}_s^{\mathrm{m}} + \int_0^t Y_s^\top \, \mathrm{d}\hat{H}_s^{\mathrm{fv}}.$$

Following (5.19) it is obvious that $\int_0^t Y_s^\top \, \mathrm{d}\hat{H}_s^{\mathrm{fv}} = \hat{I}^{\mathrm{fv}}(Y_\cdot)$. Hence, following the proof of the previous lemma, it suffices to prove that, $\bar{\mathbb{P}}$-almost surely, $\int_0^t Y_s^\top \, \mathrm{d}\hat{H}_s^{\mathrm{m}} = \hat{J}^{\mathrm{m}}(Y_\cdot)$ where $\hat{J}^{\mathrm{m}}(Y_\cdot)$ is the function defined in (5.20). Without loss of generality we assume that $m = 1$ (the general case follows by treating each of the m components in turn) and we note that we only need to prove that, for arbitrary $K > 0$, $\bar{\mathbb{P}}$-almost surely,

$$\int_0^t Y_s^K \, \mathrm{d}\hat{H}_s^{\mathrm{m}} = \hat{J}^{\mathrm{m}}(Y_\cdot^K), \tag{5.22}$$

where

$$Y_s^K = \begin{cases} Y_s & \text{if } |Y_s| \leq K \\ K & \text{otherwise.} \end{cases}$$

In turn, (5.22) follows once we prove that

$$\lim_{n \to \infty} \bar{\mathbb{E}} \left[\left(\sum_{i=0}^{n-1} \left(Y_{it/n}^K \right)^\top \left(\hat{H}_{(i+1)t/n}^{\mathrm{m}} - \hat{H}_{it/n}^{\mathrm{m}} \right) - \hat{J}^{\mathrm{m}} \left(Y_\cdot^K \right) \right)^2 \right] = 0.$$

By Fubini's theorem, using the $\bar{\mathcal{F}}$-measurability of $\hat{J}^{\mathrm{m}}(Y_\cdot^K)$ and the fact that $\hat{I}^{\mathrm{m}}(Y_\cdot^K)$ coincides with $\hat{J}^{\mathrm{m}}(Y_\cdot^K)$ on $(\Omega \backslash \mathcal{N}) \times \hat{\Omega}$ we have

$$\bar{\mathbb{E}} \left[\left(\sum_{i=0}^{n-1} \left(Y_{it/n}^K \right)^\top \left(\hat{H}_{(i+1)t/n}^{\mathrm{m}} - \hat{H}_{it/n}^{\mathrm{m}} \right) - \hat{J}^{\mathrm{m}} \left(Y_\cdot^K \right) \right)^2 \right]$$

$$= \int_{\Omega \backslash \mathcal{N}} \hat{\mathbb{E}} \left[\left(\hat{I}_n^{\mathrm{m}} \left(Y_\cdot^K(\omega) \right) - \hat{J}^{\mathrm{m}} \left(Y_\cdot^K \right) \right)^2 \right] \mathrm{d}\tilde{\mathbb{P}}(\omega)$$

$$= \int_{\Omega \backslash \mathcal{N}} \hat{\mathbb{E}} \left[\left(\hat{I}_n^{\mathrm{m}}(Y_\cdot^K(\omega)) - \hat{I}^{\mathrm{m}}(Y_\cdot^K) \right)^2 \right] \mathrm{d}\tilde{\mathbb{P}}(\omega).$$

Now since $s \mapsto Y_s^K(\omega)$ is a continuous function and $\hat{I}^{\mathrm{m}}(Y_\cdot^K(\omega))$ is a version of the stochastic integral $\int_0^t \left(Y_s^K \right)^\top (\omega) \, \mathrm{d}\hat{H}_s^{\mathrm{m}}$, it follows that

$$\lim_{n \to \infty} \hat{\mathbb{E}} \left[\left(\hat{I}_n^{\mathrm{m}}(Y_\cdot^K(\omega)) - \hat{I}^{\mathrm{m}}(Y_\cdot^K(\omega)) \right)^2 \right] = 0$$

for all $\omega \in \Omega \setminus \mathcal{N}$. Also, we have the following upper bound

$$\hat{\mathbb{E}}\left[\left(\hat{I}_n^m(Y_\cdot^K(\omega)) - \hat{I}^m(Y_\cdot^K(\omega))\right)^2\right] \leq 4K^2 \hat{\mathbb{E}}\left[(\hat{H}_t^m)^2\right] < \infty.$$

Hence, by the dominated convergence theorem,

$$\lim_{n\to\infty} \hat{\mathbb{E}}\left[\left(\sum_{i=0}^{n-1}\left(Y_{it/n}^K\right)^\top \left(\hat{H}_{(i+1)t/n}^m - \hat{H}_{it/n}^m\right) - \hat{I}^m(Y_\cdot^K)\right)^2\right]$$
$$= \int_{\Omega\setminus\mathcal{N}} \lim_{n\to\infty} \hat{\mathbb{E}}\left[\left(\hat{I}_n^m(Y_\cdot^K(\omega)) - \hat{I}^m(Y_\cdot^K(\omega))\right)^2\right] d\tilde{\mathbb{P}}(\omega) = 0.$$

$\square$

Theorem 5.12. *The random variable $\hat{f}^\varphi(Y_\cdot)$ is a version of $\pi_t(\varphi)$; that is, $\pi_t(\varphi) = \hat{f}^\varphi(Y_\cdot)$, $\mathbb{P}$-almost surely. Hence $\hat{f}^\varphi(Y_\cdot)$ is the unique robust representation of $\pi_t(\varphi)$.*

Proof. It suffices to prove that, $\mathbb{P}$-almost surely (or, equivalently, $\tilde{\mathbb{P}}$-almost surely),

$$\rho_t(\varphi) = \hat{g}^\varphi(Y_\cdot) \quad \text{and} \quad \rho_t(\mathbf{1}) = \hat{g}^\mathbf{1}(Y_\cdot).$$

We need only prove the first identity as the second is just a special case obtained by setting $\varphi = \mathbf{1}$ in the first. From the definition of abstract conditional expectation therefore it suffices to show

$$\tilde{\mathbb{E}}\left[\rho_t(\varphi)b(Y_\cdot)\right] = \tilde{\mathbb{E}}\left[\hat{g}^\varphi(Y_\cdot)b(Y_\cdot)\right], \tag{5.23}$$

where b is an arbitrary continuous bounded function $b: C_{\mathbb{R}^m}[0,t] \to \mathbb{R}$. Since X and Y are independent under $\tilde{\mathbb{P}}$, it follows that the pair processes (X, Y) under $\tilde{\mathbb{P}}$, and $(\hat{X}, Y)$ under $\bar{\mathbb{P}}$ have the same distribution. Hence, the left-hand side of (5.23) has the following representation,

$$\tilde{\mathbb{E}}\left[\rho_t(\varphi)b(Y_\cdot)\right]$$
$$= \tilde{\mathbb{E}}\left[\varphi(X_t)\exp\left(\int_0^t h(X_s)^\top dY_s - \tfrac{1}{2}\int_0^t \|h(X_s)\|^2 ds\right)b(Y_\cdot)\right]$$
$$= \bar{\mathbb{E}}\left[\hat{\varphi}(X_t)\exp\left(\int_0^t h(\hat{X}_s)^\top dY_s - \tfrac{1}{2}\int_0^t \|h(\hat{X}_s)\|^2 ds\right)b(Y_\cdot)\right]$$
$$= \bar{\mathbb{E}}\left[\varphi(\hat{X}_t)\exp\left(h(\hat{X}_t)^\top Y_t - \int_0^t Y_s^\top dh(\hat{X}_s) - \tfrac{1}{2}\int_0^t \|h(\hat{X}_s)\|^2 ds\right)b(Y_\cdot)\right].$$

On the other hand, using (5.18), the right-hand side of (5.23) has the representation

$$\tilde{\mathbb{E}}\left[\hat{g}^{\varphi}(Y.)b(Y.)\right]$$
$$= \tilde{\mathbb{E}}\left[b(Y.)\hat{\mathbb{E}}\left[\varphi(\hat{X}_t)\exp\left(h(\hat{X}_t)^{\top}Y_t - \hat{I}^{\text{fv}}(Y.) - \hat{I}^{\text{m}}(Y.)\right.\right.\right.$$
$$\left.\left.\left. - \frac{1}{2}\int_0^t \|h(\hat{X}_s)\|^2\,\mathrm{d}s\right)\right]\right]$$
$$= \tilde{\mathbb{E}}\left[b(Y.)\hat{\mathbb{E}}\left[\varphi(\hat{X}_t)\exp\left(h(X_t)^{\top}Y_t - \hat{I}^{\text{fv}}(Y.) - \hat{J}^{\text{m}}(Y.)\right.\right.\right.$$
$$\left.\left.\left. - \frac{1}{2}\int_0^t \|h(\hat{X}_s)\|^2\,\mathrm{d}s\right)\right]\right].$$

Hence by Fubini's theorem (using, again the $\bar{\mathcal{F}}$-measurability of $\hat{J}^{\text{m}}(Y.)$)

$$\tilde{\mathbb{E}}\left[\hat{g}^{\varphi}(Y.)b(Y.)\right] = \bar{\mathbb{E}}\left[\varphi(\hat{X}_t)\exp\left(h(\hat{X}_t)^{\top}Y_t - \hat{I}^{\text{fv}}(Y.) - \hat{J}^{\text{m}}(Y.)\right.\right.$$
$$\left.\left. - \frac{1}{2}\int_0^t \|h(\hat{X}_s)\|^2\,\mathrm{d}s\right)b(Y.)\right].$$

Finally, from Lemma 5.11, the two representations coincide. □

Remark 5.13. Lemma 5.11 appears to suggest a pathwise construction for the stochastic integral

$$\int_0^t h(X_s)^{\top}\,\mathrm{d}Y_s,$$

but we know that for cases such as $\int_0^t B_s\,\mathrm{d}B_s$ a stochastic integral cannot be defined pathwise (see Remark B.17). However, this apparent paradox is resolved by noting that the terms appearing in the lemma are only constructed on the space $\bar{\Omega}$.

This construction has other uses in the numerical solution of problems involving stochastic integrals. For example, *adaptive pathwise approximation* is sometimes used in numerical evaluation of stochastic integrals. Suppose we wish to evaluate the stochastic integral $\int_0^t X_s\,\mathrm{d}Y_s$ where X and Y are càdlàg processes and we assume the usual conditions on the filtration. Given $\delta > 0$, if we define stopping times $T_0^{\delta} = 0$ and

$$T_k^{\delta} = \inf\{t > T_{k-1}^{\delta} : |X_t - X_{t_{k-1}}| > \delta\},$$

then the stochastic integral may be approximated pathwise by

$$(X \cdot Y)^{(\delta)} \triangleq \sum_{k=0}^{\infty} X_{T_k^{\delta}}(Y_{T_{k+1}^{\delta}} - Y_{T_k^{\delta}}).$$

If δ_n is a sequence of values of δ which tends to zero sufficiently fast, by similar calculations to those used in the justification that I^{m} is a pathwise approximation to the stochastic integral, this series of approximations can be shown to converge $\mathbb{P}$-a.s. uniformly on a finite interval to the stochastic integral as $n \to \infty$.

5.5 Solutions to Exercises

5.1 Repeat the proof of the formula for a Borel-measurable function φ such that $\mathbb{E}\left[|\varphi(X_t)|\right] < \infty$. Alternatively use the following argument. It suffices to prove the result only for φ a non-negative Borel-measurable function such that $\mathbb{E}\left[\varphi(X_t)\right] < \infty$, as the general result follows by decomposing the function into its positive and negative parts. Consider the sequence $(\varphi_n)_{n \geq 0}$ of functions defined as

$$\varphi_n = \begin{cases} \varphi(x) & \text{if } \varphi(x) \leq n \\ n & \text{otherwise} \end{cases}.$$

Then φ_n is bounded and by the Kallianpur–Striebel formula (3.33),

$$\pi_t(\varphi_n) = \frac{\rho_t(\varphi_n)}{\rho_t(\mathbf{1})} \qquad \tilde{\mathbb{P}}(\mathbb{P})-\text{a.s.}$$

Also

$$\tilde{\mathbb{E}}\left[\varphi\left(X_t\right)\tilde{Z}_t\right] = \mathbb{E}\left[\varphi(X_t)\right] < \infty.$$

Hence, by the conditional monotone convergence theorem

$$\pi_t(\varphi) = \lim_{n \to \infty} \pi_t(\varphi_n) = \frac{1}{\rho_t(\mathbf{1})} \lim_{n \to \infty} \rho_t(\varphi_n) = \frac{\rho_t(\varphi)}{\rho_t(\mathbf{1})}.$$

5.3 Let $\hat{f}_1^\varphi$ and $\hat{f}_2^\varphi$ be two continuous functions both versions of $\pi_t(\varphi)$. Then

$$A = \left\{y_\cdot \in C_{\mathbb{R}^m}[0,t] : \hat{f}_1^\varphi(y_\cdot) \neq \hat{f}_2^\varphi(y_\cdot)\right\}$$

is an open set $C_{\mathbb{R}^m}[0,t]$. Also, from (5.7), we get that

$$\mathbb{P} \circ Y_\cdot^{-1}(A) = \mathbb{P}\left(\left\{\omega \in \Omega : \hat{f}_1^\varphi\left(Y_\cdot(\omega)\right) \neq \hat{f}_2^\varphi\left(Y_\cdot(\omega)\right)\right\}\right) = 0.$$

Since $\mathbb{P} \circ Y_\cdot^{-1}$ positively charges all non-empty open sets in $C_{\mathbb{R}^m}[0,t]$, it follows that A must be empty. Finally observe that, by Girsanov's theorem the distribution of $Y_\cdot$ under $\mathbb{P}$ is absolutely continuous with respect to the distribution of Y under $\tilde{\mathbb{P}}$. The results follows since the Wiener measure charges all open sets in $C_{\mathbb{R}^m}[0,t]$ and the Radon–Nikodym derivative $\mathrm{d}\tilde{\mathbb{P}}/\mathrm{d}\mathbb{P}$ is almost surely positive.

5.6 Bibliographic Note

The robust representation was introduced by Clark [56]. Both Clark and Kushner [179] show that the associated robust expression for the conditional distribution f^φ given by (5.10) is locally Lipschitz continuous in the observation path y. Very general robustness results have been obtained by Gyöngy [115] and Gyöngy and Krylov [114].

6
Finite-Dimensional Filters

In Section 3.5 we analyzed the case when X is a Markov process with finite state space I and associated Q-matrix Q (see Exercise 3.27). In that case, $\pi = \{\pi_t,\ t \geq 0\}$ the conditional distribution of X_t given the σ-algebra $\mathcal{Y}_t$ is a finite-dimensional process. More precisely $\pi = \{(\pi_t^i)_{i \in I},\ t \geq 0\}$, the conditional distribution of X_t given the σ-algebra $\mathcal{Y}_t$ is a process with values in $\mathbb{R}^I$ which solves the stochastic differential equation (3.53). The natural question which arises is whether the finite-dimensionality property is preserved when the signal is a diffusion process, in particular when the signal is the solution of the d-dimensional stochastic differential equation (3.9) (see Section 3.2 for details). In general, the answer to this question is negative (see, e.g. [42, 189, 231, 233]). With some notable exceptions, π is truly an infinite-dimensional stochastic process. The aim of this chapter is to study two special classes of filters for which the corresponding π is finite-dimensional: the Beneš filter (see [9]) and the linear filter, also known as the Kalman–Bucy filter ([29, 146, 147]).

6.1 The Beneš Filter

To simplify the calculations, we assume that both the signal and the observation are one-dimensional. We also assume that the signal process satisfies a stochastic differential equation with constant diffusion term and non-random initial condition; that is, X is a solution of the equation

$$X_t = x_0 + \int_0^t f(X_s)\,\mathrm{d}s + \sigma V_t. \tag{6.1}$$

In (6.1) $\sigma > 0$ is a positive constant, $x_0 \in \mathbb{R}$, V is a Brownian motion and the function $f : \mathbb{R} \to \mathbb{R}$ is differentiable, and satisfies the analogue of (3.10),

$$|f(x) - f(y)| \leq K|x - y|. \tag{6.2}$$

A. Bain, D. Crisan, *Fundamentals of Stochastic Filtering*,
DOI 10.1007/978-0-387-76896-0_6, © Springer Science+Business Media, LLC 2009

As in Chapter 3, the Lipschitz condition (6.2) is to ensure that the SDE for the signal process has a unique solution. We assume that W is a standard Brownian motion which is independent of V and that Y is the process satisfying the following evolution equation

$$Y_t = \int_0^t h(X_s)\,\mathrm{d}s + W_t. \tag{6.3}$$

In (6.3) $h : \mathbb{R} \to \mathbb{R}$ is chosen to be the linear function

$$h(x) = h_1 x + h_2, \quad x \in \mathbb{R}, \quad \text{where } h_1, h_2 \in \mathbb{R}.$$

We assume that the following condition, introduced by Beneš in [9], is satisfied

$$f'(x) + f^2(x)\sigma^{-2} + h^2(x) = P(x), \quad x \in \mathbb{R}, \tag{6.4}$$

where f' is the derivative of f and $P(x)$ is a second-order polynomial with positive leading-order coefficient.

Exercise 6.1. i. Show that if f is linear then the Beneš condition is satisfied (which establishes that the linear filter with time-independent coefficients is a Beneš filter).

ii. Show that the function f defined as

$$f(x) = \alpha\sigma \frac{\beta e^{2\alpha x/\sigma} - 1}{\beta e^{2\alpha x/\sigma} + 1}, \quad \text{where } \alpha, \beta \in \mathbb{R}$$

satisfies the Beneš condition. Thus show that $f(x) = a\sigma \tanh(ax/\sigma)$ satisfies the Beneš condition.

iii. Show that the function f defined as

$$f(x) = a\sigma \tanh(b + ax/\sigma), \quad \text{where } a, b \in \mathbb{R},$$

satisfies the Beneš condition.

6.1.1 Another Change of Probability Measure

We need to apply a change of the probability measure similar to the one detailed in Section 3.3. This time both the distribution of X and Y are affected, not just that of the observation process Y as was previously the case. Let $\check{Z} = \{\check{Z}_t,\ t > 0\}$ be the process defined by

$$\check{Z}_t \triangleq \exp\Bigg(-\int_0^t \frac{f(X_s)}{\sigma}\,\mathrm{d}V_s - \frac{1}{2}\int_0^t \frac{f(X_s)^2}{\sigma^2}\,\mathrm{d}s \\ - \int_0^t h(X_s)\,\mathrm{d}W_s - \frac{1}{2}\int_0^t h(X_s)^2\,\mathrm{d}s\Bigg), \quad t \geq 0. \tag{6.5}$$

Exercise 6.2. Show that the process $\check{Z} = \{\check{Z}_t,\ t \geq 0\}$ is an $\mathcal{F}_t$-adapted martingale under the measure $\mathbb{P}$.

Let $\hat{\mathbb{P}}$ be a new probability measure such that its Radon–Nikodym derivative with respect to $\mathbb{P}$ is

$$\left.\frac{\mathrm{d}\hat{\mathbb{P}}}{\mathrm{d}\mathbb{P}}\right|_{\mathcal{F}_t} = \check{Z}_t$$

for all $t \geq 0$. Let $\hat{V} = \{\hat{V}_t,\ t > 0\}$, be the process

$$\hat{V}_t \triangleq V_t + \int_0^t \frac{f(X_s)}{\sigma}\,\mathrm{d}s, \qquad t \geq 0.$$

Using Girsanov's theorem the pair process $(\hat{V}, Y) = \{(\hat{V}_t, Y_t),\ t > 0\}$ is a standard two-dimensional Brownian motion. Let $\hat{Z} = \{\hat{Z}_t,\ t \geq 0\}$ be the process defined as $\hat{Z}_t = \check{Z}_t^{-1}$ for $t \geq 0$. By Itô's formula, this process $\hat{Z}$ satisfies the following stochastic differential equation,

$$\mathrm{d}\hat{Z}_t = \hat{Z}_t \left(h(X_t)\,\mathrm{d}Y_t + f(X_t)\sigma^{-1}\,\mathrm{d}\hat{V}_t \right), \tag{6.6}$$

and since $\hat{Z}_0 = 1$,

$$\hat{Z}_t = \exp\left(\int_0^t \frac{f(X_s)}{\sigma}\,\mathrm{d}\hat{V}_s - \frac{1}{2}\int_0^t \frac{f(X_s)^2}{\sigma^2}\,\mathrm{d}s \right.$$
$$\left. + \int_0^t h(X_s)\,\mathrm{d}Y_s - \frac{1}{2}\int_0^t h(X_s)^2\,\mathrm{d}s \right), \qquad t \geq 0. \tag{6.7}$$

It is clear that $\hat{\mathbb{E}}\hat{Z}_t = \mathbb{E}(\hat{Z}_t \check{Z}_t) = 1$, so $\hat{Z}$ is a martingale under $\hat{\mathbb{P}}$ and we have

$$\left.\frac{\mathrm{d}\mathbb{P}}{\mathrm{d}\hat{\mathbb{P}}}\right|_{\mathcal{F}_t} = \hat{Z}_t \qquad \text{for } t \geq 0.$$

Let F be an antiderivative of f; that is, F is such that $F'(x) = f(x)$ for all $x \in \mathbb{R}$. By Itô's formula,

$$F(X_t) = F(X_0) + \int_0^t f(X_s)\sigma\,\mathrm{d}\hat{V}_s + \frac{1}{2}\int_0^t f'(X_s)\sigma^2\,\mathrm{d}s.$$

Thus from the Beneš condition (6.4) we get that, for all $t \geq 0$,

$$\check{Z}_t = \exp\left(\frac{F(X_t)}{\sigma^2} - \frac{F(x_0)}{\sigma^2} + \int_0^t h(X_s)\,\mathrm{d}Y_s - \frac{1}{2}\int_0^t P(X_s)\,\mathrm{d}s \right).$$

Exercise 6.3. Prove that, under $\hat{\mathbb{P}}$ the observation process Y is a Brownian motion independent of X, where we can write

$$X_t = X_0 + \sigma \hat{V}_t.$$

144   6 Finite-Dimensional Filters

Define $\hat{\rho}_t$ to be a measure-valued process following the definition of the unnormalised conditional expectation in Chapter 3. For every φ a bounded Borel-measurable function, it follows that $\hat{\rho}_t(\varphi)$ satisfies

$$\hat{\rho}_t(\varphi) \triangleq \hat{\mathbb{E}}[\varphi(X_t)\hat{Z}_t|\mathcal{Y}] \quad \hat{\mathbb{P}}\text{-a.s.,} \qquad (6.8)$$

where $\hat{\mathbb{E}}$ is the expectation with respect to $\hat{\mathbb{P}}$. As a consequence of Proposition 3.15, the process $\hat{\rho}(\varphi)$ is a modification of that defined with $\mathcal{Y}$ replaced by $\mathcal{Y}_t$ in (6.8).

Exercise 6.4. For every φ a bounded Borel-measurable function we have

$$\pi_t(\varphi) = \frac{\hat{\rho}_t(\varphi)}{\hat{\rho}_t(1)}, \quad \hat{\mathbb{P}}(\mathbb{P})\text{-a.s.} \qquad (6.9)$$

6.1.2 The Explicit Formula for the Beneš Filter

We aim to obtain an explicit expression of the (normalised) density of $\hat{\rho}_t$. For this we make use of the closed form expression (B.30) of the functional $I_t^{\beta,\Gamma,\delta}$ as described in equation (B.22) of the appendix. This cannot be done directly as the argument of the exponential in (B.22) contains no stochastic integral. However, similar to the analysis in Chapter 5, one can show that

$$\frac{\hat{\rho}_t(\varphi)}{\hat{\rho}_t(1)} = \lim_{n\to\infty} \frac{\hat{\rho}_t^n(\varphi)}{\hat{\rho}_t^n(1)},$$

where $\hat{\rho}_t^n$ is the measure defined as

$$\hat{\rho}_t^n(\varphi) \triangleq \hat{\mathbb{E}}\left[\varphi(X_t)\exp\left(\frac{F(X_t)}{\sigma^2} - \frac{F(x_0)}{\sigma^2} + \int_0^t h(X_s)y_s^n\,ds\right.\right.$$
$$\left.\left. -\frac{1}{2}\int_0^t P(X_s)\,ds\right)\right], \qquad (6.10)$$

for any bounded measurable function φ and $y^n = \{y_s^n,\ s\in[0,t]\}$ the piecewise constant process

$$y_s^n = \frac{Y_{(k+1)t/n} - Y_{kt/n}}{t/n}, \quad s\in[kt/n,(k+1)t/n), \quad k=0,1,\ldots,n-1.$$

As explained in Chapter 5, the expectation in (6.10) is no longer conditional. We keep y^n fixed to the observation path, or rather the approximation of its 'derivative' and integrate with respect to the law of $\hat{V}$.

Exercise 6.5. Prove that, almost surely,

$$\lim_{n\to\infty}\int_0^t \frac{\sinh(sp\sigma)}{\sinh(tp\sigma)}y_s^n\,ds = \int_0^t \frac{\sinh(sp\sigma)}{\sinh(tp\sigma)}\,dY_s,$$

and that there exists a positive random variable $c(t,Y)$ such that, uniformly in $n \geq 1$, we have
$$\left|\int_0^t \frac{\sinh(sp\sigma)}{\sinh(tp\sigma)} y_s^n \, ds\right| \leq c(t,Y).$$

In the following, we express the polynomial $P(x)$ in the form
$$P(x) = p^2 x^2 + 2qx + r,$$
where $p, q, r \in \mathbb{R}$ are arbitrary. Then we have the following.

Lemma 6.6. *For an arbitrary bounded Borel-measurable function φ, the ratio $\hat{\rho}_t^n(\varphi)/\hat{\rho}_t^n(\mathbf{1})$ has the following explicit formula*
$$\frac{\hat{\rho}_t^n(\varphi)}{\hat{\rho}_t^n(\mathbf{1})} = \frac{1}{c_t^n} \int_{-\infty}^{\infty} \varphi(x_0 + \sigma z) \exp\bigl(F(x_0 + \sigma z)\sigma^{-2} + Q_t^n(z)\bigr) \, dz, \qquad (6.11)$$
where $Q_t^n(z)$ is the second-order polynomial
$$Q_t^n(z) \triangleq z\left(\int_0^t \frac{\sinh(sp\sigma)}{\sinh(tp\sigma)} \sigma\left(h_1 y_s^n - q - p^2 x_0\right) ds\right) - \frac{p\sigma \coth(tp\sigma)}{2} z^2,$$
and c_t^n is the normalising constant
$$c_t^n \triangleq \int_{-\infty}^{\infty} \exp\bigl(F(x_0 + \sigma z)\sigma^{-2} + Q^n(z)\bigr) \, dz.$$

Proof. From (6.10), the expression for $\hat{\rho}_t^n(\varphi)$ becomes
$$\hat{\rho}_t^n(\varphi) = \lambda_t^n \hat{\mathbb{E}}\left[\varphi(x_0 + \sigma \hat{V}_t) \exp\left(F\left(x_0 + \sigma \hat{V}_t\right)\sigma^{-2} + \int_0^t \hat{V}_s \beta_s^n \, ds - \frac{1}{2}\int_0^t (p\sigma \hat{V}_s)^2 \, ds\right)\right], \qquad (6.12)$$
where
$$\lambda_t^n \triangleq \exp\left(-F(x_0)\sigma^{-2} + \int_0^t (h_1 x_0 + h_2) y_s^n \, ds - \frac{1}{2}(r + 2x_0 q + p^2 x_0^2)t\right),$$
$$\beta_s^n \triangleq \sigma(h_1 y_s^n - q - p^2 x_0).$$

If we make the definition
$$I_t^{\beta^n, p\sigma, z} \triangleq \hat{\mathbb{E}}\left[\exp\left(\int_0^t \hat{V}_s \beta_s^n \, ds - \frac{1}{2}\int_0^t (p\sigma \hat{V}_s)^2 \, ds\right) \Big| \hat{V}_t = z\right],$$
then

$$\hat{\mathbb{E}}\left[\varphi(x_0 + \sigma\hat{V}_t)\exp\left(F\left(x_0 + \sigma\hat{V}_t\right)\sigma^{-2} + \int_0^t \hat{V}_s \beta_s^n \, \mathrm{d}s \right.\right.$$
$$\left.\left. - \frac{1}{2}\int_0^t (p\sigma\hat{V}_s)^2 \, \mathrm{d}s\right)\bigg|\, \hat{V}_t = z\right]$$
$$= I_t^{\beta^n, p\sigma, z} \varphi(x_0 + \sigma z) \exp\left(\frac{F(x_0 + \sigma z)}{\sigma^2}\right). \quad (6.13)$$

Following (B.36) we get that
$$I_t^{\beta^n, p\sigma, z} = \bar{f}_t^{\beta^n, p\sigma} \exp\left(z \int_0^t \frac{\sinh(sp\sigma)}{\sinh(tp\sigma)} \beta_s^n \, \mathrm{d}s - \frac{p\sigma \coth(tp\sigma)}{2} z^2 + \frac{z^2}{2t}\right), \quad (6.14)$$

where
$$\bar{f}_t^{\beta^n, p\sigma} \triangleq \sqrt{\frac{tp\sigma}{\sinh(tp\sigma)}} \exp\left(\int_0^t \int_0^t \frac{\sinh((s-t)p\sigma)\sinh(s'p\sigma)}{2p\sigma \sinh(tp\sigma)} \beta_s^n \beta_{s'}^n \, \mathrm{d}s \, \mathrm{d}s'\right).$$

Identity (6.11) then follows from (6.12)–(6.14) by integrating over the $N(0, t)$ law of $\hat{V}_t$,
$$\hat{\mathbb{E}}(\cdot) = \frac{1}{\sqrt{2\pi t}} \int_{-\infty}^{\infty} \hat{\mathbb{E}}(\cdot \mid \hat{V}_t = z) \mathrm{e}^{-z^2/2t} \, \mathrm{d}z.$$
$\square$

Observe that the function $\bar{f}_t^{\beta^n, p\sigma}$ which is used in the above proof does not appear in the final expression for $\hat{\rho}_t^n(\varphi)/\hat{\rho}_t^n(\mathbf{1})$. We are now ready to obtain the formula for $\pi_t(\varphi)$.

Proposition 6.7. *If the Beneš condition (6.4) is satisfied then for arbitrary bounded Borel-measurable φ, it follows that $\pi_t(\varphi)$ satisfies the following explicit formula*
$$\pi_t(\varphi) = \frac{1}{c_t} \int_{-\infty}^{\infty} \varphi(z) \exp\left(F(z)\sigma^{-2} + Q_t(z)\right) \mathrm{d}z, \quad (6.15)$$

where $Q_t(z)$ is the second-order polynomial
$$Q_t(z) \triangleq z \left(h_1 \sigma \int_0^t \frac{\sinh(sp\sigma)}{\sinh(tp\sigma)} \mathrm{d}Y_s + \frac{q + p^2 x_0}{p\sigma \sinh(tp\sigma)} - \frac{q}{p\sigma} \coth(tp\sigma)\right)$$
$$- \frac{p \coth(tp\sigma)}{2\sigma} z^2,$$

and c_t is the corresponding normalising constant,
$$c_t \triangleq \int_{-\infty}^{\infty} \exp\left(F(z)\sigma^{-2} + Q_t(z)\right) \mathrm{d}z. \quad (6.16)$$

In particular, π depends only on the one-dimensional $\mathcal{Y}_t$-adapted process
$$t \mapsto \int_0^t \sinh(sp\sigma) \, \mathrm{d}Y_s.$$

Proof. Making a change of variable in (6.11), we get that

$$\frac{\hat{\rho}_t^n(\varphi)}{\hat{\rho}_t^n(1)} = \frac{1}{c_t^n}\int_{-\infty}^{\infty} \varphi(u)\exp\left(\frac{F(u)}{\sigma^2} + Q_t^n\left(\frac{u-x_0}{\sigma}\right)\right)\frac{1}{\sigma}\,du.$$

Following Exercise 6.5 we get that

$$\lim_{n\to\infty}\int_0^t \frac{\sinh(sp\sigma)}{\sinh(tp\sigma)} y_s^n\,ds = \int_0^t \frac{\sinh(sp\sigma)}{\sinh(tp\sigma)}\,dY_s,$$

hence[†]

$$\lim_{n\to\infty} Q_t^n(z) = z\sigma h_1 \int_0^t \frac{\sinh(sp\sigma)}{\sinh(tp\sigma)}\,dY_s$$
$$-\frac{q+p^2 x_0}{p}(\coth(tp\sigma) - \operatorname{csch}(tp\sigma))\,z - \frac{p\sigma \coth(tp\sigma)}{2}z^2.$$

Thus

$$Q_t(u) = \lim_{n\to\infty} Q_t^n\left(\frac{u-x_0}{\sigma}\right)$$
$$= uh_1 \int_0^t \frac{\sinh(sp\sigma)}{\sinh(tp\sigma)}\,dY_s - \frac{p\coth(tp\sigma)}{2\sigma}u^2$$
$$-\frac{q+p^2 x_0}{p\sigma}(\coth(tp\sigma) - \operatorname{csch}(tp\sigma))\,u + \frac{p\coth(tp\sigma)}{\sigma}ux_0.$$

Finally, since

$$\pi_t(\varphi) = \lim_{n\to\infty}\frac{\hat{\rho}_t^n(\varphi)}{\hat{\rho}_t^n(1)},$$

the proposition follows by the dominated convergence theorem (again use Exercise 6.5). □

Remark 6.8. For large t, as $\coth(x) \to 1$ and $\operatorname{csch}(x) \to 0$ as $x \to \infty$, it follows that $\pi_t(\varphi)$ is approximately equal to

$$\pi_t(\varphi) \simeq \frac{1}{c_t}\int_{-\infty}^{\infty}\varphi(z)\exp\left(F(z)\sigma^{-2} + \tilde{P}_t(z)\right)\,dz,$$

where $\tilde{P}_t(z)$ is the second-order polynomial

$$\tilde{P}_t(z) \triangleq \left(h_1\sigma\int_{t'}^t \frac{\sinh(sp\sigma)}{\sinh(tp\sigma)}\,dY_s - \frac{q}{\sigma p}\right)z - \frac{p}{2\sigma}z^2, \quad t' < t.$$

In particular, past observations become quickly (exponentially) irrelevant and so does the initial position of the signal x_0.

[†] Recall that $\coth(x) = \cosh(x)/\sinh(x)$ and $\operatorname{csch}(x) = 1/\sinh(x)$.

Exercise 6.9. Compute the normalising constant c_t for the linear filter and the filter given by $f(x) = a\sigma \tanh(ax/\sigma)$, which were shown to satisfy the Beneš condition described in Exercise 6.1. Hence determine an explicit expression for the density of π_t. What is the asymptotic behaviour of π_t for large t?

If the initial state of the signal X_0 is random, then the formula for $\pi_t(\varphi)$ is obtained by integrating (6.15) in the x_0 variable with respect to the law of X_0. A multidimensional version of (6.15) can be obtained by following the same procedure as above. The details of the computation of the exponential Brownian function $I_t^{\beta,\Gamma,\delta}$ are described in formula (B.22) of the appendix in the multidimensional case. Including the full form of $\pi_t(\varphi)$ in this case would make this chapter excessively long. However, the fact that such a computation is possible is fairly important, due to the scarcity of explicit expressions for π. Such explicit expressions provide benchmarks for testing numerical algorithms for computing approximations to π.

6.2 The Kalman–Bucy Filter

Let now $X = (X^i)_{i=1}^d$ be the solution of the linear SDE driven by a p-dimensional Brownian motion process $V = (V^j)_{j=1}^p$,

$$X_t = X_0 + \int_0^t (F_s X_s + f_s)\, ds + \int_0^t \sigma_s\, dV_s, \tag{6.17}$$

where, for any $s \geq 0$, F_s is a $d \times d$ matrix, σ_s is a $d \times p$ matrix and f_s is a d-dimensional vector. The functions $s \mapsto F_s$, $s \mapsto \sigma_s$ and $s \mapsto f_s$ are measurable and locally bounded.[†] Assume that $X_0 \sim N(x_0, r_0)$ is independent of V. Next assume that W is a standard $\mathcal{F}_t$-adapted m-dimensional Brownian motion on $(\Omega, \mathcal{F}, \mathbb{P})$ independent of X and let Y be the process satisfying the following evolution equation

$$Y_t = \int_0^t (H_s X_s + h_s)\, ds + W_t, \tag{6.18}$$

where, for any $s \geq 0$, H_s is a $m \times d$ matrix and h_s is an m-dimensional vector.

Remark 6.10. Let $\mathbb{I}_m$ be the $m \times m$-identity matrix and $0_{a,b}$ be the $a \times b$ matrix with all entries equal to 0. Let L_s be the $(d+m) \times (d+m)$ matrix, l_s be the $(d+m)$-dimensional vector and $\mathcal{F}_s$ be the $(d+m) \times (r+m)$ matrix given by, respectively,

$$L_s = \begin{pmatrix} F_s & \mathbb{O}_{d,m} \\ H_s & \mathbb{O}_{m,m} \end{pmatrix}, \quad l_s = \begin{pmatrix} f_s \\ h_s \end{pmatrix}, \quad \mathcal{F}_s = \begin{pmatrix} \sigma_s & \mathbb{O}_{d,m} \\ \mathbb{O}_{m,r} & \mathbb{I}_m \end{pmatrix}.$$

[†] That is, for every time t, the functions are bounded for $s \in [0, t]$.

Let $T = \{T_t,\ t > 0\}$ be the $(d+m)$-dimensional pair process (X, Y) and $U = \{U_t,\ t > 0\}$ be the $(p+m)$-dimensional Brownian motion (V, W). Then T is a solution of the linear SDE

$$T_t = T_0 + \int_0^t (L_s T_s + l_s)\,\mathrm{d}s + \int_0^t F_s\,\mathrm{d}U_s. \tag{6.19}$$

Exercise 6.11. i. Prove that T has the following representation

$$T_t = \Phi_t \left[T_0 + \int_0^t \Phi_s^{-1} l_s\,\mathrm{d}s + \int_0^t \Phi_s^{-1} F_s\,\mathrm{d}U_s \right], \tag{6.20}$$

where Φ is the unique solution of the matrix equation

$$\frac{\mathrm{d}\Phi_t}{\mathrm{d}t} = L_t \Phi_t, \tag{6.21}$$

with initial condition $\Phi_0 = \mathbb{I}_{d+m}$.

ii. Deduce from (i) that for any $n > 0$ and any $n+1$-tuple of the form

$$\left(Y_{t_1}, Y_{t_2}, \ldots, Y_{t_{n-1}}, Y_t, X_t \right),$$

where $0 \leq t_1 \leq \cdots \leq t_{n-1} \leq t$, has a $(d+nm)$-variate normal distribution.

iii. Let $K : [0, t] \to \mathbb{R}^{d \times m}$ be a measurable $(d \times m)$ matrix-valued function with all of its entries square integrable. Deduce from (ii) that the pair

$$\left(X_t, \int_0^t K_s\,\mathrm{d}Y_s \right)$$

has a $2d$-variate normal distribution.

Lemma 6.12. *In the case of the linear filter, the normalised conditional distribution π_t of X_t conditional upon $\mathcal{Y}_t$ is a multivariate normal distribution.*

Proof. Consider the orthogonal projection of the components of the signal X_t^i, $i = 1, \ldots, d$, onto the Hilbert space $\mathcal{H}_t^Y \subset L_2(\Omega)$ generated by the components of the observation process

$$\left\{ Y_s^j,\ s \in [0, t],\ j = 1, \ldots, m \right\}.$$

Using Lemma 4.3.2, page 122 in Davis [71], the elements of $\mathcal{H}_t^Y$ have the following representation

$$\mathcal{H}_t^Y = \left\{ \sum_{i=1}^m \int_0^t a_i\,\mathrm{d}Y_s^i : a_i \in L_2([0,t]),\ i = 1, \ldots, m \right\}.$$

It follows that there exists a $(d \times m)$ matrix-valued function $K : [0, t] \to \mathbb{R}^{d \times m}$ with all of its entries square integrable, and a random variable $\check{X}_t = (\check{X}_t^i)_{i=1}^d$ with entries orthogonal on $\mathcal{H}_t^Y$ such that

$$X_t = \check{X}_t + \int_0^t K_s \, dY_s.$$

In particular, as a consequence of Exercise 6.11 part (iii), $\check{X}_t$ has a Gaussian distribution. Moreover, for any $n > 0$ any n-tuple of the form

$$\left(Y_{t_1}, Y_{t_2}, \ldots, Y_{t_{n-1}}, \check{X}_t\right),$$

where $0 \leq t_1 \leq \cdots \leq t_{n-1} \leq t$ has a $(d + (n-1)m)$-variate normal distribution. Now since $\check{X}_t$ has all entries orthogonal on $\mathcal{H}_t^Y$ it follows that $\check{X}_t$ is independent of $(Y_{t_1}, Y_{t_2}, \ldots, Y_{t_{n-1}})$ and since the time instances $0 \leq t_1 \leq \cdots \leq t_{n-1} \leq t$ have been arbitrarily chosen it follows that $\check{X}_t$ is independent of $\mathcal{Y}_t$. This observation is crucial! It basically says that, in the linear/Gaussian case, the linear projection (the projection onto the linear space generated by the observation) coincides with the non-linear projection (the conditional expectation with respect to the observation σ-algebra). Hence the distribution of X_t conditional upon $\mathcal{Y}_t$ is the same as the distribution of $\check{X}_t$ shifted by the (fixed) quantity $\int_0^t K_s \, dY_s$. In particular π_t is characterized by its first and second moments alone. □

6.2.1 The First and Second Moments of the Conditional Distribution of the Signal

We know from Chapter 3 that the conditional distribution of the signal is the unique solution of the Kushner–Stratonovich equation (3.57). Unlike the model analysed in Chapter 3, the above linear filter has time-dependent coefficients. Nevertheless all the results and proofs presented there apply to the linear filter with time-dependent coefficients (see Remark 3.1). In the following we deduce the equations for the first and second moments of π. Let φ_i, φ_{ij} for $i, j = 1, \ldots, d$ be the functions

$$\varphi_i(x) = x_i, \qquad \varphi_{ij}(x) = x_i x_j, \qquad x \in \mathbb{R}$$

and let π_t^i, π_t^{ij} be the moments of π_t

$$\pi_t^i = \pi_t(\varphi_i), \qquad \pi_t^{ij} = \pi_t(\varphi_{ij}), \qquad i, j = 1, \ldots, d.$$

Exercise 6.13. i. Show that for any $t \geq 0$ and $i = 1, \ldots, d$ and $p \geq 1$, the solution of the equation (6.17) satisfies

$$\sup_{s \in [0,t]} \mathbb{E}\left[|X_s^i|^p\right] < \infty.$$

ii. Deduce from (i) that for any $t \geq 0$ and $i, j = 1, \ldots, d$

$$\sup_{s \in [0,t]} \mathbb{E}\left[(\pi_s(|\varphi_i|))^p\right] < \infty, \qquad \sup_{s \in [0,t]} \mathbb{E}\left[|\pi_s(|\varphi_{ij}|)|^p\right] < \infty.$$

In particular
$$\sup_{s\in[0,t]} \mathbb{E}\left[|\pi_s^i|^p\right] < \infty, \qquad \sup_{s\in[0,t]} \mathbb{E}\left[|\pi_s^{ij}|^p\right] < \infty.$$

In this case the innovation process $I = \{I_t,\ t \geq 0\}$ defined by (2.17) has the components

$$I_t^j = Y_t^j - \int_0^t \left(\sum_{i=1}^d H_s^{ji}\pi_s^i + h_s^j\right) \mathrm{d}s, \qquad t \geq 0,\ j = 1,\ldots,m.$$

The Kushner–Stratonovich equation (3.57) now takes the form

$$\pi_t(\varphi) = \pi_0(\varphi) + \int_0^t \pi_s(A_s\varphi)\,\mathrm{d}s + \sum_{i=1}^d \sum_{j=1}^m \int_0^t \pi_s\left(\varphi\left(\varphi^i - \pi_s^i\right)\right) H_s^{ji}\,\mathrm{d}I_s^j \quad (6.22)$$

where the time-dependent generator A_s, $s \geq 0$ is given by

$$A_s\varphi = \sum_{i,j=1}^d \left(F_s^{ij} x_j + f_s^i\right) \frac{\partial \varphi}{\partial x_i} + \frac{1}{2}\sum_{i=1}^d \sum_{j=1}^d (\sigma_s \sigma_s^\top)_{ij} \frac{\partial^2 \varphi}{\partial x_i \partial x_j},$$

and φ is chosen in the domain of A_s for any $s \in [0,t]$ such that

$$\sup_{s\in[0,t]} \|A_s\varphi\| < \infty.$$

To find the equations satisfied by π_t^i and π_t^{ij} we cannot replace φ by φ_i and φ by φ_{ij} in (6.22) because neither of them belongs to the domain of A_s (since they are unbounded). We proceed by cutting off φ_i and φ_{ij} at a fixed level which we let tend to infinity. For this let us introduce the functions $(\psi^k)_{k>0}$ defined as

$$\psi^k(x) = \psi(x/k), \qquad x \in \mathbb{R}^d, \qquad (6.23)$$

where

$$\psi(x) = \begin{cases} 1 & \text{if } |x| \leq 1 \\ \exp\left(\frac{|x|^2-1}{|x|^2-4}\right) & \text{if } 1 < |x| < 2 \\ 0 & \text{if } |x| \geq 2 \end{cases}.$$

Obviously, for all $k > 0$, $\psi^k \in C_b^\infty(\mathbb{R}^d)$ and $0 \leq I_{B(k)} \leq \psi_k \leq 1$. Also, all partial derivatives of ψ_k tend uniformly to 0. In particular

$$\lim_{k\to\infty} \|A\psi_k\|_\infty = 0, \quad \lim_{k\to\infty} \|\partial_i \psi_k\|_\infty = 0, \qquad i = 1,\ldots,d.$$

In the following we use the relations

$$\lim_{k \to \infty} \varphi_i(x)\psi^k(x) = \varphi_i(x), \qquad |\varphi_i(x)\psi^k(x)| \le |\varphi_i(x)|, \qquad (6.24)$$

$$\lim_{k \to \infty} A_s(\varphi_i \psi^k)(x) = A_s \varphi_i(x), \qquad (6.25)$$

$$\sup_{s \in [0,t]} |A_s(\varphi_i \psi^k)(x)| \le C_t \left(\sum_{i=1}^n |\varphi_i(x)| + \sum_{i,j=1}^n |\varphi_{ij}(x)| \right). \qquad (6.26)$$

Proposition 6.14. *Let $\hat{x} = \{\hat{x}_t,\ t \ge 0\}$ be the conditional mean of the signal. In other words, $\hat{x}$ is the d-dimensional process with components*

$$\hat{x}_t^i = \mathbb{E}[X_t^i | \mathcal{Y}_t] = \pi_t^i, \qquad i = 1, \ldots, d,\ t \ge 0.$$

Define $R = \{R_t,\ t \ge 0\}$ to be the conditional covariance matrix of the signal. In other words, R_t is the $d \times d$-dimensional process with components

$$\begin{aligned}R_t^{ij} &= \mathbb{E}[X_t^i X_t^j | \mathcal{Y}_t] - \mathbb{E}[X_t^i | \mathcal{Y}_t]\mathbb{E}[X_t^j | \mathcal{Y}_t] \\ &= \pi_t^{ij} - \pi_t^i \pi_t^j, \qquad i,j = 1, \ldots, d,\ t \ge 0.\end{aligned}$$

Then $\hat{x}$ satisfies the stochastic differential equation

$$\mathrm{d}\hat{x}_t = (F_t \hat{x}_t + f_t)\,\mathrm{d}t + R_t H_t^\top (\mathrm{d}Y_t - (H_t \hat{x}_t + h_t)\,\mathrm{d}t), \qquad (6.27)$$

and R satisfies the deterministic matrix Riccati equation

$$\frac{\mathrm{d}R_t}{\mathrm{d}t} = \sigma_t \sigma_t^\top + F_t R_t + R_t F_t^\top - R_t H_t^\top H_t R_t. \qquad (6.28)$$

Proof. Replacing φ by $\varphi_i \psi^k$ in (6.22) gives us

$$\begin{aligned}\pi_t(\varphi_i \psi^k) = {}& \pi_0(\varphi_i \psi^k) + \int_0^t \pi_s(A_s(\varphi_i \psi^k))\,\mathrm{d}s \\ & + \sum_{l=1}^d \sum_{j=1}^m \int_0^t \pi_s\left((\varphi_i \psi^k(\varphi^l - \pi_s^l))\right) H_s^{jl}\,\mathrm{d}I_s^j.\end{aligned} \qquad (6.29)$$

By the dominated convergence theorem (use (6.24)–(6.26)) we may pass to the limit as $k \to \infty$,

$$\lim_{k \to \infty} \pi_t(\varphi_i \psi^k) = \pi_t(\varphi_i) \qquad (6.30)$$

$$\lim_{k \to \infty} \pi_0(\varphi_i \psi^k) + \int_0^t \pi_s(A_s(\varphi_i \psi^k))\,\mathrm{d}s = \pi_0(\varphi_i) + \int_0^t \pi_s(A_s \varphi_i)\,\mathrm{d}s. \qquad (6.31)$$

Also

$$\lim_{k \to \infty} \mathbb{E}\left[\left|\int_0^t \pi_s\left((\varphi_i(\psi^k - 1)(\varphi^k - \pi_s^l))\right) H_s^{jl}\,\mathrm{d}I_s^j\right|\right] = 0.$$

Hence at least for subsequence $(k_n)_{n \ge 0}$, we have that

$$\lim_{k_n \to \infty} \sum_{l=1}^{d} \sum_{j=1}^{m} \int_0^t \pi_s \left((\varphi_i \psi^k (\varphi^l - \pi_s^l)) \right) H_s^{jl} \, \mathrm{d}I_s^j = \sum_{l=1}^{d} \sum_{j=1}^{m} \int_0^t R_s^{il} H_s^{jl} \, \mathrm{d}I_s^j. \tag{6.32}$$

By taking the limit in (6.29) along a convenient subsequence and using (6.30)–(6.32) we obtain (6.27).

We now derive the equation for the evolution of the covariance matrix R. Again we cannot apply the Kushner–Stratonovich equation directly to φ_{ij} but use first an intermediate step. We 'cut off' φ_{ij} and use the functions $(\psi^k)_{k>0}$ and take the limit as k tends to infinity. After doing that we obtain the equation for π_t^{ij} which is

$$\mathrm{d}\pi_t^{ij} = \left((\sigma_t \sigma_t^\top)^{ij} + \sum_{k=1}^{d} F_t^{ik} \pi_t^{kj} + F_t^{jk} \pi_t^{ik} + f_t^i \left(\hat{x}_t^i + \hat{x}_t^j \right) \right) \mathrm{d}t$$

$$+ \sum_{k=1}^{d} \sum_{l=1}^{m} \left(\pi_t \left(\varphi_i \varphi_j \varphi_k \right) - \pi_t^{ij} \hat{x}_t^k \right) H_t^{lk} \mathrm{d}I_t^l. \tag{6.33}$$

Observe that since π_t is normal we have the following result on the third moments of a multivariate normal distribution

$$\pi_t \left(\varphi_i \varphi_j \varphi_k \right) = \hat{x}_t^i \hat{x}_t^j \hat{x}_t^k + \hat{x}_t^i R_t^{jk} + \hat{x}_t^j R_t^{ik} + \hat{x}_t^k R_t^{ij}.$$

It is clear that
$$\mathrm{d}R_t^{ij} = \mathrm{d}\pi_t^{ij} - \mathrm{d}(\hat{x}_t^i \hat{x}_t^j), \tag{6.34}$$

where the first term is given by (6.33) and using Itô's form of the product rule to expand out the second term

$$\mathrm{d}(\hat{x}_t^i \hat{x}_t^j) = \hat{x}_t^i \mathrm{d}\hat{x}_t^j + \hat{x}_t^j \mathrm{d}\hat{x}_t^i + \mathrm{d}\langle \hat{x}^i, \hat{x}^j \rangle_t.$$

Therefore using (6.27) we can evaluate this as

$$\mathrm{d}\left(\hat{x}_t^i \hat{x}_t^j \right) = \sum_{k=1}^{d} F_t^{ik} \hat{x}_t^k \hat{x}_t^j \mathrm{d}t + F_t^{jk} \hat{x}_t^i \hat{x}_t^k \mathrm{d}t + f_t^i \left(\hat{x}_t^i + \hat{x}_t^j \right) \mathrm{d}t + \hat{x}_t^i (H_t R_t^\top \mathrm{d}I_t)^j$$

$$+ \hat{x}_t^j (H_t R_t^\top \mathrm{d}I_t)^i + \left\langle (H_t^\top R_t \mathrm{d}I_t)^i, (H_t R_t^\top \mathrm{d}I_t)^j \right\rangle. \tag{6.35}$$

For evaluating the quadratic covariation term in this expression it is simplest to work componentwise using the Einstein summation convention and use the fact that by Proposition 2.30 the innovation process I_t is a $\mathbb{P}$-Brownian motion

$$\left\langle (H_t R_t^\top \mathrm{d}I_t)^i, (H_t^\top R_t \mathrm{d}I_t)^j \right\rangle = \left\langle R_t^{il} H_t^{kl} \mathrm{d}I_t^k, R_t^{jm} H_t^{nm} \mathrm{d}I_t^n \right\rangle$$

$$= R_t^{il} H_t^{kl} R_t^{jm} H_t^{nm} \delta_{kn} \mathrm{d}t$$

$$= R_t^{il} H_t^{kl} H_t^{km} R_t^{jm} \mathrm{d}t$$

$$= (RH^\top H R^\top)^{ij} \mathrm{d}t$$

$$= (RH^\top H R)^{ij} \mathrm{d}t, \tag{6.36}$$

where the last equality follows since $R^\top = R$. Substituting (6.33), (6.35) and (6.36) into (6.34) yields the following equation for the evolution of the ijth element of the covariance matrix

$$dR_t^{ij} = \left((\sigma_t \sigma_t^\top)^{ij} + (F_t R_t)^{ij} + (R_t^\top F_t^\top)^{ij} - (R_t H_t^\top H_t R_t)^{ij}\right) dt$$
$$+ (\hat{x}_t^i R_t^{jm} + \hat{x}_t^j R_t^{jm}) H_t^{lm} \, dI_t^l - (\hat{x}_t^i R_t^{jm} + \hat{x}_t^j R_t^{jm}) H_t^{lm} \, dI_t^l.$$

Thus we obtain the final differential equation for the evolution of the conditional covariance matrix (notice that all of the stochastic terms will cancel out). □

6.2.2 The Explicit Formula for the Kalman–Bucy Filter

In the following we use the notation $R^{1/2}$ to denote the square root of the symmetric positive semi-definite matrix R; that is, the matrix $R^{1/2}$ is the (unique) symmetric positive semi-definite matrix A such that $A^2 = R$.

Theorem 6.15. *The conditional distribution of X_t given the observation σ-algebra is given by the explicit formula*

$$\pi_t(\varphi) = \frac{1}{(2\pi)^{n/2}} \int_{\mathbb{R}^d} \varphi\left(\hat{x}_t + R_t^{1/2}\zeta\right) \exp\left(-\frac{1}{2}\|\zeta\|^2\right) d\zeta$$

for any $\varphi \in \mathcal{B}\left(\mathbb{R}^d\right)$.

Proof. Immediate as π_t is a normal distribution with mean $\hat{x}_t$ and covariance matrix R_t. □

We remark that, in this case too, π is finite-dimensional as it depends only on the $(d+d^2)$-process (x,R) (its mean and covariance matrix).

Corollary 6.16. *The process ρ_t satisfying the Zakai equation (3.43) is given by*

$$\rho_t(\varphi) = \hat{Z}_t \frac{1}{(2\pi)^{n/2}} \int_{\mathbb{R}^d} \varphi\left(\hat{x}_t + R_t^{1/2}\zeta\right) \exp\left(-\frac{1}{2}\|\zeta\|^2\right) d\zeta,$$

where $\varphi \in \mathcal{B}\left(\mathbb{R}^d\right)$ and

$$\hat{Z}_t = \exp\left(\int_0^t (H\hat{x}_t + h)^\top \, dY_s - \int_0^t \|H\hat{x}_t + h\|^2 \, ds\right).$$

Proof. Immediate from Theorem 6.15 and the fact that $\rho_t(\mathbf{1})$ has the representation

$$\rho_t(\mathbf{1}) = \exp\left(\int_0^t (H\hat{x}_s + h)^\top \, dY_s - \int_0^t \|H\hat{x}_s + h\|^2 \, ds\right)$$

as proved in Exercise 3.37. □

6.3 Solutions to Exercises

6.1

i. Suppose that $f(x) = ax + b$; then $P(x) = a + (ax+b)^2 \sigma^{-2} + (h_1 x + h_2)^2$ which is a second-order polynomial with leading coefficient $a^2/\sigma^2 + h_1^2 \geq 0$. The Lipschitz condition on f is trivial.

ii. In this case $P(x) = \alpha^2 + (h_1 x + h_2)^2$ which is a second order polynomial with leading coefficient $h_1^2 \geq 0$. The case $f(x) = a\sigma \tanh(ax/\sigma)$ is obtained by taking $\alpha = a$ and $\beta = 1$. The derivative $f'(x)$ is bounded by $1/(4\beta)$, thus the function f is Lipschitz and satisfied (6.2).

iii. Use the previous result with $\alpha = a$, $\beta = e^{2b}$.

6.2
Lemma 3.9 implies that it is sufficient to show that

$$\mathbb{E}\left[\int_0^t \left(f(X_s)^2 \sigma^{-2} + h^2(X_s)\right) \mathrm{d}s\right] < \infty.$$

From the Lipschitz condition (6.2) on f, the fact that σ is constant, and that $X_0 = x_0$ is constant and thus trivially has bounded second moment, it follows from Exercise 3.11 that for $0 \leq t \leq T$, $\mathbb{E}X_t^2 < G_T < \infty$. It also follows from Exercise 3.3 that $f(X)$ has a linear growth bound $f(x) \leq \kappa(1+\|x\|)$, therefore

$$\mathbb{E}\left[\int_0^t \left(\frac{f(X_s)^2}{\sigma^2} + h(X_s)^2\right) \mathrm{d}s\right] \leq \mathbb{E}\left[\int_0^t \frac{\kappa^2}{\sigma^2}(1+|X_s|)^2 + (h_1 X_s + h_2)^2 \,\mathrm{d}s\right]$$

$$\leq 2\left(h_1^2 + \frac{\kappa^2}{\sigma^2}\right)\int_0^t \mathbb{E}|X_s|^2 \,\mathrm{d}s + \left(h_2^2 + \frac{\kappa^2}{\sigma^2}\right)t$$

$$\leq 2\left(h_1^2 + \frac{\kappa^2}{\sigma^2}\right)t G_T + \left(h_2^2 + \frac{\kappa^2}{\sigma^2}\right)t < \infty.$$

6.3
By Girsanov's theorem under $\tilde{\mathbb{P}}$, the process with components

$$X_t^1 = W_t - \left\langle W, -\int_0^t \frac{f(X_s)}{\sigma}\mathrm{d}V_s - \int_0^t h(X_s)\,\mathrm{d}W_s\right\rangle = W_t + \int_0^t h(X_s)\,\mathrm{d}s$$

and

$$X_t^2 = V_t - \left\langle V, -\int_0^t \frac{f(X_s)}{\sigma}\mathrm{d}V_s - \int_0^t h(X_s)\,\mathrm{d}W_s\right\rangle = V_t + \int_0^t \frac{F(X_s)}{\sigma}\,\mathrm{d}s$$

is a two-dimensional Brownian motion. Therefore the law of $(X_t^1, X_t^2) = (T_t, \hat{V}_t)$ is bivariate normal, so to show the components are independent it is sufficient to consider the covariation

$$\langle \hat{V}_t, Y_t\rangle = \langle V_t, W_t\rangle = 0, \quad \forall t \in [0, \infty),$$

from which we may conclude that Y is independent of $\hat{V}$, and since $X_t = X_0 + \sigma \hat{V}$, it follows that under $\hat{\mathbb{P}}$ the processes Y and X are independent.

6.4 Follow the same argument as in the proof of Proposition 3.16.

6.5 Consider t as fixed; it is then sufficient to show that uniformly in n,

$$\int_0^t \sinh(s p\sigma) y_s^n \, ds$$

$$= \sum_{k=0}^{n-1} \frac{Y_{(k+1)t/n} - Y_{kt/n}}{t/n} \int_{kt/n}^{(k+1)t/n} \sinh(s p\sigma) \, ds$$

$$= \sum_{k=0}^{n-1} \frac{\cosh((k+1) p\sigma t/n) - \cosh(k p\sigma t/n)}{p\sigma t/n} \int_{kt/n}^{(k+1)t/n} dY_s$$

$$= \int_0^t \sum_{k=0}^{n-1} \frac{\cosh((k+1) p\sigma t/n) - \cosh(k p\sigma t/n)}{p\sigma t/n} 1_{(kt/n,(k+1)t/n]}(s) \, dY_s.$$

Thus by Itô's isometry, since Y is a Brownian motion under $\hat{\mathbb{P}}$, therefore it is sufficient to show

$$\mathbb{E}\left[\int_0^t \left(\sum_{k=0}^{n-1} \frac{\cosh((k+1) p\sigma t/n) - \cosh(k p\sigma t/n)}{p\sigma t/n} 1_{(kt/n,(k+1)t/n]}(s) \right.\right.$$

$$\left.\left. - \sinh(s p\sigma)\right)^2 ds \right] \to 0.$$

Using the mean value theorem, for each interval for $k = 0, \ldots, n-1$, there exists $\xi \in [k p\sigma/n, (k+1) p\sigma/n]$ such that

$$\sinh(\xi p\sigma) = \frac{\cosh((k+1) p\sigma/n) - \cosh(k p\sigma/n)}{p\sigma/n}$$

therefore since $\sinh(x)$ is monotonic increasing for $x > 0$,

$$\mathbb{E}\left[\sum_{k=0}^{n-1} \int_{kt/n}^{(k+1)t/n} \left(\frac{\cosh((k+1) p\sigma t/n) - \cosh(k p\sigma t/n)}{p\sigma t/n} - \sinh(s p\sigma)\right)^2 ds\right]$$

$$\leq \sum_{k=0}^{n-1} \frac{t}{n} (\sinh((k+1) p\sigma t/n) - \sinh(k p\sigma t/n))^2$$

$$\leq \sum_{k=0}^{n-1} \frac{t}{n} \cosh^2((k+1) p\sigma t/n) \left(\frac{t p\sigma}{n}\right)^2$$

$$\leq t \cosh^2(t p\sigma) \frac{(t p\sigma)^2}{n^2},$$

where we use the bound for $a, x > 0$,

$$\sinh(a + x) - \sinh(a) \leq \sinh'(a + x) x = \cosh(a + x) x.$$

Thus this tends to zero as $n \to \infty$, which establishes the required convergence.

For the uniform bound, it is sufficient to show that for fixed t,

$$\mathbb{E}\left[\sum_{k=0}^{n-1}\int_{kt/n}^{(k+1)t/n} \frac{\cosh((k+1)p\sigma t/n) - \cosh(kp\sigma t/n)}{p\sigma t/n}\,\mathrm{d}Y_s\right]^2 \qquad (6.37)$$

is uniformly bounded in n. We can then use the fact that $\mathbb{E}|Z| < \sqrt{\mathbb{E}Z^2}$, to see that the modulus of the integral is bounded in the L^1 norm and hence in probability. The dependence on ω in this bound arises solely from the process Y; thus considered as a functional of Y, there is a uniform in n bound.

To complete the proof we establish a uniform in n bound on (6.37) using the Itô isometry

$$\mathbb{E}\left[\sum_{k=0}^{n-1}\int_{kt/n}^{(k+1)t/n} \frac{\cosh((k+1)p\sigma t/n) - \cosh(kp\sigma t/n)}{p\sigma t/n}\,\mathrm{d}Y_s\right]^2$$

$$\leq \sum_{k=0}^{n-1}\int_{kt/n}^{(k+1)t/n}\left(\frac{\cosh((k+1)p\sigma t/n) - \cosh(kp\sigma t/n)}{p\sigma t/n}\right)^2\,\mathrm{d}s$$

$$\leq \left(\frac{n}{p\sigma t}\right)^2 \sinh^2(tp\sigma)\left(\frac{p\sigma t}{n}\right)^2$$

$$\leq \sinh^2(tp\sigma).$$

6.9 For the linear filter take $F(x) = ax^2/2 + bx$; computing the normalising constant involves computing for $B > 0$,

$$\int_{-\infty}^{\infty} \exp(-Bx^2 + Ax)\,\mathrm{d}x = e^{A^2/(4B)}\int_{-\infty}^{\infty}\exp\left(-\left(\sqrt{B}x - \frac{A}{2\sqrt{B}}\right)^2\right)\,\mathrm{d}x$$

$$= e^{A^2/(4B)}\sqrt{\pi}/\sqrt{B}. \qquad (6.38)$$

In the case of the linear filter the coefficients $p = \sqrt{a^2/\sigma^2 + h_1^2}$, $q = ab/\sigma^2 + h_1 h_2$ and $r = a + b^2/\sigma^2 + h_2^2$. Thus from the equation for the normalising constant (6.16),

$$A_t = b/\sigma^2 + h_1\Psi_t + \frac{q + p^2 x_0}{p\sigma\sinh(tp\sigma)} - \frac{q}{p\sigma}\coth(tp\sigma),$$

where

$$\Psi_t = \int_0^t \frac{\sinh(sp\sigma)}{\sinh(tp\sigma)}\,\mathrm{d}Y_s$$

and

$$B_t = -\frac{a}{2\sigma^2} + \frac{p\coth(tp\sigma)}{2\sigma}.$$

Since $\coth(x) > 1$ for $x > 0$ and $p \geq a/\sigma$, it follows that $B > 0$ as required. Using the result (6.38) we see that the normalised conditional distribution is given by (6.15),

$$\pi_t(\varphi) = \frac{\sqrt{B_t}}{\sqrt{\pi}} \int_{-\infty}^{\infty} \varphi(x) \exp\left(-\frac{1}{2}\left(\frac{x - A_t/(2B_t)}{1/\sqrt{2B_t}}\right)^2\right) dx,$$

which corresponds to a Gaussian distribution with mean $\hat{x}_t = A_t/(2B_t)$ and variance $R_t = 1/2B_t$. Differentiating

$$\frac{dR_t}{dt} = \frac{p^2}{4\sinh^2(tp\sigma)B_t^2}$$

thus with the aid of the identity $\coth^2(x) - 1 = 1/\sinh^2(x)$, it is easy to check that

$$\frac{dR_t}{dt} = \sigma^2 + 2aR_t - R_t^2 h_1^2$$

which is the one-dimensional form of the Kalman filter covariance equation (6.28).

In one dimension the Kalman filter equation for the conditional mean is

$$d\hat{x}_t = (a\hat{x}_t + b)dt + R_t h_1 dY_t - R_t h_1 (h_1 x + h_2) dt$$

thus to verify that the mean $A_t R_t$ is a solution of this SDE we compute

$$d(A_t R_t) = \frac{A_t R_t^2 p^2}{\sinh^2(tp\sigma)} dt + R_t h_1 dY_t - R_t \coth(tp\sigma) p\sigma (A_t - b/\sigma^2) - qR_t$$

$$= R_t h_1 dY_t + (A_t R_t)\left(R_t p^2 \coth^2(tp\sigma) - p\sigma \coth(tp\sigma) - R_t p^2\right)$$

$$+ \frac{pR_t b}{\sigma} \coth(tp\sigma) - \frac{ab}{\sigma^2} R_t - h_1 h_2 R_t$$

$$= R_t h_1 dY_t - h_1 h_2 R_t + b$$

$$+ (A_t R_t)\left(R_t p^2 \coth^2(tp\sigma) - p\sigma \coth(tp\sigma) - R_t p^2\right)$$

$$= R_t h_1 dY_t - h_1 h_2 R_t + b - R_t h_1^2 (R_t A_t)$$

$$+ (A_t R_t) R_t \left(p^2 \coth^2(tp\sigma) - \frac{p\sigma}{R_t} \coth(tp\sigma) - \frac{a^2}{\sigma^2}\right)$$

$$= R_t h_1 dY_t - h_1 h_2 R_t + b - Rh_1^2(A_t R_t) + (A_t R_t) R_t$$

$$\times \left(p^2 \coth^2(tp\sigma) - p\sigma \coth(tp\sigma)\left(-\frac{a}{\sigma^2} + \frac{p}{\sigma} \coth(tp\sigma)\right) - \frac{a^2}{\sigma^2}\right)$$

$$= R_t h_1 dY_t - R_t h_1 (h_1 A_t R_t + h_2) + (A_t R_t) a + b.$$

Therefore the solution computed explicitly solves the SDEs for the one-dimensional Kalman filter.

6.3 Solutions to Exercises

In the limit as $t \to \infty$, $B_t \to -a/\sigma^2 + p/(2\sigma)$ and $A_t \simeq b/\sigma^2 + h_1\Psi_t - q/(p\sigma)$ and thus the law of the conditional distribution asymptotically for large t is given by

$$N\left(\frac{h_1\Psi_t\sigma^2 + b - q\sigma/p}{p\sigma - a}, \frac{\sigma^2}{p\sigma - a}\right).$$

For the second Beneš filter, from the solution to Exercise 6.1 $p = h_1$, $q = h_1 h_2$ and $r = h_2^2 + \alpha^2$, so

$$Q_t(x) = \left(h_1\Psi_t + \frac{h_2 + h_1 x_0}{\sigma \sinh(t p \sigma)} - \frac{h_2}{\sigma}\coth(t p \sigma)\right)x - \frac{h_1}{2\sigma}\coth(t p \sigma) x^2.$$

In the general case we can take as antiderivative to f

$$F(x) = \frac{\sigma^2}{\alpha}\log\left(e^{2\alpha x/\sigma} + 1/\beta\right) - \sigma x.$$

However, there does not seem to be an easy way to evaluate this integral in general, so consider the specific case where $\beta = 1$ and $\alpha = a$,

$$F(x) = \sigma^2 \log(\cosh(ax/\sigma));$$

thus from (6.16) the normalising constant is

$$c_t = \int_{-\infty}^{\infty} \cosh\left(\frac{ax}{\sigma}\right)$$
$$\times \exp\left(\left(h_1\Psi_t + \frac{h_2 + h_1 x_0}{\sigma \sinh(t p \sigma)} - \frac{h_2}{\sigma}\coth(t p \sigma)\right)x - \frac{h_1}{2\sigma}\coth(t p \sigma) x^2\right) dx,$$

which can be evaluated using two applications of the result (6.38), with

$$B_t \triangleq \frac{h_1}{2\sigma}\coth(t p \sigma),$$

and

$$A_t^\pm \triangleq \pm\frac{a}{\sigma} + h_1\Psi_t + \frac{h_2 + h_1 x_0}{\sigma \sinh(t p \sigma)} - \frac{h_2}{\sigma}\coth(t p \sigma).$$

Thus the normalising constant is given by

$$c_t = \frac{\sqrt{\pi}}{2\sqrt{B_t}}\left(e^{(A_t^+)^2/(4B_t)} + e^{(A_t^-)^2/(4B_t)}\right).$$

Therefore the normalised conditional distribution is given by

160   6 Finite-Dimensional Filters

$$\pi_t(\varphi) = \frac{\sqrt{B_t}}{\sqrt{\pi}} \frac{1}{e^{(A_t^+)^2/(4B_t)} + e^{(A_t^-)^2/(4B_t)}}$$

$$\times \int_{-\infty}^{\infty} \varphi(x) \exp(-B_t x^2) \left(\exp(A_t^+ x) + \exp(A_t^- x) \right) dx$$

$$= \frac{\sqrt{B_t}}{\sqrt{\pi}} \frac{1}{e^{(A_t^+)^2/(4B_t)} + e^{(A_t^-)^2/(4B_t)}}$$

$$\times \left[e^{-(A^+)^2/(4B_t)} \int_{-\infty}^{\infty} \varphi(x) \exp\left(-\frac{1}{2} \left(\frac{x - A_t^+/(2B_t)}{1/\sqrt{2B_t}} \right)^2 \right) dx \right.$$

$$\left. + e^{-(A^-)^2/(4B_t)} \int_{-\infty}^{\infty} \varphi(x) \exp\left(-\frac{1}{2} \left(\frac{x - A_t^-/(2B_t)}{1/\sqrt{2B_t}} \right)^2 \right) dx \right].$$

Thus the normalised conditional distribution is the weighted mixture of two normal distributions, with weight

$$w^{\pm} = \frac{\exp(-(A_t^{\pm})^2/(4B_t))}{\exp((A_t^+)^2/(4B_t)) + \exp((A_t^-)^2/(4B_t))}$$

on a $N(A_t^{\pm}/(2B_t), 1/(2B_t))$ distributed random variable.

In the limit as $t \to \infty$, $B_t \to h_1/(2\sigma)$ and $A_t^{\pm} \simeq \pm a/\sigma + h_1 \Psi_t - h_2/\sigma$ and the asymptotic expressions for the weights become

$$w^{\pm} = 2 \frac{\exp(\pm 2a/(h_1 \Psi_t/\sigma - h_2/\sigma^2))}{\cosh(2a/(h_1 \Psi_t/\sigma - h_2/\sigma^2))}$$

and the distributions $N(\pm a/h_1 + \sigma \Psi_t - h_2/h_1, \sigma/h_1)$.

6.11

i. Setting

$$C_t \triangleq T_0 + \int_0^t \Phi_s^{-1} l_s \, ds + \int_0^t \Phi_s^{-1} F_s \, dU_s, \quad (6.39)$$

and $A_t \triangleq \Phi_t C_t$, where Φ_t is given by (6.21), it follows by integration by parts that

$$dA_t = d\Phi_t \left[T_0 + \int_0^t \Phi_s^{-1} l_s \, dt + \int_0^t \Phi_s^{-1} F_s \, dU_s \right]$$
$$+ \Phi_t \left[\Phi_t^{-1} l_t \, dt + \Phi_t^{-1} F_t \, dU_t \right]$$
$$= L_t A_t + l_t \, dt + F_t \, dU_t$$

which is the SDE for T_t. As $\Phi_0 = \mathbb{I}_{d+m}$, it follows that $A_0 = T_0$. Thus T_t has the representation (6.20).

ii. In this part we continue to use the notation for the process C_t introduced above. It is clearly sufficient to show that $(T_{t_1}, \ldots, T_{t_{n-1}}, T_n)$ has a multivariate-normal distribution, since $T = (X, Y)$. Note that the process

Φ_t is a deterministic matrix-valued process, thus if for fixed t, C_t has a multivariate normal distribution then so does $\Phi_t C_t$.

Since X_0 has a multivariate normal distribution and $Y_0 = 0$, T_0 has a multivariate normal distribution. From the SDE (6.39) it follows that $C_{t_1}, C_{t_2} - C_{t_1}, \ldots, C_t - C_{t_{n-1}}$ are independent random variables, each of which has a multivariate-normal distribution. The result now follows since

$$T_{t_1} = \Phi_{t_1} C_{t_1}$$
$$T_{t_2} = \Phi_{t_2}(C_{t_1} + (C_{t_2} - C_{t_1}))$$
$$\vdots \quad \vdots$$
$$T_t = \Phi_t(C_{t_1} + \cdots + (C_{t_{n-2}} - C_{t_{n-1}}) + (C_t - C_{t_{n-1}})).$$

iii. It follows from (ii) and the fact that the image under a linear map of a multivariate-normal distribution is also multivariate-normal, that for any n and fixed times $0 \leq t_1 \leq \cdots \leq t_{n-1} \leq t$,

$$\left(X_t, \sum_{i=0}^{n-2} K_{t_i} \left(Y_{t_{i+1}} - Y_{t_i} \right) + K_{t_{n-1}} \left(Y_t - Y_{t_{n-1}} \right) \right)$$

has a multivariate-normal distribution. By the usual Itô isometry argument as the mesh of the partition tends to zero, this term converges in L^2 and thus in probability to

$$\left(X_t, \int_0^t K_s \, \mathrm{d}Y_s \right).$$

By a standard result on weak convergence (e.g. Theorem 4.3 of [19]) this convergence in probability implies that the sequence converges weakly; consequently the characteristic functions must also converge. As each element of the sequence is multivariate normal it follows that the limit must be multivariate normal.

6.13

i. The first part follows using the SDE for X and Itô's formula using the local boundedness of f_s, F_s, σ_s. In the case $p = 1$ local boundedness of σ_s implies that the stochastic integral is a martingale, thus using the notation

$$\|F\|_{[0,t]} \triangleq \sup_{0 \leq s \leq t} \max_{i,j=1,\ldots,d} |F_s^{ij}| < \infty, \qquad \|f\|_{[0,t]} \triangleq \sup_{0 \leq s \leq t} \max_{i=1,\ldots,d} |f_s^i| < \infty,$$

we can obtain the following bound

$$\mathbb{E}\|X_t\| \leq \mathbb{E}\|X_0\| + \mathbb{E}\left[\left\| \int_0^t F_s X_s + f_s \, \mathrm{d}s \right\| \right]$$
$$\leq x_0 + td\|f\|_{[0,t]} + d\|F\|_{[0,t]} \int_0^t \|X_s\| \, \mathrm{d}s.$$

Thus from Corollary A.40 to Gronwall's lemma

$$\mathbb{E}\|X_t\| \leq \left(x_0 + td\|f_s\|_{[0,t]}\right)\exp\left(td\|F\|_{[0,t]}\right).$$

Similarly for $p = 2$, use $f(x) = x^\top x$,

$$\mathrm{d}\|X_t\|^2 = 2X_s^\top(F_sX_s + f_s)\mathrm{d}s + 2X_s^\top\sigma\mathrm{d}V_s + \mathrm{tr}(\sigma^\top\sigma)\mathrm{d}s.$$

Let T_n be a reducing sequence for the stochastic integral, which is a local martingale (see Exercise 3.10 for more details). Then

$$\mathbb{E}\|X_{t\wedge T_n}\|^2 = \mathbb{E}\|X_0\|^2 + \mathbb{E}\left[\int_0^{t\wedge T_n} 2X_s^\top(F_sX_s + f_s) + \mathrm{tr}(\sigma^\top\sigma)\,\mathrm{d}s\right]$$

$$\leq \mathbb{E}\|X_0\|^2 + 2d^2\|F\|_{[0,t]}\int_0^t \mathbb{E}[\|X_s\|^2]\,\mathrm{d}s$$

$$+ dt\|f\|_{[0,t]}\sup_{0\leq s\leq t}\mathbb{E}\|X_s\| + td\|\sigma\|_{[0,t]}^2.$$

Using the first moment bound, Gronwall's inequality yields a bound independent of n, thus as $n \to \infty$ Fatou's lemma implies that

$$\sup_{0\leq s\leq t}\mathbb{E}\|X_s\|^2 < \infty.$$

We can proceed by induction to the general case for the pth moment. Apply Itô's formula to $f(x) = x^{p/2}$ for $p \geq 3$ to obtain the pth moment bound; thus

$$\mathrm{d}\|X_t\|^p = p\|X\|^{p-2}\left(2X_t^\top(F_tX_t + f_t)\mathrm{d}t + \mathrm{tr}(\sigma^\top\sigma)\,\mathrm{d}s + 2X_t^\top\sigma\mathrm{d}V_t\right)$$

$$+ \frac{p(p-1)}{2}\|X_t\|^{p-4}(X_t^\top\sigma\sigma^\top X_t)\mathrm{d}t.$$

The stochastic integral is a local martingale and so a reducing sequence T_n can be found. The other terms involve moments of order p, $p-1$ and $p-2$, so the result follows as in the case above from the inductive hypotheses, Gronwall's lemma followed by Fatou's lemma and the fact that all moments of the initial X_0 are finite since it is normally distributed.

ii. For any $s \in [0,\infty)$,

$$\mathbb{E}\left[(\pi_s(|\varphi_i|))^p\right] = \mathbb{E}\left[(\mathbb{E}\left[|\varphi_i(X_s)| \mid \mathcal{Y}_s\right])^p\right] \leq \mathbb{E}\left[\mathbb{E}\left[|\varphi_i(X_s)|^p \mid \mathcal{Y}_s\right]\right]$$
$$= \mathbb{E}\left[|\varphi_i(X_s)|^p\right],$$

where the inequality follows from the conditional form of Jensen's inequality. Therefore from part (i),

$$\sup_{s\in[0,t]}\mathbb{E}\left[(\pi_s(|\varphi_i|))^p\right] \leq \sup_{s\in[0,t]}\mathbb{E}\left[|X_s^i|^p\right] < \infty.$$

For the product term

$$\mathbb{E}\left[(\pi_s(|\varphi_{ij}|))^p\right] = \mathbb{E}\left[\left(\mathbb{E}\left[|X_s^i X_s^j| \mid \mathcal{Y}_s\right]\right)^p\right] \leq \mathbb{E}\left[\mathbb{E}\left[|X_s^i|^p |X_s^j|^p \mid \mathcal{Y}_s\right]\right]$$
$$= \mathbb{E}\left[|X_s^i|^p |X_s^j|^p\right]$$
$$\leq \sqrt{\mathbb{E}[|X_s^i|^{2p}]\, \mathbb{E}\left[|X_s^j|^{2p}\right]} < \infty.$$

7

The Density of the Conditional Distribution of the Signal

The question which we consider in this chapter is whether π_t, the conditional distribution of X_t given the observation σ-algebra $\mathcal{Y}_t$, has a density with respect to a reference measure, in particular with respect to Lebesgue measure. We prove that, under fairly mild conditions, the unnormalised conditional distribution ρ_t, which is the unique solution of the Zakai equation (3.43), has a square integrable density with respect to Lebesgue measure. This automatically implies that π_t has the same property. There are various approaches to answer this question. The approach presented here is that adopted by Kurtz and Xiong in [174]. In the second part of the chapter we discuss the smoothness properties (i.e. the differentiability) of the density of ρ. Finally we show the existence of the dual of the solution of the Zakai equation (see (7.30) below). The dual of ρ plays an important rôle in establishing the rates of convergence of particle approximations to π and ρ which are discussed in more detail in Chapter 9.

In the following, we take the signal X to be the solution of the stochastic differential equation (3.9); that is, $X = (X^i)_{i=1}^d$ is the solution of the stochastic differential equation

$$dX_t = f(X_t)dt + \sigma(X_t)\,dV_t, \tag{7.1}$$

where $f : \mathbb{R}^d \to \mathbb{R}^d$ and $\sigma : \mathbb{R}^d \to \mathbb{R}^{d \times p}$ are bounded and globally Lipschitz (i.e. they satisfy the conditions (3.10)) and $V = (V^j)_{j=1}^p$ is a p-dimensional Brownian motion. The observation process is the solution of the evolution equation (3.5). That is, Y is an m-dimensional stochastic process which satisfies

$$dY_t = h(X_t)\,dt + dW_t,$$

where $h = (h_i)_{i=1}^m : \mathbb{R}^d \to \mathbb{R}^m$ is a bounded measurable function and W is a standard m-dimensional Brownian motion which is independent of X.

In the following we make use of an embedding theorem which we state below. In order to state this theorem, we need a few notions related to Sobolev spaces. Further details on this topic can be found, for example, in Adams [1].

A. Bain, D. Crisan, *Fundamentals of Stochastic Filtering*,
DOI 10.1007/978-0-387-76896-0_7, © Springer Science+Business Media, LLC 2009

7.1 An Embedding Theorem

Let $\alpha = (\alpha_1, \ldots, \alpha_d) \in \mathbb{N}^d$ be an arbitrary multi-index. Given two functions f and $g \in L^p(\mathbb{R}^d)$, we say that $\partial^\alpha f = g$ in the weak sense if for all $\psi \in C_0^\infty(\mathbb{R}^d)$ we have

$$\int_{\mathbb{R}^d} f(x) \partial^\alpha \psi(x) \, dx = (-1)^{|\alpha|} \int_{\mathbb{R}^d} g(x) \psi(x) \, dx. \tag{7.2}$$

We immediately see that if the partial derivative of a function exists in the conventional sense and is continuous up to order $|\alpha|$, integration by parts will yield (7.2). The converse is not true; to see this one can consider, for example, the function $\exp(i/|x|^n)$. Let k be a non-negative integer. The Sobolev space, denoted $W_k^p(\mathbb{R}^d)$, is the space of all functions $f \in L^p(\mathbb{R}^d)$ such that the partial derivatives $\partial^\alpha f$ exist in the weak sense and are in $L^p(\mathbb{R}^d)$ whenever $|\alpha| \leq k$, where α is a multi-index. We endow $W_k^p(\mathbb{R}^d)$ with the norm

$$\|f\|_{k,p} = \left(\sum_{|\alpha| \leq k} \|\partial^\alpha f\|_p^p \right)^{1/p}, \tag{7.3}$$

where $\partial^0 f = f$ and the norms on the right are the usual norms in $L^p(\mathbb{R}^d)$. Then $W_k^p(\mathbb{R}^n)$ is complete with respect to the norm defined by (7.3); hence it is a Banach space. In the following, we make use, without proof, of the following Sobolev-type embedding theorem (for a proof see Adams [1], Saloff-Coste [252], or Stein [256]).

Theorem 7.1. *If $k > d/p$ then there exists a modification of $f \in W_k^p(\mathbb{R}^d)$ on a set of zero Lebesgue measure so that the resulting function is continuous.*

In the following we work mostly with the space $W_k^2(\mathbb{R}^d)$. This space is a Hilbert space with the inner product

$$\langle f, g \rangle_{W_2^p(\mathbb{R}^d)} = \sum_{|\alpha| \leq k} \langle \partial^\alpha f, \partial^\alpha g \rangle,$$

where $\langle \cdot, \cdot \rangle$ is the usual inner product on $L^2(\mathbb{R}^d)$

$$\langle f, g \rangle = \int_{\mathbb{R}^d} f(x) g(x) \, dx.$$

Exercise 7.2. Let $\{\varphi_i\}_{i>0}$ be an orthonormal basis of $L^2(\mathbb{R}^d)$ with the property that $\varphi_i \in C_b(\mathbb{R}^d)$ for all $i > 0$. Let $\mu \in \mathcal{M}(\mathbb{R}^d)$ be a finite measure. Show that if

$$\sum_{i=1}^\infty \mu(\varphi_i)^2 < \infty,$$

then μ is absolutely continuous with respect to Lebesgue measure. Moreover if $g_\mu : \mathbb{R}^d \to \mathbb{R}$ is the density of μ with respect to Lebesgue measure then $g_\mu \in L^2(\mathbb{R}^d)$.

7.1 An Embedding Theorem

The results from below make use of the *regularisation method*. Let ψ be the kernel for the heat equation $(\partial_t u = 1/2 \sum_{i=1}^{d} \partial_i \partial_i u)$, viz

$$\psi_\varepsilon(x) \triangleq (2\pi\varepsilon)^{-d/2} \exp\bigl(-\|x\|^2/2\varepsilon\bigr),$$

and define the convolution operator

$$T_\varepsilon \colon \mathcal{B}(\mathbb{R}^d) \to \mathcal{B}(\mathbb{R}^d)$$
$$T_\varepsilon f(x) \triangleq \int_{\mathbb{R}^d} \psi_\varepsilon(x-y) f(y) \, \mathrm{d}y, \quad x \in \mathbb{R}^d. \tag{7.4}$$

Also define the corresponding operator on the space of finite measures

$$T_\varepsilon \colon \mathcal{M}(\mathbb{R}^d) \to \mathcal{M}(\mathbb{R}^d)$$
$$T_\varepsilon \mu(f) \triangleq \mu(T_\varepsilon f)$$
$$= \int_{\mathbb{R}^d} \int_{\mathbb{R}^d} \psi_\varepsilon(x-y) f(y) \, \mathrm{d}y \, \mu(\mathrm{d}x)$$
$$= \int_{\mathbb{R}^d} f(y) T_\varepsilon \mu(y) \, \mathrm{d}y,$$

where $y \mapsto T_\varepsilon \mu(y)$ is the density of the measure $T_\varepsilon \mu$ with respect to Lebesgue measure, which by the above exists even if μ is not absolutely continuous with respect to Lebesgue measure; furthermore the density is given by

$$T_\varepsilon \mu(y) = \int_{\mathbb{R}^d} \psi_\varepsilon(x-y) \mu(\mathrm{d}x), \quad y \in \mathbb{R}^d.$$

In the following, we use the same notation $T_\varepsilon \mu$ for the regularized measure and its density.

Exercise 7.3. Let μ be a finite measure on $\mathbb{R}^d$ and $|\mu| \in \mathcal{M}(\mathbb{R}^d)$ be its total variation measure. Show that:

i. For any $\varepsilon > 0$ and $g \in L^2(\mathbb{R}^d)$, $\|T_\varepsilon g\|_2 \le \|g\|_2$.
ii. For any $\varepsilon > 0$, $T_\varepsilon \mu \in W_k^2(\mathbb{R}^d)$.
iii. For any $\varepsilon > 0$, $\|T_{2\varepsilon}\mu\|_2 \le \|T_\varepsilon |\mu|\|_2$.

Let μ be a finite signed measure on $\mathbb{R}^d$; then for $f \in C_b(\mathbb{R}^d)$, denote by $f\mu$ the finite signed measure on $\mathbb{R}^d$ which is absolutely continuous with respect to μ and whose density with respect to μ is f.

Exercise 7.4. Let μ be a finite (signed) measure on $\mathbb{R}^d$ and $|\mu| \in \mathcal{M}(\mathbb{R}^d)$ be its total variation measure. Also let $f \in C_b(\mathbb{R}^d)$ be a Lipschitz continuous and bounded function. Denote by $k_f \triangleq \sup_{x \in \mathbb{R}^d} |f(x)|$ and let k_f' be the Lipschitz constant of f. Show that:

i. For any $\varepsilon > 0$, $\|T_\varepsilon f\mu\|_2 \le k_f \|T_\varepsilon |\mu|\|_2$.
ii. For any $\varepsilon > 0$ and $i = 1, \ldots, d$, we have $\bigl|\langle T_\varepsilon \mu, f \partial^i T_\varepsilon \mu\rangle\bigr| \le \tfrac{1}{2} k_f' \|T_\varepsilon |\mu|\|_2^2$.
iii. For any $\varepsilon > 0$ and $i = 1, \ldots, d$, we have $\|f \partial^i T_\varepsilon \mu - \partial^i T_\varepsilon f\mu\|_2 \le 2^{d/2+2} k_f' \|T_{2\varepsilon} |\mu|\|_2$.

7.2 The Existence of the Density of ρ_t

In this section we prove that the unnormalised conditional distribution ρ_t is absolutely continuous with respect to Lebesgue measure and its density is square integrable. We start with two technical lemmas.

We require a set of functions $\{\varphi_i\}_{i\geq 1}$, where $\varphi \in C_b^2(\mathbb{R}^d)$, such that these functions form an orthonormal basis of the space $L^2(\mathbb{R}^d)$. There are many methods to construct such a basis. One of the most straightforward ones is to use wavelets (see, e.g. [224]). For any orthonormal basis of $L^2(\mathbb{R}^d)$ and arbitrary $f \in L^2(\mathbb{R}^d)$,

$$f = \sum_{i=1}^{\infty} \langle f, \varphi_i \rangle \varphi_i,$$

so

$$\|f\|_2^2 = \sum_{i=1}^{\infty} \langle f, \varphi_i \rangle^2 \|\varphi_i\|_2^2 = \sum_{i=1}^{\infty} \langle f, \varphi_i \rangle^2.$$

The function $\psi_\varepsilon(x)$ decays to zero as $\|x\| \to \infty$, therefore for $\varphi \in C_b^1(\mathbb{R}^d)$, using the symmetry of $\psi_\varepsilon(x-y)$ and integration by parts

$$\partial^i T_\varepsilon \varphi = \frac{\partial}{\partial x_i} \int_{\mathbb{R}^d} \psi_\varepsilon(x-y) \varphi(y) \, dy = \int_{\mathbb{R}^d} \frac{\partial}{\partial x_i} \psi_\varepsilon(x-y) \varphi(y) \, dy$$

$$= -\int_{\mathbb{R}^d} \frac{\partial}{\partial y_i} \psi_\varepsilon(x-y) \varphi(y) \, dy = \int_{\mathbb{R}^d} \psi_\varepsilon(x-y) \frac{\partial \varphi(y)}{\partial y^i} \, dy$$

$$= T_\varepsilon(\partial^i \varphi).$$

Lemma 7.5. *Let A be a generator of the form*

$$A\varphi = \sum_{i,j=1}^{d} a^{ij} \frac{\partial^2 \varphi}{\partial x^i \partial x^j} + \sum_{i=1}^{d} f^i \frac{\partial \varphi}{\partial x^i}, \quad \varphi \in \mathcal{D}(A) \subset C_b(\mathbb{R}^d), \tag{7.5}$$

where the matrix a is defined as in (3.12); that is, $a = \frac{1}{2}\sigma\sigma^\top$. Let $\{\varphi_i\}_{i>0}$ be any orthonormal basis of $L^2(\mathbb{R}^d)$ with the property that $\varphi_i \in C_b^2(\mathbb{R}^d)$ for all $i > 0$. Then

$$\sum_{k=1}^{\infty} \rho_s(A(T_\varepsilon \varphi_k))^2 \leq d \sum_{i=1}^{d} \|\partial^i T_\varepsilon(f^i \rho_s)\|_2^2 + d^2 \sum_{i,j=1}^{d} \|\partial^i \partial^j T_\varepsilon(a^{ij} \rho_s)\|_2^2. \tag{7.6}$$

In particular, if

$$k_f = \max_{i=1,\ldots,d} \sup_{x \in \mathbb{R}^d} |f^i(x)| < \infty$$

$$k_a = \max_{i,j=1,\ldots,d} \sup_{x \in \mathbb{R}^d} |a^{ij}(x)| < \infty,$$

then there exists a constant $k = k(f, a, \varepsilon, d)$ such that

7.2 The Existence of the Density of ρ_t

$$\sum_{k=1}^{\infty} \rho_s(A(T_\varepsilon \varphi_k))^2 \le k \|T_\varepsilon \rho_s\|_2^2.$$

Proof. For any $i \ge 0$, for $\varphi \in C_b^2(\mathbb{R}^d)$, integration by parts yields

$$\rho_s(f^i \partial^i T_\varepsilon \varphi) = \rho_s(f^i T_\varepsilon \partial^i \varphi) = (f^i \rho_s)(T_\varepsilon \partial^i \varphi) = \langle \partial^i \varphi, T_\varepsilon(f^i \rho_s) \rangle$$
$$= -\langle \varphi, \partial^i T_\varepsilon(f^i \rho_s) \rangle \qquad (7.7)$$

and

$$\rho_s(a^{ij} \partial^i \partial^j T_\varepsilon \varphi) = \rho_s(a^{ij} T_\varepsilon \partial^i \partial^j \varphi) = (a^{ij} \rho_s)(T_\varepsilon \partial^i \partial^j \varphi) = \langle \partial^i \partial^j \varphi, T_\varepsilon(a^{ij} \rho_s) \rangle$$
$$= \langle \varphi, \partial^i \partial^j T_\varepsilon(a^{ij} \rho_s) \rangle. \qquad (7.8)$$

Thus using (7.7) and (7.8),

$$\rho_s(A(T_\varepsilon \varphi_k)) = -\sum_{i=1}^d \langle \varphi_k, \partial^i T_\varepsilon(f^i \rho_s) \rangle + \sum_{i,j=1}^d \langle \varphi_k, \partial^i \partial^j T_\varepsilon(a^{ij} \rho_s) \rangle, \qquad (7.9)$$

from which inequality (7.6) follows. Then

$$|\partial^i T_\varepsilon(f^i \rho_s)(x)| \le \left| \int_{\mathbb{R}^d} \frac{|x_i - y_i|}{\varepsilon} \psi_\varepsilon(x-y)(f^i \rho_s)(dy) \right|$$
$$\le 2^{d/2} k_f \int_{\mathbb{R}^d} \frac{|x_i - y_i|}{\varepsilon} \exp\left(-\frac{\|x-y\|^2}{4\varepsilon}\right) \psi_{2\varepsilon}(x-y) \rho_s(dy)$$
$$\le \frac{2^{d/2} k_f}{\sqrt{\varepsilon}} T_{2\varepsilon} \rho_s(x),$$

where the last inequality follows as $\sup_{t \ge 0} t \exp(-t^2/4) = \sqrt{2/e} < 1$. For the second term in (7.9) we can construct a similar bound

$$|\partial^i \partial^j T_\varepsilon(a^{ij} \rho_s)(x)| \le \left| \int_{\mathbb{R}^d} \left(\frac{(x_i - y_i)(x_j - y_j)}{\varepsilon^2} - \frac{1_{i=j}}{\varepsilon} \right) \psi_\varepsilon(x-y)(a^{ij} \rho_s)(dy) \right|$$
$$\le 2^{d/2} k_a \int_{\mathbb{R}^d} \left(\frac{\|x-y\|^2}{\varepsilon^2} + \frac{1}{\varepsilon} \right)$$
$$\times \exp\left(-\frac{\|x-y\|^2}{4\varepsilon}\right) \psi_{2\varepsilon}(x-y) \rho_s(dy)$$
$$\le 2^{d/2} k_a (2 + 1/\varepsilon) T_{2\varepsilon} \rho_s(x),$$

where we used the fact that $\sup_{t \ge 0} t e^{-t/4} = 4/e < 2$. The lemma then follows using part iii. of Exercise 7.3. $\square$

Lemma 7.6. *Let k'_σ be the Lipschitz constant of the function σ, where $a = \frac{1}{2} \sigma \sigma^\top$. Then we have*

$$\sum_{i,j=1}^{d} \langle T_\varepsilon \rho_s, \partial^i \partial^j T_\varepsilon(a^{ij}\rho_s)\rangle + \frac{1}{2}\sum_{k=1}^{p}\left\|\sum_{i=1}^{d}\partial^i T_\varepsilon(\sigma^{ik}\rho_s)\right\|_2^2$$
$$\leq 2^{d/2+3}d^2 p(k'_\sigma)^2 \|T_\varepsilon \rho_s\|_2^2. \quad (7.10)$$

Proof. First let us note that

$$\langle T_\varepsilon \rho_s, \partial^i \partial^j T_\varepsilon(a^{ij}\rho_s)\rangle$$
$$= \int_{\mathbb{R}^d}\int_{\mathbb{R}^d}\psi_\varepsilon(x-y)\rho_s(\mathrm{d}y)\int_{\mathbb{R}^d}\frac{\partial^2}{\partial x_i \partial x_j}\psi_\varepsilon(x-z)a^{ij}(z)\rho_s(\mathrm{d}z)\,\mathrm{d}x$$
$$= \int_{\mathbb{R}^d}\int_{\mathbb{R}^d}\Theta(y,z)a^{ij}(z)\rho_s(\mathrm{d}y)\rho_s(\mathrm{d}z)$$
$$= \int_{\mathbb{R}^d}\int_{\mathbb{R}^d}\Theta(y,z)\frac{a^{ij}(z)+a^{ij}(y)}{2}\rho_s(\mathrm{d}y)\rho_s(\mathrm{d}z), \quad (7.11)$$

where the last equality follows from the symmetry in z and y, and where

$$\Theta(y,z) \triangleq \int_{\mathbb{R}^d}\psi_\varepsilon(x-y)\frac{\partial^2}{\partial x_i \partial x_j}\psi_\varepsilon(x-z)\,\mathrm{d}x$$
$$= \frac{\partial^2}{\partial z_i \partial z_j}\int_{\mathbb{R}^d}\psi_\varepsilon(x-z)\psi_\varepsilon(x-y)\,\mathrm{d}x$$
$$= \frac{\partial^2}{\partial z_i \partial z_j}\psi_{2\varepsilon}(z-y)$$
$$= \left(\frac{(z_i-y_i)(z_j-y_j)}{4\varepsilon^2} - \frac{1_{\{i=j\}}}{2\varepsilon}\right)\psi_{2\varepsilon}(z-y).$$

Then by integration by parts and the previous calculation we get that

$$\langle \partial^i T_\varepsilon(\sigma^{ik}\rho_s), \partial^j T_\varepsilon(\sigma^{jk}\rho_s)\rangle$$
$$= -\langle T_\varepsilon(\sigma^{ik}\rho_s), \partial^i \partial^j T_\varepsilon(\sigma^{jk}\rho_s)\rangle$$
$$= -\int_{\mathbb{R}^d}\int_{\mathbb{R}^d}\Theta(y,z)\frac{\sigma^{ik}(y)\sigma^{jk}(z)+\sigma^{ik}(z)\sigma^{jk}(y)}{2}\rho_s(\mathrm{d}y)\rho_s(\mathrm{d}z). \quad (7.12)$$

Combining (7.11) and (7.12) summing over all the indices, and using the fact that $a = \sigma\sigma^\top$, the left-hand side of (7.10) is equal to

$$\frac{1}{2}\int_{\mathbb{R}^d}\int_{\mathbb{R}^d}\Theta(y,z)\sum_{k=1}^{p}\sum_{i,j=1}^{d}\left(\sigma^{ik}(y)-\sigma^{ik}(z)\right)\left(\sigma^{jk}(y)-\sigma^{jk}(z)\right)\rho_s(\mathrm{d}y)\rho_s(\mathrm{d}z)$$

and hence using the Lipschitz property of σ,

$$\sum_{i,j=1}^{d} \langle T_\varepsilon \rho_s, \partial^i \partial^j T_\varepsilon(a^{ij}\rho_s) \rangle + \sum_{k=1}^{p} \left\| \sum_{i=1}^{d} \partial^i T_\varepsilon(\sigma^{ik}\rho_s) \right\|_2^2$$
$$\leq \frac{d^2 p}{2}(k'_\sigma)^2 \int_{\mathbb{R}^d} \int_{\mathbb{R}^d} \|y - z\|^2 \Theta(y, z) \rho_s(\mathrm{d}y) \rho_s(\mathrm{d}z).$$

It then follows that

$$\|y - z\|^2 |\Theta(y, z)| \leq 2^{d/2} \|y - z\|^2 \psi_{4\varepsilon}(z - y)$$
$$\times \left(\frac{\|z - y\|^2}{4\varepsilon^2} + \frac{1}{2\varepsilon} \right) \exp\left(-\frac{\|z - y\|^2}{8\varepsilon} \right)$$
$$\leq 2^{d/2+5} \psi_{4\varepsilon}(z - y),$$

where the final inequality follows by setting $x = \|y-z\|^2/(2\varepsilon)$ in the inequality

$$\sup_{x \geq 0}(x^2 + x)\exp(-x/4) < 2^5.$$

Hence the left-hand side of (7.10) is bounded by

$$2^{d/2+3} d^2 p(k'_\sigma)^2 \|T_{2\varepsilon}\rho_s\|_2^2 \leq 2^{d/2+3} d^2 p(k'_\sigma)^2 \|T_\varepsilon \rho_s\|_2^2,$$

the final inequality being a consequence of Exercise 7.3, part (iii). □

Proposition 7.7. *If the function h is uniformly bounded, then there exists a constant c depending only on the functions f, σ and h and such that for any $\varepsilon > 0$ and $t \geq 0$ we have*

$$\tilde{\mathbb{E}}\left[\|T_\varepsilon \rho_t\|_2^2 \right] \leq \|T_\varepsilon \pi_0\|_2^2 + c \int_0^t \tilde{\mathbb{E}}\left[\|T_\varepsilon \rho_s\|_2^2 \right] \mathrm{d}s.$$

Proof. For any $t \geq 0$ and φ_i an element of an orthonormal basis of $L^2(\mathbb{R}^d)$ chosen so that $\varphi_i \in C_b(\mathbb{R}^d)$ we have from the Zakai equation using the fact that $\rho_t(T_\varepsilon \varphi_i) = T_\varepsilon \rho_t(\varphi_i)$,

$$T_\varepsilon \rho_t(\varphi_i) = T_\varepsilon \pi_0(\varphi_i) + \int_0^t \rho_s(A(T_\varepsilon \varphi_i))\,\mathrm{d}s + \sum_{j=1}^m \int_0^t \rho_s(h^j T_\varepsilon \varphi_i)\,\mathrm{d}Y_s^j$$

and by Itô's formula

$$(T_\varepsilon \rho_t(\varphi_i))^2 = (T_\varepsilon \pi_0(\varphi_i))^2 + 2\int_0^t T_\varepsilon \rho_s(\varphi_i) \rho_s(A(T_\varepsilon \varphi_i))\,\mathrm{d}s$$
$$+ 2\sum_{j=1}^m \int_0^t T_\varepsilon \rho_s(\varphi_i) \rho_s(h^j T_\varepsilon \varphi_i)\,\mathrm{d}Y_s^j$$
$$+ \sum_{j=1}^m \int_0^t \left(\rho_s(h^j T_\varepsilon \varphi_i) \right)^2 \mathrm{d}s.$$

The stochastic integral term in the above identity is a martingale, hence its expectation is 0. By taking expectation and using Fatou's lemma we get that

$$\tilde{\mathbb{E}}\left[\|T_\varepsilon\rho_t\|_2^2\right] \le \liminf_{n\to\infty} \tilde{\mathbb{E}}\left[\sum_{i=1}^n (T_\varepsilon\rho_t(\varphi_i))^2\right]$$

$$\le \|T_\varepsilon\pi_0\|_2^2$$

$$+ \liminf_{n\to\infty} \sum_{i=1}^n \tilde{\mathbb{E}}\left[\int_0^t \left(2T_\varepsilon\rho_s(\varphi_i)\rho_s\left(A\left(T_\varepsilon\varphi_i\right)\right)\right.\right.$$

$$\left.\left.+ \sum_{j=1}^m \left(\rho_s(h^j T_\varepsilon\varphi_i)\right)^2\right) ds\right]. \quad (7.13)$$

By applying the inequality $|ab| \le (a^2 + b^2)/2$,

$$\sum_{i=1}^n \tilde{\mathbb{E}}\left[\int_0^t |T_\varepsilon\rho_s(\varphi_i)\rho_s(A\left(T_\varepsilon\varphi_i\right))| \, ds\right]$$

$$\le \frac{1}{2}\tilde{\mathbb{E}}\left[\int_0^t \sum_{i=1}^n (T_\varepsilon\rho_s(\varphi_i))^2 \, ds\right] + \frac{1}{2}\tilde{\mathbb{E}}\left[\int_0^t \sum_{i=1}^n (\rho_s(A(T_\varepsilon\varphi_i)))^2 \, ds\right].$$

Thus using the bound of Lemma 7.5, it follows that uniformly in $n \ge 0$,

$$\sum_{i=1}^n \tilde{\mathbb{E}}\left[\int_0^t |T_\varepsilon\rho_s(\varphi_i)\rho_s(A\left(T_\varepsilon\varphi_i\right))| \, ds\right] \le \frac{1+k}{2}\int_0^t \tilde{\mathbb{E}}\left[\|T_\varepsilon\rho_s\|_2^2\right] ds.$$

For the second part of the last term on the right-hand side of (7.13) for any $n \ge 0$,

$$\sum_{i=1}^n \tilde{\mathbb{E}}\left[\sum_{j=1}^m \int_0^t \left(\rho_s(h^j T_\varepsilon\varphi_i)\right)^2 ds\right] \le mk_h^2 \int_0^t \tilde{\mathbb{E}}\left[\|T_\varepsilon\rho_s\|_2^2\right] ds,$$

where

$$k_h \triangleq \max_{j=1,\dots,m} \sup_{x\in\mathbb{R}^d} |h^j(x)|.$$

As a consequence, there exists a constant $\bar{k} = \bar{k}(f, a, h, \varepsilon, d, m)$ such that

$$\tilde{\mathbb{E}}\left[\|T_\varepsilon\rho_t\|_2^2\right] \le \|T_\varepsilon\pi_0\|_2^2 + \bar{k}\int_0^t \tilde{\mathbb{E}}\left[\|T_\varepsilon\rho_s\|_2^2\right] ds;$$

hence by Corollary A.40 to Gronwall's lemma

$$\tilde{\mathbb{E}}\left[\|T_\varepsilon\rho_t\|_2^2\right] \le \|T_\varepsilon\pi_0\|_2^2 e^{\bar{k}t},$$

thus

$$\int_0^t \tilde{\mathbb{E}}\left[\|T_\varepsilon \rho_s\|_2^2\right] \, \mathrm{d}s \le \frac{\|T_\varepsilon \pi_0\|_2^2}{\bar{k}} e^{\bar{k}t} < \infty,$$

where we used Exercise 7.3 part (ii) to see that $\|T_\varepsilon \pi_0\|_2^2 < \infty$. Thus as a consequence of the dominated convergence theorem in (7.13) the limit can be exchanged with the integral and expectation (which is a double integral). From (7.9), using $\langle f, g \rangle = \sum_{i=1}^\infty \langle f, \varphi_i \rangle \langle g, \varphi_i \rangle$, we then get that

$$\tilde{\mathbb{E}}\left[\|T_\varepsilon \rho_t\|_2^2\right] \le \|T_\varepsilon \pi_0\|_2^2 + 2\sum_{i=1}^d \int_0^t \tilde{\mathbb{E}}\left[\langle T_\varepsilon \rho_s, \partial^i T_\varepsilon f^i \rho_s \rangle\right] \, \mathrm{d}s$$

$$+ \sum_{i,j=1}^d \int_0^t \tilde{\mathbb{E}}\left[\langle T_\varepsilon \rho_s, \partial^i \partial^j T_\varepsilon a^{ij} \rho_s \rangle\right] \, \mathrm{d}s$$

$$+ \sum_{j=1}^m \int_0^t \tilde{\mathbb{E}}\left[\|T_\varepsilon h^j \rho_s\|_2^2\right] \, \mathrm{d}s. \quad (7.14)$$

From Exercise 7.4 parts (ii) and (iii), we obtain

$$|\langle T_\varepsilon \rho_s, \partial^i T_\varepsilon f^i \rho_s \rangle| \le |\langle T_\varepsilon \rho_s, f^i \partial^i T_\varepsilon \rho_s \rangle| + |\langle T_\varepsilon \rho_s, \partial^i T_\varepsilon(f^i \rho_s) - f^i \partial^i T_\varepsilon \rho_s \rangle|$$

$$\le \frac{1}{2} k'_f \|T_\varepsilon \rho_s\|_2^2 + 2^{d/2+2} k'_f \|T_\varepsilon \rho_s\|_2 \|T_{2\varepsilon} \rho_s\|_2. \quad (7.15)$$

Since the function h is uniformly bounded, it follows that

$$\|T_\varepsilon(h^j \rho_t)\|_2^2 \le k_h^2 \|T_\varepsilon \rho_t\|_2^2, \quad j = 1, \ldots, m. \quad (7.16)$$

The proposition follows now by bounding the terms on the right-hand side of (7.14) using (7.10) for the third term, (7.15) for the second term and (7.16) for the fourth term. □

Theorem 7.8. *If π_0 is absolutely continuous with respect to Lebesgue measure with a density which is in $L^2(\mathbb{R}^d)$ and the sensor function h is uniformly bounded, then almost surely ρ_t has a density with respect to Lebesgue measure and this density is square integrable.*

Proof. In view of Exercise 7.2, it is sufficient to show that

$$\tilde{\mathbb{E}}\left[\sum_{i=1}^\infty \rho_t(\varphi_i)^2\right] < \infty,$$

where $\{\varphi_i\}_{i>0}$ is an orthonormal basis of $L^2(\mathbb{R}^d)$ with the property that $\varphi_i \in C_b(\mathbb{R}^d)$ for all $i > 0$. From Proposition 7.7, Corollary A.40 to Gronwall's lemma and Exercise 7.3 part (iii) we get that,

$$\sup_{\varepsilon > 0} \tilde{\mathbb{E}}\left[\|T_\varepsilon \rho_t\|_2^2\right] \le e^{ct} \|\pi_0\|_2^2. \quad (7.17)$$

Hence, by Fatou's lemma

$$\tilde{\mathbb{E}}\left[\sum_{i=1}^{\infty}(\rho_t(\varphi_i))^2\right] = \tilde{\mathbb{E}}\left[\lim_{\varepsilon \to 0}\sum_{i=1}^{\infty}(T_\varepsilon \rho_t(\varphi_i))^2\right]$$
$$\leq \liminf_{\varepsilon \to 0}\tilde{\mathbb{E}}\left[\|T_\varepsilon \rho_t\|_2^2\right]$$
$$\leq e^{ct}\|\pi_0\|_2^2 < \infty,$$

hence the result. □

Corollary 7.9. *If π_0 is absolutely continuous with respect to Lebesgue measure with a density which is in $L^2(\mathbb{R}^d)$ and the sensor function h is uniformly bounded, then almost surely π_t has a density with respect to Lebesgue measure and this density is square integrable.*

Proof. Immediate from Theorem 7.8 and the fact that π_t is the normalised version of ρ_t. □

7.3 The Smoothness of the Density of ρ_t

So far we have proved that ρ_t has a density in $L^2(\mathbb{R}^d)$. The above proof has the advantage that the conditions on the coefficients are fairly minimal. In particular, the diffusion matrix a is not required to be strictly positive. From (7.17) we get that

$$\sup_{\varepsilon > 0}\tilde{\mathbb{E}}\left[\|T_\varepsilon \rho_t\|_2\right] < \infty.$$

Since, for example, the sequence $(\|T_{2^{-n}}\rho_t\|_2)_{n>0}$ is non-decreasing (see part (iii) of Exercise 7.3), by Fatou's lemma, this implies that

$$\sup_{n > 0}\|T_{2^{-n}}\rho_t\|_2 < \infty.$$

This implies that $T_{2^{-n}}\rho_t$ belongs to a finite ball in $L^2(\mathbb{R}^d)$. But $L^2(\mathbb{R}^d)$ and in general any Sobolev space $W_k^p(\mathbb{R}^d)$ with $p \in (1, \infty)$ has the property that its balls are weakly sequentially compact (as Banach spaces, they are reflexive; see, for instance, Adams [1]). In particular, this implies that the sequence $T_{2^{-n}}\rho_t$ has a weakly convergent subsequence. So ρ_t, the (weak) limit of the convergent subsequence of $T_{2^{-n}}\rho_t$ must be in $L^2(\mathbb{R}^d)$ almost surely. Similarly, if we can prove the stronger result

$$\sup_{\varepsilon > 0}\tilde{\mathbb{E}}\left[\|T_\varepsilon \rho_t\|_{W_k^p(\mathbb{R}^d)}\right] < \infty, \qquad (7.18)$$

then, by the same argument, we can get that the density of ρ_t belongs to $W_k^p(\mathbb{R}^d)$. Moreover by Theorem 7.1, if $k > d/p$ then the density of ρ_t is continuous (more precisely it has a continuous modification with which we can identify it) and bounded. Furthermore, if $k > d/p+n$, not just the density of ρ_t but also all of its partial derivatives up to order n are continuous and bounded. To obtain (7.18) we require additional smoothness conditions imposed on the coefficients f, σ and h and we also need π_0 to have a density that belongs to $W_k^p(\mathbb{R}^d)$. We need to analyse the evolution equation not just of $T_\varepsilon \rho_t$ but also that of all of its partial derivatives up to the required order k. Unfortunately, the analysis becomes too involved to be covered here. The following exercise should provide a taster of what would be involved if we were to take this route.

Exercise 7.10. Consider the case where $d = m = 1$ and let $\{z_t^\varepsilon, t \geq 0\}$ be the measure-valued process (signed measures) whose density is the spatial derivative of $T_\varepsilon \rho_t$. Show that

$$\tilde{\mathbb{E}}\left[\|z_t^\varepsilon\|_2^2\right] \leq \|(T_\varepsilon \pi_0)'\|_2^2 - 2\int_0^t \tilde{\mathbb{E}}\left[\langle z_s^\varepsilon, (T_\varepsilon f \rho_s)''\rangle\right] ds$$
$$- \int_0^t \tilde{\mathbb{E}}\left[\langle z_s^\varepsilon, (T_\varepsilon a \rho_s)'''\rangle\right] ds + \int_0^t \tilde{\mathbb{E}}\left[\|(T_\varepsilon h \rho_s)'\|_2^2\right] ds.$$

A much cleaner approach, but just as lengthy, is to recast the Zakai equation in its strong form. Heuristically, if the unconditional distribution of the signal ρ_t has a density p_t with respect to Lebesgue measure for all $t \geq 0$ and p_t is 'sufficiently nice' then from (3.43) we get that

$$\rho_t(\varphi) = \int_{\mathbb{R}^d} \varphi(x) p_t(x) \, dx$$
$$= \int_{\mathbb{R}^d} \varphi(x) \left(p_0(x) + \int_0^t A^* p_s(x) \, ds + \int_0^t h^\top(x) p_s(x) \, dY_s \right) dx. \quad (7.19)$$

In (7.19), φ is a bounded function of compact support with bounded first and second derivatives and A^* is the adjoint of the operator A, where

$$A\varphi = \sum_{i,j=1}^d a^{ij} \frac{\partial^2 \varphi}{\partial x^i \partial x^j} + \sum_{i=1}^d f^i \frac{\partial \varphi}{\partial x^i}$$

$$A^*\varphi = \sum_{i,j=1}^d \frac{\partial^2}{\partial x_i \partial x_j}(a^{ij}\varphi) - \sum_{i=1}^d \frac{\partial}{\partial x_i}(f^i \varphi)$$

and for suitably chosen functions ψ, φ (e.g. $\psi, \varphi \in W_2^2(\mathbb{R}^d)$),[†]

$$\langle A^*\psi, \varphi\rangle = \langle \psi, A\varphi\rangle.$$

It follows that it is natural to look for a solution of the stochastic partial differential equation

---
[†] We also need f to be differentiable and a to be twice differentiable.

$$p_t(x) = p_0(x) + \int_0^t A^* p_s(x) \, ds + \int_0^t h^\top(x) p_s(x) \, dY_s, \qquad (7.20)$$

in a suitably chosen function space. It turns out that a suitable function space within which we can study (7.20) is the Hilbert space $W_k^2(\mathbb{R}^d)$. A multitude of difficulties arise when studying (7.20): the stochastic integral in (7.20) needs to be redefined as a Hilbert space operator, the operator A^* has to be rewritten in its divergence form and the solution of (7.20) needs further explanations in terms of measurability, continuity and so on. A complete analysis of (7.20) is contained in Rozovskii [250]. The following two results are immediate corollaries of Theorem 1, page 155 and, respectively, Corollary 1, page 156 in [250] (see also Section 6.2, page 229). We need to assume the following.

C1. The matrix-valued function a is uniformly strictly elliptic. That is, there exists a constant c such that $\xi^\top a \xi \geq c \|\xi\|^2$ for any $x, \xi \in \mathbb{R}^d$ such that $\xi \neq 0$.
C2. For all $i, j = 1, \ldots, d$, $a_{ij} \in C_b^{k+2}(\mathbb{R}^d)$, $f_i \in C_b^{k+1}(\mathbb{R}^d)$ and for all $i = 1, \ldots, m$, we have $h_i \in C_b^{k+1}(\mathbb{R}^d)$.
C3. $p_0 \in W_k^r(\mathbb{R}^d)$, $r \geq 2$.

Theorem 7.11. *Under the assumptions C1–C3 there exists a unique $\mathcal{Y}_t$-adapted process $p = \{p_t, \, t \geq 0\}$, such that $p_t \in W_k^2(\mathbb{R}^d)$ and p is a solution of the stochastic PDE (7.20). Moreover there exists a constant $c = c(k, r, t)$ such that*

$$\tilde{\mathbb{E}}\left[\sup_{0 \leq s \leq t} \|p_s\|_{W_k^r(\mathbb{R}^d)}^{r'}\right] \leq c \|p_0\|_{W_k^r(\mathbb{R}^d)}^{r'}, \qquad (7.21)$$

where r' can be chosen to be either 2 or r.

Theorem 7.12. *Under the assumptions C1–C3, if $n \in \mathbb{N}$ is given and $(k-n)r > d$, then $p = \{p_t, \, t \geq 0\}$; the solution of (7.20) has a unique modification with the following properties.*

1. *For every $x \in \mathbb{R}^d$, $p_t(x)$ is a real-valued $\mathcal{Y}_t$-adapted process.*
2. *Almost surely, $(t, x) \to p_t(x)$ is jointly continuous over $[0, \infty) \times \mathbb{R}^d$ and is continuously differentiable up to order n in the space variable. Both p_t and its partial derivatives are continuous bounded functions.*
3. *There exists a constant $c = c(k, n, r, t)$ such that*

$$\tilde{\mathbb{E}}\left[\sup_{s \in [0,t]} \|p_s\|_{n,\infty}^r\right] \leq c \|p_0\|_{W_k^r(\mathbb{R}^d)}^r. \qquad (7.22)$$

Remark 7.13. The inequality (7.21) implies that, almost surely, p_t belongs to the subspace $W_k^r(\mathbb{R}^d)$ or $W_k^2(\mathbb{R}^d)$. However, the definition of the solution of (7.20) requires the Hilbert space structure of $W_k^2(\mathbb{R}^d)$ which is why the conclusion of Theorem 7.11 is that p is a $W_k^2(\mathbb{R}^d)$-valued process.

Let now $\tilde{\rho}_t$ be the measure which is absolutely continuous with respect to Lebesgue measure with density p_t. For the following exercise, use the fact that the stochastic integral appearing on the right-hand side of the stochastic partial differential equation (7.20) is defined as the unique $L^2(\mathbb{R}^d)$-valued stochastic process $M = \{M_t,\ t \geq 0\}$ satisfying

$$\langle M_t, \varphi \rangle = \int_0^t \langle p_s h^\top, \varphi \rangle \, dY_s, \quad t \geq 0 \tag{7.23}$$

for any $\varphi \in L^2(\mathbb{R}^d)$ (see Chapter 2 in Rozovskii [250] for details).

Exercise 7.14. Show that $\tilde{\rho} = \{\tilde{\rho}_t,\ t \geq 0\}$ satisfies the Zakai equation (3.43); that is for any test function $\varphi \in C_k^2(\mathbb{R}^d)$,

$$\tilde{\rho}_t(\varphi) = \pi_0(\varphi) + \int_0^t \tilde{\rho}_s(A\varphi) \, ds + \int_0^t \tilde{\rho}_s(\varphi h^\top) \, dY_s. \tag{7.24}$$

Even though we proved that $\tilde{\rho}$ satisfies the Zakai equation we cannot conclude that it must be equal to ρ based on the uniqueness theorems proved in Chapter 4. This is because the measure-valued process $\tilde{\rho}$ does not a priori belong to the class of processes within which we proved uniqueness for the solution of the Zakai equation. In particular, we do not know if $\tilde{\rho}$ has finite mass (i.e. $\tilde{\rho}(\mathbf{1})$ may be infinite), so the required inequalities (4.4), or (4.37) may not be satisfied. Instead we use the same approach as that adopted in Section 4.1.

Exercise 7.15. Let $\varepsilon_t \in S_t$ where S_t is the set defined in Corollary B.40; that is,

$$\varepsilon_t = \exp\left(i \int_0^t r_s^\top \, dY_s + \frac{1}{2} \int_0^t \|r_s\|^2 \, ds \right),$$

where $r \in C_b^p([0,t], \mathbb{R}^m)$. Then show that

$$\tilde{\mathbb{E}}[\varepsilon_t \tilde{\rho}_t(\varphi_t)] = \pi_0(\varphi_0) + \tilde{\mathbb{E}}\left[\int_0^t \varepsilon_s \tilde{\rho}_s \left(\frac{\partial \varphi_s}{\partial s} + A\varphi_s + i\varphi_s h^\top r_s \right) ds \right], \tag{7.25}$$

for any $\varphi \in C_b^{1,2}([0,t] \times \mathbb{R}^d)$, such that for any $t \geq 0$, $\varphi \in W_2^2(\mathbb{R}^d)$ and

$$\sup_{s \in [0,t]} \|\varphi_s\|_{W_2^2(\mathbb{R}^d)} < \infty. \tag{7.26}$$

Proposition 7.16. *Under the assumptions C1–C3, for any $\psi \in C_k^\infty(\mathbb{R}^d)$ we have, almost surely,*

$$\tilde{\rho}_t(\psi) = \rho_t(\psi), \quad \tilde{\mathbb{P}}\text{-a.s.}$$

Proof. Since all coefficients are now bounded and a is not degenerate there exists a (unique) function $\varphi \in C_b^{1,2}([0,t] \times \mathbb{R}^d)$ which solves the parabolic PDE (4.14); that is,

$$\frac{\partial \varphi_s}{\partial s} + A\varphi_s + i\varphi_s h^\top r_s = 0, \qquad s \in [0,t]$$

with final condition $\varphi_t = \psi$. The compact support of ψ ensures that (7.26) is also satisfied. From (7.25) we obtain that

$$\tilde{\mathbb{E}}[\varepsilon_t \tilde{\rho}_t(\psi)] = \pi_0(\varphi_0).$$

As the same identity holds for $\rho_t(\psi)$ the conclusion follows since the set S_t is total. □

Theorem 7.17. *Under the assumptions C1–C3, the unnormalised conditional distribution of the signal has a density with respect to Lebesgue measure and its density is the process $p = \{p_t,\ t \geq 0\}$ which is the unique solution of the stochastic PDE (7.20).*

Proof. Similar to Exercise 4.1, choose $(\varphi_i)_{i \geq 0}$ to be a sequence of $C_k^\infty(\mathbb{R}^d)$ functions dense in the set of all continuous functions with compact support. Then choose a common null set for all the elements of the sequence outside which $\rho_t(\varphi_i) = \tilde{\rho}_t(\varphi_i)$ for all $i \geq 0$ and by a standard approximation argument one shows that outside this null set

$$\rho_t(A) = \tilde{\rho}_t(A)$$

for any ball $A = B(x,r)$ for arbitrary $x \in \mathbb{R}^d$ and $r > 0$, hence the two measures must coincide. □

The following corollary identifies the density of the conditional distribution of the signal (its existence follows from Corollary 7.9). Denote the density of π_t by $\tilde{\pi}_t \in L^2(\mathbb{R}^d)$.

Corollary 7.18. *Under the assumptions C1–C3, the conditional distribution of the signal has a density with respect to Lebesgue measure and its density is the normalised version of process $p = \{p_t,\ t \geq 0\}$ which is the solution of the stochastic PDE (7.20). In particular, $\tilde{\pi}_t \in W_k^2(\mathbb{R}^d)$ and there exists a constant $c = c(k,r,t)$ such that*

$$\tilde{\mathbb{E}}\left[\sup_{0 \leq s \leq t} \|\tilde{\pi}_s\|_{W_k^r(\mathbb{R}^d)}^{r'}\right] \leq c \|p_0\|_{W_k^r(\mathbb{R}^d)}^{r'}, \qquad (7.27)$$

where r' can be chosen to be either 1 or $r/2$.

Proof. The first part of the corollary is immediate from Theorem 7.11 and Theorem 7.17. Inequality (7.27) follows from (7.21) and the Cauchy–Schwarz inequality

$$\tilde{\mathbb{E}}\left[\sup_{0 \leq s \leq t} \|\tilde{\pi}_s\|_{W_k^r(\mathbb{R}^d)}^{r'}\right] \leq \sqrt{\tilde{\mathbb{E}}\left[\sup_{0 \leq s \leq t} \rho_s^{-2r'}(\mathbf{1})\right] \tilde{\mathbb{E}}\left[\sup_{0 \leq s \leq t} \|p_s\|_{W_k^r(\mathbb{R}^d)}^{2r'}\right]}.$$

Exercise 9.16 establishes the finiteness of the term $\tilde{\mathbb{E}}[\sup_{0 \leq s \leq t} \rho_s^{-2r'}(\mathbf{1})]$.

7.3 The Smoothness of the Density of ρ_t

Additional smoothness properties of π follow in a similar manner from Theorem 7.12. Following the Kushner–Stratonovich equation (see Theorem 3.30), the density of π satisfies the following non-linear stochastic PDE

$$\tilde{\pi}_t(x) = \tilde{\pi}_0(x) + \int_0^t A^* \tilde{\pi}_s(x)\,\mathrm{d}s + \int_0^t \tilde{\pi}_s(x)(h^\top(x) - \tilde{\pi}_s(h^\top))\,(\mathrm{d}Y_s - \tilde{\pi}_s(h)\,\mathrm{d}s). \tag{7.28}$$

It is possible to recast the SPDE for the density p into a form in which there are no stochastic integral terms. This form can be analysed; for example, Baras et al. [7] treat the one-dimensional case in this way, establishing the existence of a fundamental solution to this form of the Zakai equation. They then use this fundamental solution to prove existence and uniqueness results for the solution to the Zakai equation without requiring bounds on the sensor function h.

Theorem 7.19. *If we write*

$$R_t \triangleq \exp\left(-Y_t^\top h(x) + \frac{1}{2}\|h(x)\|^2 t\right) \tag{7.29}$$

and define $\tilde{p}_t(x) \triangleq R_t(x)p_t(x)$ then this satisfies the following partial differential equation with stochastic coefficients

$$\mathrm{d}\tilde{p}_t = R_t A^*(R_t^{-1}\tilde{p}_t)\,\mathrm{d}t$$

with initial condition $\tilde{p}_0(x) = p_0(x)$.

Proof. Clearly

$$\mathrm{d}R_t = R_t\left(-h^\top(x)\mathrm{d}Y_t + \frac{1}{2}\|h(x)\|^2\,\mathrm{d}t + \frac{1}{2}\|h(x)\|^2\,\mathrm{d}\langle Y\rangle_t\right)$$
$$= R_t\left(-h^\top(x)\mathrm{d}Y_t + \|h(x)\|^2\,\mathrm{d}t\right).$$

Therefore using (7.20) for $\mathrm{d}p_t$ it follows by Itô's formula that

$$\begin{aligned}\mathrm{d}\tilde{p}_t(x) &= \mathrm{d}(R_t(x)p_t(x)) \\ &= R_t A^* p_t(x)\mathrm{d}t + R_t(x)h^\top(x)p_t(x)\,\mathrm{d}Y_t \\ &\quad + p_t(x)R_t(x)(-h^\top(x)\,\mathrm{d}Y_t + \|h(x)\|^2\mathrm{d}t) - p_t(x)R_t\|h(x)\|^2\,\mathrm{d}t \\ &= R_t A^* p_t(x)\,\mathrm{d}t \\ &= R_t A^*(R_t(x)^{-1}\tilde{p}_t(x))\,\mathrm{d}t.\end{aligned}$$

The initial condition result follows from the fact that $R_0(x) = 1$. □

7.4 The Dual of ρ_t

A result similar to Theorem 7.12 justifies the existence of a function dual for the unnormalised conditional distribution of the signal. Theorem 7.20 stated below is an immediate corollary of Theorem 7.12 using a straightforward time-reversal argument. Choose a fixed time horizon $t > 0$ and let $\mathcal{Y}^t = \{\mathcal{Y}_s^t, s \in [0, t]\}$, be the backward filtration

$$\mathcal{Y}_s^t = \sigma(Y_t - Y_r, \ r \in [s, t]).$$

Theorem 7.20. *Let $m > 2$ be an integer such that $(m-2)p > d$. Then under the assumptions C1 – C2, for any bounded $\varphi \in W_p^m(\mathbb{R}^d)$ there exists a unique function-valued process $\psi^{t,\varphi} = \{\psi_s^{t,\varphi}, \ s \in [0,t]\}$:*

1. *For every $x \in \mathbb{R}^d$, $\psi_s^{t,\varphi}(x)$ is a real-valued process measurable with respect to the backward filtration $\mathcal{Y}_s^t$.*
2. *Almost surely, $\psi_s^{t,\varphi}(x)$ is jointly continuous over $(s,x) \in [0,\infty) \times \mathbb{R}^d$ and is twice differentiable in the spatial variable x. Both $\psi_s^{t,\varphi}$ and its partial derivatives are continuous bounded functions.*
3. *$\psi^{t,\varphi}$ is a (classical) solution of the following backward stochastic partial differential equation,*

$$\psi_s^{t,\varphi}(x) = \varphi(x) - \int_s^t A\psi_p^{t,\varphi}(x) \, \mathrm{d}p$$
$$- \int_s^t \psi_p^{t,\varphi}(x) h^\top(x) \, \bar{\mathrm{d}} Y_p, \qquad 0 \leq s \leq t, \tag{7.30}$$

where $\int_s^t \psi_p^{t,\varphi} h^\top \bar{\mathrm{d}} Y_p^k$ is a backward Itô integral.
4. *There exists a constant $c = c(m,p)$ independent of φ such that*

$$\tilde{\mathbb{E}}\left[\sup_{s \in [0,t]} \|\psi_s^{t,\varphi}\|_{2,\infty}^p\right] \leq c_1^{m,p} \|\varphi\|_{m,p}^p. \tag{7.31}$$

Exercise 7.21. *If $\varphi \in W_p^m(\mathbb{R}^d)$ as above, prove that for $0 \leq r \leq s \leq t$ we have*

$$\psi_r^{s, \psi_s^{t,\varphi}} = \psi_r^{t,\varphi}.$$

Theorem 7.22. *The process $\psi^{t,\varphi} = \{\psi_s^{t,\varphi}, \ s \in [0,t]\}$ is the dual of the solution of the Zakai equation. That is, for any $\varphi \in W_p^m(\mathbb{R}^d) \cap B(\mathbb{R}^d)$, the process*

$$s \mapsto \rho_s\left(\psi_s^{t,\varphi}\right), \qquad s \in [0,t]$$

is almost surely constant.

Proof. Let $\varepsilon_t \in S_t$ where S_t is the set defined in Corollary B.40; that is,

$$\varepsilon_t = \exp\left(i\int_0^t r_s^\top \, \mathrm{d}Y_s + \frac{1}{2}\int_0^t \|r_s\|^2 \, \mathrm{d}s\right),$$

where $r \in C_b^m([0,t], \mathbb{R}^m)$. Then for any $\varphi \in C_b^{1,2}([0,t] \times \mathbb{R}^d)$, the identity (4.13) gives

$$\tilde{\mathbb{E}}\left[\varepsilon_t \rho_t(\varphi_t)\right] = \tilde{\mathbb{E}}\left[\varepsilon_r \rho_r(\varphi_r)\right]$$
$$+ \tilde{\mathbb{E}}\left[\int_r^t \varepsilon_s \rho_s \left(\frac{\partial \varphi_s}{\partial s} + A\varphi_s + i\varphi_s h^\top r_s\right) ds\right]. \quad (7.32)$$

Let

$$\tilde{\varepsilon}_s = \exp\left(i \int_s^t r_u^\top dY_u + \frac{1}{2} \int_s^t \|r_u\|^2 du\right);$$

then for $s \in [0,t]$, it is immediate that

$$\tilde{\mathbb{E}}\left[\psi_s^{t,\varphi} \varepsilon_t \mid \mathcal{Y}_s\right] = \varepsilon_s \tilde{\mathbb{E}}\left[\psi_s^{t,\varphi} \tilde{\varepsilon}_s \mid \mathcal{Y}_s\right].$$

Since $\psi_s^{t,\varphi}$ and $\tilde{\varepsilon}_s$ are both $\mathcal{Y}_s^t$-measurable, it follows that they are independent of $\mathcal{Y}_s$; thus defining $\Xi = \{\Xi_s,\ s \in [0,t]\}$ to be given by $\Xi_s = \tilde{\mathbb{E}}[\psi_s^{t,\varphi} \tilde{\varepsilon}_s]$, it follows that

$$\tilde{\mathbb{E}}\left[\psi_s^{t,\varphi} \varepsilon_t \mid \mathcal{Y}_s\right] = \varepsilon_s \Xi_s.$$

Since $\tilde{\varepsilon} = \{\tilde{\varepsilon}_s,\ s \in [0,t]\}$ is a solution of the backward stochastic differential equation:

$$\tilde{\varepsilon}_s = 1 - i \int_s^t \tilde{\varepsilon}_u r_u^\top \bar{d}Y_u, \quad 0 \le s \le t.$$

It follows by stochastic integration by parts using the SDE (7.30) that

$$d(\psi_p^{t,\varphi} \tilde{\varepsilon}_p) = -i\psi_p^{t,\varphi} \tilde{\varepsilon}_p r_p^\top \bar{d}Y_p + \tilde{\varepsilon}_p A\psi_p^{t,\varphi} dp + \tilde{\varepsilon}_p \psi_p^{t,\varphi} h^\top \bar{d}Y_p + i\tilde{\varepsilon}_p h^\top r_p \psi_p^{t,\varphi} dp$$

and taking expectation and using the fact that $\psi_t^{t,\varphi} = \varphi$, and $\tilde{\varepsilon}_t = 1$,

$$\Xi_s = \varphi - \tilde{\mathbb{E}}\left[\int_s^t \tilde{\varepsilon}_p A\psi_p^{t,\varphi} dp\right] - i\tilde{\mathbb{E}}\left[\int_s^t h^\top r_p \psi_p^{t,\varphi} dp\right], \quad 0 \le s \le t;$$

using the boundedness properties of ψ, a, f, h and r we see that

$$\tilde{\mathbb{E}}\left[\int_s^t \tilde{\varepsilon}_p A\psi_p^{t,\varphi} dp\right] = \int_s^t A\Xi_p\, dp,$$
$$\tilde{\mathbb{E}}\left[\int_s^t \tilde{\varepsilon}_p h^\top r_p \psi_p^{t,\varphi} dp\right] = \int_s^t h^\top r_p \Xi_p\, dp,$$

hence

$$\Xi_s = \varphi - \int_s^t A\Xi_p\, dp - i \int_s^t h^\top r_p \Xi_p\, dp, \quad 0 \le s \le t;$$

in other words $\Xi = \{\Xi_s,\ s \in [0,t]\}$ is the unique solution of the the parabolic PDE (4.14), therefore $\Xi \in C_b^{1,2}([0,t] \times \mathbb{R}^d)$. Hence from (7.32), for arbitrary $r \in [0,t]$

$$\tilde{\mathbb{E}}[\varepsilon_t \rho_t(\varphi)] = \tilde{\mathbb{E}}[\varepsilon_t \rho_t(\Xi_t)] = \tilde{\mathbb{E}}[\rho_r(\varepsilon_r \Xi_r)] = \tilde{\mathbb{E}}[\varepsilon_r \Xi_r]$$
$$= \tilde{\mathbb{E}}\left[\varepsilon_r \tilde{\mathbb{E}}\left[\psi_r^{t,\varphi} \tilde{\varepsilon}_r \mid \mathcal{Y}_r\right]\right] = \tilde{\mathbb{E}}\left[\tilde{\mathbb{E}}\left[\varepsilon_r \tilde{\varepsilon}_r \psi_r^{t,\varphi} \mid \mathcal{Y}_r\right]\right]$$
$$= \tilde{\mathbb{E}}\left[\varepsilon_t \psi_r^{t,\varphi}\right] = \tilde{\mathbb{E}}\left[\varepsilon_t \tilde{\mathbb{E}}\left[\psi_r^{t,\varphi} \mid \mathcal{Y}_r\right]\right] = \tilde{\mathbb{E}}\left[\varepsilon_t \rho_r(\psi_r^{t,\varphi})\right],$$

where the penultimate equality uses the fact that $\psi_r^{t,\varphi}$ is $\mathcal{Y}_r^t$-adapted and hence independent of $\mathcal{Y}_r$. The conclusion of the theorem then follows since this holds for any $\varepsilon_t \in S_t$ and the set S_t is total, thus $\rho_r(\psi_r^{t,\varphi}) = \rho_t(\varphi)$ $\mathbb{P}$-a.s., and as t is fixed this implies that $\rho_r(\psi_r^{t,\varphi})$ is a.s. constant. □

Remark 7.23. Theorem 7.22 with $r = 0$ implies that

$$\rho_t(\varphi) = \pi_0\left(\psi_0^{t,\varphi}\right), \qquad \tilde{\mathbb{P}}\text{-a.s.},$$

hence the solution of the Zakai equation is unique (up to indistinguishability).

We can represent $\psi^{t,\varphi}$ by using the following version of the Feynman–Kac formula (see Pardoux [238])

$$\psi_s^{t,\varphi}(x) = \tilde{\mathbb{E}}\left[\varphi\left(X_t(x)\right) a_s^t(X(x), Y) \mid \mathcal{Y}\right], \; s \in [0, t], \qquad (7.33)$$

where

$$a_s^t(X(x), Y) = \exp\left(\int_s^t h^\top(X_s(x))\,\mathrm{d}Y_s - \frac{1}{2}\int_s^t \|h(X_s(x))\|^2\,\mathrm{d}s\right), \qquad (7.34)$$

and $X_t(x)$ follows the law of the signal starting from x, viz

$$X_t = x + \int_s^t \tilde{f}(X_s)\,\mathrm{d}s + \int_s^t \sigma(X_s)\,\mathrm{d}V_s + \int_s^t \bar{\sigma}(X_s)\,\mathrm{d}W_s. \qquad (7.35)$$

The same formula appears in Rozovskii [250] (formula (0.3), page 176) under the name of the averaging over the characteristics (AOC) formula. Using (7.33) we can prove that if φ is a non-negative function, then so is $\psi_s^{t,\varphi}$ for any $s \in [0, t]$ (see also Corollary 5, page 192 of Rozovskii [250]). We can also use (7.33) to define the dual $\psi^{t,\varphi}$ of ρ for φ in a larger class than $W_p^m(\mathbb{R}^d)$, for example, for $B(\mathbb{R}^d)$. For these classes of φ, Rozovskii's result no longer applies: the dual may not be differentiable and may not satisfy an inequality similar to (7.31). However, if φ has higher derivatives, one can use Kunita's theory of stochastic flows (see Kunita [164]) to prove that $\psi^{t,\varphi}$ is differentiable.

7.5 Solutions to Exercises

7.2 Let $\bar{g}_\mu : \mathbb{R}^d \to \mathbb{R}$ be defined as

$$\bar{g}_\mu = \sum_{i=1}^\infty \mu(\varphi_i)\varphi_i.$$

Then $\bar{g}_\mu \in L^2(\mathbb{R}^d)$. Let $\bar{\mu}$ be a measure absolutely continuous with respect to Lebesgue measure with density $\bar{g}_\mu$. Then $\mu(\varphi_i) = \bar{\mu}(\varphi_i)$, since

$$\bar{\mu}(\varphi_i) = \int_{\mathbb{R}^d} \varphi_i \bar{g}_\mu \, dx = \left\langle \sum_{j=1}^\infty \mu(\varphi_j)\varphi_j, \varphi_i \right\rangle = \mu(\varphi_i);$$

hence via an approximation argument $\mu(A) = \bar{\mu}(A)$ for any ball A of arbitrary center and radius. Hence $\mu = \bar{\mu}$ and since $\bar{\mu}$ is absolutely continuous with respect to Lebesgue measure the result follows.

7.3

i. First we show that if for $p, q \geq 1$, $1/p + 1/q = 1 + 1/r$ then $\|f \star g\|_r \leq \|f\|_p \|g\|_q$, where $f \star g$ denotes the convolution of f and g. Then choosing $p = 2$, $q = 1$, and $r = 2$, we see that for $g \in L^2(\mathbb{R}^d)$, using the fact that the L^1 norm of the heat kernel is unity,

$$\|\psi_\varepsilon g\|_2 = \|\psi_\varepsilon \star g\|_2 \leq \|\psi_\varepsilon\|_1 \|g\|_2 = \|g\|_2.$$

We now prove the result for convolution. Consider f, g non-negative; let $1/p' + 1/p = 1$ and $1/q + 1/q' = 1$. Since $1/p' + 1/q' + 1/r = 1$ we may apply Hölder's inequality,

$$f \star g(x) = \int_{\mathbb{R}^d} f(y) g(x-y) \, dy$$

$$= \int_{\mathbb{R}^d} f(y)^{p/r} g(x-y)^{q/r} f(y)^{1-p/r} g(x-y)^{1-q/r} \, dy$$

$$\leq \left(\int_{\mathbb{R}^d} f(y)^p g(x-y)^q \, dy \right)^{1/r} \left(\int_{\mathbb{R}^d} f(y)^{(1-p/r)q'} \, dy \right)^{1/q'}$$

$$\times \left(\int_{\mathbb{R}^d} g(x-y)^{(1-q/r)p'} \, dy \right)^{1/p'}$$

$$= \left(\int_{\mathbb{R}^d} f(y)^p g(x-y)^q \, dy \right)^{1/r} \left(\int_{\mathbb{R}^d} f(y)^p \, dy \right)^{1/q'}$$

$$\times \left(\int_{\mathbb{R}^d} g(y)^q \, dy \right)^{1/p'}.$$

Therefore

$$(f \star g)^r(x) \leq (f^p \star g^q)(x) \|f\|_p^{pr/q'} \|g\|_q^{rq/p'},$$

so by Fubini's theorem

$$\|f \star g\|_r^r \leq \|f\|_p^{r-p} \|g\|_q^{r-q} \int_{\mathbb{R}^d} \int_{\mathbb{R}^d} f^p(y) g^q(x-y) \, dy \, dx$$

$$\leq \|f\|_p^{r-p} \|g\|_q^{r-q} \int_{\mathbb{R}^d} f^p(y) \int_{\mathbb{R}^d} g^q(x-y) \, dx \, dy$$

$$\leq \|f\|_p^{r-p} \|g\|_q^{r-q} \|f\|_p^p \|g\|_q^q = \|f\|_p^r \|g\|_q^r.$$

ii. The function $\psi_{2\varepsilon}(x)$ is bounded by $1/(2\pi\varepsilon)^{d/2}$, therefore

$$\|T_\varepsilon\mu\|_2^2 = \int_{\mathbb{R}^d}\int_{\mathbb{R}^d}\int_{\mathbb{R}^d} \psi_\varepsilon(x-y)\psi_\varepsilon(x-z)\mu(\mathrm{d}y)\,\mu(\mathrm{d}z)\,\mathrm{d}x$$

$$= \int_{\mathbb{R}^d}\int_{\mathbb{R}^d} \psi_{2\varepsilon}(y-z)\mu(\mathrm{d}y)\,\mu(\mathrm{d}z)$$

$$\le \left(\frac{1}{4\pi\varepsilon}\right)^{d/2} \int_{\mathbb{R}^d}\int_{\mathbb{R}^d} |\mu|(\mathrm{d}y)|\mu(\mathrm{d}z)|$$

$$\le \left(\frac{1}{4\pi\varepsilon}\right)^{d/2} \left(|\mu|(\mathbb{R}^d)\right)^2 < \infty.$$

Also

$$\|\partial^i T_\varepsilon\mu\|_2^2 = \int_{\mathbb{R}^d}\int_{\mathbb{R}^d}\int_{\mathbb{R}^d} \frac{(x_i-y_i)}{\varepsilon}\psi_\varepsilon(x-y)$$

$$\times \frac{(x_i-z_i)}{\varepsilon}\psi_\varepsilon(x-z)\,\mu(\mathrm{d}y)\,\mu(\mathrm{d}z)\,\mathrm{d}x$$

$$= 2^d \int_{\mathbb{R}^d}\int_{\mathbb{R}^d}\int_{\mathbb{R}^d} \frac{(x_i-y_i)}{\varepsilon}\psi_{2\varepsilon}(x-y)\exp\left(-\frac{\|x-y\|^2}{4\varepsilon}\right)$$

$$\times \frac{(x_i-z_i)}{\varepsilon}\psi_{2\varepsilon}(x-z)\exp\left(-\frac{\|x-z\|^2}{4\varepsilon}\right)\mu(\mathrm{d}y)\,\mu(\mathrm{d}z)\,\mathrm{d}x$$

$$\le \frac{2^d}{\varepsilon}\int_{\mathbb{R}^d}\int_{\mathbb{R}^d}\int_{\mathbb{R}^d} \psi_{2\varepsilon}(x-y)\psi_{2\varepsilon}(x-z)\,\mu(\mathrm{d}y)\,\mu(\mathrm{d}z)\,\mathrm{d}x$$

$$\le \frac{2^d}{\varepsilon}\int_{\mathbb{R}^d}\int_{\mathbb{R}^d} \psi_{4\varepsilon}(y-z)\,\mu(\mathrm{d}y)\,\mu(\mathrm{d}z)$$

$$\le \frac{2^d}{\varepsilon}\left(\frac{1}{8\pi\varepsilon}\right)^{d/2} \left(|\mu|(\mathbb{R}^d)\right)^2 < \infty.$$

In the above the bound $\sup_{t\ge 0} te^{-t^2/4} < 1$ was used twice. Similar bounds hold for higher-order derivatives and are proved in a similar manner.

iii. From part (ii) $T_\varepsilon\mu \in L^2(\mathbb{R})$, thus by part (i),

$$\|T_{2\varepsilon}\mu\|_2^2 = \|T_\varepsilon(T_\varepsilon\mu)\|_2^2 \le \|T_\varepsilon\mu\|_2^2.$$

7.4

i. Immediate from

$$|T_\varepsilon f\mu(x)| = \left|\int_{\mathbb{R}^d} \psi_\varepsilon(x-y)f(y)\mu(\mathrm{d}y)\right| \le k_f T_\varepsilon|\mu|(x).$$

ii. Assuming first that $f \in C_b^1(\mathbb{R}^d)$, integration by parts yields

$$\langle T_\varepsilon\mu, f\partial_i T_\varepsilon\mu\rangle = \frac{1}{2}\int_{\mathbb{R}^d} f(x)\partial^i\left((T_\varepsilon\mu(x))^2\right)\,\mathrm{d}x$$

$$= -\frac{1}{2}\int_{\mathbb{R}^d} (T_\varepsilon\mu(x))^2\partial^i f(x)\,\mathrm{d}x.$$

Thus
$$|\langle T_\varepsilon\mu, f\partial_i T_\varepsilon\mu\rangle| \leq \tfrac{1}{2}k'_f \|T_\varepsilon\mu\|_2^2,$$
which implies (ii) for $f \in C_b^1(\mathbb{R}^d)$. The general result follows via a standard approximation argument.

iii.
$$\begin{aligned}
&\left|f\partial^i T_\varepsilon\mu(x) - \partial^i T_\varepsilon(f\mu)(x)\right| \\
&= \left|\int_{\mathbb{R}^d}(f(x)-f(y))\partial^i\psi_\varepsilon(x-y)\mu(\mathrm{d}y)\right| \\
&\leq k'_f\left|\int_{\mathbb{R}^d}\|x-y\|\frac{|x_i-y_i|}{\varepsilon}\psi_\varepsilon(x-y)|\mu|(\mathrm{d}y)\right| \\
&\leq 2^{d/2}k'_f\left|\int_{\mathbb{R}^d}\frac{\|x-y\|^2}{\varepsilon}\exp\left(-\frac{\|x-y\|^2}{4\varepsilon}\right)\psi_{2\varepsilon}(x-y)|\mu|(\mathrm{d}y)\right| \\
&\leq 2^{d/2+1}k'_f T_{2\varepsilon}|\mu|(x),
\end{aligned}$$

where the final inequality follows as a consequence of the fact that $\sup_{t\geq 0}(t\exp(-t/4)) < 2$.

7.10 Using primes to denote differentiation with respect to the spatial variable, from the Zakai equation,
$$T_\varepsilon\rho(\varphi') = T_\varepsilon\pi_0(\varphi') + \int_0^t \rho_s(AT_\varepsilon\varphi')\,\mathrm{d}s + \int_0^t \rho_s(hT_\varepsilon\varphi')\,\mathrm{d}Y_s.$$

By Itô's formula, setting $z_t^\varepsilon = (T_\varepsilon\rho)'$,
$$\begin{aligned}
(z_t^\varepsilon(\varphi))^2 &= (T_\varepsilon\pi_0)'\varphi + 2\int_0^t z_t^\varepsilon(\varphi)\rho_s(AT_\varepsilon\varphi')\,\mathrm{d}s + 2\int_0^t z_t^\varepsilon(\varphi)\,\mathrm{d}Y_s \\
&\quad + \int_0^t (\rho_s(hT_\varepsilon\varphi'))^2\,\mathrm{d}s.
\end{aligned}$$

Taking expectation and using Fatou's lemma,
$$\begin{aligned}
\tilde{\mathbb{E}}\left(z_t^\varepsilon(\varphi)\right)^2 &\leq \tilde{\mathbb{E}}\left[(T_\varepsilon\pi_0)'(\varphi)\right] + 2\tilde{\mathbb{E}}\int_0^t z_t^\varepsilon(\varphi)\rho_s(AT_\varepsilon\varphi')\,\mathrm{d}s \\
&\quad + \tilde{\mathbb{E}}\int_0^t (\rho_s(hT_\varepsilon\varphi'))^2\,\mathrm{d}s.
\end{aligned}$$

For the final term
$$\rho_s(hT_\varepsilon\varphi') = (h\rho)(T_\varepsilon\varphi') = \langle\varphi, T_\varepsilon(h\rho)\rangle;$$
using this and the result (7.9) of Lemma 7.5 it follows that

$$\tilde{\mathbb{E}}\left(z_t^\varepsilon(\varphi)\right)^2 \leq \tilde{\mathbb{E}}\left[(T_\varepsilon \pi_0)'(\varphi)\right] + 2\tilde{\mathbb{E}} \int_0^t z_t^\varepsilon(\phi)\langle \varphi', (T_\varepsilon f\rho)'\rangle \, ds$$

$$+ 2\tilde{\mathbb{E}} \int_0^t z_t^\varepsilon(\phi)\langle \varphi', (T_\varepsilon a\rho)''\rangle \, ds + \tilde{\mathbb{E}} \int_0^t \langle \varphi', T_\varepsilon(h\rho)\rangle^2 \, ds.$$

Therefore integrating by parts yields,

$$\tilde{\mathbb{E}}\left(z_t^\varepsilon(\varphi)\right)^2 \leq \tilde{\mathbb{E}}\left[(T_\varepsilon \pi_0)'(\varphi)\right] + 2\tilde{\mathbb{E}} \int_0^t z_t^\varepsilon(\phi)\langle \varphi, (T_\varepsilon f\rho)''\rangle \, ds$$

$$+ 2\tilde{\mathbb{E}} \int_0^t z_t^\varepsilon(\phi)\langle \varphi, (T_\varepsilon a\rho)'''\rangle \, ds + \tilde{\mathbb{E}} \int_0^t \langle \varphi, T_\varepsilon(h\rho)'\rangle^2 \, ds. \quad (7.36)$$

Now let φ range over an orthonormal basis of $L^2(\mathbb{R}^d)$, and bound

$$\lim_{n\to\infty} \sum_{i=1}^n (z_t^\varepsilon(\varphi_i))^2$$

using the result (7.36) applied to each term. By the dominated convergence theorem the limit can be exchanged with the integrals and the result is obtained.

7.14 By Fubini and integration by parts (use the bound (7.21) to prove the integrability of $\int_0^t A^* p_s(x) \, ds$),

$$\langle \int_0^t A^* p_s \, ds, \varphi \rangle = \int_0^t \tilde{\rho}_s(A\varphi) \, ds.$$

Next using the definition 7.23 of the stochastic integral appearing in the stochastic partial differential equation (7.20),

$$\langle \int_0^t h^\top(x) p_s(x) \, dY_s, \varphi \rangle = \int_0^t \tilde{\rho}_s(\varphi h^\top) \, dY_s.$$

Hence the result.

7.15 This proof requires that we repeat, with suitable modifications, the proof of Lemma 4.8 and Exercise 4.9. In the earlier proofs, (4.4) was used for two purposes, firstly in the proof of Lemma 4.8 to justify via dominated convergence interchange of limits and integrals, and secondly in the solution to Exercise 4.9 to show that the various stochastic integrals are martingales. The condition (7.26) must be used instead.

First for the analogue of Lemma 4.9 we show that (7.24) also holds for $\varphi \in W_2^2(\mathbb{R}^d)$, by considering a sequence $\varphi^n \in C_k^2(\mathbb{R}^d)$ converging to φ in the $\|\cdot\|_{2,2}$ norm. From Theorem 7.11 with $k=0$,

$$\tilde{\mathbb{E}}\left[\sup_{0\leq s\leq t} \|p_s\|_2^2\right] \leq c\|p_0\|_2^2 < \infty,$$

since we assumed the initial state density was in $L^2(\mathbb{R}^d)$; thus $\sup_{0 \le s \le t} \|p_s\|_2^2 < \infty$ $\tilde{\mathbb{P}}$-a.s. Therefore by the Cauchy–Schwartz inequality

$$\int_0^t \tilde{\rho}_s(\varphi)\,ds = \int_0^t \langle p_s, \varphi \rangle\,ds \le \int_0^t \|p_s\|_2 \|\varphi\|_2 \,ds$$

$$\le \|\varphi\|_2 \int_0^t \|p_s\|_2 \,ds \le t \|\varphi\|_2 \sup_{0 \le s \le t} \|p_s\|_2 < \infty \qquad \tilde{\mathbb{P}}\text{-a.s.}$$

and similarly

$$\int_0^t \tilde{\rho}_s(\partial^i \varphi)\,ds \le \|\partial^i \varphi\|_2 \int_0^t \|p_s\|_2 \,ds < \infty \qquad \tilde{\mathbb{P}}\text{-a.s.},$$

and

$$\int_0^t \tilde{\rho}_s(\partial^i \partial^j \varphi)\,ds \le t \|\partial^i \partial^j \varphi\|_2 \int_0^t \|p_s\|_2 \,ds < \infty \qquad \tilde{\mathbb{P}}\text{-a.s.}$$

Thus using the boundedness (from C2) of the a_{ij} and f_i, it follows from the dominated convergence theorem that

$$\lim_{n \to \infty} \int_0^t \tilde{\rho}_s(A\varphi^n)\,ds = \int_0^t \tilde{\rho}_s(A\varphi)\,ds.$$

From the boundedness of h, and Cauchy–Schwartz

$$\lim_{n \to \infty} \int_0^t [\tilde{\rho}_s(h_i \varphi^n) - \tilde{\rho}_s(h_i \varphi)]^2 \,ds \le \|h\|_\infty^2 \int_0^t \langle p_s, \varphi^n - \varphi \rangle^2 \,ds$$

$$\le \|h\|_\infty^2 \sup_{0 \le s \le t} \|p_t\|^2 t \|\varphi^n - \varphi\|_2^2 = 0,$$

so by Itô's isometry

$$\lim_{n \to \infty} \int_0^t \tilde{\rho}_s(h^\top \varphi^n)\,dY_s = \int_0^t \tilde{\rho}_s(h^\top \varphi)\,dY_s.$$

Thus from these convergence results (7.24) is satisfied for any $\varphi \in W_2^2$. The result can then be extended to time-dependent φ, which is uniformly bounded in W_2^2 over $[0,t]$ by piecewise approximation followed by the dominated convergence theorem using the bounds just derived. Thus for any $\varphi \in C_b^{1,2}([0,t] \times \mathbb{R}^d)$ such that $\varphi_t \in W_2^2$,

$$\tilde{\rho}_t(\varphi_t) = \tilde{\rho}_0(\varphi_0) + \int_0^t \tilde{\rho}_s\left(\frac{\partial \varphi_s}{\partial s} + A\varphi_s\right)ds + \int_0^t \tilde{\rho}_s(\varphi_s h^\top)\,dY_s.$$

For the second part of the proof, apply Itô's formula to $\varepsilon_t \rho_t(\varphi_t)$ and then take expectation. In order to show that the stochastic integrals are martingales and therefore have zero expectation, we may use the bound

$$\tilde{\mathbb{E}}\left[\int_0^t \varepsilon_s^2 \left(\tilde{\rho}_s(\varphi_s)\right)^2 ds\right] \le e^{\|r\|_\infty^2 t}\tilde{\mathbb{E}}\left[\int_0^t \left(\tilde{\rho}_s(\varphi_s)\right)^2 ds\right]$$

$$\le \tilde{\mathbb{E}}\left[\int_0^t \langle p_s, \varphi \rangle^2 ds\right]$$

$$\le \tilde{\mathbb{E}}\left[\int_0^t \|\varphi_s\|_2^2 \|p_s\|_2^2 ds\right]$$

$$\le t\left(\sup_{0\le s\le t}\|\varphi_s\|_2\right)^2 \tilde{\mathbb{E}}\left[\sup_{0\le s\le t}\|p_s\|_2^2\right] < \infty.$$

Consequently since the stochastic integrals are all martingales, we obtain

$$\tilde{\mathbb{E}}\left[\varepsilon_t \tilde{\rho}_t(\varphi_t)\right] = \pi_0(\varphi_0) + \tilde{\mathbb{E}}\left[\int_0^t \varepsilon_s \tilde{\rho}_s\left(\frac{\partial \varphi_s}{\partial s} + A\varphi_s + i\varphi_s h^\top r_s\right) ds\right].$$

7.21 It is immediate from (7.30) that

$$\psi_s^{s,\psi_s^{t,\varphi}} = \psi_s^{t,\varphi};$$

thus by subtraction of (7.30) at times s and r, for $0 \le r \le s \le t$, we obtain

$$\psi_r^{t,\varphi} = \psi_s^{t,\varphi} - \int_r^s A\psi_p^{t,\varphi}\, dp - \int_r^s \psi_p^{t,\varphi} h^\top\, \bar{d}Y_p$$

and this is the same as the evolution equation for $\psi_r^{s,\psi_s^{t,\varphi}}$. Therefore by the uniqueness of its solution (Theorem 7.20), $\psi_r^{t,\varphi} = \psi_r^{s,\psi_s^{t,\varphi}}$ for $r \in [0,s]$.

Part II

Numerical Algorithms

8
Numerical Methods for Solving the Filtering Problem

This chapter contains an overview of six classes of numerical methods for solving the filtering problem. For each of the six classes, we give a brief description of the ideas behind the methods and state some related results. The last class of methods presented here, particle methods, is developed and studied in depth in Chapter 9 for the continuous time framework and in Chapter 10 for the discrete one.

8.1 The Extended Kalman Filter

This approximation method is based on a natural extension of the exact computation of the conditional distribution for the linear/Gaussian case. Recall from Chapter 6, that in the linear/Gaussian framework the pair (X,Y) satisfies the $(d+m)$-dimensional system of linear stochastic differential equations (6.17) and (6.18); that is,

$$\begin{aligned} \mathrm{d}X_t &= (F_t X_t + f_t)\,\mathrm{d}t + \sigma_t \mathrm{d}V_t \\ \mathrm{d}Y_t &= (H_t X_t + h_t)\,\mathrm{d}t + \mathrm{d}W_t. \end{aligned} \qquad (8.1)$$

In (8.1), the pair (V,W) is a $(d+m)$-dimensional standard Brownian motion. Also $Y_0 = 0$ and X_0 has a Gaussian distribution, $X_0 \sim N(x_0, p_0)$, and is independent of (V,W). The functions

$$\begin{aligned} F &: [0,\infty) \to \mathbb{R}^{d\times d}, & f &: [0,\infty) \to \mathbb{R}^d \\ H &: [0,\infty) \to \mathbb{R}^{d\times m}, & h &: [0,\infty) \to \mathbb{R}^m \end{aligned}$$

are locally bounded, measurable functions. Then π_t, the conditional distribution of the signal X_t, given the observation σ-algebra $\mathcal{Y}_t$ is Gaussian. Therefore π_t is uniquely identified by its mean and covariance matrix. Let $\hat{x} = \{\hat{x}_t,\ t \geq 0\}$ be the conditional mean of the signal; that is, $\hat{x}_t^i = \mathbb{E}[X_t^i|\mathcal{Y}_t]$. Then $\hat{x}$ satisfies the stochastic differential equation (6.27), that is,

A. Bain, D. Crisan, *Fundamentals of Stochastic Filtering*,
DOI 10.1007/978-0-387-76896-0_8, © Springer Science+Business Media, LLC 2009

$$d\hat{x}_t = (F_t \hat{x}_t + f_t)\,dt + R_t H_t^\top (dY_t - (H_t \hat{x}_t + h_t)\,dt),$$

and $R = \{R_t,\ t \geq 0\}$ satisfies the deterministic matrix Riccati equation (6.28),

$$\frac{dR_t}{dt} = \sigma_t \sigma_t^\top + F_t R_t + R_t F_t^\top - R_t H_t^\top H_t R_t.$$

We note that $R = \{R_t,\ t \geq 0\}$ is the conditional covariance matrix of the signal; that is, $R_t = (R_t^{ij})_{i,j=1}^d$ has components

$$R_t^{ij} = \mathbb{E}[X_t^i X_t^j | \mathcal{Y}_t] - \mathbb{E}[X_t^i | \mathcal{Y}_t]\mathbb{E}[X_t^j | \mathcal{Y}_t], \qquad i,j = 1,\ldots,d,\ t \geq 0.$$

Therefore, in this particular case, the conditional distribution of the signal is explicitly described by a finite set of parameters ($\hat{x}_t$ and R_t) which, in turn, are easy to compute numerically. The conditional mean $\hat{x}_t$ satisfies a stochastic differential equation driven by the observation process Y and is computed online, in a recursive fashion, updating it as new observation values become available. However R_t is independent of Y and can be computed offline, i.e., before any observation is obtained.

Some of the early applications of the linear/Gaussian filter, known as the Kalman–Bucy filter, date back to the early 1960s. They include applications to space navigation, aircraft navigation, anti-submarine warfare and calibration of inertial navigation systems. Notably, the Kalman–Bucy filter was used to guide Rangers VI and VII in 1964 and the Apollo space missions. See Bucy and Joseph [31] for details and a list of early references. For a recent self-contained treatment of the Kalman–Bucy filter and a number of applications to mathematical finance, genetics and population modelling, see Aggoun and Elliott [2] and references therein.

The result obtained for the linear filtering problem (8.1) can be generalized as follows. Let (X,Y) be the solution of the following $(d+m)$-dimensional system of stochastic differential equations

$$dX_t = (F(t,Y)X_t + f(t,Y))\,dt + \sigma(t,Y)\,dV_t$$
$$+ \sum_{i=1}^m (G_i(t,Y)X_t + g_i(t,Y))dY_t^i \qquad (8.2)$$
$$dY_t = (H(t,Y)X_t + h(t,Y))\,dt + dW_t,$$

where $F, \sigma, G_1, \ldots, G_n : [0,\infty) \times \Omega \to \mathbb{R}^{n \times n}$, $f, g_1, \ldots, g_n : [0,\infty) \times \Omega \to \mathbb{R}^n$, $H : [0,\infty) \times \Omega \to \mathbb{R}^{n \times m}$ and $h : [0,\infty) \times \Omega \to \mathbb{R}^m$ are progressively measurable[†] locally bounded functions. Then, as above, π_t is Gaussian with mean $\hat{x}_t$, and variance R_t which satisfy the following equations

---

[†] If $(\Omega, \mathcal{F}, \mathcal{F}_t, \mathbb{P})$ is a filtered probability space, then we say that $a : [0,\infty) \times \Omega \to \mathbb{R}^N$ is a progressively measurable function if, for all $t \geq 0$, its restriction to $[0,t] \times \Omega$ is $\mathcal{B}([0,t]) \times \mathcal{F}_t$-measurable, where $\mathcal{B}([0,t])$ is the Borel σ-algebra on $[0,t]$).

$$d\hat{x}_t = \left(F(t,Y)\hat{x}_t + f(t,Y) + \sum_{i=1}^{m} G_i(t,Y)R_t H_i^\top(t,Y) \right) dt$$
$$+ \sum_{i=1}^{m} (G_i(t,Y)\hat{x}_t + g_i(t,Y))\, dY_t^i$$
$$+ R_t H^\top(t,Y)\, (dY_t - (H_t(t,Y)\hat{x}_t + h_t(t,Y))\, dt) \quad (8.3)$$

$$dR_t = \Big(F(t,Y)R_t + R_t F(t,Y) + \sigma(t,Y)\sigma^\top(t,Y)$$
$$+ \sum_{i=1}^{m} G_i(t,Y)R_t G_i^\top(t,Y) \Big) dt - R_t H^\top(t,Y)H(t,Y)R_t\, dt$$
$$+ \sum_{i=1}^{m} (G_i(t,Y)R_t + R_t G_i^\top(t,Y))\, dY_t^i. \quad (8.4)$$

The above formulae can be used to estimate π_t for more general classes of filtering problems, which are non-linear. This will lead to the well-known extended Kalman filter (EKF for short). The following heuristic justification of the EKF follows that given in Pardoux [238].

Let (X,Y) be the solution of the following $(d+m)$-dimensional system of *non-linear* stochastic differential equations

$$dX_t = f(X_t)\, dt + \sigma(X_t)\, dV_t + g(X_t)\, dW_t$$
$$dY_t = h(X_t)\, dt + dW_t, \quad (8.5)$$

and assume that $(X_0, Y_0) = (x_0, 0)$, where $x_0 \in \mathbb{R}^d$. Define $\bar{x}_t$ to be the solution of the ordinary differential equation

$$\frac{d\bar{x}_t}{dt} = f(\bar{x}_t), \quad \bar{x}_0 = x_0.$$

The contribution of the two stochastic terms in (8.5) remains small, at least within a small window of time $[0,\varepsilon]$, so a trajectory $t \mapsto X_t$ may be viewed as being a perturbation from the (deterministic) trajectory $t \to \bar{x}_t$. Therefore the following Taylor-like expansion is expected

$$dX_t \simeq (f'(\bar{x}_t)(X_t - \bar{x}_t) + f(\bar{x}_t))\, dt + \sigma(\bar{x}_t)\, dV_t + g(\bar{x}_t)\, dW_t$$
$$dY_t \simeq (h'(\bar{x}_t)(X_t - \bar{x}_t) + h(\bar{x}_t))\, dt + dW_t.$$

In the above equation, '$\simeq$' means approximately equal, although one can not attach a rigorous mathematical meaning to it. Here f' and h' are the derivatives of f and h. In other words, for a small time window, the equation satisfied by the pair (X,Y) is nearly linear. By analogy with the generalized linear filter (8.2), we can 'conclude' that π_t is 'approximately' normal with mean $\hat{x}_t$ and with covariance R_t which satisfy (cf. (8.3) and (8.4))

$$d\hat{x}_t = [(f' - gh')(\bar{x}_t)\hat{x}_t + (f - gh)(\bar{x}_t) - (f' - gh')(\bar{x}_t)\bar{x}_t] dt$$
$$+ g(\bar{x}_t)dY_t + R_t h'^\top(\bar{x}_t)[dY_t - (h'(\bar{x}_t)\hat{x}_t + h(\bar{x}_t) - h'(\bar{x}_t)\bar{x}_t) dt]$$
$$\frac{dR_t}{dt} = (f' - gh')(\bar{x}_t)R_t + R_t(f' - gh')^\top(\bar{x}_t) + \sigma\sigma^\top(\bar{x}_t) - R_t h'^\top h'(\bar{x}_t)R_t$$

with $\hat{x}_0 = x_0$ and $R_0 = p_0$. Hence, we can estimate the position of the signal by using $\hat{x}_t$ as computed above. We can use the same procedure, but instead of $\bar{x}_t$ we can use any $\mathcal{Y}_t$-adapted 'estimator' process m_t. Thus, we obtain a mapping Λ from the set of $\mathcal{Y}_t$-adapted 'estimator' processes into itself

$$m_t \xrightarrow{\Lambda} \hat{x}_t.$$

The extended Kalman filter (EKF) is the fixed point of Λ; that is, the solution of the following system

$$d\hat{x}_t = (f - gh)(\hat{x}_t)dt + g(\hat{x}_t)dY_t + R_t h'^\top(\hat{x}_t)[dY_t - h(\hat{x}_t)dt]$$
$$\frac{dR_t}{dt} = (f' - gh')(\hat{x}_t)R_t + R_t(f' - gh')^\top(\hat{x}_t) + \sigma\sigma^\top(\hat{x}_t) - R_t h'^\top h'(\hat{x}_t)R_t.$$

Although this method is not mathematically justified, it is widely used in practice. The following is a minute sample of some of the more recent applications of the EKF.

- In Bayro-Corrochano et al. [8], a variant of the EKF is used for the motion estimation of a visually guided robot operator.
- In Kao et al. [148], the EKF is used to optimise a model's physical parameters for the simulation of the evolution of a shock wave produced through a high-speed flyer plate.
- In Mangold et al. [202], the EKF is used to estimate the state of a molten carbonate fuel cell.
- In Ozbek and Efe [235], the EKF is used to estimate the state and the parameters for a model for the ingestion and subsequent metabolism of a drug in an individual.

The EKF will give a good estimate if the initial position of the signal is well approximated (p_0 is 'small'), the coefficients f and g are only 'slightly' non-linear, h is injective and the system is stable. Theorem 8.5 (below) gives a result of this nature. The result requires a number of definitions.

Definition 8.1. *The family of function $f^\varepsilon : [0, \infty) \times \mathbb{R}^d \to \mathbb{R}^d$, $\varepsilon \geq 0$, is said to be almost linear if there exists a family of matrix-valued functions $F_t : \mathbb{R}^d \to \mathbb{R}^{d \times d}$ such that, for any $t \geq 0$ and $x, y \in \mathbb{R}^d$, we have*

$$|f^\varepsilon(t, x) - f^\varepsilon(t, y) - F_t(x - y)| \leq \mu_\varepsilon |x - y|,$$

for some family of numbers μ_ε converging to 0 as ε converges to 0.

Definition 8.2. *The function* $f^\varepsilon : [0, \infty) \times \mathbb{R}^d \to \mathbb{R}^d$ *is said to be* strongly injective *if there exists a constant* $c > 0$ *such that*

$$|f(t, x) - f(t, y)| \geq c|x - y|$$

for any $x, y \in \mathbb{R}^d$.

Definition 8.3. *A family of stochastic processes* $\{\xi_t^\varepsilon, \ t \geq 0\}$, $\varepsilon > 0$, *is said to be bounded in* $L^{\infty-}$ *if, for any* $q < \infty$ *there exists* $\varepsilon^q > 0$ *such that* $\|\xi_t^\varepsilon\|_q$ *is bounded uniformly for* $(t, \varepsilon) \in [0, \infty) \times [0, \varepsilon_q]$.

Definition 8.4. *The family* ξ_t^ε, $\varepsilon > 0$, *is said to be of order* ε^α *for some* $\alpha > 0$ *if* $\varepsilon^{-\alpha}\xi_t^\varepsilon$ *is bounded in* $L^{\infty-}$.

Assume that the pair $(X^\varepsilon, Y^\varepsilon)$ satisfies the following system of SDEs,

$$dX_t^\varepsilon = \beta^\varepsilon(t, X_t^\varepsilon)dt + \sqrt{\varepsilon}\sigma(t, X_t^\varepsilon)dW_t + \sqrt{\varepsilon}\gamma(t, X_t^\varepsilon)dB_t$$
$$dY_t^\varepsilon = h^\varepsilon(t, X_t^\varepsilon)dt + \sqrt{\varepsilon}dB_t.$$

The following theorem is proved in Picard [240].

Theorem 8.5. *Assume that* $p_0^{-1/2}(X_0^\varepsilon - \hat{x}_0)$ *is of order* $\sqrt{\varepsilon}$ *and the following conditions are satisfied.*

- σ *and* γ *are bounded.*
- β^ε *and* h^ε *are continuously. differentiable and almost linear.*
- h *is strongly injective and* $\sigma\sigma^\top$ *is uniformly elliptic.*
- *The ratio of the largest and smallest eigenvalues of* P_0 *is bounded.*

Then $(R_t^\varepsilon)^{-1/2}(X_t^\varepsilon - \hat{x}_t^\varepsilon)$ *is of order* $\sqrt{\varepsilon}$.

Hence the EKF works well under the conditions described above. If any of these conditions are not satisfied, the approximation can be very bad. The following two examples, again taken from [240], show this fact.

Suppose first that X^ε and Y^ε are one-dimensional and satisfy

$$dX_t^\varepsilon = (2\arctan X_t^\varepsilon - X_t^\varepsilon)dt + \sqrt{\varepsilon}dW_t$$
$$dY_t^\varepsilon = HX_t^\varepsilon dt + \sqrt{\varepsilon}dB_t,$$

where H is a positive real number. In particular, the signal's drift is no longer almost linear. The deterministic dynamical system associated with X^ε (obtained for $\varepsilon = 0$) has two stable points of equilibrium denoted by $x_0 > 0$ and $-x_0$. The point 0 is an unstable equilibrium point.

The EKF performs badly in this case. For instance, it cannot be used to detect phase transitions of the signal. More precisely, suppose that the signal starts from x_0. Then, for all ε, X_t^ε will change sign with probability one. In fact, one can check that

$$\alpha_0 = \lim_{\varepsilon \to 0} \varepsilon \log(\mathbb{E}\left[\inf\{t > 0; \ X_t^\varepsilon < 0\}\right])$$

exists and is finite. We choose $\alpha_1 > \alpha_0$ and $t_1 \triangleq \exp(\alpha_1/\varepsilon)$. One can prove that
$$\lim_{\varepsilon \to 0} \mathbb{P}\left[(X_{t_1}^\varepsilon < 0)\right] = \frac{1}{2},$$
but on the other hand,
$$\lim_{\varepsilon \to 0} \mathbb{P}\left[(\hat{x}_{t_1} > x_0 - \delta)\right] = 1$$
for small $\delta > 0$. Hence $X_t^\varepsilon - \hat{x}_t^\varepsilon$ does not converge to 0 in probability as ε tends to 0.

In the following example the EKF does not work because the initial condition of the signal is imprecisely known. Assume that X^ε is one-dimensional, Y^ε is two-dimensional, and that they satisfy the system of SDEs,
$$\begin{aligned} dX_t^\varepsilon &= \sqrt{\varepsilon}\,dW_t \\ dY_t^{\varepsilon,1} &= X_t^\varepsilon + \sqrt{\varepsilon}\,dB_t^1 \\ dY_t^{\varepsilon,2} &= 2|X_t^\varepsilon| + \sqrt{\varepsilon}\,B_t^2, \end{aligned}$$
and $X_0^\varepsilon \sim N(-2, 1)$. In this case $X_t^\varepsilon - \hat{x}_t^\varepsilon$ does not converge to 0. To be precise,
$$\liminf_{\varepsilon \to 0} \mathbb{P}\left(\inf_{s \le t} X_s^\varepsilon \ge 1, \sup \hat{x}_t^\varepsilon \le -1\right) > 0.$$
For further results and examples see Bensoussan [12], Bobrovsky and Zakai [21], Fleming and Pardoux [97] and Picard [240, 243].

8.2 Finite-Dimensional Non-linear Filters

We begin by recalling the explicit expression of the conditional distribution of the Beneš filter as presented in Chapter 6. Let X and Y be one-dimensional processes satisfying the system of stochastic differential equations (6.1) and (6.3); that is,
$$\begin{aligned} dX_t &= f(X_t)\,dt + \sigma\,dV_t \\ dY_t &= (h_1 X_t + h_2)\,dt + dW_t \end{aligned} \qquad (8.6)$$
with $(X_0, Y_0) = (x_0, 0)$, where $x_0 \in \mathbb{R}$. In (8.6), the pair process (V, W) is a two-dimensional Brownian motion, $h_1, h_2, \sigma \in \mathbb{R}$ are constants with $\sigma > 0$, and $f : \mathbb{R} \to \mathbb{R}$ is differentiable with bounded derivative (Lipschitz) satisfying the Beneš condition
$$f'(x) + f^2(x)\sigma^{-2} + (h_1 x + h_2)^2 = p^2 x^2 + 2qx + r, \qquad x \in \mathbb{R},$$
where $p, q, r \in \mathbb{R}$ are arbitrary. Then π_t satisfies the explicit formula (6.15); that is,

8.2 Finite-Dimensional Non-linear Filters

$$\pi_t(\varphi) = \frac{1}{c_t} \int_{-\infty}^{\infty} \varphi(z) \exp\bigl(F(z)\sigma^{-2} + Q_t(z)\bigr) \, dz, \tag{8.7}$$

where F is an antiderivative of f, φ is an arbitrary bounded Borel-measurable function, $Q_t(z)$ is the second-order polynomial

$$Q_t(z) \triangleq z \left(h_1 \sigma \int_0^t \frac{\sinh(s p \sigma)}{\sinh(t p \sigma)} \, dY_s + \frac{q + p^2 x_0}{p\sigma \sinh(tp\sigma)} - \frac{q}{p\sigma} \coth(tp\sigma) \right)$$
$$- \frac{p \coth(tp\sigma)}{2\sigma} z^2$$

and c_t is the corresponding constant,

$$c_t \triangleq \int_{-\infty}^{\infty} \exp\bigl(F(z)\sigma^{-2} + Q_t(z)\bigr) \, dz. \tag{8.8}$$

In particular, π only depends on the one-dimensional $\mathcal{Y}_t$-adapted process

$$t \mapsto \psi_t = \int_0^t \sinh(sp\sigma) \, dY_s.$$

The explicit formulae (8.7) and (8.8) are very convenient. If the observations arrive at the given times $(t_i)_{i\geq 0}$, then ψ_{t_i} can be recursively approximated using, for example, the Euler method

$$\psi_{t_{i+1}} = \psi_{t_i} + \sinh(t_{i+1} p \sigma)(Y_{t_{i+1}} - Y_{t_i})$$

and provided the constant c_t and the antiderivative F can be computed this gives an explicit approximation of the density of π_t. Chapter 6 gives some examples where this is possible. If c_t and F are not available in closed form then they can be approximated via a Monte Carlo method for c and numerical integration for F.

The following extension to the d-dimensional case (see Beneš [9] for details) is valid. Let $f : \mathbb{R}^d \to \mathbb{R}^d$ be an irrotational vector field; that is, there exists a scalar function F such that $f = \nabla F$ and assume that the signal and the observation satisfy

$$dX_t = f(X_t)dt + dV_t, \qquad X_0 = x \tag{8.9}$$
$$dY_t = X_t dt + W_t, \qquad Y_0 = 0, \tag{8.10}$$

and further assume that F satisfies the following condition

$$\nabla^2 F + |\nabla F|^2 + |z|^2 = z^\top Q z + q^\top Z + c, \tag{8.11}$$

where $Q \geq 0$ and $Q = Q^\top$. Let T be an orthogonal matrix such that $TQT^\top = \Lambda$, where Λ is the diagonal matrix of (nonnegative) eigenvalues λ_i of Q and $b = Tq$. Let $k = (\sqrt{\lambda_1}, \ldots, \sqrt{\lambda_d})$, $u^\top = (0, 1, -1, 0, 1, -1, \ldots$ repeated d times) and m be the $3d$-dimensional solution of the equation

$$\frac{dm}{dt} = Am, \tag{8.12}$$

where $m(0) = (x_1, 0, 0, x_2, 0, 0, \ldots, x_d, 0, 0)$ and

$$A = \begin{bmatrix} A_1 & & & 0 \\ & A_2 & & \\ & & \ddots & \\ 0 & & & A_d \end{bmatrix}, \quad A_i = \begin{bmatrix} -k_i & & 0 & 0 \\ 0 & & & 0 & 0 \\ k_i(Ty)_i - b_i/2 & 0 & 0 \end{bmatrix}.$$

Let also R be the $3d \times 3d$ matrix-valued solution of

$$\frac{dR}{dt} = \bar{Y} + RA^* + AR,$$

where

$$R = \begin{bmatrix} R_1 & & & 0 \\ & R_2 & & \\ & & \ddots & \\ 0 & & & R_d \end{bmatrix}, \quad \bar{Y} = \begin{bmatrix} \bar{Y}_1 & & & 0 \\ & \bar{Y}_2 & & \\ & & \ddots & \\ 0 & & & \bar{Y}_d \end{bmatrix},$$

$$\bar{Y}_i = \begin{pmatrix} 1 \\ (TY_t)_i \\ 0 \end{pmatrix} (1, (TY_t)_i, 0).$$

Then we have the following theorem (see Beneš [9] for details).

Theorem 8.6. *If condition (8.11) is satisfied, then π_t satisfies the explicit formula*

$$\pi_t(\varphi) = \frac{1}{c_t} \int_{\mathbb{R}^d} \varphi(z) \exp(F(z) + U_t(z)) \, dz,$$

where φ is an arbitrary bounded Borel-measurable function, $U_t(z)$ is the second-order polynomial

$$U_t(z) = z^\top Y_t + \frac{1}{2} z^\top Q^{1/2} z - \frac{1}{2} (Tz + Ru - m)^\top R^{-1} (Tz + Ru - m), \quad z \in \mathbb{R}^d$$

and c_t is the corresponding normalising constant

$$c_t = \int_{\mathbb{R}^d} \exp(F(z) + U_t(z)) \, dz.$$

As in the one-dimensional case, this filter is finite-dimensional. The conditional distribution of the signal π_t depends on the triplet (Y, m, R), which can be recursively computed/approximated. Again, as long as the normalising constant c_t and the antiderivative F can be computed we have an explicit approximation of the density of π_t and if c_t and F are not available in closed form they can be approximated via a Monte Carlo method and numerical integration, respectively.

The above filter is equivalent to the Kalman–Bucy filter: one can be obtained from the other via a certain space transformation. This in turn induces a homeomorphism which makes the Lie algebras associated with the two filters equivalent (again see Beneš [9] for details). However in [10], Beneš has extended the above class of finite-dimensional non-linear filters to a larger class with corresponding Lie algebras which are no longer homeomorphic to the Lie algebra associated with the Kalman–Bucy filter. Further work on finite-dimensional filters and numerical schemes based on approximation using these classes of filter can be found in Cohen de Lara [58, 59], Daum [69, 70], Schmidt [253] and the references therein. See also Darling [68] for another related approach.

8.3 The Projection Filter and Moments Methods

The projection filter (see Brigo et al. [24] and the references therein) is an algorithm which provides an approximation of the conditional distribution of the signal in a systematic way, the method being based on the differential geometric approach to statistics. The algorithm works well in some cases, for example, the cubic sensor example discussed below, but no general convergence theorem is known.

Let $S \triangleq \{p(\cdot, \theta), \ \theta \in \Theta\}$ be a family of probability densities on $\mathbb{R}^d$, where $\Theta \subseteq \mathbb{R}^n$ is an open set of parameters and let

$$S^{1/2} \triangleq \{\sqrt{p(\cdot, \theta)}, \ \theta \in \Theta\} \in L^2(\mathbb{R}^d)$$

be the corresponding set of square roots of densities. We assume that for all $\theta \in \Theta$,

$$\left\{\frac{\partial \sqrt{p(\cdot, \theta)}}{\partial \theta_1}, \ldots, \frac{\partial \sqrt{p(\cdot, \theta)}}{\partial \theta_n}\right\}$$

are independent vectors in $L^2(\mathbb{R}^d)$, i.e., that $S^{1/2}$ is an n-dimensional submanifold of $L^2(\mathbb{R}^d)$, The tangent vector space at $\sqrt{p(\cdot, \theta)}$ to $S^{1/2}$ is

$$L_{\sqrt{p(\cdot,\theta)}} S^{1/2} = \operatorname{span}\left\{\frac{\partial \sqrt{p(\cdot, \theta)}}{\partial \theta_1}, \ldots, \frac{\partial \sqrt{p(\cdot, \theta)}}{\partial \theta_n}\right\}.$$

The L^2-inner product of any two elements of the basis is defined as

$$\left\langle \frac{\partial \sqrt{p(\cdot, \theta)}}{\partial \theta_i}, \frac{\partial \sqrt{p(\cdot, \theta)}}{\partial \theta_j} \right\rangle = \frac{1}{4} \int_{\mathbb{R}^d} \frac{1}{p(x, \theta)} \frac{\partial p(x, \theta)}{\partial \theta_i} \frac{\partial p(x, \theta)}{\partial \theta_j} \, \mathrm{d}x = \frac{1}{4} g_{ij}(\theta),$$

where $g(\theta) = (g_{ij}(\theta))$ is called the Fisher information matrix and following normal tensorial convention, its inverse is denoted by $g^{-1}(\theta) = (g^{ij}(\theta))$.

8 Numerical Methods for Solving the Filtering Problem

In the following we choose S to be an exponential family, i.e.,

$$S = \{p(x,\theta) = \exp\left(\theta^\top c(x) - \psi(\theta)\right) : \theta \in \Theta\},$$

where $c_1, \ldots, c_n$ are scalar functions such that $\{1, c_1, \ldots, c_n\}$ are linearly independent. We also assume that $\Theta \subseteq \Theta_0$ where

$$\Theta_0 = \left\{\theta \in \mathbb{R}^n : \psi(\theta) \triangleq \log \int e^{\theta^\top c(x)} \, dx < \infty\right\}$$

and that Θ_0 has non-empty interior. Let X and Y be the solution of the following system of SDEs,

$$\begin{aligned} dX_t &= f(t, X_t) \, dt + \sigma(t, X_t) \, dW_t \\ dY_t &= h(t, X_t) \, dt + dV_t. \end{aligned}$$

The density $\pi_t(z)$ of the conditional distribution of the signal satisfies the Stratonovich SDE,

$$\begin{aligned} d\pi_t(z) &= A^* \pi_t(z) dt - \tfrac{1}{2} \pi_t(z)(\|h(z)\|^2 - \pi_t(\|h\|^2)) \\ &\quad + \pi_t(z)(h^\top(z) - \pi_t(h^\top)) \circ dY_t, \end{aligned} \tag{8.13}$$

where $\circ$ is used to denote Stratonovich integration and A^* is the operator which is the formal adjoint of A,

$$A^*\varphi \triangleq -\sum_{i=1}^d \frac{\partial}{\partial x_i}(f^i \varphi) + \tfrac{1}{2} \sum_{i,j=1}^d \frac{\partial^2}{\partial x_i \partial x_j}\left(\varphi \sum_{k=1}^d \sigma_{ik} \sigma_{jk}\right).$$

By using the Stratonovich chain rule, we get from (8.13) that

$$d\sqrt{\pi_t} = \frac{1}{2\sqrt{\pi_t}} \circ d\pi_t = R_t(\sqrt{\pi_t}) dt - Q_t^0(\sqrt{\pi_t}) dt + \sum_{k=1}^m Q_t^k(\sqrt{\pi_t}) \circ dY_t^k,$$

where R_t and $(Q_t^k)_{k=0}^m$ are the following non-linear time-dependent operators

$$R_t(\sqrt{p}) \triangleq \frac{A^* p}{2\sqrt{p}}$$

$$Q_t^0(\sqrt{p}) \triangleq \frac{\sqrt{p}}{4} \left(\|h\|^2 - \pi_t\left(\|h\|^2\right)\right)$$

$$Q_t^k(\sqrt{p}) \triangleq \frac{\sqrt{p}}{2} \left(\|h\|^k - \pi_t\left(\|h\|^k\right)\right).$$

Assume now that for all $\theta \in \Theta$ and all $t \geq 0$

$$\mathbb{E}_{p(\cdot,\theta)}\left[\left(\frac{A^* p(\cdot,\theta)}{p(\cdot,\theta)}\right)^2\right] < \infty$$

8.3 The Projection Filter and Moments Methods

and $\mathbb{E}_{p(\cdot,\theta)}[|h|^4] < \infty$. This implies that $R_t(\sqrt{p(\cdot,\theta)})$ and $Q_t^k(\sqrt{p(\cdot,\theta)})$, for $k = 0, 1, \ldots, m$, are vectors in $L^2(\mathbb{R}^d)$.

We define the *exponential projection filter* for the exponential family S to be the solution of the stochastic differential equation

$$\mathrm{d}\sqrt{p(\cdot,\theta_t)} = \Lambda_{\theta_t} \circ R_t(\sqrt{p(\cdot,\theta_t)})\,\mathrm{d}t - \Lambda_{\theta_t} \circ Q_t^0(\sqrt{p(\cdot,\theta_t)})\,\mathrm{d}t$$
$$+ \sum_{k=1}^{m} \Lambda_{\theta_t} \circ Q_t^k(\sqrt{p(\cdot,\theta_t)}) \circ Y_t^k,$$

where $\Lambda_{\theta_t} : L^2 \to L_{\sqrt{p(\cdot,\theta)}} S^{1/2}$ is the orthogonal projection

$$v \stackrel{\Lambda_{\theta_t}}{\mapsto} \sum_{i=1}^{n} \left[\sum_{j=1}^{n} 4g^{ij}(\theta) \left\langle v, \frac{\partial \sqrt{p(\cdot,\theta)}}{\partial \theta_j} \right\rangle \right] \frac{\partial \sqrt{p(\cdot,\theta)}}{\partial \theta_i}.$$

In other words, $\sqrt{p(\cdot,\theta_t)}$ satisfies a differential equation whose driving vector fields are the projections of the corresponding vector fields appearing in the equation satisfied by $\sqrt{\pi_t}$ onto the tangent space of the manifold $S^{1/2}$, and therefore, $p(\cdot,\theta_t)$ is a natural candidate for an approximation of the conditional distribution of the signal at time t, when the approximation is sought among the elements of S.

One can prove that for the exponential family

$$p(x,\theta) = \exp\left[\theta^\top c(x) - \psi(\theta)\right],$$

the projection filter density R_t^π is equal to $p(\cdot,\theta_t)$, where the parameter θ_t satisfies the stochastic differential equation

$$\mathrm{d}\theta_t = g^{-1}(\theta_t)\left(\bar{\mathbb{E}}\left[Ac - \tfrac{1}{2}\|h\|^2(c - \bar{\mathbb{E}}[c])\right]\,\mathrm{d}t \right.$$
$$\left. + \sum_{k=1}^{m} \bar{\mathbb{E}}[h_t^k(c - \bar{\mathbb{E}}[c])] \circ Y_t^k \right), \tag{8.14}$$

where $\bar{\mathbb{E}}[\cdot] = \mathbb{E}_{p(\cdot,\theta_t)}[\cdot]$. Therefore, in order to approximate π_t, solve (8.14) and then compute the density corresponding to its solution.

Example 8.7. We consider the cubic sensor, i.e., the following problem

$$\mathrm{d}X_t = \sigma\,\mathrm{d}W_t$$
$$\mathrm{d}Y_t = X_t^3\,\mathrm{d}t + \mathrm{d}V_t.$$

We choose now S to be the following family of densities

$$S = \left\{ p(x,\theta) = \exp\left(\sum_{i=1}^{6} \theta_i x^i - \psi(\theta)\right) : \theta \in \Theta \subset \mathbb{R}^6,\ \theta_6 < 0 \right\}.$$

Let $\eta_k(\theta)$ be the kth moment of the probability with density $p(\cdot,\theta)$, i.e., $\eta_k(\theta) \triangleq \int_{-\infty}^{\infty} x^k p(x,\theta)\,dx$; clearly $\eta_0(\theta) = 1$. It is possible to show that the following recurrence relation holds

$$\eta_{6+i}(\theta) = -\frac{1}{6\theta_6}\left[(i+1)\eta_i(\theta) + \sum_{j=1}^{6}\theta_j\eta_{i+j}(\theta)\right], \quad i \geq 0,$$

and therefore we only need to compute $\eta_1(\theta), \ldots, \eta_5(\theta)$ in order to compute all the moments. The entries of the Fisher information matrix $g_{ij}(\theta)$ are given by

$$g_{ij}(\theta) = \frac{\partial^2 \psi(\theta)}{\partial \theta_i \partial \theta_j} = \eta_{i+j}(\theta) - \eta_i(\theta)\eta_j(\theta)$$

and (8.14) reduces to the SDE,

$$d\theta_t = g^{-1}(\theta_t)\gamma_\bullet(\theta_t)dt - \lambda_\bullet^0\,dt + \lambda_\bullet dY_t,$$

where

$$\lambda_\bullet^0 = (0,0,0,0,0,1/2)^\top$$
$$\lambda_\bullet = (0,0,1,0,0,0)^\top$$
$$\gamma_\bullet = \tfrac{1}{2}\sigma^2(0, 2\eta_0(\theta), 6\eta_1(\theta), 12\eta_2(\theta), 2 - \eta_3(\theta), 30\eta_4(\theta))^\top.$$

See Brigo et al. [24] for details of the numerical implementation of the projection filter in this case.

The idea of fixing the form of the approximating conditional density and then evolving it by imposing appropriate constraints on the parameters was first introduced by Kushner in 1967 (see [177]). In [183], the same method is used to produce approximations for the filtering problem with a continuous time signal and discrete time observations.

8.4 The Spectral Approach

The spectral approach for the numerical estimation of the conditional distribution of the signal was introduced by Lototsky, Mikulevicius and Rozovskii in 1997 (see [197] for details). Further developments on spectral methods can be found in [195, 198, 199]. For a recent survey see [196]. This section follows closely the original approach and the results contained in [197] (see also [208]).

Let us begin by recalling from Chapter 7 that $p_t(z)$, the density of the unnormalised conditional distribution of the signal, is the (unique) solution of the stochastic partial differential equation (7.20),

$$p_t(x) = p_0(x) + \int_0^t A^* p_s(x)\,ds + \int_0^t h^\top(x) p_s(x)\,dY_s,$$

8.4 The Spectral Approach

in a suitably chosen function space (e.g. $L_k^2(\mathbb{R}^d)$). The spectral approach is based on decomposing p_t into a sum of the form

$$p_t(z) = \sum_\alpha \frac{1}{\sqrt{\alpha!}} \varphi_\alpha(t,z) \xi_\alpha(Y), \qquad (8.15)$$

where $\xi_\alpha(Y)$ are certain polynomials (see below) of Wiener integrals with respect to Y and $\varphi_\alpha(t,z)$ are deterministic Hermite–Fourier coefficients in the Cameron–Martin orthogonal decomposition of $p_t(z)$. This expansion separates the parameters from the observations: the Hermite–Fourier coefficients are determined only by the coefficients of the signal process, its initial distribution and the observation function h, whereas the polynomials $\xi_\alpha(Y)$ are completely determined by the observation process.

A collection $\alpha = (\alpha_k^l)_{1 \leq l \leq d, k \geq 1}$ of nonnegative integers is called a d-dimensional multi-index if only finitely many of α_k^l are different from zero. Let J be the set of all d-dimensional multi-indices. For $\alpha \in J$ we define:

$$|\alpha| \triangleq \sum_{l,k} \alpha_k^l \ : \ \text{the length of } \alpha$$

$$d(\alpha) \triangleq \max\left\{k \geq 1 : \alpha_k^l > 0 \text{ for some } 1 \leq l \leq d\right\} \ : \ \text{the order of } \alpha$$

$$\alpha! \triangleq \prod_{k,l} \alpha_k^l!.$$

Let $\{m_k\} = \{m_k(s)\}_{k \geq 1}$ be an orthonormal system in the space $L^2([0,t])$ and $\xi_{k,l}$ be the following random variables

$$\xi_{k,l} = \int_0^t m_k(s) \, \mathrm{d}Y^l(s).$$

Under the new probability measure $\tilde{\mathbb{P}}$, $\xi_{k,l}$ are i.i.d. Gaussian random variables (as $Y = (Y^l)$ is a standard Brownian motion under $\tilde{\mathbb{P}}$). Let also $(H_n)_{n \geq 1}$ be the Hermite polynomials

$$H_n(x) \triangleq (-1)^n e^{x^2/2} \frac{\mathrm{d}^2}{\mathrm{d}x^n} e^{-x^2/2}$$

and $(\xi_\alpha)_\alpha$ be the Wick polynomials

$$\xi_\alpha \triangleq \prod_{k,l} \left(\frac{H_{\alpha_k^l}(\xi_{k,l})}{\sqrt{\alpha_k^l!}} \right).$$

Then $(\xi_\alpha)_\alpha$ form a complete orthonormal system in $L^2(\Omega, \mathcal{Y}_t, \tilde{\mathbb{P}})$. Their corresponding coefficients in the expansion (8.15) satisfy the following system of deterministic partial differential equations

$$\frac{d\varphi_t^\alpha(z)}{dt} = A^*\varphi_\alpha(t,z) + \sum_{k,l} \alpha_k^l m_k(t) h^l(z) \varphi_{\alpha(k,l)}(t,z) \tag{8.16}$$

$$\varphi_0^\alpha(z) = \pi_0(z) 1_{\{|\alpha|=0\}},$$

where
$$\alpha = (\alpha_k^l)_{1 \le l \le d, k \ge 1} \in J$$

and $\alpha(i,j)$ stands for the multi-index $(\tilde{\alpha}_k^l)_{1 \le l \le d, k \ge 1}$ with

$$\tilde{\alpha}_k^l = \begin{cases} \alpha_k^l & \text{if } k \ne i \text{ or } \ell \ne j \text{ or both} \\ \max(0, \alpha_i^j - 1) & \text{if } k = i \text{ and } \ell = j \end{cases}.$$

Theorem 8.8. *Under certain technical assumptions (given in Lototsky et al. [197]), the series*

$$\sum_\alpha \frac{1}{\sqrt{\alpha!}} \varphi_t^\alpha(z) \xi_\alpha$$

converges in $L^2(\Omega, \tilde{\mathbb{P}})$ and in $L^1(\Omega, \mathbb{P})$ and we have

$$p_t(z) = \sum_\alpha \frac{1}{\sqrt{\alpha!}} \varphi_\alpha(t,z) \xi_\alpha, \qquad \mathbb{P}\text{-a.s.} \tag{8.17}$$

Also the following Parseval's equality holds

$$\tilde{\mathbb{E}}[|p_t(z)|^2] = \sum_\alpha \frac{1}{\alpha!} |\varphi_\alpha(t,z)|^2.$$

For computational purposes one needs to truncate the sum in the expansion of p_t. Let J_N^n be the following finite set of indices

$$J_N^n = \{\alpha : |\alpha| \le N,\ d(\alpha) \le n\}$$

and choose the following deterministic basis

$$m_1(s) = \frac{1}{\sqrt{t}}; \quad m_k(s) = \sqrt{\frac{2}{t}} \cos\left(\frac{\pi(k-1)s}{t}\right), \quad k \ge 1, 0 \le s \le t.$$

Then, again under some technical assumptions, we have the following.

Theorem 8.9. *If $p_t^{n,N}(z) \triangleq \sum_{\alpha \in J_N^n} (1/\sqrt{\alpha!}) \varphi_\alpha(t,z) \xi_\alpha$, then*

$$\tilde{\mathbb{E}}[\|p_t^{n,N} - p_t\|_{L_2}^2] \le \frac{C_t^1}{(N+1)!} + \frac{C_t^2}{n},$$

$$\sup_{z \in \mathbb{R}^d} \tilde{\mathbb{E}}[|p_t^{n,N}(z) - p_t(z)|^2] \le \frac{\bar{C}_t^1}{(N+1)!} + \frac{\bar{C}_t^2}{n},$$

where the constants C_t^1, C_t^2, $\bar{C}_t^1$, and $\bar{C}_t^2$ are independent of n and N.

One can also construct a recursive version of the expansion (8.17) (see [197] for a discussion of the method based on the above approximation). Let $0 = t_0 < t_1 < \cdots < t_M = T$ be a uniform partition of the interval $[0, T]$ with step Δ ($t_i = i\Delta$, $i = 0, \ldots, M$). Let $m_k^i = \{m_k^i(s)\}$ be a complete orthonormal system in $L^2([t_{i-1}, t_i])$. We define the random variables

$$\xi_{k,l}^i = \int_{t_{i-1}}^{t_i} m_k^i(s)\,\mathrm{d}Y^l(s), \qquad \xi_\alpha^i = \prod_{k,l}\left(\frac{H_{\alpha_k^l}(\xi_{k,l}^i)}{\sqrt{(\alpha_k^l)!}}\right),$$

where H_n is the nth Hermite polynomial. Consider the following system of deterministic partial differential equations

$$\frac{\mathrm{d}\varphi_\alpha^i(t,z,g)}{\mathrm{d}t} = A^*\varphi_\alpha^i(t,z,g)$$
$$+ \sum_{k,l} \alpha^{l,k} m_\alpha^i(t) h^l(z) \varphi_{\alpha(k,l)}^i(t,z,g), \quad t \in [t_{i-1}, t_i] \quad (8.18)$$

$$\varphi_\alpha^i(t_{i-1}, z, g) = g(z)\mathbf{1}_{\{|\alpha|=0\}}.$$

We observe that, for each $i = 1, \ldots, M$, the system (8.18) is similar to (8.16), the difference being that the initial time is no longer zero and we allow for an arbitrary initial condition which may be different for different is. The following is the recursive version of Theorem 8.8.

Theorem 8.10. *If $p_0(z) = \pi_0(z)$, then for each $z \in \mathbb{R}^d$ and each t_i, $i = 1,\ldots,M$, the unnormalised conditional distribution of the signal is given by*

$$p_{t_i}(z) = \sum_\alpha \frac{1}{\sqrt{\alpha!}} \varphi_\alpha^i(t_i, z, p_{t_{i-1}}(\cdot)) \xi_\alpha^i \qquad (\mathbb{P}\text{-a.s.}). \qquad (8.19)$$

The series converges in $L^2(\Omega, \mathcal{Y}_t, \tilde{\mathbb{P}})$ and $L^1(\Omega, \mathcal{Y}_t, \mathbb{P})$ and the following Parseval's equality holds,

$$\tilde{\mathbb{E}}[|p_{t_i}(z)|^2] = \sum_\alpha \frac{1}{\alpha!} |\varphi_\alpha^i(t_i, z, p_{t_{i-1}}(\cdot))|^2.$$

For computational purposes we truncate (8.19). We introduce the following basis

$$m_k^i(t) = m_k(t - t_{i-1}), \qquad t_{i-1} \le t \le t_i,$$
$$m_1(t) = \frac{1}{\sqrt{\Delta}},$$
$$m_k(t) = \sqrt{\frac{2}{\Delta}} \cos\left(\frac{\pi(k-1)t}{\Delta}\right), \qquad k \ge 1,\ t \in [0, \Delta],$$
$$m_k(t) = 0, \qquad k \ge 1,\ t \notin [0, \Delta].$$

Theorem 8.11. *If* $p_0^{n,N}(z) = \pi_0(z)$ *and*

$$p_{t_i}^{n,N}(z) = \sum_{\alpha \in J_N^n} \frac{1}{\sqrt{\alpha!}} \varphi_\alpha^i(\Delta, z) \xi_\alpha^i,$$

where $\varphi_\alpha^i(\Delta, z)$ *are the solutions of the system*

$$\frac{\mathrm{d}\varphi_\alpha^i(t,z)}{\mathrm{d}t} = A^* \varphi_\alpha^i(t,z) + \sum_{k,l} \alpha^{l,k} m_\alpha^i(t) h^l(z) \varphi_{\alpha(k,l)}^i(t,z), \quad t \in [0, \Delta]$$

$$\varphi_\alpha^i(0, z) = p_{t_{i-1}}^{n,N}(z) 1_{\{|\alpha|=0\}},$$

then

$$\max_{1 \le i \le M} \tilde{\mathbb{E}}[\|p_{t_i}^{n,N} - p_{t_i}\|_{L_2}^2] \le B e^{BT} \left(\frac{(C\Delta)^N}{(N+1)!} + \frac{\Delta^2}{n} \right),$$

$$\max_{1 \le i \le M} \sup_z \tilde{\mathbb{E}}[|p_{t_i}^{n,N}(z) - p_{t_i}(z)|^2] \le \bar{B} e^{\bar{B}T} \left(\frac{(\bar{C}\Delta)^N}{(N+1)!} + \frac{\Delta^2}{n} \right),$$

where the constants B, C, $\bar{B}$ *and* $\bar{C}$ *are independent of* n, N, Δ *and* T.

8.5 Partial Differential Equations Methods

This type of method uses the fact that $p_t(z)$, the density of the unnormalised conditional distribution of the signal, is the solution of a partial differential equation, albeit a stochastic one. Therefore classical PDE methods may be applied to this stochastic PDE to obtain an approximation to the density p_t. These methods are very successful in low-dimensional problems, but cannot be applied in high-dimensional problems as they require the use of a space grid whose size increases exponentially with the dimension of the state space of the signal. This section follows closely the description of the method given in Cai et al. [37]. The first step is to apply the splitting-up algorithm (see [186, 187] for results and details) to the Zakai equation

$$\mathrm{d}p_t(z) = A^* p_t(z) \, \mathrm{d}t + p_t(z) h^\top(z) \, \mathrm{d}Y_t.$$

Let $0 = t_0 < t_1 < \cdots < t_n < \cdots$ be a uniform partition of the interval $[0, \infty)$ with time step $\Delta = t_n - t_{n-1}$. Then the density $p_{t_n}(z)$ will be approximated by $p_n^\Delta(z)$, where the transition from $p_{n-1}^\Delta(z)$ to $p_n^\Delta(z)$ is divided into the following two steps.

- The first step, called the *prediction* step, consists in solving the following Fokker–Planck equation for the time interval $[t_{n-1}, t_n]$,

$$\frac{\partial p_t^n}{\partial t} = A^* p_t(z)$$

$$p_{t_{n-1}}^n = p_{n-1}^\Delta$$

and we denote the prior estimate by $\bar{p}_n^\Delta \triangleq p_{t_n}^n$. The Fokker–Planck equation is solved by using the implicit Euler scheme, i.e., we solve

$$\bar{p}_n^\Delta - \Delta A^* \bar{p}_n^\Delta = p_{n-1}^\Delta. \tag{8.20}$$

- The second step, called the *correction* step, uses the new observation Y_{t_n} to update $\bar{p}_n^\Delta$. Define

$$z_n^\Delta \triangleq \frac{1}{\Delta}\left(Y_{t_n} - Y_{t_{n-1}}\right) = \frac{1}{\Delta}\int_{t_{n-1}}^{t_n} h(X_s)\,\mathrm{d}s + \frac{1}{\Delta}\left(W_{t_n} - W_{t_{n-1}}\right).$$

Using the Kallianpur–Striebel formula, define $p_n^\Delta(z)$ for $z \in \mathbb{R}^d$ as

$$p_n^\Delta(z) \triangleq c_n \psi_n^\Delta(z) \bar{p}_n^\Delta(z),$$

where $\psi_n^\Delta(z) \triangleq \exp(-\frac{1}{2}\Delta \|z_n^\Delta - h(z)\|^2)$ and c_n is a normalisation constant chosen such that

$$\int_{\mathbb{R}^d} p_n^\Delta(z)\,\mathrm{d}z = 1.$$

Assume that the infinitesimal generator of the signal is the following second-order differential operator

$$A = \sum_{i,j=1}^{d} a_{ij}(\cdot)\frac{\partial^2}{\partial x_i \partial x_j} + \sum_{i=1}^{d} f_i(\cdot)\frac{\partial}{\partial x_i}.$$

We can approximate the solution to equation (8.20) by using a finite difference scheme on a given d-dimensional regular grid Ω^h with mesh $h = (h_1, \ldots, h_m)$ in order to approximate the differential operator A. The scheme approximates first-order derivatives evaluated at x as (e_i is the unit vector in the ith coordinate)

$$\left.\frac{\partial \varphi}{\partial x_i}\right|_x \simeq \begin{cases} \dfrac{\varphi(x+e_i h_i) - \varphi(x)}{h_i} & \text{if } f_i(x) \geq 0 \\[2mm] \dfrac{\varphi(x) - \varphi(x - e_i h_i)}{h_i} & \text{if } f_i(x) < 0 \end{cases}$$

and the second-order derivatives as

$$\left.\frac{\partial^2 \varphi}{\partial x_i^2}\right|_x \simeq \frac{\varphi(x+e_i h_i) - 2\varphi(x) + \varphi(x - e_i h_i)}{h_i^2}$$

and

$$\left.\frac{\partial^2 \varphi}{\partial x_i \partial x_j}\right|_x \simeq \begin{cases} \frac{1}{2h_i}\left(\frac{\varphi(x+e_i h_i + e_j h_j) - \varphi(x+e_i h_i)}{h_j} - \frac{\varphi(x+e_j h_j) - \varphi(x)}{h_j}\right. \\ \left.\quad + \frac{\varphi(x) - \varphi(x - e_j h_j)}{h_j} - \frac{\varphi(x - e_i h_i) - \varphi(x - e_i h_i - e_j h_j)}{h_j}\right) & \text{if } a_{ij} \geq 0, \\[2mm] \frac{1}{2h_i}\left(\frac{\varphi(x+e_i h_i) - \varphi(x+e_i h_i - e_j h_j)}{h_j} - \frac{\varphi(x) - \varphi(x - e_j h_j)}{h_j}\right. \\ \left.\quad + \frac{\varphi(x+e_j h_j) - \varphi(x)}{h_j} - \frac{\varphi(x - e_i h_i + e_j h_j) - \varphi(x - e_i h_i)}{h_j}\right) & \text{if } a_{ij} < 0. \end{cases}$$

For each grid point $x \in \Omega^h$ define the set V^h to be the set of points accessible from x, that is,

$$V^h(x) \triangleq \{x + \varepsilon_i e_i h_i + \varepsilon_j e_j h_j, \ \forall \ \varepsilon_i, \varepsilon_j \in \{-1, 0, +1\}, i \neq j\}$$

and the set $N^h(x) \supset V^h(x)$ to be the set of nearest neighbors of x, including x itself

$$N^h(x) \triangleq \{x + \varepsilon_1 e_1 h_1 + \cdots + \varepsilon_d e_d h_d, \ \forall \ \varepsilon_1, \ldots, \varepsilon_d \in \{-1, 0, +1\}\}.$$

The operator A is approximated by A^h, where A^h is the operator

$$A^h \varphi(x) \triangleq \sum_{y \in V^h(x)} A^h(x, y) \varphi(y)$$

with coefficients[†] given for each $x \in \Omega^h$ by

$$A^h(x, x) = -\sum_{i=1}^{d} \left[\frac{1}{h_i^2} a_{ii}(x) - \sum_{j:\, j \neq i} \frac{1}{2 h_i h_j} |a_{ij}(x)| \right] - \sum_{i=1}^{d} \frac{1}{h_i} |f_i(x)|$$

$$A^h(x, x \pm e_i h_i) = \frac{1}{2 h_i^2} a_{ii}(x) - \sum_{j:\, j \neq i} |a_{ij}(x)| + \frac{1}{h_i} f_i^{\pm}(x)$$

$$A^h(x, x + e_i h_i \pm e_j h_j) = \frac{1}{2 h_i h_j} a_{ij}^{\pm}(x)$$

$$A^h(x, x - e_i h_i \mp e_j h_j) = \frac{1}{2 h_i h_j} a_{ij}^{\pm}(x)$$

$$A^h(x, y) = 0, \quad \text{otherwise}$$

for all $i, j = 1, \ldots, d$, $i \neq j$. One can check that, for all $x \in \bar{\Omega}^h$, where

$$\bar{\Omega}^h \triangleq \bigcup_{x \in \Omega^h} N^h(x),$$

it holds that

$$\sum_{y \in V^h(x)} A^h(x, y) = 0.$$

If for all $x \in \mathbb{R}^d$ and $i = 1, \ldots, d$, the condition

$$\frac{1}{h_i^2} a_{ii}(x) - \sum_{j:\, j \neq i} \frac{1}{2 h_i h_j} |a_{ij}(x)| \geq 0, \qquad (8.21)$$

is satisfied then

$$A^h(x, x) \leq 0 \qquad A^h(x, y) \geq 0 \quad \forall x \in \Omega^h, \forall y \in \Omega^h(x) \setminus x.$$

[†] The notation x^+ denotes $\max(x, 0)$ and x^- denotes $\min(x, 0)$.

Condition (8.21) ensures that A^h can be interpreted as the generator of a pure jump Markov process taking values in the discretisation grid Ω^h. As a consequence the solution of the resulting approximation of the Fokker–Planck equation $\bar{p}_n^\Delta$ will always be a discrete probability distribution.

For recent results regarding the splitting-up algorithm see the work of Gyöngy and Krylov in [118, 119]. The method described above can be refined to permit better approximations of p_t by using composite or adaptive grids (see Cai et al. [37] for details). See also Kushner and Dupuis [181], Lototsky et al. [194], Sun and Glowinski [263], Beneš [9] and Florchingen and Le Gland [101] for related results.

For a general framework for proving convergence results for this class of methods, see Chapter 7 of the monograph by Kushner [182] and the references contained therein. See also Kushner and Huang [184] for further convergence results.

8.6 Particle Methods

Particle methods[†] are algorithms which approximate the stochastic process π_t with discrete random measures of the form

$$\sum_i a_i(t) \delta_{v_i(t)},$$

in other words, with empirical distributions associated with sets of randomly located particles of stochastic masses $a_1(t), a_2(t), \ldots$, which have stochastic positions $v_1(t), v_2(t), \ldots$ where $v_i(t) \in \mathbb{S}$. Particle methods are currently among the most successful and versatile methods for numerically solving the filtering problem and are discussed in depth in the following two chapters.

The basis of this class of numerical method is the representation of π_t given by the Kallianpur–Striebel formula (3.33). That is, for any φ a bounded Borel-measurable function, we have

$$\pi_t(\varphi) = \frac{\rho_t(\varphi)}{\rho_t(\mathbf{1})},$$

where ρ_t is the unnormalised conditional distribution of X_t

$$\rho_t(\varphi) = \tilde{\mathbb{E}}\left[\varphi(X_t) \tilde{Z}_t \,\middle|\, \mathcal{Y}_t \right], \quad (8.22)$$

and

$$\tilde{Z}_t = \exp\left(\int_0^t h(X_s)^\top \, dY_s - \frac{1}{2} \int_0^t \|h(X_s)\|^2 \, ds \right).$$

[†] Also known as *particle filters* or *sequential Monte Carlo methods*.

The expectation in (8.22) is taken with respect to the probability measure $\tilde{\mathbb{P}}$ under which the process Y is a Brownian motion independent of X (see Section 3.3 for details).

One can then use a Monte Carlo approximation for $\tilde{\mathbb{E}}[\varphi(X_t)\tilde{Z}_t \mid \mathcal{Y}_t]$. That is, a large number of independent realisations of the signal are produced (say n) and, for each of them, the corresponding expression $\varphi(X_t)\tilde{Z}_t$ is computed. Then, by taking the average of all the resulting values, one obtains an approximation of $\tilde{\mathbb{E}}[\varphi(X_t)\tilde{Z}_t \mid \mathcal{Y}_t]$. To be more precise, let v_j, $j = 1, \ldots, n$ be n mutually independent stochastic processes and independent of Y, each of them being a solution of the martingale problem for (A, π_0). In other words the pairs (v_j, Y), $j = 1, \ldots, n$ are identically distributed and have the same distribution as the pair (X, Y) (under $\tilde{\mathbb{P}}$). Also let a_j, $j = 1, \ldots, n$ be the following exponential martingales

$$a_j(t) = 1 + \int_0^t a_j(s) h(v_j(s))^\top \, dY_s, \quad t \geq 0. \tag{8.23}$$

In other words

$$a_j(t) = \exp\left(\int_0^t h(v_j(s))^\top \, dY_s - \frac{1}{2} \int_0^t \|h(v_j(s))\|^2 \, ds \right), \quad t \geq 0.$$

Hence, the triples (v_j, a_j, Y), $j = 1, \ldots, n$ are identically distributed and have the same distribution as the triple $(X, \tilde{Z}, Y)$ (under $\tilde{\mathbb{P}}$).

Exercise 8.12. Show that the pairs $(v_j(t), a_j(t))$, $j = 1, \ldots, n$ are mutually independent conditional upon the observation σ-algebra $\mathcal{Y}_t$.

Let $\rho^n = \{\rho_t^n, \ t \geq 0\}$ and $\pi^n = \{\pi_t^n, \ t \geq 0\}$ be the following sequences of measure-valued processes

$$\rho_t^n \triangleq \frac{1}{n} \sum_{j=1}^n a_j(t) \delta_{v_j(t)}, \quad t \geq 0 \tag{8.24}$$

$$\pi_t^n \triangleq \frac{\rho_t^n}{\rho_t^n(1)}, \quad t \geq 0$$

$$= \sum_{j=1}^n \bar{a}_j^n(t) \delta_{v_j(t)}, \quad t \geq 0, \tag{8.25}$$

where the normalised weights $\bar{a}_j^n$ have the form

$$\bar{a}_j^n(t) = \frac{a_j(t)}{\sum_{k=1}^n a_k(t)}, \quad j = 1, \ldots, n, \ t \geq 0.$$

That is, ρ_t^n is the empirical measure of n (random) particles with positions $v_j(t)$, $j = 1, \ldots, n$ and weights $a_j(t)/n$, $j = 1, \ldots, n$ and π_t^n is its normalised version. We have the following.

Lemma 8.13. *For any $\varphi \in B(\mathbb{S})$ we have*

$$\tilde{\mathbb{E}}[(\rho_t^n(\varphi) - \rho_t(\varphi))^2 \mid \mathcal{Y}_t] = \frac{c_{1,\varphi}(t)}{n}, \qquad (8.26)$$

where $c_{1,\varphi}(t) \triangleq \tilde{\mathbb{E}}[(\varphi(X_t)\tilde{Z}_t - \rho_t(\varphi))^2 \mid \mathcal{Y}_t]$. Moreover

$$\tilde{\mathbb{E}}\left[(\rho_t^n(\varphi) - \rho_t(\varphi))^4 \mid \mathcal{Y}_t\right] \leq \frac{c_{2,\varphi}(t)}{n^2}, \qquad (8.27)$$

where $c_{2,\varphi}(t) \triangleq 6\tilde{\mathbb{E}}[(\varphi(X_t)\tilde{Z}_t - \rho_t(\varphi))^4 \mid \mathcal{Y}_t]$.

Proof. Observe that since the triples (v_j, a_j, Y), $j = 1, \ldots, n$ are identically distributed and have the same distribution as the triple $(X, \tilde{Z}, Y)$, we have for $j = 1, \ldots, m$,

$$\tilde{\mathbb{E}}\left[\varphi(v_j(t))a_j(t) \mid \mathcal{Y}_t\right] = \tilde{\mathbb{E}}\left[\varphi(X_t)\tilde{Z}_t \mid \mathcal{Y}_t\right] = \rho_t(\varphi).$$

In particular

$$\tilde{\mathbb{E}}\left[\rho_t^n(\varphi) \mid \mathcal{Y}_t\right] = \rho_t(\varphi)$$

and the random variables ξ_j^φ, $j = 1, \ldots, n$ defined by

$$\xi_j^\varphi \triangleq \varphi(v_j(t))\, a_j(t) - \rho_t(\varphi), \qquad j = 1, \ldots, n,$$

have zero mean and the same distribution as $\varphi(X_t)\tilde{Z}_t - \rho_t(\varphi)$. It then follows that

$$\frac{1}{n}\sum_{j=1}^n \xi_j^\varphi = \rho_t^n(\varphi) - \rho_t(\varphi).$$

Since the pairs $(v_i(t), a_i(t))$ and $(v_j(t), a_j(t))$ for $i \neq j$, conditional upon $\mathcal{Y}_t$ are independent, it follows that the random variables ξ_j^φ, $j = 1, \ldots, n$ are mutually independent conditional upon $\mathcal{Y}_t$. It follows immediately that

$$\begin{aligned}
\tilde{\mathbb{E}}\left[(\rho_t^n(\varphi) - \rho_t(\varphi))^2 \Big| \mathcal{Y}_t\right] &= \frac{1}{n^2} \tilde{\mathbb{E}}\left[\left(\sum_{j=1}^n \xi_j^\varphi\right)^2 \Big| \mathcal{Y}_t\right] \\
&= \frac{1}{n^2} \sum_{j=1}^n \tilde{\mathbb{E}}\left[(\xi_j^\varphi)^2 \big| \mathcal{Y}_t\right] \\
&= \frac{1}{n^2} \sum_{j=1}^n \tilde{\mathbb{E}}\left[(\varphi(v_j(t))a_j(t) - \rho_t(\varphi))^2 \Big| \mathcal{Y}_t\right] \\
&= \frac{c_{1,\varphi}(t)}{n}.
\end{aligned}$$

Similarly

$$\tilde{\mathbb{E}}\left[(\rho_t^n(\varphi)-\rho_t(\varphi))^4\big|\mathcal{Y}_t\right] = \frac{1}{n^4}\tilde{\mathbb{E}}\left[\left(\sum_{j=1}^n \xi_j^\varphi\right)^4\bigg|\mathcal{Y}_t\right]$$

$$= \frac{1}{n^4}\sum_{j=1}^n \tilde{\mathbb{E}}\left[(\xi_j^\varphi)^4\big|\mathcal{Y}_t\right]$$

$$+ \frac{12}{n^4}\sum_{1\le j_1<j_2\le n} \tilde{\mathbb{E}}\left[(\xi_{j_1}^\varphi)^2\big|\mathcal{Y}_t\right]\tilde{\mathbb{E}}\left[(\xi_{j_2}^\varphi)^2\big|\mathcal{Y}_t\right]$$

$$\le \frac{\tilde{\mathbb{E}}[(\varphi(X_t)\tilde{Z}_t-\rho_t(\varphi))^4]}{n^3} + \frac{6n(n-1)}{n^4}(c_{1,\varphi}(t))^2$$

and the claim follows since, by Jensen's inequality, we have

$$(c_{1,\varphi}(t))^2 \le \tilde{\mathbb{E}}[(\varphi(X_t)\tilde{Z}_t-\rho_t(\varphi))^4].$$

□

Remark 8.14. More generally one can prove that for any integer p and any $\varphi \in B(\mathbb{S})$,

$$\tilde{\mathbb{E}}[(\rho_t^n(\varphi)-\rho_t(\varphi))^{2p}\big|\mathcal{Y}_t] \le \frac{c_{p,\varphi}(t)}{n^p}, \tag{8.28}$$

where

$$c_{p,\varphi}(t) = k_p \tilde{\mathbb{E}}\left[(\varphi(X_t)\tilde{Z}_t-\rho_t(\varphi))^{2p}\big|\mathcal{Y}_t\right], \tag{8.29}$$

where k_p is some universal constant.

Of course, Lemma 8.13 and Remark 8.14 are of little use if the random variables $c_{p,\varphi}(t)$ are not finite a.s. In the following we assume that they are. Under this condition the lemma implies that $\rho_t^n(\varphi)$ converges in expectation to $\rho_t(\varphi)$ for any $\varphi \in B(\mathbb{S})$ with the rate of convergence of order $1/\sqrt{n}$.

Exercise 8.15. Let $c_{p,\varphi}(t)$ be the $\mathcal{Y}_t$-adapted random variable defined in (8.29). Show that if $\tilde{\mathbb{E}}[\tilde{Z}_t^{2p}] < \infty$, then $\tilde{\mathbb{E}}[c_{p,\varphi}(t)] < \infty$, hence the random variable $c_{p,\varphi}(t)$ is finite $\tilde{\mathbb{P}}$-almost surely for any $\varphi \in B(\mathbb{S})$. In particular, show that if the function h is bounded, then $c_{p,\varphi}(t) < \infty$, $\tilde{\mathbb{P}}$-almost surely for any $\varphi \in B(\mathbb{S})$.

The convergence of $\rho_t^n(\varphi)$ to $\rho_t(\varphi)$ is valid for larger classes of function φ (not just bounded functions) provided that $\varphi(X_t)\tilde{Z}_t$ is $\tilde{\mathbb{P}}$-integrable. Moreover, the existence of higher moments of $\varphi(X_t)\tilde{Z}_t$ ensures a control on the rate of convergence. However, in the following we restrict ourselves to just bounded test functions.

Proposition 8.16. *If $\tilde{\mathbb{E}}[\tilde{Z}_t^{2p}] < \infty$, then for any $\varphi \in B(\mathbb{S})$, there exists a finite $\mathcal{Y}_t$-adapted random variable $\bar{c}_{p,\varphi}(t)$ such that for any $\varphi \in B(\mathbb{S})$,*

$$\tilde{\mathbb{E}}[(\pi_t^n(\varphi)-\pi_t(\varphi))^{2p}\big|\mathcal{Y}_t] \le \frac{\bar{c}_{p,\varphi}(t)}{n^p}. \tag{8.30}$$

Proof. Observe that

$$\pi_t^n(\varphi) - \pi_t(\varphi) = \frac{\rho_t^n(\varphi)}{\rho_t^n(\mathbf{1})} \frac{1}{\rho_t(\mathbf{1})}(\rho_t(\mathbf{1}) - \rho_t^n(\mathbf{1})) + \frac{1}{\rho_t(\mathbf{1})}(\rho_t^n(\varphi) - \rho_t(\varphi)),$$

hence, since $|\rho_t^n(\varphi)| \le \|\varphi\|_\infty \rho_t^n(\mathbf{1})$, we have

$$|\pi_t^n(\varphi) - \pi_t(\varphi)| \le \frac{\|\varphi\|_\infty}{\rho_t(\mathbf{1})}|\rho_t^n(\mathbf{1}) - \rho_t(\mathbf{1})| + \frac{1}{\rho_t(\mathbf{1})}|\rho_t^n(\varphi) - \rho_t(\varphi)| \quad (8.31)$$

and, by the triangle inequality,

$$\tilde{\mathbb{E}}[(\pi_t^n(\varphi) - \pi_t(\varphi))^{2p} | \mathcal{Y}_t]^{1/2p} \le \frac{\|\varphi\|_\infty}{\rho_t(\mathbf{1})}\tilde{\mathbb{E}}\left[(\rho_t^n(\mathbf{1}) - \rho_t^n(\mathbf{1}))^{2p} | \mathcal{Y}_t\right]^{1/2p}$$
$$+ \frac{1}{\rho_t(\mathbf{1})}\tilde{\mathbb{E}}\left[(\rho_t^n(\varphi) - \rho_t(\varphi))^{2p} | \mathcal{Y}_t\right]^{1/2p}.$$

Remark 8.14 and Exercise 8.15 imply that there exists a finite $\mathcal{Y}_t$-adapted random variable such that for any $\varphi \in B(\mathbb{S})$ we have

$$\tilde{\mathbb{E}}[(\rho_t^n(\varphi) - \rho_t(\varphi))^{2p} | \mathcal{Y}_t] \le \frac{c_{p,\varphi}(t)}{n^p};$$

hence (8.30) holds with $\bar{c}_{p,\varphi}(t)$ being the $\mathcal{Y}_t$-adapted random variable

$$\bar{c}_{p,\varphi}(t) \triangleq \frac{\left(c_{p,\varphi}(t)^{1/2p} + \|\varphi\|_\infty c_{p,\mathbf{1}}(t)^{1/2p}\right)^{2p}}{\rho_t(\mathbf{1})^{2p}}. \quad (8.32)$$

$\square$

Lemma 8.13 shows the convergence of $\rho_t^n(\varphi)$ to $\rho_t(\varphi)$ when conditioned with respect to the observation σ-algebra $\mathcal{Y}_t$. It also implies the convergence in expectation,[†] and the almost sure convergence of ρ_t^n to ρ_t.

Theorem 8.17. *If $\tilde{\mathbb{E}}[\tilde{Z}_t^2] < \infty$, then for any $\varphi \in B(\mathbb{S})$ we have*

$$\tilde{\mathbb{E}}[|\rho_t^n(\varphi) - \rho_t(\varphi)|] \le \frac{\tilde{c}_1(t)}{\sqrt{n}}\|\varphi\|_\infty,$$

where $\tilde{c}_1(t) \triangleq \sqrt{\tilde{\mathbb{E}}[\tilde{Z}_t^2]}$. In particular e $\lim_{n\to\infty} \rho_t^n = \rho_t$. Moreover, if $\tilde{\mathbb{E}}[\tilde{Z}_t^{2p}] < \infty$, for $p \ge 2$ then for any $\varepsilon \in (0, 1/2 - 1/(2p))$ and $\varphi \in B(\mathbb{S})$ there exists a positive random variable $\tilde{c}_{\varepsilon,p,\varphi}(t)$ which is almost surely finite such that

$$|\rho_t^n(\varphi) - \rho_t(\varphi)| \le \frac{\tilde{c}_{\varepsilon,p,\varphi}(t)}{n^\varepsilon}. \quad (8.33)$$

In particular, ρ_t^n converges to ρ_t, $\tilde{\mathbb{P}}$-almost surely.

[†] Recall that $\rho_t^n \to \rho_t$ in expectation if $\lim_{n\to\infty} \tilde{\mathbb{E}}[|\rho_t^n f - \rho_t f|] = 0$ for all $f \in C_b(\mathbb{S})$. See Section A.10 for the definition of convergence in expectation.

214   8 Numerical Methods for Solving the Filtering Problem

Proof. From Lemma 8.13 we get, using Jensen's inequality, that

$$\tilde{\mathbb{E}}[|\rho_t^n(\varphi) - \rho_t(\varphi)|] \leq \sqrt{\tilde{\mathbb{E}}[(\rho_t^n(\varphi) - \rho_t(\varphi))^2]} = \frac{\sqrt{\tilde{\mathbb{E}}[c_{1,\varphi}(t)]}}{\sqrt{n}},$$

hence the first claim is true since

$$\tilde{\mathbb{E}}[c_{1,\varphi}(t)] = \tilde{\mathbb{E}}[(\varphi(X_t)\tilde{Z}_t - \rho_t(\varphi))^2]$$
$$= \tilde{\mathbb{E}}\left[\varphi(X_t)^2 \tilde{Z}_t^2 - 2\rho_t(\varphi)\tilde{Z}_t\varphi(X_t) + \rho_t(\varphi)^2\right]$$
$$= \tilde{\mathbb{E}}\left[\varphi(X_t)^2 \tilde{Z}_t^2 - \rho_t(\varphi)^2\right]$$
$$\leq \|\varphi\|_\infty^2 \tilde{\mathbb{E}}[\tilde{Z}_t^2].$$

Similarly

$$\tilde{\mathbb{E}}[(\rho_t^n(\varphi) - \rho_t(\varphi))^{2p}] \leq \frac{\tilde{\mathbb{E}}[c_{p,\varphi}(t)]}{n^p},$$

where $c_{p,\varphi}(t)$ is the random variable defined in (8.29), which implies (8.33) and the almost sure convergence of ρ_t^n to ρ follows as a consequence of Remark A.38 in the appendix. □

Let us turn our attention to the convergence of π_t^n. The almost sure convergence of π_t^n to π_t holds under the same conditions as the convergence of ρ_t^n to ρ_t. However, the convergence in expectation of π_t^n to π_t requires an additional integrability condition on $\rho_t^{-1}(\mathbf{1})$.

Theorem 8.18. *If $\tilde{\mathbb{E}}[\tilde{Z}_t^2] < \infty$ and $\tilde{\mathbb{E}}\left[\rho_t^{-2}(\mathbf{1})\right] < \infty$, then for any $\varphi \in B(\mathbb{S})$, we have*

$$\tilde{\mathbb{E}}[|\pi_t^n(\varphi) - \pi_t(\varphi)|] \leq \frac{\hat{c}_1(t)}{\sqrt{n}} \|\varphi\|, \tag{8.34}$$

where $\hat{c}_1(t) = 2\sqrt{\tilde{\mathbb{E}}[\tilde{Z}_t^2]\tilde{\mathbb{E}}[\rho_t^{-2}(\mathbf{1})]}$. In particular π_t^n converges to π_t in expectation. Moreover, if $\tilde{\mathbb{E}}[\tilde{Z}_t^{2p}] < \infty$, for $p \geq 2$ then for any $\varepsilon \in (0, 1/2 - 1/(2p))$ there exists a positive random variable $\hat{c}_{\varepsilon,p,\varphi}(t)$ almost surely finite such that

$$|\pi_t^n(\varphi) - \pi_t(\varphi)| \leq \frac{\hat{c}_{\varepsilon,p,\varphi}(t)}{n^\varepsilon}. \tag{8.35}$$

In particular, π_t^n converges to π_t, $\tilde{\mathbb{P}}$-almost surely.

Proof. By inequality (8.31) and the Cauchy–Schwartz inequality we get that

$$\tilde{\mathbb{E}}[|\pi_t^n(\varphi) - \pi_t(\varphi)|] \leq \|\varphi\| \sqrt{\tilde{\mathbb{E}}\left[\rho_t^{-2}(\mathbf{1})\right] \tilde{\mathbb{E}}\left[|\rho_t^n(\mathbf{1}) - \rho_t(\mathbf{1})|^2\right]}$$
$$+ \sqrt{\tilde{\mathbb{E}}\left[\rho_t^{-2}(\mathbf{1})\right] \tilde{\mathbb{E}}\left[|\rho_t^n(\varphi) - \rho_t(\varphi)|^2\right]}.$$

Moreover, for any $\varphi \in B(\mathbb{S})$, from the proof of Theorem 8.17, it follows that

$$\tilde{\mathbb{E}}[(\rho_t^n(\varphi) - \rho_t(\varphi))^2] \leq \frac{1}{n} \|\varphi\|^2 \tilde{\mathbb{E}}[\tilde{Z}_t^2];$$

hence the first claim is true.

For the almost sure convergence result observe that inequalities (8.31) and (8.33) imply that

$$|\pi_t^n(\varphi) - \pi_t(\varphi)| \leq \frac{\|\varphi\|_\infty}{\rho_t(\mathbf{1})} |\rho_t^n(\mathbf{1}) - \rho_t(\mathbf{1})| + \frac{1}{\rho_t(\mathbf{1})} |\rho_t^n(\varphi) - \rho_t(\varphi)|$$

$$\leq \frac{\|\varphi\|_\infty}{\rho_t(\mathbf{1})} \frac{\tilde{c}_{\varepsilon,p,1}(t)}{n^\varepsilon} + \frac{1}{\rho_t(\mathbf{1})} \frac{\tilde{c}_{\varepsilon,p,\varphi}(t)}{n^\varepsilon}$$

and the claim follows with

$$\hat{c}_{\varepsilon,p,\varphi}(t) = \frac{\|\varphi\|_\infty \tilde{c}_{\varepsilon,p,1}(t) + \tilde{c}_{\varepsilon,p,\varphi}(t)}{\rho_t(\mathbf{1})}.$$

□

Exercise 8.19. Show that $\tilde{\mathbb{E}}[\rho_t^{-2}(\mathbf{1})] < \infty$ if the function h is bounded.

Exercise 8.20. Show that if $\tilde{\mathbb{E}}[\tilde{Z}_t^{2p}] < \infty$, then there exists a positive constant $\tilde{c}_p(t)$ such that for any $\varphi \in B(\mathbb{S})$ we have

$$\tilde{\mathbb{E}}[|\rho_t^n(\varphi) - \rho_t(\varphi)|^p] \leq \frac{\tilde{c}_p(t)}{n^{p/2}} \|\varphi\|_\infty^p.$$

Similarly, show that if $\tilde{\mathbb{E}}[\tilde{Z}_t^{2p}] < \infty$ and $\tilde{\mathbb{E}}[\rho_t^{-2p}(\mathbf{1})] < \infty$, then for any $\varphi \in B(\mathbb{S})$, we have

$$\tilde{\mathbb{E}}[|\pi_t^n(\varphi) - \pi_t(\varphi)|^p] \leq \frac{\hat{c}_p(t)}{n^{p/2}} \|\varphi\|_\infty^p.$$

Let $\mathcal{M} = \{\varphi_i, i \geq 0\}$, where $\varphi_i \in C_b(\mathbb{S})$ be a countable convergence determining set such that $\|\varphi_i\|_\infty \leq 1$ for any $i \geq 0$ and let $d_\mathcal{M}$ be the metric on $\mathcal{M}(\mathbb{S})$ (see Section A.10 for additional details):

$$d_\mathcal{M} : \mathcal{M}(\mathbb{S}) \times \mathcal{M}(\mathbb{S}) \to [0, \infty), \qquad d(\mu, \nu) = \sum_{i=0}^\infty \frac{1}{2^i} |\mu\varphi_i - \nu\varphi_i|.$$

Theorems 8.17 and 8.18 give the following corollary.

Corollary 8.21. *If $\tilde{\mathbb{E}}[\tilde{Z}_t^2] < \infty$, then*

$$\tilde{\mathbb{E}}[d_\mathcal{M}(\rho_t^n, \rho_t)] \leq \frac{2\sqrt{\tilde{\mathbb{E}}[\tilde{Z}_t^2]}}{\sqrt{n}}. \qquad (8.36)$$

Similarly if $\tilde{\mathbb{E}}[\tilde{Z}_t^2] < \infty$ and $\tilde{\mathbb{E}}\left[\rho_t^{-2}(\mathbf{1})\right] < \infty$, then for any $\varphi \in B(\mathbb{S})$, we have

$$\tilde{\mathbb{E}}[d_\mathcal{M}(\pi^n_t, \pi_t)] \leq \frac{4\sqrt{\tilde{\mathbb{E}}[\tilde{Z}^2_t]\tilde{\mathbb{E}}\left[\rho_t^{-2}(\mathbf{1})\right]}}{\sqrt{n}}. \qquad (8.37)$$

Moreover, if $\tilde{\mathbb{E}}[\tilde{Z}^{2p}_t] < \infty$, for $p \geq 2$ then for any $\varepsilon \in (0, 1/2 - 1/(2p))$ there exists a positive random variable $\tilde{c}_\varepsilon(t)$ which is almost surely finite such that

$$d_\mathcal{M}(\rho^n_t, \rho_t) \leq \frac{\tilde{c}_\varepsilon(t)}{n^\varepsilon}, \qquad d_\mathcal{M}(\pi^n_t, \pi) \leq \frac{\tilde{c}_\varepsilon(t)}{n^\varepsilon}. \qquad (8.38)$$

Proof. From Theorem 8.17 we get, using the fact that $\|\varphi_i\|_\infty \leq 1$, that

$$\tilde{\mathbb{E}}[d_\mathcal{M}(\rho^n_t, \rho_t)] \leq \sum_{i=0}^\infty \frac{1}{2^i} \tilde{\mathbb{E}}[|\rho^n_t(\varphi_i) - \rho_t(\varphi_i)|]$$

$$\leq \frac{\tilde{c}_1(t)}{\sqrt{n}} \sum_{i=0}^\infty \frac{1}{2^i} \leq \frac{2\tilde{c}_1(t)}{\sqrt{n}},$$

which establishes (8.36). Inequality (8.37) follows by a similar argument. By the triangle inequality and Exercise 8.20 it follows that

$$\tilde{\mathbb{E}}[d_\mathcal{M}(\rho^n_t, \rho_t)^p]^{1/p} \leq \sum_{i=0}^\infty \frac{1}{2^i} \tilde{\mathbb{E}}[|\rho^n_t(\varphi_i) - \rho_t(\varphi_i)|^p]^{1/p} \leq \frac{\tilde{c}_p(t)}{n^{p/2}} \sum_{i=0}^\infty \frac{1}{2^i}.$$

The first inequality in (8.38) then follows from Remark A.38 in the appendix. The second inequality in (8.38) follows in a similar manner. □

Corollary 8.21 states that both ρ^n_t converges to ρ_t, and π^n_t converges to π_t in expectation with the rate $1/\sqrt{n}$. It also states that the corresponding rate for the almost sure convergence is slightly lower than $1/\sqrt{n}$.

The above analysis requires the existence of higher moments of the martingale $\tilde{Z}$. Of course, the question arises as to what happens if they do not exist and we only know that $\tilde{Z}$ is integrable. In this case π^n_t still converges almost surely to π_t for fixed observation paths $s \mapsto Y_s$ as a consequence of the strong law of large numbers. To state this precisely, it is necessary to use an explicit description of the underlying probability space Ω as a product space, where the processes (X, Y) live on one component and the processes v_j, $j = 1, 2, \ldots$ live on another. The details and the ensuing analysis is cumbersome, so we do not include the details. Moreover, in this case, the random measure π^n_t will not converge to the random measure π_t (over the product space) and no convergence rates may be available. In Chapter 10 we discuss the convergence for fixed observations for the discrete time framework.

Theorems 8.17 and 8.18 and Corollary 8.21 show that the Monte Carlo method will produce approximations for ρ_t, respectively π_t, provided enough particles (independent realizations of the signal) are used. The number of particles depends upon the magnitude of the constants appearing in the upper bounds of the rates of convergence, which in turn depend on the magnitude of the higher moments of the exponential martingale $\tilde{Z}$. This is bad news, because these higher moments of the exponential martingale $\tilde{Z}$ increase very rapidly as functions of time.

The particle picture makes the reason for the deterioration in the accuracy of the approximations with time clearer. Each particle has a trajectory which is independent of the signal trajectory, and its corresponding weight depends on how close its trajectory is to the signal trajectory: the weight is the likelihood of the trajectory given the observation.[†] Typically, most particles' trajectories diverge very quickly from the signal trajectory, with a few 'lucky' ones remaining close to the signal. Therefore the majority of the weights decrease to zero, while a small minority become very large. As a result only the 'lucky' particles will contribute significantly to the sums (8.24) and (8.25) giving the approximations for ρ_t, respectively, π_t. The convergence of the Monte Carlo method is therefore very slow as a large number of particles is needed in order to have a sufficient number of particles in the right area (with correspondingly large weights).

To solve this problem, a wealth of methods have been proposed. In filtering theory, the generic name for these methods is particle filters or sequential Monte Carlo methods. These methods use a correction mechanism that culls particles with small weights and multiplies particles with large weights. The correction procedure depends on the trajectory of the particle and the observation data. This is effective as particles with small weights (i.e. particles with unlikely trajectories/positions) are not carried forward uselessly whereas the most probable regions of the signal state space are explored more thoroughly. The result is a cloud of particles, with those surviving to the current time providing an estimate for the conditional distribution of the signal.

In the following two chapters we study this class of methods in greater detail. In Chapter 9 we discuss such a particle method for the continuous time framework together with corresponding convergence results. In Chapter 10, we look at particle methods for solving the filtering problem in the discrete framework.

8.7 Solutions to Exercises

8.12 It is enough to show that the stochastic integrals

$$I_j = \int_0^t h(v_j(s))^\top \, dY_s, \quad j = 1, \ldots, n$$

are mutually independent given $\mathcal{Y}_t$. This follows immediately from the fact that the random variables

$$I_j^m = \sum_{i=1}^m h(v_j(it/m))^\top (Y_{(i+1)t/m} - Y_{it/m}), \quad j = 1, \ldots, n$$

[†] In Chapter 10 we make this statement precise for the discrete time framework.

are mutually independent given $\mathcal{Y}_t$, hence by the bounded convergence theorem

$$\tilde{\mathbb{E}}\left[\prod_{j=1}^{n}\exp(i\lambda_j I_j)\bigg|\mathcal{Y}_t\right] = \lim_{k\to\infty}\tilde{\mathbb{E}}\left[\prod_{j=1}^{n}\exp(i\lambda_j I_j^{m_k})\bigg|\mathcal{Y}_t\right]$$

$$= \lim_{k\to\infty}\prod_{j=1}^{n}\tilde{\mathbb{E}}\left[\exp(i\lambda_j I_j^{m_k})\big|\mathcal{Y}_t\right]$$

$$= \prod_{j=1}^{n}\tilde{\mathbb{E}}\left[\exp(i\lambda_j I_j)\big|\mathcal{Y}_t\right] \quad (8.39)$$

for any λ_j, $j=1,\ldots,n$. In (8.39), $(I_j^{m_k})_{k>0}$ is a suitably chosen subsequence of $(I_j^m)_{m>0}$ so that $I_j^{m_k}$ converges to I_j almost surely.

8.15 From (8.29) and the inequality $(a+b)^k \leq 2^{k-1}(a^k+b^k)$,

$$\tilde{\mathbb{E}}[c_{p,\varphi}] = k_p\tilde{\mathbb{E}}\left[\left(\varphi(X_t)\tilde{Z}_t - \rho_t(\varphi)\right)^{2p}\right]$$

$$\leq 2^{2p-1}k_p\tilde{\mathbb{E}}\left[(\varphi(X_t)\tilde{Z}_t)^{2p} + (\rho_t(\varphi))^{2p}\right]$$

$$\leq 2^{2p-1}k_p\|\varphi\|_\infty^{2p}\left(\tilde{\mathbb{E}}[\tilde{Z}_t^{2p}] + \tilde{\mathbb{E}}[(\rho_t(\mathbf{1}))^{2p}]\right).$$

The first term is bounded by the assumption $\tilde{\mathbb{E}}[\tilde{Z}^{2p}] < \infty$; for the second term use the conditional form of Jensen

$$\tilde{\mathbb{E}}\left[(\rho_t(\mathbf{1}))^{2p}\right] = \tilde{\mathbb{E}}\left[(\tilde{\mathbb{E}}[\tilde{Z}_t\mid\mathcal{Y}_t])^{2p}\right] \leq \tilde{\mathbb{E}}\left[\tilde{\mathbb{E}}[\tilde{Z}_t^{2p}\mid\mathcal{Y}_t]\right] = \tilde{\mathbb{E}}[\tilde{Z}_t^{2p}] < \infty.$$

Therefore $\mathbb{E}[c_{p,\varphi}] < \infty$, which implies that $c_{p,\varphi} < \infty$ $\tilde{\mathbb{P}}$-a.s.

For the second part, where h is bounded, use the explicit form

$$\tilde{Z}_t^{2p} = \exp\left(2p\sum_{i=1}^{m}\int_0^t h^i\,\mathrm{d}Y_s^i - p\sum_{i=1}^{m}\int_0^t h^i(X_s)^2\,\mathrm{d}s\right)$$

$$\leq \exp((2p^2-p)mt\|h\|_\infty^2)\Theta_t,$$

where $\Theta = \{\Theta_t,\ t\geq 0\}$ is the exponential martingale

$$\Theta_t \triangleq \exp\left(2p\sum_{i=1}^{m}\int_0^t h^i\,\mathrm{d}Y_s^i - \frac{(2p)^2}{2}\sum_{i=1}^{m}\int_0^t h^i(X_s)^2\,\mathrm{d}s\right).$$

The boundedness of h implies that Θ is a genuine martingale via Novikov's condition (see Theorem B.34). Taking expectations, we see that $\tilde{\mathbb{E}}[\tilde{Z}_t^{2p}]$ is bounded by $\exp((2p^2-p)mt\|h\|_\infty^2)$.

8.19 By Jensen's inequality

$$\tilde{\mathbb{E}}[\rho_t^{-2}(\mathbf{1})] = \tilde{\mathbb{E}}\left[\tilde{\mathbb{E}}\left[\tilde{Z}_t \mid \mathcal{Y}_t\right]^{-2}\right] \le \tilde{\mathbb{E}}\left[\tilde{\mathbb{E}}\left[\tilde{Z}_t^{-2} \mid \mathcal{Y}_t\right]\right] = \tilde{\mathbb{E}}\left[\tilde{Z}_t^{-2}\right]$$

and from the explicit form for $\tilde{Z}_t$,

$$\tilde{Z}_t^{-2} = \exp\left(-2\int_0^t h(X_s)^\top \, \mathrm{d}Y_s + \int_0^t \|h(X_s)\|^2 \, \mathrm{d}s\right)$$
$$\le \exp(3mt\|h\|_\infty^2)\bar{\Theta}_t,$$

where $\bar{\Theta} = \{\bar{\Theta}_t, t \ge 0\}$ is the exponential martingale

$$\bar{\Theta}_t \triangleq \exp\left(-2\sum_{i=1}^m \int_0^t h^i \, \mathrm{d}Y_s^i - 2\sum_{i=1}^m \int_0^t h^i(X_s)^2 \, \mathrm{d}s\right).$$

The boundedness of h implies that $\bar{\Theta}$ is a genuine martingale via Novikov's condition (see Theorem B.34). Taking expectations, we see that $\tilde{\mathbb{E}}[\rho_t^{-2}(\mathbf{1})]$ is bounded by $\exp(3mt\|h\|_\infty^2)$.

8.20 By Jensen's inequality and (8.28)

$$\tilde{\mathbb{E}}\left[|\rho_t^n(\varphi) - \rho(\varphi)|^p\right] \le \sqrt{\tilde{\mathbb{E}}\left[(\rho_t^n(\varphi) - \rho(\varphi))^{2p}\right]}$$
$$\le \sqrt{\tilde{\mathbb{E}}\left[\tilde{\mathbb{E}}\left[(\rho_t^n(\varphi) - \rho(\varphi))^{2p} \mid \mathcal{Y}_t\right]\right]}$$
$$\le \frac{\sqrt{\tilde{\mathbb{E}}[c_{p,\varphi}(t)]}}{n^{p/2}}.$$

From the computations in Exercise 8.15,

$$\tilde{\mathbb{E}}[c_{p,\varphi}(t)] \le K_p(t)\|\varphi\|_\infty^{2p},$$

where

$$K_p(t) = 4^p k_p \tilde{\mathbb{E}}\left[\tilde{Z}_t^{2p}\right],$$

thus

$$\tilde{\mathbb{E}}\left[|\rho_t^n(\varphi) - \rho(\varphi)|^p\right] \le \frac{\sqrt{K_p(t)}\|\varphi\|_\infty^p}{n^{p/2}}.$$

Therefore the result follows with $\tilde{c}_p(t) = \sqrt{K_p(t)}$.

For the second part, from (8.31) and the inequality $(a+b)^p < 2^{p-1}(a^p+b^p)$,

$$|\pi_t^n(\varphi) - \pi(\varphi)|^p \le 2^{p-1}\frac{\|\varphi\|_\infty^p}{\rho_t(\mathbf{1})^p}|\rho_t^n(\mathbf{1}) - \rho_t(\mathbf{1})|^p + \frac{2^{p-1}}{\rho_t(\mathbf{1})^p}|\rho_t^n(\varphi) - \rho_t(\varphi)|^p,$$

so by Cauchy–Schwartz

$$\tilde{\mathbb{E}}\left[|\pi_t^n(\varphi) - \pi(\varphi)|^p\right] \leq 2^{p-1} \|\varphi\|_\infty^p \sqrt{\tilde{\mathbb{E}}\left[\rho_t(1)^{-2p}\right] \tilde{\mathbb{E}}\left[\frac{c_{p,1}}{n^p}\right]}$$

$$+ 2^{p-1} \sqrt{\tilde{\mathbb{E}}\left[\rho_t(1)^{-2p}\right] \tilde{\mathbb{E}}\left[\frac{c_{p,\varphi}}{n^p}\right]}$$

$$\leq 2^{p-1} \frac{\sqrt{\tilde{\mathbb{E}}\left[\rho_t(1)^{-2p}\right]}}{n^{p/2}} \left(\|\varphi\|_\infty^p \sqrt{\tilde{\mathbb{E}}\left[c_{p,1}\right]} + \sqrt{\tilde{\mathbb{E}}\left[c_{p,\varphi}\right]}\right)$$

$$\leq 2^{p-1} \frac{\sqrt{\tilde{\mathbb{E}}\left[\rho_t(1)^{-2p}\right]}}{n^{p/2}} \|\varphi\|_\infty^p 2\sqrt{K_p(t)},$$

so the result follows with

$$\hat{c}_p(t) = 2^p \sqrt{K_p(t)} \sqrt{\tilde{\mathbb{E}}\left[\rho_t(1)^{-2p}\right]}.$$

9
A Continuous Time Particle Filter

9.1 Introduction

Throughout this chapter, we take the signal X to be the solution of (3.9); that is, $X = (X^i)_{i=1}^d$ is the solution of the stochastic differential equation

$$dX_t = f(X_t)dt + \sigma(X_t)\,dV_t, \tag{9.1}$$

where $f : \mathbb{R}^d \to \mathbb{R}^d$ and $\sigma : \mathbb{R}^d \to \mathbb{R}^{d \times p}$ are bounded and globally Lipschitz functions and $V = (V^j)_{j=1}^p$ is a p-dimensional Brownian motion. As discussed in Section 3.2, the generator A associated with the process X is the second-order differential operator,

$$A = \sum_{i=1}^d f^i \frac{\partial}{\partial x_i} + \sum_{i,j=1}^d a^{ij} \frac{\partial^2}{\partial x_i \partial x_j},$$

where $a = \frac{1}{2}\sigma\sigma^\top$. Since both f and a are bounded, the domain of the generator A, $\mathcal{D}(A)$ is $C_b^2(\mathbb{R}^d)$, the space of bounded twice continuously differentiable functions with bounded first and second partial derivatives; for any $\varphi \in C_b^2(\mathbb{R}^d)$, the process $M^\varphi = \{M_t^\varphi,\, t \geq 0\}$ defined by[†]

$$M_t^\varphi \triangleq \varphi(X_t) - \varphi(X_0) - \int_0^t A\varphi(X_s)\,ds,$$

$$= \int_0^t ((\nabla\varphi)^\top \sigma)(X_s)\,dV_s,\ t \geq 0$$

is an $\mathcal{F}_t$-adapted martingale.

The observation process is the solution of the evolution equation (3.5); that is, Y is an m-dimensional stochastic process that satisfies

$$dY_t = h(X_t)\,dt + dW_t,$$

[†] In the following $(\nabla\varphi)^\top$ is the row vector $(\partial_1\varphi, \ldots, \partial_d\varphi)$.

A. Bain, D. Crisan, *Fundamentals of Stochastic Filtering*,
DOI 10.1007/978-0-387-76896-0_9, © Springer Science+Business Media, LLC 2009

where $h = (h_i)_{i=1}^m : \mathbb{R}^d \to \mathbb{R}^m$ is a bounded measurable function and W is a standard m-dimensional Brownian motion independent of X. Since h is bounded, condition (3.25) is satisfied. Hence the process $Z = \{Z_t,\ t > 0\}$ defined by

$$Z_t \triangleq \exp\left(-\int_0^t h(X_s)^\top \, dW_s - \frac{1}{2}\int_0^t \|h(X_s)\|^2 \, ds\right), \qquad t \geq 0, \qquad (9.2)$$

is a genuine martingale and the probability $\tilde{\mathbb{P}}$ whose Radon–Nikodym derivative with respect to $\mathbb{P}$ is given on $\mathcal{F}_t$ by Z_t, viz

$$\left.\frac{d\tilde{\mathbb{P}}}{d\mathbb{P}}\right|_{\mathcal{F}_t} = Z_t,$$

is well defined (see Section 3.3 for details, also Theorem B.34 and Corollary B.31). As was shown in Chapter 3, under $\tilde{\mathbb{P}}$, the process Y is a Brownian motion independent of X. Then the Kallianpur–Striebel formula (3.33) states that

$$\pi_t(\varphi) = \frac{\rho_t(\varphi)}{\rho_t(\mathbf{1})}, \qquad \tilde{\mathbb{P}}(\mathbb{P})\text{-a.s.},$$

where ρ_t is the unnormalized conditional distribution of X, which satisfies

$$\rho_t(\varphi) = \tilde{\mathbb{E}}\left[\varphi(X_t)\tilde{Z}_t \mid \mathcal{Y}_t\right]$$

for any bounded Borel-measurable function φ and

$$\tilde{Z}_t = \exp\left(\int_0^t h(X_s)^\top \, dY_s - \frac{1}{2}\int_0^t \|h(X_s)\|^2 \, ds\right). \qquad (9.3)$$

Similar to the Monte Carlo method which is described in Section 8.6, the particle filter presented below produces a measure-valued process $\pi^n = \{\pi_t^n,\ t \geq 0\}$ which represents the empirical measure of n (random) particles with varying weights

$$\pi_t^n \triangleq \sum_{j=1}^n \bar{a}_j^n(t)\delta_{v_j^n(t)}, \qquad t \geq 0.$$

The difference between the Monte Carlo method described earlier and the particle filter which we are about to describe is the presence of an additional correction procedure, which is applied at regular intervals to the system of particles. At the correction times, each particle is replaced by a random number of particles (possibly zero). We say that the particles *branch* into a random number of offspring. This is done in a consistent manner so that particles with small weights have no offspring (i.e. are killed), and particles with large weights are replaced by several offspring.

The chapter is organised as follows. In the following section we describe in detail the particle filter and some of its properties. In Section 9.3 we review the dual of the process ρ, which was introduced in Chapter 7, and give a number of preliminary results. The convergence results are proved in Section 9.4.

9.2 The Approximating Particle System

The particle system at time 0 consists of n particles all with equal weights $1/n$, and positions $v_j^n(0)$, for $j = 1, \ldots, n$. We choose the initial positions of the particles to be independent, identically distributed random variables with common distribution π_0, for $j, n \in \mathbb{N}$. Hence the approximating measure at time 0 is

$$\pi_0^n = \frac{1}{n} \sum_{j=1}^n \delta_{v_j^n(0)}.$$

The time interval $[0, \infty)$ is partitioned into sub-intervals of equal length δ. During the time interval $[i\delta, (i+1)\delta)$, the particles all move with the same law as the signal X; that is, for $t \in [i\delta, (i+1)\delta)$,

$$v_j^n(t) = v_j^n(i\delta) + \int_{i\delta}^t f(v_j^n(s))\,\mathrm{d}s + \int_{i\delta}^t \sigma(v_j^n(s))\,\mathrm{d}V_s^{(j)}, \quad j = 1,\ldots,n, \quad (9.4)$$

where $(V^{(j)})_{j=1}^n$ are mutually independent $\mathcal{F}_t$-adapted p-dimensional Brownian motions which are independent of Y, and independent of all other random variables in the system. The notation $V^{(j)}$ is used to make it clear that these are not the components of each p-dimensional Brownian motion. The weights $\bar{a}_j^n(t)$ are of the form

$$\bar{a}_j^n(t) \triangleq \frac{a_j^n(t)}{\sum_{k=1}^n a_k^n(t)},$$

where

$$a_j^n(t) = 1 + \sum_{k=1}^m \int_{i\delta}^t a_j^n(s) h^k(v_j^n(s))\,\mathrm{d}Y_s^k; \quad (9.5)$$

in other words

$$a_j^n(t) = \exp\left(\int_{i\delta}^t h(v_j^n(s))^\top \,\mathrm{d}Y_s - \frac{1}{2}\int_{i\delta}^t \|h(v_j^n(s))\|^2\,\mathrm{d}s\right). \quad (9.6)$$

For $t \in [i\delta, (i+1)\delta)$, define

$$\pi_t^n \triangleq \sum_{j=1}^n \bar{a}_j^n(t) \delta_{v_j^n(t)}.$$

At the end of the interval, each particle branches into a random number of particles. Each offspring particle initially inherits the spatial position of its parent. After branching all the particles are reindexed (from 1 to n) and all of the (unnormalized) weights are reinitialised back to 1. When necessary, we use the notation $j' = 1, 2, \ldots, n$ to denote the particle index prior to the branching event, to distinguish it from the index after the branching event which we denote by $j = 1, 2, \ldots, n$. Let $o_{j'}^{n,(i+1)\delta}$ be the number of offspring produced by the j'th particle at time $(i+1)\delta$ in the n-particle approximating system. Then $o_{j'}^{n,(i+1)\delta}$ is $\mathcal{F}_{(i+1)\delta}$-adapted and[†]

$$o_{j'}^{n,(i+1)\delta} \triangleq \begin{cases} \left[n\bar{a}_{j'}^{n,(i+1)\delta}\right] & \text{with prob. } 1 - \{n\bar{a}_{j'}^{n,(i+1)\delta}\} \\ \left[n\bar{a}_{j'}^{n,(i+1)\delta}\right] + 1 & \text{with prob. } \{n\bar{a}_{j'}^{n,(i+1)\delta}\}, \end{cases} \quad (9.7)$$

where $\bar{a}_{j'}^{n,(i+1)\delta}$ is the value of the particle's weight immediately prior to the branching; in other words,

$$\bar{a}_{j'}^{n,(i+1)\delta} = \bar{a}_{j'}^{n}((i+1)\delta-) = \lim_{t \nearrow (i+1)\delta} \bar{a}_{j'}^{n}(t). \quad (9.8)$$

Hence if $\mathcal{F}_{(i+1)\delta-}$ is the σ-algebra of events up to time $(i+1)\delta$, viz

$$\mathcal{F}_{(i+1)\delta-} = \sigma(\mathcal{F}_s, \ s < (i+1)\delta),$$

then from (9.7),

$$\mathbb{E}\left[o_{j'}^{n,(i+1)\delta} \mid \mathcal{F}_{(i+1)\delta-}\right] = n\bar{a}_{j'}^{n,(i+1)\delta}, \quad (9.9)$$

and the conditional variance of the number of offspring is

$$\mathbb{E}\left[\left(o_{j'}^{n,(i+1)\delta}\right)^2 \mid \mathcal{F}_{(i+1)\delta-}\right] - \left(\mathbb{E}\left[o_{j'}^{n,(i+1)\delta} \mid \mathcal{F}_{(i+1)\delta-}\right]\right)^2$$
$$= \{n\bar{a}_{j'}^{n,(i+1)\delta}\}\left(1 - \{n\bar{a}_{j'}^{n,(i+1)\delta}\}\right).$$

Exercise 9.1. Let $a > 0$ be a positive constant and $\mathcal{A}_a$ be the set of all integer-valued random variables ξ such that $\mathbb{E}[\xi] = a$, viz

$$\mathcal{A}_a \triangleq \{\xi : \Omega \to \mathbb{N} \mid \mathbb{E}[\xi] = a\}.$$

Let $\operatorname{var}(\xi) = \mathbb{E}[\xi^2] - a^2$ be the variance of an arbitrary random variable $\xi \in \mathcal{A}_a$. Show that there exists a random variable $\xi^{\min} \in \mathcal{A}_a$ with minimal variance. That is, $\operatorname{var}(\xi^{\min}) \leq \operatorname{var}(\xi)$ for any $\xi \in \mathcal{A}_a$. Moreover show that

$$\xi^{\min} = \begin{cases} [a] & \text{with prob. } 1 - \{a\} \\ [a] + 1 & \text{with prob. } \{a\} \end{cases} \quad (9.10)$$

[†] In the following, $[x]$ is the largest integer smaller than x and $\{x\}$ is the fractional part of x; that is, $\{x\} = x - [x]$.

and $\operatorname{var}(\xi^{\min}) = \{a\}(1-\{a\})$. More generally show that $\mathbb{E}[\varphi(\xi^{\min})] \leq \mathbb{E}[\varphi(\xi)]$ for any convex function $\varphi : \mathbb{R} \to \mathbb{R}$.

Remark 9.2. Following Exercise 9.1, we deduce that the random variables $o_{j'}^{n,(i+1)\delta}$ defined by (9.7) have conditional minimal variance in the set of all integer-valued random variables ξ such that $\mathbb{E}[\xi \mid \mathcal{F}_{(i+1)\delta-}] = n\bar{a}_{j'}^{n,(i+1)\delta}$ for $j = 1, \ldots, n$. This property is important as it is the variance of the random variables o_j^n that influences the speed of convergence of the corresponding algorithm.

9.2.1 The Branching Algorithm

We wish to control the branching process so that the number of particles in the system remains constant at n; that is, we require that for each i,

$$\sum_{j'=1}^{n} o_{j'}^{n,(i+1)\delta} = n,$$

which implies that the random variables $o_{j'}^{n,(i+1)\delta}$, $j' = 1, \ldots, n$ will be correlated.

Let $u_{j'}^{n,(i+1)\delta}$, $j' = 1, \ldots, n-1$ be $n-1$ mutually independent random variables, uniformly distributed on $[0,1]$, which are independent of all other random variables in the system. To simplify notation in the statement of the algorithm, we omit the superscript $(i+1)\delta$ in the notation for $o_{j'}^{n,(i+1)\delta}$, $\bar{a}_{j'}^{n,(i+1)\delta}$ and $u_{j'}^{n,(i+1)\delta}$. The following algorithm is then applied.

```
g := n      h := n
for j' := 1 to n − 1
    if {nā_{j'}^n} + {g − nā_{j'}^n} < 1 then
        if u_{j'}^n < 1 − ({nā_{j'}^n}/{g}) then
            o_{j'}^n := [nā_{j'}^n]
        else
            o_{j'}^n := [nā_{j'}^n] + (h − [g])
        end if
    else
        if u_{j'}^n < 1 − (1 − {nā_{j'}^n})/(1 − {g}) then
            o_{j'}^n := [nā_{j'}^n] + 1
        else
            o_{j'}^n := [nā_{j'}^n] + (h − [g])
        end if
    end if
    g := g − nā_{j'}^n
    h := h − o_{j'}^n
end for
o_n^n := h
```

Some of the properties of the random variables $\{o_{j'}^n, j' = 1, \ldots, n\}$ are given by the following proposition. Since there is no risk of confusion, in the statement and proof of this proposition, the primes on the indices are omitted and thus the variables are denoted $\{o_j^n\}_{j=1}^n$.

Proposition 9.3. *The random variables o_j^n for $j = 1, \ldots, n$ have the following properties.*

a. $\sum_{j=1}^n o_j^n = n$.
b. *For any $j = 1, \ldots, n$ we have $\mathbb{E}[o_j^n] = n\bar{a}_j^n$.*
c. *For any $j = 1, \ldots, n$, o_j^n has minimal variance, specifically*

$$\mathbb{E}[(o_j^n - n\bar{a}_j^n)^2] = \{n\bar{a}_j^n\}(1 - \{n\bar{a}_j^n\}).$$

d. *For any $k = 1, \ldots, n-1$, the random variables $o_{1:k}^n = \sum_{j=1}^k o_j^n$, and $o_{k+1:n}^n = \sum_{j=k+1}^n o_j^n$ have variance*

$$\mathbb{E}[(o_{1:k}^n - n\bar{a}_{1:k}^n)^2] = \{n\bar{a}_{1:k}^n\}\left(1 - \{n\bar{a}_{1:k}^n\}\right).$$
$$\mathbb{E}[(o_{k+1:n}^n - n\bar{a}_{k+1:n}^n)^2] = \{n\bar{a}_{k+1:n}^n\}\left(1 - \{n\bar{a}_{k+1:n}^n\}\right),$$

where $\bar{a}_{1:k}^n = \sum_{j=1}^k \bar{a}_j^n$ and $\bar{a}_{k+1:n}^n = \sum_{j=k+1}^n \bar{a}_j^n$.
e. *For $1 \le i < j \le n$, the random variables o_i^n and o_j^n are negatively correlated. That is,*

$$\mathbb{E}[(o_i^n - n\bar{a}_i^n)(o_j^n - n\bar{a}_j^n)] \le 0.$$

Proof. Property (a) follows immediately from the fact that o_n^n is defined as

$$o_n^n = n - \sum_{j'=1}^{n-1} o_{j'}^n.$$

For properties (b), (c) and (d), we proceed by induction. First define the sequence of σ-algebras

$$\mathcal{U}_k = \sigma(\{u_j^n, \ j = 1, \ldots, k\}), \quad k = 1, \ldots, n-1,$$

where u_j^n, $j = 1, \ldots, n-1$ are the random variables used to construct the $o_j^{n'}$s. Then from the algorithm,

$$o_1^n = [n\bar{a}_1^n] + 1_{[0, \{n\bar{a}_1^n\}]}(u_1^n);$$

hence o_1^n has mean $n\bar{a}_1^n$ and minimal variance from Exercise 9.1. As a consequence of property (a), it also holds that $o_{2:n}^n$ has minimal variance. The induction step follows from the fact that h stores the number of offspring which are not yet assigned and g stores the sum of their corresponding means. In other words at the kth iteration for $k \ge 2$, $h = o_{k:n}^n = n - o_{1:k-1}^n$ and $g = n\bar{a}_{k:n}^n = n - n\bar{a}_{1:k-1}^n$. It is clear that $\{n\bar{a}_k^n\} + \{n\bar{a}_{k+1:n}^n\}$ is either equal

9.2 The Approximating Particle System

to $\{n\bar{a}^n_{k:n}\}$ or $\{n\bar{a}^n_{k:n}\} + 1$. In the first of these cases, from the algorithm it follows that for $k \geq 2$,

$$o^n_k = [n\bar{a}^n_k] + (o^n_{k:n} - [n\bar{a}^n_{k:n}])\, 1_{[1-\{n\bar{a}^n_k\}/\{n\bar{a}^n_{k:n}\},1]}(u^n_k), \qquad (9.11)$$

from which it follows from the fact that $o^n_{k+1:n} + o^n_k = o^n_{k:n}$, that

$$o^n_{k+1:n} = [n\bar{a}^n_{k+1:n}] + (o^n_{k:n} - [n\bar{a}^n_{k:n}])\, 1_{[0,1-\{n\bar{a}^n_k\}/\{n\bar{a}^n_{k:n}\}]}(u^n_k); \qquad (9.12)$$

hence, using the fact that $o^n_{k:n}$ is $\mathcal{U}_{k-1}$-measurable and u^n_k is independent of $\mathcal{U}_{k-1}$, we get from (9.11) that

$$\mathbb{E}\left[(o^n_k - n\bar{a}^n_k) \mid \mathcal{U}_{k-1}\right] = -\{n\bar{a}^n_k\} + (o^n_{k:n} - [n\bar{a}^n_{k:n}]) \frac{\{n\bar{a}^n_k\}}{\{n\bar{a}^n_{k:n}\}}$$

$$= \frac{\{n\bar{a}^n_k\}}{\{n\bar{a}^n_{k:n}\}} (o^n_{k:n} - n\bar{a}^n_{k:n})$$

$$= (o^n_{k:n} - n\bar{a}^n_{k:n}) \frac{\{n\bar{a}^n_k\}}{\{n\bar{a}^n_{k:n}\}} \qquad (9.13)$$

and by a similar calculation

$$\mathbb{E}[(o^n_k - n\bar{a}^n_k)^2 \mid \mathcal{U}_{k-1}]$$

$$= (o^n_{k:n} - n\bar{a}^n_{k:n})^2 \frac{\{n\bar{a}^n_k\}}{\{n\bar{a}^n_{k:n}\}} + (\{n\bar{a}^n_{k:n}\} - \{n\bar{a}^n_k\})\{n\bar{a}^n_k\}$$

$$+ 2(o^n_{k:n} - n\bar{a}^n_{k:n})(\{n\bar{a}^n_{k:n}\} - \{n\bar{a}^n_k\}) \frac{\{n\bar{a}^n_k\}}{\{n\bar{a}^n_{k:n}\}}. \qquad (9.14)$$

The identities (9.13), (9.14) and the corresponding identities derived from (9.12), viz:

$$\mathbb{E}\left[o_{k+1:n} - n\bar{a}^n_{k+1:n} \mid \mathcal{U}_{k-1}\right] = (o_{k:n} - n\bar{a}^n_{k:n})\left(1 - \frac{\{n\bar{a}^n_k\}}{\{n\bar{a}^n_{k:n}\}}\right)$$

and

$$\mathbb{E}[(o^n_{k+1:n} - n\bar{a}^n_{k+1:n})^2 \mid \mathcal{U}_{k-1}] = (o^n_{k:n} - n\bar{a}^n_{k:n})^2 \left(1 - \frac{\{n\bar{a}^n_k\}}{\{n\bar{a}^n_{k:n}\}}\right)$$

$$+ 2(o^n_{k:n} - n\bar{a}^n_{k:n})\{n\bar{a}^n_k\}\left(1 - \frac{\{n\bar{a}^n_k\}}{\{n\bar{a}^n_{k:n}\}}\right)$$

$$+ (\{n\bar{a}^n_{k:n}\} - \{n\bar{a}^n_k\})\{n\bar{a}_k\}$$

which give the induction step for properties (b), (c) and (d).

For example, in the case of (b), taking expectation over (9.13) we see that

$$\mathbb{E}\left[o^n_k - n\bar{a}^n_k\right] = \frac{\{n\bar{a}^n_k\}}{\{n\bar{a}^n_{k:n}\}} \mathbb{E}\left[o^n_{k:n} - n\bar{a}^n_{k:n}\right]$$

and the right-hand side is zero by the inductive hypothesis. The case $\{n\bar{a}_k^n\} + \{n\bar{a}_{k+1:n}^n\} = \{n\bar{a}_{k:n}^n\} + 1$ is treated in a similar manner. Finally, for the proof of property (e) one shows first that for $j > i$,

$$\mathbb{E}\left[(o_j^n - n\bar{a}_j^n) \mid \mathcal{U}_i\right] = c_{i:j}\left(o_{i+1:n}^n - n\bar{a}_{i+1:n}^n\right)$$

$$c_{i:j} = p_j \prod_{k=i}^{j-2} q_k \geq 0,$$

where we adopt the convention $\prod_{k=i}^{j-2} q_k = 1$ if $i = j-1$, and where

$$p_j = \begin{cases} \{n\bar{a}_j^n\}/\{n\bar{a}_{j:n}^n\} & \text{if } \{n\bar{a}_j^n\} + \{n\bar{a}_{j+1:n}^n\} = \{n\bar{a}_{j:n}^n\} \\ (1 - \{n\bar{a}_j^n\})/(1 - \{n\bar{a}_{j:n}^n\}) & \text{if } \{n\bar{a}_j^n\} + \{n\bar{a}_{j+1:n}^n\} = \{n\bar{a}_{j:n}^n\} + 1 \end{cases}$$

$$q_k = \begin{cases} \{n\bar{a}_{k:n}^n\}/\{n\bar{a}_{k-1:n}^n\} & \text{if } \{n\bar{a}_{k-1}^n\} + \{n\bar{a}_{k:n}^n\} = \{n\bar{a}_{k-1:n}^n\} \\ (1 - \{n\bar{a}_{k:n}^n\})/(1 - \{n\bar{a}_{k-1:n}^n\}) & \text{otherwise.} \end{cases}$$

Then, for $j > i$

$$\mathbb{E}\left[(o_i^n - n\bar{a}_i^n)(o_j^n - n\bar{a}_j^n)\right] = c_{i:j}\mathbb{E}\left[(o_i^n - n\bar{a}_i^n)(o_{i+1:n}^n - n\bar{a}_{i+1:n}^n)\right]$$
$$= -r_i c_{i:j},$$

where

$$r_i = \begin{cases} \{n\bar{a}_i^n\}\{n\bar{a}_{i+1:n}^n\} & \text{if } \{n\bar{a}_i^n\} + \{n\bar{a}_{i+1:n}^n\} = \{n\bar{a}_{i:n}^n\} \\ (1 - \{n\bar{a}_i^n\})(1 - \{n\bar{a}_{i+1:n}^n\}) & \text{if } \{n\bar{a}_i^n\} + \{n\bar{a}_{i+1:n}^n\} = \{n\bar{a}_{i:n}^n\} + 1. \end{cases}$$

As $r_i > 0$ and $c_{i:j} > 0$, it follows that

$$\mathbb{E}\left[(o_i^n - n\bar{a}_i^n)(o_j^n - n\bar{a}_j^n)\right] < 0.$$

□

Remark 9.4. Proposition 9.3 states that the algorithm presented above produces an n-tuple of integer-valued random variables o_j^n for $j = 1, \ldots, n$ with minimal variance, negatively correlated and whose sum is always n. Moreover, not only do the individual o_j^ns have minimal variance, but also any sum of the form $\sum_{j=1}^{k} o_j^n$ or $\sum_{j=k}^{n} o_j^n$ is an integer-valued random variable with minimal variance for any $k = 1, \ldots, n$. This additional property can be interpreted as a further restriction on the random perturbation introduced by the branching correction.

Remark 9.5. Since the change of measure from $\mathbb{P}$ to $\tilde{\mathbb{P}}$ does not affect the distribution of the random variables $u_{j'}^n$, for $j' = 1, \ldots, n-1$, all the properties stated in Proposition 9.3 hold true under $\tilde{\mathbb{P}}$ as well.

Lemma 9.6. *The process* $\pi^n = \{\pi^n_t, \ t \geq 0\}$ *is a probability measure-valued process with càdlàg paths. In particular, π^n is continuous on any interval $[i\delta, (i+1)\delta)$, $i \geq 0$. Also, for any $i > 0$ we have*

$$\mathbb{E}[\pi^n_{i\delta} \mid \mathcal{F}_{i\delta-}] = \lim_{t \nearrow i\delta} \pi^n_t. \tag{9.15}$$

The same identity holds true under the probability measure $\tilde{\mathbb{P}}$. That is,

$$\tilde{\mathbb{E}}[\pi^n_{i\delta} \mid \mathcal{F}_{i\delta-}] = \lim_{t \nearrow i\delta} \pi^n_t.$$

Proof. Since the pair processes $(\bar{a}^n_j(t), v^n_j(t))$, $j = 1, 2, \ldots, n$ are continuous in the interval $[i\delta, (i+1)\delta)$ it follows that for any $\varphi \in C_b(\mathbb{R}^d)$ the function

$$\pi^n_t(\varphi) = \sum_{j=1}^n \bar{a}^n_j(t) \varphi(v^n_j(t))$$

is continuous for $t \in (i\delta, (i+1)\delta)$. Hence π^n is continuous with respect to the weak topology on $\mathcal{M}(\mathbb{R}^d)$ for $t \in (i\delta, (i+1)\delta)$, for each $i \geq 0$. By the same argument, π^n is right continuous and has left limits at $i\delta$ for any $i > 0$. For any $t \geq 0$,

$$\pi^n_t(\mathbf{1}) = \sum_{j=1}^n \bar{a}^n_j(t) = 1,$$

therefore π^n is probability measure-valued.

The identity (9.15) follows by observing that at the time $i\delta$ the weights are reset to one; thus for $\varphi \in B(\mathbb{R}^d)$, it follows that

$$\pi^n_{i\delta}(\varphi) = \frac{1}{n} \sum_{j'=1}^n o^{n,i\delta}_{j'} \varphi(v^n_{j'}(i\delta))$$

and from (9.8) and (9.9), we have

$$\mathbb{E}\left[\pi^n_{i\delta}(\varphi) \mid \mathcal{F}_{i\delta-}\right] = \frac{1}{n} \sum_{j'=1}^n \mathbb{E}[o^{n,i\delta}_{j'} \mid \mathcal{F}_{i\delta-}] \varphi\left(v^n_{j'}(i\delta)\right)$$

$$= \sum_{j'=1}^n \bar{a}^{n,i\delta}_{j'} \varphi\left(v^n_{j'}(i\delta)\right)$$

$$= \lim_{t \nearrow i\delta} \sum_{j'=1}^n \bar{a}^n_{j'}(t) \varphi\left(v^n_{j'}(t)\right).$$

Finally, from Remark 9.5, since the change of measure from $\mathbb{P}$ to $\tilde{\mathbb{P}}$ does not affect the distribution of the random variables $u^n_{j'}$, for $j' = 1, \ldots, n-1$, it follows that

$$\tilde{\mathbb{E}}\left[o^{n,i\delta}_{j'} \mid \mathcal{F}_{(i+1)\delta-}\right] = n\bar{a}^{n,i\delta}_{j'},$$

hence also $\tilde{\mathbb{E}}[\pi^n_{i\delta} \mid \mathcal{F}_{i\delta-}] = \lim_{t \nearrow i\delta} \pi^n_t$. □

If the system does not undergo any corrections, that is, $\delta = \infty$, then the above method is simply the Monte Carlo method described in Section 8.6. The convergence of the Monte Carlo approximation is very slow as the particles wander away from the signal's trajectory forcing the unnormalised weights to become infinitesimally small. Consequently the branching correction procedure is introduced to cull the unlikely particles and multiply those situated in the right areas.

However, the branching procedure introduces randomness into the system as it replaces each weight with a random number of offspring. As such, the distribution of the number of offspring has to be chosen with great care to minimise this effect. The random number of offspring should have minimal variance. That is, as the mean number of offspring is pre-determined, we should choose the o_j^ns to have the smallest possible variance amongst all integer-valued random variables with the given mean $n\bar{a}_j^n$. It is easy to check that if the o_j^ns have the distribution described by (9.7) then they have minimal variance.

In [66], Crisan and Lyons describe a generic way to construct n-tuples of integer-valued random variables with the minimal variance property and the total sum equal to n. This is done by means of an associated binary tree, hence the name Tree-based branching Algorithms (which are sometimes abbreviated as TBBAs). The algorithm presented above is a specific example of the class described in [66]. To the authors' knowledge only one other alternative algorithm is known that produces n-tuples which satisfy the minimal variance property. It was introduced by Whitley [268] and independently by Carpenter, Clifford and Fearnhead [39]. Further remarks on the branching algorithm can be found at the end of Chapter 10.

9.3 Preliminary Results

The following proposition gives us the evolution equation for the approximating measure-valued process π^n.

Proposition 9.7. *The probability measure-valued process $\pi^n = \{\pi_t^n, \, t \geq 0\}$ satisfies the following evolution equation*

$$\pi_t^n(\varphi) = \pi_0^n(\varphi) + \int_0^t \pi_s^n(A\varphi)\,\mathrm{d}s + S_t^{n,\varphi} + M_{[t/\delta]}^{n,\varphi}$$

$$+ \sum_{k=1}^m \int_0^t (\pi_s^n(h_k\varphi) - \pi_s^n(h_k)\pi_s^n(\varphi))\left(\mathrm{d}Y_s^k - \pi_s^n(h_k)\,\mathrm{d}s\right), \quad (9.16)$$

for any $\varphi \in C_b^2(\mathbb{R}^d)$, where $S^{n,\varphi} = \{S_t^{n,\varphi}, \, t \geq 0\}$ is the $\mathcal{F}_t$-adapted martingale

$$S_t^{n,\varphi} = \frac{1}{n}\sum_{i=0}^\infty \sum_{j=1}^n \int_{i\delta \wedge t}^{(i+1)\delta \wedge t} \bar{a}_j^n(s)(\nabla \varphi)^\top \sigma)(v_j^n(s))\,\mathrm{d}V_s^{(j)},$$

9.3 Preliminary Results

and $M^{n,\varphi} = \{M_k^{n,\varphi}, \ k > 0\}$ is the discrete parameter martingale

$$M_k^{n,\varphi} = \frac{1}{n}\sum_{i=1}^{k}\sum_{j'=1}^{n}(o_{j'}^n(i\delta) - n\bar{a}_{j'}^{n,i\delta})\varphi(v_{j'}^n(i\delta)), \quad k > 0. \tag{9.17}$$

Proof. Let $\mathcal{F}_{k\delta-} = \sigma(\mathcal{F}_s, \ 0 \le s < k\delta)$ be the σ-algebra of events up to time $k\delta$ (the time of the kth-branching) and $\pi_{k\delta-}^n = \lim_{t \nearrow k\delta} \pi_t^n$. For $t \in [i\delta,(i+1)\delta)$, we have[†] for $\varphi \in C_b^2(\mathbb{R}^d)$,

$$\pi_t^n(\varphi) = \pi_0^n(\varphi) + M_i^{n,\varphi} + \sum_{k=1}^{i}\left(\pi_{k\delta-}^n(\varphi) - \pi_{(k-1)\delta}^n(\varphi)\right)$$
$$+ (\pi_t^n(\varphi) - \pi_{i\delta}^n(\varphi)), \tag{9.18}$$

where $M^{n,\varphi} = \{M_j^{n,\varphi}, \ j \ge 0\}$ is the process defined as

$$M_j^{n,\varphi} = \sum_{k=1}^{j}\left(\pi_{k\delta}^n(\varphi) - \pi_{k\delta-}^n(\varphi)\right), \quad \text{for } j \ge 0.$$

The martingale property of $M^{n,\varphi}$ follows from (9.15) and the explicit expression (9.17) from the fact that $\pi_{k\delta}^n = (1/n)\sum_{j'=1}^{n}o_{j'}^{n,k\delta}\delta_{v_{j'}^n(k\delta)}$ and $\pi_{k\delta-}^n = \sum_{j'=1}^{n}\bar{a}_{j'}^{n,k\delta}\delta_{v_{j'}^n(k\delta)}$.

We now find an expression for the third and fourth terms on the right-hand side of (9.18). From Itô's formula using (9.4), (9.5) and the independence of Y and V, it follows that

$$d\left(a_j^n(t)\varphi(v_j^n(t))\right) = a_j^n(t)A\varphi(v_j^n(t))\,dt$$
$$+ a_j^n(t)((\nabla\varphi)^\top\sigma)(v_j^n(t))\,dV_t^{(j)}$$
$$+ a_j^n(t)\varphi(v_j^n(t))h^\top(v_j^n(t))\,dY_t,$$

and

$$d\left(\sum_{k=1}^{n}a_k^n(t)\right) = \sum_{k=1}^{n}a_k^n(t)h^\top(v_k^n(t))\,dY_t,$$

for any $\varphi \in C_b^2(\mathbb{R}^d)$. Hence for $t \in [k\delta,(k+1)\delta)$ and $k = 0,1,\ldots,i$, we have

---
[†] We use the standard convention $\sum_{k=1}^{0} = 0$.

$$\pi_t^n(\varphi) - \pi_{(k-1)\delta}^n(\varphi) = \int_{(k-1)\delta}^t d\left(\sum_{j=1}^n \bar{a}_j^n \varphi(v_j^n(s))\right) \quad (9.19)$$

$$= \int_{(k-1)\delta}^t \sum_{j=1}^n d\left(\frac{a_j^n(s)\varphi(v_j^n(s))}{\sum_{p=1}^n a_p^n(s)}\right)$$

$$= \int_{(k-1)\delta}^t \pi_s^n(A\varphi) \, ds$$

$$+ \sum_{r=1}^m \int_{(k-1)\delta}^t (\pi_s^n(h_r\varphi) - \pi_s^n(h_r)\pi_s^n(\varphi))$$

$$\times (dY_s^r - \pi_s^n(h_r) \, ds)$$

$$+ \sum_{j=1}^n \int_{(k-1)\delta}^t \bar{a}_j^n(s)((\nabla\varphi)^\top \sigma)(v_j^n(s)) \, dV_s^{(j)}. \quad (9.20)$$

Taking the limit as $t \nearrow k\delta$ yields,

$$\pi_{k\delta-}^n(\varphi) - \pi_{(k-1)\delta}^n(\varphi) = \int_{(k-1)\delta}^{k\delta} \pi_s^n(A\varphi) \, ds$$

$$+ \sum_{j=1}^n \int_{(k-1)\delta}^{k\delta} \bar{a}_j^n(s)((\nabla\varphi)^\top \sigma)(v_j^n(s)) \, dV_s^{(j)}$$

$$+ \sum_{r=1}^m \int_{(k-1)\delta}^{k\delta} (\pi_s^n(h_r\varphi) - \pi_s^n(h_r)\pi_s^n(\varphi))$$

$$\times (dY_s^r - \pi_s^n(h_r) \, ds). \quad (9.21)$$

Finally, (9.18), (9.20) and (9.21) imply (9.16). □

In the following we choose a fixed time horizon $t > 0$ and let $\mathcal{Y}^t = \{\mathcal{Y}_s^t, \, s \in [0,t]\}$ be the backward filtration

$$\mathcal{Y}_s^t = \sigma(Y_t - Y_r, \, r \in [s,t]).$$

Recall that $C_b^m(\mathbb{R}^d)$ is the set of all bounded, continuous functions with bounded partial derivatives up to order m on which we define the norm

$$\|\varphi\|_{m,\infty} = \sum_{|\alpha| \le m} \sup_{x \in \mathbb{R}^d} |D_\alpha \varphi(x)|, \qquad \varphi \in C_b^m(\mathbb{R}^d),$$

where $\alpha = (\alpha^1, \ldots, \alpha^d)$ is a multi-index and $D_\alpha \varphi = (\partial_1)^{\alpha^1} \cdots (\partial_d)^{\alpha^d} \varphi$. Also recall that $W_p^m(\mathbb{R}^d)$ is the set of all functions with generalized partial derivatives up to order m with both the function and all its partial derivatives being p-integrable on which we define the Sobolev norm

$$\|\varphi\|_{m,p} = \left(\sum_{|\alpha| \leq m} \int_{\mathbb{R}^d} |D_\alpha \varphi(x)|^p \, dx \right)^{1/p}.$$

In the following we impose conditions under which the dual of the solution of the Zakai equation exists (see Chapter 7 for details). We assume that the matrix-valued function a is uniformly strictly elliptic. We also assume that there exists an integer $m > 2$ and a positive constant $p > \max(d/(m-2), 2)$ such that for all $i, j = 1, \ldots, d$, $a_{ij} \in C_b^{m+2}(\mathbb{R}^d)$, $f_i \in C_b^{m+1}(\mathbb{R}^d)$ and for all $i = 1, \ldots, m$ we have $h_i \in C_b^{m+1}(\mathbb{R}^d)$. Under these conditions, for any bounded $\varphi \in W_p^m(\mathbb{R}^d)$ there exists a function-valued process $\psi^{t,\varphi} = \{\psi_s^{t,\varphi}, \, s \in [0,t]\}$ which is the dual of the measure-valued process $\rho = \{\rho_s, \, s \in [0,t]\}$ (the solution of the Zakai equation) in the sense of Theorem 7.22. That is, for any $\varphi \in W_p^m(\mathbb{R}^d) \cap B(\mathbb{R}^d)$, the process

$$s \mapsto \rho_s\left(\psi_s^{t,\varphi}\right), \quad s \in [0,t]$$

is almost surely constant. We recall below the properties of the dual as described in Chapter 7.

1. For every $x \in \mathbb{R}^d$, $\psi_s^{t,\varphi}(x)$ is a real-valued process measurable with respect to the backward filtration $\mathcal{Y}^t$.
2. Almost surely, $\psi^{t,\varphi}$ is jointly continuous over $[0,\infty) \times \mathbb{R}^d$ and is twice differentiable in the spatial variable. Both $\psi_s^{t,\varphi}$ and its partial derivatives are continuous bounded functions.
3. $\psi^{t,\varphi}$ is a solution of the following backward stochastic partial differential equation which is identical to (7.30):

$$\psi_s^{t,\varphi}(x) = \varphi(x) - \int_s^t A\psi_p^{t,\varphi}(x) \, dp$$
$$- \int_s^t \psi_p^{t,\varphi}(x) h^\top(x) \, \bar{d}Y_p, \quad 0 \leq s \leq t, x \in \mathbb{R}^d,$$

where $\int_s^t \psi_p^{t,\varphi} h^\top \, \bar{d}Y_p$ is a backward Itô integral.
4. There exists a constant $c = c(p)$ independent of φ such that

$$\tilde{\mathbb{E}}\left[\sup_{s \in [0,t]} \|\psi_s^{t,\varphi}\|_{2,\infty}^p \right] \leq c\|\varphi\|_{m,p}^p. \tag{9.22}$$

As mentioned in Chapter 7, the dual $\psi^{t,\varphi}$ can be defined for a larger class of test functions φ than $W_p^m(\mathbb{R}^d)$, using the representation (7.33). We can rewrite (7.33) in the following form,

$$\psi_s^{t,\varphi}(x) = \tilde{\mathbb{E}}\left[\varphi(v(t)) a_s^t(v, Y) \mid \mathcal{Y}_t, v(s) = x \right], \tag{9.23}$$

for any $\varphi \in B(\mathbb{R}^d)$. In (9.23), $v = \{v(s), \, s \in [0,t]\}$ is an $\mathcal{F}_s$-adapted Markov process, independent of Y that satisfies the same stochastic differential equation as the signal; that is,

$$dv(t) = f(v(t)) \, dt + \sigma(v(t)) \, dV_t$$

and

$$a_s^t(v, Y) = \exp\left(\int_s^t h(v(r))^\top \, dY_r - \frac{1}{2}\int_s^t \|h(v(r))\|^2 \, dr\right).$$

Lemma 9.8. *For $s \in [0, t]$ and $\varphi \in B(\mathbb{R}^d)$, we have*

$$\psi_s^{t,\varphi}(v(s)) = \tilde{\mathbb{E}}\left[\varphi(v(t)) a_s^t(v, Y) \mid \mathcal{F}_s \vee \mathcal{Y}_t\right].$$

Proof. From (9.23) and the properties of the conditional expectation

$$\psi_s^{t,\varphi}(v(s)) = \tilde{\mathbb{E}}\left[\varphi(v(t)) a_s^t(v, Y) \mid \mathcal{Y}_t \vee \sigma(v(s))\right]$$

and the claim follows by the Markov property of the process v and its independence from $\mathcal{Y}_t$. □

Lemma 9.9. *For any $\varphi \in B(\mathbb{R}^d)$ and any $k < [t/\delta]$, the real-valued process*

$$s \in [k\delta, (k+1)\delta \wedge t] \mapsto \psi_s^{t,\varphi}(v_j^n(s)) a_j^n(s)$$

is an $\mathcal{F}_s \vee \mathcal{Y}_t$-adapted martingale. Moreover, if $\varphi \in W_p^m(\mathbb{R}^d) \cap B(\mathbb{R}^d)$ where $m > 2$ and $(m-2)p > d$

$$\psi_s^{t,\varphi}(v_j^n(s)) a_j^n(s) = \psi_{k\delta}^{t,\varphi}\left(v_j^n(k\delta)\right)$$
$$+ \int_{k\delta}^s a_j^n(p)((\nabla \psi_p^{t,\varphi})^\top \sigma)\left(v_j^n(p)\right) dV_p^{(j)}, \quad (9.24)$$

for $s \in [k\delta, (k+1)\delta \wedge t]$ and $j = 1, \ldots, n$.

Proof. For the first part of the proof we cannot simply use the fact that $\psi^{t,\varphi}$ is a (classical) solution of the backward stochastic partial differential equation (7.30) as the test function φ does not necessarily belong to $W_p^m(\mathbb{R}^d)$. However, from Lemma 9.8 it follows that

$$\psi_s^{t,\varphi}(v_j^n(s)) = \tilde{\mathbb{E}}\left[\varphi\left(v_j^n(t)\right) a_s^t(v_j^n, Y) \mid \mathcal{F}_s \vee \mathcal{Y}_t\right], \quad (9.25)$$

where for $j = 1, \ldots, n$, following (9.6),

$$a_s^t(v_j^n, Y) = \exp\left(\int_s^t h\left(v_j^n(r)\right)^\top dY_r - \frac{1}{2}\int_s^t \|h(v_j^n(r))\|^2 \, dr\right)$$

and $v_j^n(s)$ is given by

$$v_j^n(s) = v_j^n(k\delta) + \int_{k\delta}^s f(v_j^n(r)) \, dr + \int_{k\delta}^s \sigma(v_j^n(r)) \, dV_r^{(j)}, \quad j = 1, \ldots, n, \quad (9.26)$$

which is taken as the definition for $s \in [k\delta, t]$. Comparing this with (9.4) it is clear that if $(k+1)\delta < t$, then this $v_j^n(s)$ may not agree with the previous definition on $((k+1)\delta, t]$. Observe that $a_s^t(v_j^n, Y) = a_j^n(t)/a_j^n(s)$ where $a_j^n(s)$ is given for $s \in [k\delta, t]$ by

$$a_j^n(s) = \exp\left(\int_{k\delta}^s h(v_j^n(p))^\top \, dY_p - \frac{1}{2}\int_{k\delta}^s \|h(v_j^n(p))\|^2 \, dp\right); \tag{9.27}$$

since $a_j^n(s)$ is $\mathcal{F}_s$-adapted it is also $\mathcal{F}_s \vee \mathcal{Y}_t$-adapted, thus

$$\psi_s^{t,\varphi}(v_j^n(s))a_j^n(s) = \tilde{\mathbb{E}}[\varphi\left(v_j^n(t)\right)a_j^n(t) \mid \mathcal{F}_s \vee \mathcal{Y}_t]. \tag{9.28}$$

Since $s \mapsto \tilde{\mathbb{E}}[\varphi\left(v_j^n(t)\right)a_j^n(t) \mid \mathcal{F}_s \vee \mathcal{Y}_t]$ is an $\mathcal{F}_s \vee \mathcal{Y}_t$-adapted martingale for $s \in [0, t]$, so is $s \mapsto \psi_s^{t,\varphi}(v_j^n(s))a_j^n(s)$. This completes the proof of the first part of the lemma.

For the second part of the lemma, as $\varphi \in W_p^m(\mathbb{R}^d)$, it is now possible to use properties 1–4 of the dual process $\psi^{t,\varphi}$, in particular the fact that $\psi^{t,\varphi}$ is differentiable. The stochastic integral on the right-hand side of (9.24) is well defined as the Brownian motion $V^{(j)} = \{V_s^{(j)}, \; s \in [k\delta, (k+1)\delta \wedge t)\}$ is $\mathcal{F}_s \vee \mathcal{Y}_t$-adapted ($V^{(j)}$ is independent of Y) and so is the integrand

$$s \in [k\delta, (k+1)\delta \wedge t) \mapsto a_j^n(p)((\nabla \psi_p^{t,\varphi})^\top \sigma)\left(v_j^n(p)\right).$$

Moreover, the stochastic integral on the right-hand side of (9.24) is a genuine martingale since its quadratic variation process $Q = \{Q_s, \; s \in [k\delta, (k+1)\delta \wedge t)\}$ satisfies the inequality

$$\tilde{\mathbb{E}}[Q_s] \leq K_\sigma^2 \int_{k\delta}^s \tilde{\mathbb{E}}\left[\|\psi_p^{t,\varphi}\|_{1,\infty}^2\right] \tilde{\mathbb{E}}\left[(a_j^n(p))^2\right] dp < \infty. \tag{9.29}$$

In (9.29) we used the fact that $\|\psi_p^{t,\varphi}\|_{1,\infty}^2$ and $a_j^n(p)$ are mutually independent and that σ is uniformly bounded by K_σ. We cannot prove (9.24) by applying Itô's formula directly: $\psi_p^{t,\varphi}$ is $\mathcal{Y}_p^t$-measurable, whereas $a_j^n(p)$ is $\mathcal{F}_p$-measurable. Instead, we use a density argument.

Since all terms appearing in (9.24) are measurable with respect to the σ-algebra $\mathcal{F}_{k\delta} \vee \mathcal{Y}_{k\delta}^t \vee (\mathcal{V}^j)_{k\delta}^t$, where $\mathcal{Y}_{k\delta}^t = \sigma(Y_r - Y_{k\delta}, \; r \in [k\delta, t])$ and $(\mathcal{V}^j)_{k\delta}^t = \sigma(V_r^j - V_{k\delta}^j, \; r \in [k\delta, t])$, it suffices to prove that

$$\tilde{\mathbb{E}}\left[\chi\left(\psi_s^{t,\varphi}(v_j^n(s))a_j^n(s) - \psi_{k\delta}^{t,\varphi}\left(v_j^n(k\delta)\right)\right)\right]$$
$$= \tilde{\mathbb{E}}\left[\chi \int_{k\delta}^s a_j^n(p)((\nabla \psi_p^{t,\varphi})^\top \sigma)\left(v_j^n(p)\right) dV_p^{(j)}\right], \tag{9.30}$$

where χ is any bounded $\mathcal{F}_{k\delta} \vee \mathcal{Y}_{k\delta}^t \vee (\mathcal{V}^j)_{k\delta}^t$-measurable random variable. It is sufficient to work with a much smaller class of bounded $\mathcal{F}_{k\delta} \vee \mathcal{Y}_{k\delta}^t \vee (\mathcal{V}^j)_{k\delta}^t$-measurable random variables. Let $b : [k\delta, t] \to \mathbb{R}^m$ and $c : [k\delta, t] \to \mathbb{R}^d$ be bounded, Borel-measurable functions and let θ^b and θ^c be the following (bounded) processes

$$\theta_r^b \triangleq \exp\left(i \int_{k\delta}^r b_p^\top \, dY_p + \frac{1}{2} \int_{k\delta}^r \|b_p\|^2 \, dp\right), \tag{9.31}$$

and

$$\theta_r^c \triangleq \exp\left(i \int_{k\delta}^r c_p^\top \, dV_p^{(j)} + \frac{1}{2} \int_{k\delta}^r \|c_p\|^2 \, dp\right). \tag{9.32}$$

Then it is sufficient to show that (9.30) holds true for χ of the form $\chi = \zeta \theta_t^b \theta_t^c$, for any choice of b in (9.31) and c in (9.32) and any bounded $\mathcal{F}_{k\delta}$-measurable random variable ζ (see Corollary B.40 for a justification of the above). For $s \in [k\delta, (k+1)\delta \wedge t)$,

$$\tilde{\mathbb{E}}\left[\psi_s^{t,\varphi}(v_j^n(s))a_j^n(s)\zeta\theta_t^b\theta_t^c \mid \mathcal{F}_{k\delta} \vee \mathcal{Y}_{k\delta}^s \vee (\mathcal{V}^j)_{k\delta}^s\right]$$
$$= \Xi_s(v_j^n(s))a_j^n(s)\zeta\theta_s^b\theta_s^c, \tag{9.33}$$

where $\Xi = \{\Xi_s(\cdot), \ s \in [k\delta, (k+1)\delta \wedge t]\}$ is given by

$$\Xi_s(\cdot) \triangleq \tilde{\mathbb{E}}\left[\psi_s^{t,\varphi}(\cdot)\tilde{\theta}_s^b \mid \mathcal{F}_{k\delta} \vee \mathcal{Y}_{k\delta}^s \vee (\mathcal{V}^j)_{k\delta}^s\right],$$

and

$$\tilde{\theta}_s^b \triangleq \frac{\theta_t^b}{\theta_s^b} = \exp\left(i \int_s^t b_p^\top \, dY_p + \frac{1}{2} \int_s^t \|b_p\|^2 \, dp\right).$$

Both $\psi_s^{t,\varphi}$ and $\tilde{\theta}_s^b$ are measurable with respect to the σ-algebra $\mathcal{Y}_s^t$, which is independent of $\mathcal{F}_{k\delta} \vee \mathcal{Y}_{k\delta}^s \vee (\mathcal{V}^j)_{k\delta}^s$, hence $\Xi_s(\cdot) = \tilde{\mathbb{E}}[\psi_s^{t,\varphi}(\cdot)\tilde{\theta}_s^b]$. As in the proof of Theorem 7.22 it follows that for any $r \in C_b^m([0,\infty), \mathbb{R}^d)$ and any $x \in \mathbb{R}^d$,

$$\Xi_s(x) = \varphi(x) - \int_s^t A\Xi_p(x) \, dp - i \int_s^t h^\top(x) r_p \Xi_p(x) \, dp, \quad 0 \le s \le t. \tag{9.34}$$

Equivalently $\Xi(\cdot) = \{\Xi_s(\cdot), \ s \in [0,t]\}$ is the unique solution of the parabolic PDE (4.14) with final time condition $\Xi_t(\cdot) = \varphi(\cdot)$. From the Sobolev embedding theorem as a consequence of the condition $(m-2)p > d$, it follows that φ has a modification on a set of null Lebesgue measure which is in $C_b(\mathbb{R}^d)$, therefore the solution to the PDE $\Xi \in C_b^{1,2}([0,t] \times \mathbb{R}^d)$. From (9.33) it follows that

$$\tilde{\mathbb{E}}\left[\left(\psi_s^{t,\varphi}(v_j^n(s))a_j^n(s) - \psi_{k\delta}^{t,\varphi}(v_j^n(k\delta))\right)\chi\right]$$
$$= \tilde{\mathbb{E}}\left[\zeta\left(\Xi_s(v_j^n(s))a_j^n(s)\theta_s^b\theta_s^c - \Xi_{k\delta}(v_j^n(k\delta))\right)\right]. \tag{9.35}$$

As Ξ is the solution of a deterministic PDE with deterministic initial condition, it follows that $\Xi_s(v_j^n(s))$ is $\mathcal{F}_s$-measurable. Thus as all the terms are now measurable with respect to the same filtration, it is possible to apply Itô's rule and use the PDE (9.34) to obtain

$$\tilde{\mathbb{E}}\left[\zeta\left(\Xi_s(v_j^n(s))a_j^n(s)\theta_s^b\theta_s^c - \Xi_{k\delta}\left(v_j^n(k\delta)\right)a_j^n(k\delta)\theta_{k\delta}^b\theta_{k\delta}^c\right)\right]$$

$$= \tilde{\mathbb{E}}\left[\zeta\int_{k\delta}^s \mathrm{d}\left(a_j^n(p)\Xi_p(v_j^n(p))\theta_p^b\theta_p^c\right)\right]$$

$$= \tilde{\mathbb{E}}\left[\zeta\int_{k\delta}^s \left(a_j^n(p)\theta_p^b\theta_p^c\left(A\Xi_p(v_j^n(p)) + i\Xi_p(v_j^n(p))h^\top(v_j^n(p))b_p\right.\right.\right.$$
$$\left.\left.\left.+ \frac{\partial \Xi_p}{\partial p}(v_j^n(p))\right) + i(\nabla\Xi)^\top \sigma c_p\theta_p^b\theta_p^c\right)\mathrm{d}p\right]$$

$$= \tilde{\mathbb{E}}\left[i\zeta\int_{k\delta}^s a_j^n(p)(\nabla\Xi^\top \sigma)c_p\theta_p^b\theta_p^c\,\mathrm{d}p\right]$$

$$= \tilde{\mathbb{E}}\left[i\zeta\int_{k\delta}^s a_j^n(p)\left(\nabla\Xi_p^\top \sigma\right)(v_j^n(p))c_p\theta_p^b\theta_p^c\,\mathrm{d}p\right]. \quad (9.36)$$

A second similar application of Itô's formula using (9.32) yields

$$\tilde{\mathbb{E}}\left[\zeta\theta_t^b\theta_t^c\int_{k\delta}^s a_j^n(p)((\nabla\psi_p^{t,\varphi})^\top \sigma)(v_j^n(p))\,\mathrm{d}V_p^j\,\Big|\,\mathcal{F}_{k\delta}\vee\mathcal{Y}_{k\delta}^t\right]$$

$$= \zeta\theta_t^b\tilde{\mathbb{E}}\left[\int_{k\delta}^s \mathrm{d}\left(\theta_t^c\int_{k\delta}^s a_j^n(p)((\nabla\psi_p^{t,\varphi})^\top \sigma)(v_j^n(p))\,\mathrm{d}V_p^j\right)\,\Big|\,\mathcal{F}_{k\delta}\vee\mathcal{Y}_{k\delta}^t\right]$$

$$= i\zeta\theta_t^b\tilde{\mathbb{E}}\left[\int_{k\delta}^s a_j^n(p)\left((\nabla\psi_p^{t,\varphi})^\top \sigma\right)(v_j^n(p))c_p\theta_p^c\,\mathrm{d}p\,\Big|\,\mathcal{F}_{k\delta}\vee\mathcal{Y}_{k\delta}^t\right]. \quad (9.37)$$

Use of Fubini's theorem and the tower property of conditional expectation gives

$$\tilde{\mathbb{E}}\left[\zeta\int_{k\delta}^s a_j^n(p)\left((\nabla\psi_p^{t,\varphi})^\top \sigma\right)(v_j^n(p))c_p\theta_t^b\theta_p^c\,\mathrm{d}p\right]$$

$$= \int_{k\delta}^s \tilde{\mathbb{E}}\left[\zeta a_j^n(p)\left(\nabla(\psi_p^{t,\varphi})^\top \sigma\right)(v_j^n(p))c_p\theta_t^b\theta_p^c\right]\,\mathrm{d}p$$

$$= \int_{k\delta}^s \tilde{\mathbb{E}}\left[\tilde{\mathbb{E}}\left[\zeta a_j^n(p)\left(\nabla(\psi_p^{t,\varphi})^\top \sigma\right)(v_j^n(p))c_p\theta_t^b\theta_p^c\,\Big|\,\mathcal{F}_{k\delta}\vee\mathcal{Y}_{k\delta}^p\vee(\mathcal{V}^j)_{k\delta}^p\right]\right]\,\mathrm{d}p$$

$$= \int_{k\delta}^s \tilde{\mathbb{E}}\left[\zeta\theta_p^c\theta_p^b a_j^n(p)c_p\tilde{\mathbb{E}}\left[\left(\nabla(\psi_p^{t,\varphi})^\top \sigma\right)(v_j^n(p))\tilde{\theta}_p^b\,\Big|\,\mathcal{F}_{k\delta}\vee\mathcal{Y}_{k\delta}^p\vee(\mathcal{V}^j)_{k\delta}^p\right]\right]\,\mathrm{d}p$$

$$= \int_{k\delta}^s \tilde{\mathbb{E}}\left[\zeta\theta_p^c\theta_p^b a_j^n(p)c_p\tilde{\mathbb{E}}\left[\left(\nabla(\psi_p^{t,\varphi})^\top\right)(v_j^n(p))\tilde{\theta}_p^b\right]\sigma(v_j^n(p))\right]\,\mathrm{d}p$$

$$= \int_{k\delta}^s \tilde{\mathbb{E}}\left[\zeta\theta_p^c\theta_p^b a_j^n(p)c_p\nabla\tilde{\mathbb{E}}\left[((\psi_p^{t,\varphi})^\top)(v_j^n(p))\tilde{\theta}_p^b\right]\sigma(v_j^n(p))\right]\,\mathrm{d}p$$

$$= \tilde{\mathbb{E}}\left[\zeta\int_{k\delta}^s a_j^n(p)\left(\nabla\Xi_p^\top \sigma\right)(v_j^n(p))c_p\theta_p^b\theta_p^c\,\mathrm{d}p\right].$$

Using this result and (9.37) it follows that

$$\tilde{\mathbb{E}}\left[\zeta\theta_t^b\theta_t^c\int_{k\delta}^s a_j^n(p)\left((\nabla\psi_p^{t,\varphi})^\top\sigma\right)(v_j^n(p))\,\mathrm{d}V_p^j\right]$$
$$=\tilde{\mathbb{E}}\left[i\zeta\int_{k\delta}^s a_j^n(p)\left(\nabla\Xi_p^\top\sigma\right)(v_j^n(p))c_p\theta_p^b\theta_p^c\,\mathrm{d}p\right]. \quad (9.38)$$

From (9.35), (9.36) and (9.38) we deduce (9.30) and hence the result of the lemma. □

To show that $\psi_s^{t,\varphi}$ is dual to ρ_s for arbitrary $\varphi \in B(\mathbb{R}^d)$, use the fact that $(v_j^n(s), a_j^n(s))$ have the same law as $(X,\tilde{Z})$ and (9.28),

$$\begin{aligned}
\rho_s\left(\psi_s^{t,\varphi}\right) &= \tilde{\mathbb{E}}[\tilde{Z}_s\psi_s^{t,\varphi}(X_s)\mid \mathcal{Y}_s] \\
&= \tilde{\mathbb{E}}[\tilde{Z}_s\psi_s^{t,\varphi}(X_s)\mid \mathcal{Y}_t] \\
&= \tilde{\mathbb{E}}[\psi_s^{t,\varphi}(v_j^n(s))a_j^n(s)\mid \mathcal{Y}_t] \\
&= \tilde{\mathbb{E}}\left[\tilde{\mathbb{E}}\left[\varphi\left(v_j^n(t)\right)a_j^n(t)\mid \mathcal{F}_s\vee\mathcal{Y}_t\right]\mid \mathcal{Y}_t\right] \\
&= \tilde{\mathbb{E}}\left[\varphi\left(v_j^n(t)\right)a_j^n(t)\mid \mathcal{Y}_t\right] \\
&= \tilde{\mathbb{E}}\left[\varphi(X_t)\tilde{Z}_t\mid \mathcal{Y}_t\right] \\
&= \rho_t(\varphi).
\end{aligned}$$

Define the following $\mathcal{F}_t$-adapted martingale $\xi^n = \{\xi_t^n,\ t\geq 0\}$ by

$$\xi_t^n \triangleq \left(\prod_{i=1}^{[t/\delta]}\frac{1}{n}\sum_{j=1}^n a_j^{n,i\delta}\right)\left(\frac{1}{n}\sum_{j=1}^n a_j^n(t)\right).$$

Exercise 9.10. Prove that for any $t\geq 0$ and $p\geq 1$, there exist two constants $c_1^{t,p}$ and $c_2^{t,p}$ which depend only on $\max_{k=1,\dots,m}\|h_k\|_{0,\infty}$ such that

$$\sup_{n\geq 0}\sup_{s\in[0,t]}\mathbb{E}\left[(\xi_s^n)^p\right]\leq c_1^{t,p}, \quad (9.39)$$

and

$$\max_{j=1,\dots,n}\sup_{n\geq 0}\sup_{s\in[0,t]}\tilde{\mathbb{E}}\left[\left(\xi_s^n a_j^n(s)\right)^p\right]\leq c_2^{t,p}. \quad (9.40)$$

We use the martingale ξ_t^n to linearize π_t^n in order to make it easier to analyze the convergence of π^n. Let $\rho^n = \{\rho_t^n,\ t\geq 0\}$ be the measure-valued process defined by

$$\rho_t^n \triangleq \xi_t^n \pi_t^n = \frac{\xi_{[t/\delta]\delta}^n}{n}\sum_{j=1}^n a_j^n(t)\delta_{v_j^n(t)}.$$

Exercise 9.11. Show that $\rho^n = \{\rho_t^n,\ t\geq 0\}$ is a measure-valued process which satisfies the following evolution equation

9.3 Preliminary Results

$$\rho_t^n(\varphi) = \pi_0^n(\varphi) + \int_0^t \rho_s^n(A\varphi)\,ds + \bar{S}_t^{n,\varphi} + \bar{M}_{[t/\delta]}^{n,\varphi}$$

$$+ \sum_{k=1}^m \int_0^t \rho_s^n(h_k\varphi)\,dY_s^k, \tag{9.41}$$

for any $\varphi \in C_b^2(\mathbb{R}^d)$. In (9.41), $\bar{S}^{n,\varphi} = \{\bar{S}_t^{n,\varphi},\ t \geq 0\}$ is an $\mathcal{F}_t$-adapted martingale

$$\bar{S}_t^{n,\varphi} = \frac{1}{n}\sum_{i=0}^\infty \sum_{j=1}^n \int_{i\delta\wedge t}^{(i+1)\delta\wedge t} \xi_{i\delta}^n a_j^n(s)((\nabla\varphi)^\top \sigma)(v_j^n(s))\,dV_s^j$$

and $\bar{M}^{n,\varphi} = \{\bar{M}_k^{n,\varphi},\ k > 0\}$ is the discrete martingale

$$\bar{M}_k^{n,\varphi} = \frac{1}{n}\sum_{i=1}^k \xi_{i\delta}^n \sum_{j'=1}^n (o_{j'}^n(i\delta) - n\bar{a}_{j'}^{n,i\delta})\varphi(v_{j'}^n(i\delta)), \quad k > 0.$$

Proposition 9.12. *For any $\varphi \in B(\mathbb{R}^d)$, the real-valued process $\rho_\cdot^n(\psi_\cdot^{t,\varphi}) = \{\rho_s^n(\psi_s^{t,\varphi}),\ s \in [0,t]\}$ is an $\mathcal{F}_s \vee \mathcal{Y}_t$-adapted martingale.*

Proof. From Lemma 9.9 we deduce that for $s \in [[t/\delta]\delta, t]$, we have

$$\tilde{\mathbb{E}}\left[a_j^n(t)\varphi(v_j^n(t)) \mid \mathcal{F}_s \vee \mathcal{Y}_t\right] = a_j^n(s)\psi_s^{t,\varphi}(v_j^n(s))$$

which implies, in particular that

$$\tilde{\mathbb{E}}[a_{j'}^{n,k\delta}\psi_{k\delta}^{t,\varphi}(v_{j'}^n(k\delta)) \mid \mathcal{F}_s \vee \mathcal{Y}_t] = a_{j'}^n(s)\psi_s^{t,\varphi}(v_{j'}^n(s))$$

for any $s \in [(k-1)\delta, k\delta)$. Hence

$$\tilde{\mathbb{E}}\left[\rho_t^n(\varphi) \mid \mathcal{F}_s \vee \mathcal{Y}_t\right] = \frac{\xi_{[t/\delta]\delta}^n}{n}\sum_{j=1}^n \tilde{\mathbb{E}}[a_j^n(t)\varphi\left(v_j^n(t)\right) \mid \mathcal{F}_s \vee \mathcal{Y}_t]$$

$$= \rho_s^n\left(\psi_s^{t,\varphi}\right), \quad \text{for } [t/\delta]\delta \leq s \leq t \tag{9.42}$$

and, for $s \in [(k-1)\delta, k\delta)$,

$$\tilde{\mathbb{E}}\left[\rho_{k\delta-}^n(\psi_{k\delta-}^{t,\varphi}) \mid \mathcal{F}_s \vee \mathcal{Y}_t\right] = \frac{\xi_{(k-1)\delta}^n}{n}\sum_{j'=1}^n \tilde{\mathbb{E}}\left[a_{j'}^{n,k\delta}\psi_{k\delta}^{t,\varphi}(v_{j'}^n(k\delta)) \mid \mathcal{F}_s \vee \mathcal{Y}_t\right]$$

$$= \rho_s^n(\psi_s^{t,\varphi}). \tag{9.43}$$

Finally

$$\tilde{\mathbb{E}}[\rho_{k\delta}^n\left(\psi_{k\delta}^{t,\varphi}\right) \mid \mathcal{F}_{k\delta-} \vee \mathcal{Y}_t] = \frac{\xi_{k\delta}^n}{n}\sum_{j'=1}^n \frac{a_{j'}^{n,k\delta}}{\sum_{k'=1}^n a_{k'}^{n,k\delta}/n}\psi_{k\delta}^{t,\varphi}(v_{j'}^n(k\delta))$$

$$= \rho_{k\delta-}^n(\psi_{k\delta-}^{t,\varphi}). \tag{9.44}$$

The proposition now follows from (9.42), (9.43) and (9.44). □

Proposition 9.13. *For any $\varphi \in W_p^m(\mathbb{R}^d) \cap B(\mathbb{R}^d)$, the real-valued process $\rho_\cdot^n(\psi_\cdot^{t,\varphi}) = \{\rho_s^n(\psi_s^{t,\varphi}), \, s \in [0,t]\}$ has the representation*

$$\rho_t^n(\varphi) = \pi_0^n(\psi_0^{t,\varphi}) + \hat{S}_t^{n,\varphi} + \hat{M}_{[t/\delta]}^{n,\varphi}. \tag{9.45}$$

In (9.45), $\hat{S}^{n,\varphi} = \{\hat{S}_s^{n,\varphi}, \, s \in [0,t]\}$ is the $\mathcal{F}_s \vee \mathcal{Y}_t$-adapted martingale

$$\hat{S}_s^{n,\varphi} \triangleq \sum_{i=0}^{\infty} \sum_{j=1}^{n} \frac{\xi_{i\delta}^n}{n} \int_{i\delta \wedge s}^{(i+1)\delta \wedge s} a_j^n(p)((\nabla \psi_p^{t,\varphi})^\top \sigma)(v_j^n(p)) \, dV_p^{(j)}$$

and $\hat{M}^{n,\varphi} = \{\hat{M}_k^{n,\varphi}, \, k > 0\}$ is the discrete martingale

$$\hat{M}_k^{n,\varphi} \triangleq \sum_{i=1}^{k} \frac{\xi_{i\delta}^n}{n} \sum_{j=1}^{n} (o_j^n(i\delta) - n\bar{a}_j^n(i\delta)) \psi_{i\delta}^{t,\varphi}(v_j^n(i\delta)), \quad k > 0.$$

Proof. As in (9.18), we have for $t \in [i\delta, (i+1)\delta)$ that

$$\rho_t^n(\varphi) = \rho_t^n(\psi_t^{t,\varphi})$$
$$= \pi_0^n(\psi_0^{t,\varphi}) + \hat{M}_i^{n,\varphi} + \sum_{k=1}^{i} \left(\rho_{k\delta-}^n(\psi_{k\delta-}^{t,\varphi}) - \rho_{(k-1)\delta}^n(\psi_{(k-1)\delta}^{t,\varphi}) \right)$$
$$+ (\rho_t^n(\psi_t^{t,\varphi}) - \rho_{i\delta}^n(\psi_{i\delta}^{t,\varphi})), \tag{9.46}$$

where $\hat{M}^{n,\varphi} = \{\hat{M}_i^{n,\varphi}, \, i \geq 0\}$ is the process defined as (note that $\psi_{k\delta-}^{t,\varphi} = \psi_{k\delta}^{t,\varphi}$)

$$\hat{M}_i^{n,\varphi} = \sum_{k=1}^{i} (\rho_{k\delta}^n(\psi_{k\delta}^{t,\varphi}) - \rho_{k\delta-}^n(\psi_{k\delta-}^{t,\varphi}))$$
$$= \sum_{k=1}^{i} \xi_{k\delta}^n (\pi_{k\delta}^n(\psi_{k\delta}^{t,\varphi}) - \pi_{k\delta-}^n(\psi_{k\delta}^{t,\varphi}))$$
$$= \frac{1}{n} \sum_{k=1}^{i} \xi_{k\delta}^n \sum_{j'=1}^{n} (o_{j'}^{n,k\delta} - n\bar{a}_{j'}^{n,k\delta}) \psi_{k\delta}^{t,\varphi}(v_{j'}^n(k\delta)), \quad \text{for } i \geq 0. \tag{9.47}$$

The random variables $o_{j'}^{n,k\delta}$ are independent of $\mathcal{Y}_{k\delta}^t$ since they are $\mathcal{F}_{k\delta}$-adapted. Then (9.9) implies

$$\tilde{\mathbb{E}}\left[o_{j'}^{n,k\delta} \mid \mathcal{F}_{k\delta-} \vee \mathcal{Y}_{k\delta}^t \right] = \tilde{\mathbb{E}}\left[o_{j'}^{n,k\delta} \mid \mathcal{F}_{k\delta-} \right] = n\bar{a}_{j'}^{n,k\delta},$$

whence the martingale property of $\hat{M}^{n,\varphi}$. Finally, from the representation (9.24) we deduce that for $t \in [i\delta, (i+1)\delta)$,

$$\rho_t^n(\psi_t^{t,\varphi}) = \frac{\xi_{i\delta}^n}{n} \sum_{j=1}^n a_j^n(t)\psi_t^{t,\varphi}\left(v_j^n(t)\right)$$

$$= \frac{\xi_{i\delta}^n}{n} \sum_{j=1}^n \psi_{i\delta}^{t,\varphi}\left(v_j^n(i\delta)\right)$$

$$+ \frac{\xi_{i\delta}^n}{n} \sum_{j=1}^n \int_{i\delta}^t a_j^n(p)((\nabla\psi_p^{t,\varphi})^\top \sigma)\left(v_j^n(p)\right) \, \mathrm{d}V_p^{(j)},$$

hence

$$\rho_t^n(\psi_t^{t,\varphi}) - \rho_{i\delta}^n(\psi_{i\delta}^{t,\varphi}) = \frac{\xi_{i\delta}^n}{n} \sum_{j=1}^n \int_{i\delta}^t a_j^n(p)((\nabla\psi_p^{t,\varphi})^\top \sigma)\left(v_j^n(p)\right) \, \mathrm{d}V_p^{(j)}.$$

Similarly

$$\rho_{k\delta-}^n(\psi_{k\delta-}^{t,\varphi}) - \rho_{(k-1)\delta}^n(\psi_{(k-1)\delta}^{t,\varphi})$$

$$= \frac{\xi_{(k-1)\delta}^n}{n} \sum_{j=1}^n \int_{(k-1)\delta}^{k\delta} a_j^n(p)((\nabla\psi_p^{t,\varphi})^\top \sigma)\left(v_j^n(p)\right) \, \mathrm{d}V_p^{(j)},$$

which completes the proof of the representation (9.45). □

9.4 The Convergence Results

In this section we begin by showing that $\rho_t^n(\varphi)$ converges to $\rho_t(\varphi)$ in Proposition 9.14 and that $\pi_t^n(\varphi)$ converges to $\pi_t(\varphi)$ in Theorem 9.15 for any $\varphi \in C_b(\mathbb{R}^d)$. These results imply that ρ_t^n converges to ρ_t and π_t^n converges to π_t as measure-valued random variables (Corollary 9.17). Proposition 9.14 and Theorem 9.15 are then used to prove two stronger results, namely that the process $\rho_\cdot^n(\varphi)$ converges to $\rho_\cdot(\varphi)$ in Proposition 9.18 and that the process $\pi_\cdot^n(\varphi)$ converges to $\pi_\cdot(\varphi)$ in Theorem 9.19 for any $\varphi \in C_b^2(\mathbb{R}^d)$.[†] These imply in turn, by Corollary 9.20, that the measure-valued process $\rho_\cdot^n$ converges to $\rho_\cdot$ and that the probability measure-valued process $\pi_\cdot^n$ converges to $\pi_\cdot$. Bounds on the rates of convergence are also obtained.

Proposition 9.14. *If the coefficients σ, f and h are bounded and Lipschitz, then for any $T \geq 0$, there exists a constant c_3^T independent of n such that for any $\varphi \in C_b(\mathbb{R}^d)$, we have*

$$\tilde{\mathbb{E}}[(\rho_t^n(\varphi) - \rho_t(\varphi))^2] \leq \frac{c_3^T}{n}\|\varphi\|_{0,\infty}^2, \quad t \in [0,T]. \tag{9.48}$$

In particular, for all $t \geq 0$, ρ_t^n converges in expectation to ρ_t.

[†] Note the smaller class of test functions for which results 9.18 and 9.19 hold true.

Proof. It suffices to prove (9.48) for any non-negative $\varphi \in C_b(\mathbb{R}^d)$. Obviously, we have

$$\rho_t^n(\varphi) - \rho_t(\varphi) = \left(\rho_t^n(\varphi) - \rho_{[t/\delta]\delta}^n(\psi_{[t/\delta]\delta}^{t,\varphi})\right) + \sum_{k=1}^{[t/\delta]} \left(\rho_{k\delta}^n(\psi_{k\delta}^{t,\varphi}) - \rho_{k\delta-}^n(\psi_{k\delta-}^{t,\varphi})\right)$$

$$+ \sum_{k=1}^{[t/\delta]} \left(\rho_{k\delta-}^n(\psi_{k\delta-}^{t,\varphi}) - \rho_{(k-1)\delta}^n(\psi_{(k-1)\delta}^{t,\varphi})\right)$$

$$+ \left(\pi_0^n\left(\psi_0^{t,\varphi}\right) - \pi_0\left(\psi_0^{t,\varphi}\right)\right). \tag{9.49}$$

We must bound each term on the right-hand side individually. For the first term, using the martingale property of $\rho^n(\psi^{t,\varphi})$ and the fact that the random variables $v_j^n(t)$ for $j = 1, 2, \ldots, n$ are mutually independent conditional upon $\mathcal{F}_{[t/\delta]\delta} \vee \mathcal{Y}_t$ (since the generating Brownian motions $V^{(j)}$, for $j = 1, 2, \ldots, n$ are mutually independent), we have

$$\tilde{\mathbb{E}}\left[(\rho_t^n(\varphi) - \rho_{[t/\delta]\delta}^n(\psi_{[t/\delta]\delta}^{t,\varphi}))^2 \mid \mathcal{F}_{[t/\delta]\delta} \vee \mathcal{Y}_t\right]$$

$$= \tilde{\mathbb{E}}[(\rho_t^n(\varphi) - \tilde{\mathbb{E}}[\rho_t^n(\varphi) \mid \mathcal{F}_{[t/\delta]\delta} \vee \mathcal{Y}_t])^2 \mid \mathcal{F}_{[t/\delta]\delta} \vee \mathcal{Y}_t]$$

$$= \frac{(\xi_{[t/\delta]\delta}^n)^2}{n^2} \tilde{\mathbb{E}}\left[\left(\sum_{j=1}^n \varphi(v_j^n(t))a_j^n(t)\right)^2 \Bigg| \mathcal{F}_{[t/\delta]\delta} \vee \mathcal{Y}_t\right]$$

$$- \frac{(\xi_{[t/\delta]\delta}^n)^2}{n^2} \left(\sum_{j=1}^n \tilde{\mathbb{E}}\left[\varphi(v_j^n(t))a_j^n(t) \mid \mathcal{F}_{[t/\delta]\delta} \vee \mathcal{Y}_t\right]\right)^2$$

$$\leq \frac{(\xi_{[t/\delta]\delta}^n)^2}{n^2} \|\varphi\|_{0,\infty}^2 \sum_{j=1}^n \tilde{\mathbb{E}}[a_j^n(t)^2 \mid \mathcal{F}_{[t/\delta]\delta} \vee \mathcal{Y}_t]. \tag{9.50}$$

By taking expectation on both sides of (9.50) and using (9.40) for $p = 2$, we obtain

$$\tilde{\mathbb{E}}\left[\left(\rho_t^n(\varphi) - \rho_{[t/\delta]\delta}^n(\psi_{[t/\delta]\delta}^{t,\varphi})\right)^2\right] \leq \frac{\|\varphi\|_{0,\infty}^2}{n^2} \sum_{j=1}^n \tilde{\mathbb{E}}[(\xi_{[t/\delta]\delta}^n)^2 a_j^n(t)^2]$$

$$\leq \frac{c_2^{t,2}}{n} \|\varphi\|_{0,\infty}^2. \tag{9.51}$$

Similarly (although in this case we do not have the uniform bound on $\psi_{k\delta}^{t,\varphi}$ which was used with $\psi_t^{t,\varphi}$),

$$\tilde{\mathbb{E}}\left[\left(\rho_{k\delta-}^n(\psi_{k\delta-}^{t,\varphi}) - \rho_{(k-1)\delta}^n(\psi_{(k-1)\delta}^{t,\varphi})\right)^2\right]$$

$$\leq \frac{1}{n^2} \sum_{j'=1}^n \tilde{\mathbb{E}}\left[(\xi_{(k-1)\delta}^n a_{j'}^{n,k\delta})^2 \psi_{k\delta}^{t,\varphi}(v_{j'}^n(k\delta))^2\right]. \tag{9.52}$$

From (9.25) we deduce that
$$\psi_{k\delta}^{t,\varphi}(v_{j'}^n(k\delta)) = \tilde{\mathbb{E}}\left[\varphi(v_j^n(t))a_{k\delta}^t(v_j^n, Y) \mid \mathcal{F}_{k\delta} \vee \mathcal{Y}_t\right];$$

hence by Jensen's inequality
$$\tilde{\mathbb{E}}\left[(\psi_s^{t,\varphi}(v_{j'}^n(k\delta)))^p\right] \leq \tilde{\mathbb{E}}\left[\tilde{\mathbb{E}}\left[\varphi(v_j^n(t))a_{k\delta}^t(v_j^n, Y) \mid \mathcal{F}_{k\delta} \vee \mathcal{Y}_t\right]^p\right]$$
$$= \tilde{\mathbb{E}}\left[(\varphi(v_j^n(t))a_{k\delta}^t(v_j^n, Y))^p\right].$$

Therefore
$$\tilde{\mathbb{E}}\left[(\psi_s^{t,\varphi}(v_{j'}^n(k\delta)))^p\right]$$
$$\leq \|\varphi\|_{0,\infty}^p \tilde{\mathbb{E}}\left[\exp\left(\int_{k\delta}^t ph\left(v_{j'}^n(r)\right)^\top dY_r - \frac{1}{2}\int_{k\delta}^t p^2\|h(v_{j'}^n(r))\|^2 dr\right)\right.$$
$$\left.\times \exp\left(\frac{p^2-p}{2}\int_{k\delta}^t \|h(v_{j'}^n(r))\|^2 dr\right)\right]$$
$$\leq \exp\left(\frac{1}{2}m(p^2-p)t \max_{k=1,\ldots,m}\|h_k\|_{0,\infty}^2\right) \|\varphi\|_{0,\infty}^p. \quad (9.53)$$

Using this upper bound with $p=4$, the bound (9.40) and the Cauchy–Schwarz inequality on the right-hand side of (9.52),
$$\tilde{\mathbb{E}}\left[\left(\rho_{k\delta-}^n(\psi_{k\delta-}^{t,\varphi}) - \rho_{(k-1)\delta}^n(\psi_{(k-1)\delta}^{t,\varphi})\right)^2\right]$$
$$\leq \sqrt{c_2^{t,4}} \exp\left(3mt \max_{k=1,\ldots,m}\|h_k\|_{0,\infty}^2\right) \frac{\|\varphi\|_{0,\infty}^2}{n}. \quad (9.54)$$

For the second term on the right-hand side of (9.49), observe that
$$\tilde{\mathbb{E}}\left[(\rho_{k\delta}^n(\psi_{k\delta}^{t,\varphi}) - \rho_{k\delta-}^n(\psi_{k\delta-}^{t,\varphi}))^2 \mid \mathcal{F}_{k\delta-} \vee \mathcal{Y}_t\right]$$
$$= \frac{\xi_{k\delta}^2}{n^2} \sum_{j',l'=1}^n \tilde{\mathbb{E}}\left[\left(o_{j'}^{n,k\delta} - n\bar{a}_{j'}^{n,k\delta}\right)\left(o_{l'}^{n,k\delta} - n\bar{a}_{l'}^{n,k\delta}\right) \mid \mathcal{F}_{k\delta-} \vee \mathcal{Y}_t\right]$$
$$\times \psi_{k\delta}^{t,\varphi}(v_{j'}^n(k\delta))\psi_{k\delta}^{t,\varphi}(v_{l'}^n(k\delta)).$$

Since the test function φ was chosen to be non-negative, and the random variables $\{o_{j'}^{n,k\delta}, j'=1,\ldots,n\}$ are negatively correlated (see Proposition 9.3 part e.) it follows that
$$\tilde{\mathbb{E}}\left[(\rho_{k\delta}^n(\psi_{k\delta}^{t,\varphi}) - \rho_{k\delta-}^n(\psi_{k\delta-}^{t,\varphi}))^2 \mid \mathcal{F}_{k\delta-} \vee \mathcal{Y}_t\right]$$
$$\leq \frac{\xi_{k\delta}^2}{n^2} \sum_{j'=1}^n \tilde{\mathbb{E}}\left[\left(o_{j'}^{n,k\delta} - n\bar{a}_{j'}^{n,k\delta}\right)^2 \mid \mathcal{F}_{k\delta-} \vee \mathcal{Y}_t\right] \psi_{k\delta}^{t,\varphi}(v_{j'}^n(k\delta))^2$$
$$\leq \frac{\xi_{k\delta}^2}{n^2} \sum_{j'=1}^n \{n\bar{a}_{j'}^{n,k\delta}\}\left(1 - \{n\bar{a}_{j'}^{n,k\delta}\}\right) \psi_{k\delta}^{t,\varphi}(v_{j'}^n(k\delta))^2.$$

Finally using the inequality $q(1-q) \leq \frac{1}{4}$ for $q = \{n\bar{a}_{j'}^{n,k\delta}\}$ and (9.53) with $p = 2$, it follows that

$$\tilde{\mathbb{E}}\left[(\rho_{k\delta}^n(\psi_{k\delta}^{t,\varphi}) - \rho_{k\delta-}^n(\psi_{k\delta-}^{t,\varphi}))^2\right]$$
$$\leq \frac{1}{4n}\exp\left(mt \max_{k=1,\ldots,m}\|h_k\|_{0,\infty}^2\right)\|\varphi\|_{0,\infty}^2. \quad (9.55)$$

For the last term, note that $\psi_0^{t,\varphi}$ is $\mathcal{Y}_t$-measurable, therefore using the mutual independence of the initial points $v_j^n(0)$, and the fact that

$$\tilde{\mathbb{E}}[\psi_0^{t,\varphi}(v_j^n(0)) \mid \mathcal{Y}_t] = \pi_0(\psi_0^{t,\varphi}),$$

we obtain

$$\tilde{\mathbb{E}}\left[\left(\pi_0^n(\psi_0^{t,\varphi}) - \pi_0(\psi_0^{t,\varphi})\right)^2 \mid \mathcal{Y}_t\right]$$
$$= \frac{1}{n^2}\sum_{j=1}^n \tilde{\mathbb{E}}\left[(\psi_0^{t,\varphi}(v_j^n(0)))^2 \mid \mathcal{Y}_t\right] - \left(\pi_0(\psi_0^{t,\varphi})\right)^2$$
$$\leq \frac{1}{n^2}\sum_{j=1}^n \tilde{\mathbb{E}}\left[(\psi_0^{t,\varphi}(v_j^n(0)))^2 \mid \mathcal{Y}_t\right].$$

Hence using the result (9.53) with $p = 2$,

$$\tilde{\mathbb{E}}\left[\left(\pi_0^n(\psi_0^{t,\varphi}) - \pi_0(\psi_0^{t,\varphi})\right)^2\right] \leq \frac{1}{n^2}\sum_{j=1}^n \tilde{\mathbb{E}}[\psi_0^{t,\varphi}(v_j^n(0))^2]$$
$$\leq \frac{1}{n}\exp\left(mt \max_{k=1,\ldots,m}\|h_k\|_{0,\infty}^2\right)\|\varphi\|_{0,\infty}^2. \quad (9.56)$$

The bounds on individual terms (9.51), (9.54), (9.55) and (9.56) substituted into (9.49) yields the result (9.48). □

Theorem 9.15. *If the coefficients σ, f and h are bounded and Lipschitz, then for any $T \geq 0$, there exists a constant c_4^T independent of n such that for any $\varphi \in C_b(\mathbb{R}^d)$, we have*

$$\tilde{\mathbb{E}}\left[|\pi_t^n(\varphi) - \pi_t(\varphi)|\right] \leq \frac{c_4^T}{\sqrt{n}}\|\varphi\|_{0,\infty}, \quad t \in [0,T]. \quad (9.57)$$

In particular, for all $t \geq 0$, π_t^n converges in expectation to π_t.

Proof. Since $\pi_t^n(\varphi)\rho_t^n(\mathbf{1}) = \xi_t^n \pi_t^n(\varphi) = \rho_t^n(\varphi)$

$$\pi_t^n(\varphi) - \pi_t(\varphi) = (\rho_t^n(\varphi) - \rho_t(\varphi))(\rho_t(\mathbf{1}))^{-1}$$
$$- \pi_t^n(\varphi)(\rho_t^n(\mathbf{1}) - \rho_t(\mathbf{1}))(\rho_t(\mathbf{1}))^{-1}.$$

Define
$$m_t \triangleq \sqrt{\tilde{\mathbb{E}}\left[(\rho_t(1))^{-2}\right]}.$$

Following Exercise 9.16 below, $m_t < \infty$, hence by Cauchy–Schwartz

$$\tilde{\mathbb{E}}\left[|\pi_t^n(\varphi) - \pi_t(\varphi)|\right] \leq m_t \sqrt{\tilde{\mathbb{E}}\left[(\rho_t^n(\varphi) - \rho_t(\varphi))^2\right]}$$
$$+ m_t \|\varphi\|_{0,\infty} \sqrt{\tilde{\mathbb{E}}\left[(\rho_t^n(1) - \rho_t(1))^2\right]}, \quad (9.58)$$

and the result follows by applying Proposition 9.14 to the two expectations on the right-hand side of (9.58). □

Exercise 9.16. Prove that $\tilde{\mathbb{E}}[\sup_{t \in [0,T]}(\rho_t(1))^{-2}] < \infty$ for any $T \geq 0$.

Let $\mathcal{M} = \{\varphi_i, \, i \geq 0\} \in C_b(\mathbb{R}^d)$ be a countable convergence determining set such that $\|\varphi_i\| \leq 1$ for any $i \geq 0$ and $d_\mathcal{M}$ be the metric on $\mathcal{M}_F(\mathbb{R}^d)$ (see Section A.10 for additional details)

$$d_\mathcal{M} : \mathcal{M}_F(\mathbb{R}^d) \times \mathcal{M}_F(\mathbb{R}^d) \to [0, \infty), \qquad d(\mu, \nu) = \sum_{i=0}^{\infty} \frac{|\mu\varphi_i - \nu\varphi_i|}{2^i}.$$

Proposition 9.14 and Theorem 9.15 give the following corollary.

Corollary 9.17. *If the coefficients σ, f and h are bounded and Lipschitz, then*

$$\sup_{t \in [0,T]} \tilde{\mathbb{E}}[d_\mathcal{M}(\rho_t^n, \rho_t)] \leq \frac{2\sqrt{c_3^T}}{\sqrt{n}}, \qquad \sup_{t \in [0,T]} \tilde{\mathbb{E}}[d_\mathcal{M}(\pi_t^n, \pi_t)] \leq \frac{2c_4^T}{\sqrt{n}}. \quad (9.59)$$

Thus ρ_t^n converges to ρ_t in expectation and π_t^n converges to π_t in expectation. In the following, we prove a stronger convergence result.

Proposition 9.18. *If the coefficients σ, f and h are bounded and Lipschitz, then for any $T \geq 0$, there exists a constant c_5^T independent of n such that*

$$\tilde{\mathbb{E}}\left[\sup_{t \in [0,T]}(\rho_t^n(\varphi) - \rho_t(\varphi))^2\right] \leq \frac{c_5^T}{n}\|\varphi\|_{2,\infty}^2 \quad (9.60)$$

for any $\varphi \in C_b^2(\mathbb{R}^d)$.

Proof. Again, it suffices to prove (9.60) for any non-negative $\varphi \in C_b^2(\mathbb{R}^d)$. Following Exercise 9.11 we have that

$$\rho_t^n(\varphi) - \rho_t(\varphi) = (\pi_0^n(\varphi) - \pi_0(\varphi)) + \int_0^t (\rho_s^n(A\varphi) - \rho_s(A\varphi))\,ds + \bar{S}_t^{n,\varphi}$$
$$+ \bar{M}_{[t/\delta]}^{n,\varphi} + \sum_{k=1}^m \int_0^t (\rho_s^n(h_k\varphi) - \rho_s(h_k\varphi))\,dY_s^k, \quad (9.61)$$

where $\bar{S}^{n,\varphi} = \{\bar{S}_t^{n,\varphi},\ t \geq 0\}$ is the martingale

$$\bar{S}_t^{n,\varphi} \triangleq \frac{1}{n} \sum_{i=0}^{\infty} \sum_{j=1}^{n} \int_{i\delta \wedge t}^{(i+1)\delta \wedge t} \xi_{i\delta}^n a_j^n(s)((\nabla\varphi)^\top \sigma)(v_j^n(s)) \mathrm{d}V_s^{(j)},$$

and $\bar{M}^{n,\varphi} = \{\bar{M}_k^{n,\varphi},\ k > 0\}$ is the discrete parameter martingale

$$\bar{M}_k^{n,\varphi} \triangleq \frac{1}{n} \sum_{i=1}^{k} \xi_{i\delta}^n \sum_{j'=1}^{n} (o_{j'}^n(i\delta) - n\bar{a}_{j'}^{n,i\delta}) \varphi(v_{j'}^n(i\delta)), \quad k > 0.$$

We show that each of the five terms on the right-hand side of (9.61) satisfies an inequality of the form (9.60). For the first term, using the mutual independence of the initial locations of the particles $v_j^n(0)$, we obtain

$$\tilde{\mathbb{E}}\left[(\pi_0^n(\varphi) - \pi_0(\varphi))^2\right] = \frac{1}{n}\left(\pi_0(\varphi^2) - \pi_0(\varphi)^2\right) \leq \frac{1}{n}\|\varphi\|_{0,\infty}^2. \tag{9.62}$$

For the second term, by Cauchy–Schwartz

$$\tilde{\mathbb{E}}\left[\sup_{t\in[0,T]}\left(\int_0^t (\rho_s^n(A\varphi) - \rho_s(A\varphi))\mathrm{d}s\right)^2\right]$$

$$\leq \tilde{\mathbb{E}}\left[\sup_{t\in[0,T]} t \int_0^t (\rho_s^n(A\varphi) - \rho_s(A\varphi))^2\,\mathrm{d}s\right]$$

$$= \tilde{\mathbb{E}}\left[T \int_0^T (\rho_s^n(A\varphi) - \rho_s(A\varphi))^2\,\mathrm{d}s\right]. \tag{9.63}$$

By Fubini's theorem and (9.48), we obtain

$$\tilde{\mathbb{E}}\left[\int_0^T (\rho_s^n(A\varphi) - \rho_s(A\varphi))^2\,\mathrm{d}s\right] \leq \frac{c_3^T T}{n}\|A\varphi\|_{0,\infty}^2. \tag{9.64}$$

From the boundedness of σ and f since there exists $c_6 = c_6(\|\sigma\|_{0,\infty}, \|f\|_{0,\infty})$ such that

$$\|A\varphi\|_{0,\infty}^2 \leq c_6 \|\varphi\|_{2,\infty}^2,$$

from (9.63) and (9.64) that

$$\tilde{\mathbb{E}}\left[\sup_{t\in[0,T]}\left(\int_0^t (\rho_s^n(A\varphi) - \rho_s(A\varphi))\,\mathrm{d}s\right)^2\right] \leq \frac{c_3^T c_6 T^2}{n}\|\varphi\|_{2,\infty}^2. \tag{9.65}$$

For the third term, we use the Burkholder–Davis–Gundy inequality (Theorem B.36). If we denote by C the constant in the Burkholder–Davis–Gundy inequality applied to $F(x) = x^2$, then

$$\tilde{\mathbb{E}}\left[\sup_{t\in[0,T]}(\bar{S}_t^{n,\varphi})^2\right] \leq C\tilde{\mathbb{E}}\left[\langle \bar{S}^{n,\varphi}\rangle_T\right]$$

$$= \frac{C}{n^2}\sum_{j=1}^n \int_0^T \tilde{\mathbb{E}}\left[(\xi_{[s/\delta]\delta}^n a_j^n(s))^2((\nabla\varphi)^\top\sigma\sigma^\top\nabla\varphi)(v_j^n(s))\right]ds. \quad (9.66)$$

From (9.40) and the fact that σ is bounded, we deduce that there exists a constant c_7^T such that

$$\tilde{\mathbb{E}}[(\xi_{[s/\delta]\delta}^n a_j^n(s))^2((\nabla\varphi)^\top\sigma\sigma^\top\nabla\varphi)(v_j^n(s))] \leq c_7^T \|\varphi\|_{2,\infty}^2, \quad (9.67)$$

for any $s\in[0,T]$. From (9.66) and (9.67)

$$\tilde{\mathbb{E}}\left[\sup_{t\in[0,T]}(\bar{S}_t^{n,\varphi})^2\right] \leq \frac{Cc_7^T T}{n}\|\varphi\|_{2,\infty}^2. \quad (9.68)$$

For the fourth term on the right-hand side of (9.61), by Doob's maximal inequality

$$\tilde{\mathbb{E}}\left[\max_{k=1,\ldots,[T/\delta]}\left(\bar{M}_k^{n,\varphi}\right)^2\right] \leq 4\tilde{\mathbb{E}}\left[\left(\bar{M}_{[T/\delta]}^{n,\varphi}\right)^2\right]. \quad (9.69)$$

Since φ is non-negative and the offspring numbers, $o_{j'}^n(i\delta)$ for $j'=1,\ldots,n$, are negatively correlated, from the orthogonality of martingale increments

$$\tilde{\mathbb{E}}\left[\left(\bar{M}_{[T/\delta]}^{n,\varphi}\right)^2\right]$$

$$\leq \frac{1}{n^2}\sum_{i=1}^{[T/\delta]}\sum_{j=1}^n \tilde{\mathbb{E}}\left[(\xi_{i\delta}^n)^2\{n\bar{a}_j^n(i\delta)\}\left(1-\{n\bar{a}_j^n(i\delta)\}\right)\left(\varphi\left(v_j^n(i\delta)\right)\right)^2\right]$$

$$\leq \frac{\|\varphi\|_{0,\infty}^2}{4n^2}\sum_{i=1}^{[T/\delta]}\sum_{j=1}^n \tilde{\mathbb{E}}\left[(\xi_{i\delta}^n)^2\right]. \quad (9.70)$$

Then, from (9.39), (9.69) and (9.70) there exists a constant $c_8^T = c_1^{T,2}[T/\delta]/4$ independent of n such that

$$\tilde{\mathbb{E}}\left[\max_{k=1,\ldots,[T/\delta]}\left(\bar{M}_k^{n,\varphi}\right)^2\right] \leq \frac{c_8^T}{n}\|\varphi\|_{0,\infty}^2. \quad (9.71)$$

To bound the last term, we use the Burkholder–Davis–Gundy inequality (Theorem B.36), Fubini's theorem and the conclusion of Proposition 9.14 (viz equation (9.48)) to obtain

$$\tilde{\mathbb{E}}\left[\sup_{t\in[0,T]}\left(\int_0^t (\rho_s^n(h_k\varphi) - \rho_s(h_k\varphi))\,\mathrm{d}Y_s^k\right)^2\right]$$

$$\leq C\tilde{\mathbb{E}}\left[\int_0^T (\rho_s^n(h_k\varphi) - \rho_s(h_k\varphi))^2\,\mathrm{d}s\right]$$

$$\leq C\int_0^T \tilde{\mathbb{E}}\left[(\rho_s^n(h_k\varphi) - \rho_s(h_k\varphi))^2\right]\,\mathrm{d}s$$

$$\leq \frac{Cc_3^T T\|h_k\|_{0,\infty}}{n}\|\varphi\|_{0,\infty}^2. \tag{9.72}$$

The bounds (9.62), (9.65), (9.68), (9.71) and (9.72) together imply (9.60). □

Theorem 9.19. *If the coefficients σ, f and h are bounded and Lipschitz, then for any $T \geq 0$, there exists a constant c_9^T independent of n such that*

$$\tilde{\mathbb{E}}\left[\sup_{t\in[0,T]} |\pi_t^n(\varphi) - \pi_t(\varphi)|\right] \leq \frac{c_9^T}{\sqrt{n}}\|\varphi\|_{2,\infty} \tag{9.73}$$

for any $\varphi \in C_b^2(\mathbb{R}^d)$.

Proof. As in the proof of Theorem 9.15,

$$\tilde{\mathbb{E}}\left[\sup_{t\in[0,T]} |\pi_t^n(\varphi) - \pi_t(\varphi)|\right] \leq \bar{m}_T \sqrt{\tilde{\mathbb{E}}\left[\sup_{t\in[0,T]} (\rho_t^n(\varphi) - \rho_t(\varphi))^2\right]}$$

$$+ \bar{m}_T \|\varphi\|_{0,\infty} \sqrt{\tilde{\mathbb{E}}\left[\sup_{t\in[0,T]} (\rho_t^n(\mathbf{1}) - \rho_t(\mathbf{1}))^2\right]},$$

where, following Exercise 9.16,

$$\bar{m}_T \triangleq \sqrt{\tilde{\mathbb{E}}\left[\sup_{t\in[0,T]} (\rho_t(\mathbf{1}))^{-2}\right]} < \infty$$

and the result follows from Proposition 9.18. □

Let $\bar{\mathcal{M}} = \{\varphi_i, i \geq 0\}$ where each $\varphi_i \in C_b^2(\mathbb{R}^d)$ be a countable convergence determining set such that $\|\varphi_i\|_\infty \leq 1$ and $\|\varphi\|_{2,\infty} \leq 1$ for any $i \geq 0$ and $d_{\bar{\mathcal{M}}}$ be the corresponding metric on $\mathcal{M}_F(\mathbb{R}^d)$ as defined in Section A.10. The following corollary of Proposition 9.18 and Theorem 9.19 is then immediate.

Corollary 9.20. *If the coefficients σ, f and h are bounded and Lipschitz, then we have*

$$\tilde{\mathbb{E}}\left[\sup_{t\in[0,T]} d_{\bar{\mathcal{M}}}(\rho_t^n, \rho_t)\right] \leq \frac{2\sqrt{c_5^T}}{\sqrt{n}}, \qquad \tilde{\mathbb{E}}\left[\sup_{t\in[0,T]} d_{\bar{\mathcal{M}}}(\pi_t^n, \pi_t)\right] \leq \frac{2c_9^T}{\sqrt{n}} \tag{9.74}$$

for any $T \geq 0$.

9.5 Other Results

The particle filter described above merges the weighted approximation approach, as presented in Kurtz and Xiong [171, 174] for a general class of non-linear stochastic partial differential equations (to which the Kushner–Stratonovich equation belongs) with the branching corrections approach introduced by Crisan and Lyons in [65]. The convergence of the resulting approximation follows from Theorem 9.15 under fairly mild conditions on the coefficients. The convergence results described above can be extended to the correlated noise framework. See Section 3.8 for a description of this framework and Crisan [61] for details of the proofs in this case. More refined convergence results require the use of the decomposition (9.61). For this we make use of the properties of the dual of ρ supplied by the theory of stochastic evolution systems (cf. Rozovskii [250]; see also Veretennikov [267] for a direct approach to establishing the dual property of $\psi^{t,\varphi}$).

The decomposition (9.61) is very important. It will lead to an exact rate of convergence, that is, to computing the limit

$$\lim_{n\to\infty} n\tilde{\mathbb{E}}\left[(\rho^n_t(\varphi) - \rho_t(\varphi))^2\right]$$

and also to a central limit theorem (note that the three terms on the right-hand side of (9.61) are mutually orthogonal). For this we need to understand the limiting behaviour of the covariance matrix of the random variables $\{o^n_j,\ j = 1,\ldots,n\}$. This has yet to be achieved.

In the last ten years we have witnessed a rapid development of the theory of particle approximations to the solution of non-linear filtering, and implicitly to solving SPDEs similar to the filtering equations. The discrete time framework has been extensively studied and a multitude of convergence and stability results have been proved. A comprehensive description of these developments in the wider context of approximations of Feynman–Kac formulae can be found in Del Moral [216] and the references therein. See also Del Moral and Jacod [217] for a result involving discrete observations but a continuous signal.

Results concerning particle approximations for the continuous time filtering problem are far fewer that their discrete counterparts. The development of particle filters for continuous time problems started in the mid-1990s. In Crisan and Lyons [64], the particle construction of a superprocess is extended to the case of a branching measure-valued process in a random environment. When averaged, the particle system used in the construction is shown to converge to the solution of the Zakai equation. In Crisan et al. [63], the idea of minimal variance branching is introduced (instead of fixed variance branching) with the resulting particle system shown to converge to the solution of the Zakai equation. Finally, in Crisan and Lyons [65], a direct approximation of π_t is produced by using a normalised branching approach. In Crisan et al. [62], an alternative approximation to the Kushner–Stratonovich equation (3.57) is given where the branching step is replaced by a correction procedure using multinomial resampling. The multinomial resampling procedure produces conditionally independent approximate samples from the conditional distribution of the signal, thus facilitating the analysis of the corresponding algorithms. It is, however, suboptimal. For a heuristic explanation, assume that between two consecutive correction steps, the information we receive on the signal is 'bad' (the signal-to-noise ratio is small). Consequently the corresponding weights will all be (roughly) equal: that is, all the particles are equally likely. The correction procedure should leave the particles untouched in this case as there is no reason to cull or multiply any of the particles. This is exactly what the minimal branching step does: each particle has exactly one offspring. The multinomial resampling correction will not do this: some particle will be resampled more than others thus introducing an unnecessary random perturbation to the system. For theoretical results related to the suboptimality of the multinomial resampling procedure, see e.g. Crisan and Lyons [66] and Chopin [51]. Even if one uses the minimal variance branching correction, additional randomness is still introduced in the system, which can affect the convergence rates (see Crisan [60]). It remains an open question as to when and how often should one use the correction procedure.

On a parallel approach, Del Moral and Miclo [218] produced a particle filter using the pathwise approach of Davis [74]. The idea is to recast the equations of non-linear filtering in a form in which no stochastic integration is required. Then one can apply Del Moral's general method of approximating Feynman–Kac formulae. This approach is important as it emphasises the robustness of the particle filter, although it requires that the observation noise and signal noise are independent. While it cannot be applied to the correlated noise framework, it is nevertheless a very promising approach and we expect further research to show its full potential.

9.6 The Implementation of the Particle Approximation for π_t

In the following we give a brief description of the implementation of the particle approximation analysed in this chapter. We start by choosing parameters n, δ and m. We use n particles and we apply the correction (branching) procedure at times $k\delta$, for $i > 1$, divide the inter branching intervals $[(k-1)\delta, k\delta]$ into m subintervals of length δ/m and apply the Euler method to generate the trajectories of the particles. The following is the initialization step.

Initialization

For $j := 1, \ldots, n$
    Sample $v_j(0)$ from π_0.
    $a_j(0) := 1$.
end for

$\pi_0 := \frac{1}{n}\sum_{j=1}^{n} \delta_{v_j(0)}$
Assign value $t := 0$

The standard sampling procedure can be replaced by any alternative method that produces an approximation for π_0. For example, a stratified sampling procedure, if available, will produce a better approximation. In the special case where π_0 is a Dirac measure concentrated at $x_0 \in \mathbb{R}^d$, the value x_0 is assigned to all initial positions $v_j(0)$ of the particles. The following is the (two-step) iteration procedure.

Iteration $[i\delta \text{ to } (i+1)\delta]$

1. Evolution of the particles

for $l := 0$ to $m-1$
    for $j := 1$ to n
        Generate the Gaussian random vector ΔV.
        $v_j(t+\delta/m) := v_j(t) + f(v_j(t))\delta/m + \sigma(v_j(t))\Delta V \sqrt{\delta/m}$.
        $b_j(t+\delta/m) := h(v_j(t))^\top (Y_{t+\delta/m} - Y_t) - (\delta/2m)\|h(v_j(t))\|^2$
        $a_j(t+\delta/m) := a_j(t)\exp(b_j(t+\delta/m))$
    end for
    $t := t + \delta/m$
    $\Sigma(t) := \sum_{j=1}^{n} a_j(t)$
    $\pi_t^n := \frac{1}{\Sigma(t)} \sum_{j=1}^{n} a_j(t) \delta_{v_j(t)}$.
end for

In the above $\Delta V = (\Delta V_1, \Delta V_2, \ldots, \Delta V_p)^\top$ is a p-dimensional random vector with independent identically distributed entries $\Delta V_i \sim N(0,1)$ for all $i = 1, \ldots, p$.

The Euler method used above can be replaced by any other weak approximation method for the solution of the stochastic differential equation satisfied by the signal (see for example Kloeden and Platen [151] for alternative approximation methods). The choice of the parameters δ and m depends on the frequency of the arrivals of the new observations Y_t. We have assumed that the observation Y_t is available for all time instants t which are integer multiples of δ/m. There are no theoretical results as to what is the right balance between the size of the intervals between corrections and the number of steps used to approximate the law of the signal, in other words what is the optimal choice of parameters δ and m.

2. Branching procedure

for $j := 1$ to n
    $\bar{a}_j(t) := a_j(t)/\Sigma(t)$
end for
for $j' := 1$ to n

Calculate the number of offspring $o_{j'}^n(t)$ for the j'th particle in the system of particles with weights/positions $(\bar{a}_j(t), v_j(t))$ using the algorithm described in Section 9.2.1.
end for
We have now n particles with positions

$$(\underbrace{v_1(t), v_1(t), \ldots, v_1(t)}_{o_1(t)}, \underbrace{v_2(t), v_2(t), \ldots, v_2(t)}_{o_2(t)}, \ldots) \quad (9.75)$$

Reindex the positions of the particles as $v_1(t), v_2(t), \ldots, v_n(t)$.
for $j := 1, \ldots, n$
$\quad a_j(t) := 1$
end for

The positions of the particles with no offspring will no longer appear among those described by the formula (9.75). Alternatives to the branching procedure are described in Section 10.5. For example, one can use the sampling with replacement method. In this case Step 2 is replaced by the following.

2'. Resampling procedure

for $j := 1$ to n
$\quad \bar{a}_j(t) := a_j(t)/\Sigma(t)$.
end for
for $j := 1$ to n
$\quad$ Pick $v_j(t)$ by sampling with replacement from the set of particle positions $(v_1(t), v_2(t), \ldots, v_n(t))$ according to the probability vector of normalized weights $(\bar{a}_1(t), \bar{a}_2(t), \ldots, \bar{a}_n(t))$.
end for
Reindex the positions of the particles as $v_1(t), v_2(t), \ldots, v_n(t)$.
for $j := 1, \ldots, n$
$\quad a_j(t) := 1$
end for

However, the resampling procedure generates a multinomial offspring distribution which is known to be suboptimal. In particular, it does not have the minimal variance property enjoyed by the offspring distribution produced by the algorithm described in Section 9.2.1 (see Section 10.5 for details).

9.7 Solutions to Exercises

9.1 In the case where a is an integer it is immediate that taking $\xi^{\min} = a$ achieves the minimal variance of zero, and by Jensen's inequality for any convex function φ, for $\xi \in \mathcal{A}_a$, $\mathbb{E}[\varphi(\xi)] \geq \varphi(\mathbb{E}(\xi)) = \varphi(a) = \mathbb{E}[\varphi(\xi^{\min})]$ thus $\mathbb{E}[\varphi(\xi^{\min})] \leq \mathbb{E}[\varphi(\xi)]$ for any $\xi \in \mathcal{A}_a$.

For the more general case, let $\xi \in \mathcal{A}_a$. Suppose that the law of ξ assigns non-zero probability mass to two integers which are not adjacent. That is, we can find k, l such that $\mathbb{P}(\xi = k) > 0$ and $\mathbb{P}(\xi = l) > 0$ and $k + 1 \leq l - 1$.

We construct a new random variable ζ from ξ by moving some probability mass $\beta > 0$ from k to $k+1$ and some from l to $l-1$. Let $U \subset \{\omega : \xi(\omega) = k\}$ and $D \subset \{\omega : \xi(\omega) = l\}$, be such that $\mathbb{P}(U) = \mathbb{P}(D) = \beta$; then define

$$\zeta \triangleq \xi + 1_U - 1_D.$$

Thus by direct computation, $\mathbb{E}[\zeta] = a + \beta - \beta$, so $\zeta \in \mathcal{A}_a$; secondly

$$\text{var}(\zeta) = \mathbb{E}[\zeta^2] - a^2 = \mathbb{E}[\xi^2] + 2\beta(1 + k - l) - a^2$$
$$= \text{var}(\xi) + 2\beta(1 + k - l).$$

As we assumed that $k + 1 \leq l - 1$, it follows that $\text{var}(\zeta) < \text{var}(\xi)$. Consequently the variance minimizing element of $\mathcal{A}_a$ can only have non-zero probability mass on two adjacent negative integers, and then the condition on the expectation ensures that this must be $\xi^{\min}$ given by (9.10).

Now consider φ a convex function, we use the same argument

$$\mathbb{E}[\varphi(\zeta)] = \mathbb{E}[\varphi(\xi)] + \beta\left(\varphi(k+1) - \varphi(k) + \varphi(l-1) - \varphi(l)\right).$$

Now we use that fact that if φ is a convex function for any points $a < b < c$, since the graph of φ lies below the chord $(a, \varphi(a))$–$(c, \varphi(c))$,

$$\varphi(b) \leq \varphi(a)\frac{c-b}{c-a} + \varphi(c)\frac{b-a}{c-a},$$

which implies that

$$\frac{\varphi(b) - \varphi(a)}{b - a} \leq \frac{\varphi(c) - \varphi(b)}{c - b}.$$

If $k + 1 = l - 1$ we can apply this result directly to see that $\varphi(k+1) - \varphi(k) \leq \varphi(l) - \varphi(l-1)$, otherwise we use the result twice, for $k < k+1 < l-1$ and for $k+1 < l-1 < l$, to obtain

$$\varphi(k+1) - \varphi(k) \leq \frac{\varphi(l-1) - \varphi(k+1)}{k - l - 2} \leq \varphi(l) - \varphi(l-1)$$

thus

$$\mathbb{E}[\varphi(\zeta)] \leq \mathbb{E}[\varphi(\xi)].$$

This inequality will be strict unless φ is linear between k and l. If it is strict, then we can argue as before that $\mathbb{E}[\varphi(\zeta)] < \mathbb{E}[\varphi(\zeta)]$. It is therefore clear that if we can find a non-adjacent pair of integers k and l, such that φ is not linear between k and l then the random variable ξ cannot minimize $\mathbb{E}[\varphi(\xi)]$. Consequently, a ξ which minimizes $\mathbb{E}[\varphi(\xi)]$ can either assign strictly positive mass to a single pair of adjacent integers, or it can assign strictly positive probability to any number of integers, provided that they are all contained in a single interval of $\mathbb{R}$ where the function $\phi(x)$ is linear.

In the second case where $\xi \in \mathcal{A}_a$ only assigns non-negative probability to integers in an interval where φ is linear, it is immediate that $\mathbb{E}[\varphi(\xi)] = \varphi(\mathbb{E}[\xi]) = \varphi(a)$, thus as a consequence of Jensen's inequality such a ξ achieves the minimum value of $\mathbb{E}[\varphi(\xi)]$ over $\xi \in \mathcal{A}_a$. Since $\xi \in \mathcal{A}_a$ satisfies $\mathbb{E}[\xi] = a$, the region where φ is linear must include the integers $[a]$ and $[a]+1$, therefore with $\xi^{\min}$ defined by (9.10), $\mathbb{E}[\varphi(\xi^{\min})] = \varphi(\mathbb{E}[a])$.

It therefore follows that in either case, the minimum value is uniquely attained by $\xi^{\min}$ unless φ is linear in which case $\mathbb{E}[\varphi(\xi)]$ is constant for any $\xi \in \mathcal{A}_a$. $\mathbb{E}[\varphi(\xi^{\min})] \le \mathbb{E}[\varphi(\xi)]$ for any $\xi \in \mathcal{A}_a$.

9.10 We have for $t \in [k\delta, (k+1)\delta]$

$$\begin{aligned}(a_j^n(t))^p &= \exp\left(p\int_{k\delta}^t h(v_j^n(s))^\top \, dY_s - \frac{p}{2}\int_{k\delta}^t \|h(v_j^n(s))\|^2 \, ds\right) \\ &= M_p(t)\exp\left(\frac{p^2-p}{2}\int_{k\delta}^t \|h(v_j^n(s))\|^2 \, ds\right) \\ &\le M_p(t)\exp\left(\frac{p^2-p}{2}\sum_{i=1}^m \|h^i\|_\infty^2 (t-k\delta)\right),\end{aligned}$$

where $M_p = \{M_p(t),\ t \in [k\delta,(k+1)\delta]\}$ is the exponential martingale defined as

$$M_p(t) \triangleq \exp\left(p\int_{k\delta}^t h(v_j^n(s))^\top \, dY_s - \frac{p^2}{2}\int_{k\delta}^t \|h(v_j^n(s))\|^2 \, ds\right).$$

Hence

$$\tilde{\mathbb{E}}\left[(a_j^n(t))^p \mid \mathcal{F}_{k\delta}\right] \le \exp\left(\frac{p^2-p}{2}\sum_{i=1}^m \|h^i\|_\infty^2 (t-k\delta)\right),$$

which, in turn, implies that

$$\tilde{\mathbb{E}}\left[\left(\frac{1}{n}\sum_{j=1}^n a_j^n(t)\right)^p \bigg| \mathcal{F}_{k\delta}\right] \le \exp\left(\frac{p^2-p}{2}\sum_{i=1}^m \|h^i\|_\infty^2 (t-k\delta)\right). \tag{9.76}$$

Therefore

$$\begin{aligned}\tilde{\mathbb{E}}\left[(\xi_t^n)^p \mid \mathcal{F}_{[t/\delta]\delta}\right] &= \left(\xi_{[t/\delta]\delta}^n\right)^p \tilde{\mathbb{E}}\left[\left(\frac{1}{n}\sum_{j=1}^n a_j^n(t)\right)^p \bigg| \mathcal{F}_{[t/\delta]\delta}\right] \\ &\le \left(\xi_{[t/\delta]\delta}^n\right)^p \exp\left(\frac{(p^2-p)(t-k\delta)}{2}\sum_{i=1}^m \|h^i\|_\infty^2\right). \tag{9.77}\end{aligned}$$

Also from (9.76) one proves that

$$\mathbb{E}\left[(\xi_{k\delta}^n)^p \,|\, \mathcal{F}_{(k-1)\delta}\right] \leq \left(\xi_{(k-1)\delta}^n\right)^p \exp\left(\frac{p^2-p}{2} \sum_{i=1}^m \|h^i\|_\infty^2 \delta\right)$$

hence, by induction,

$$\mathbb{E}[(\xi_{k\delta}^n)^p] \leq \exp\left(\frac{p^2-p}{2} \sum_{i=1}^m \|h^i\|_\infty^2 k\delta\right). \tag{9.78}$$

Finally from (9.76), (9.77) and (9.78) we get (9.39). The bound (9.40) follows in a similar manner.

9.11 We follow the proof of Proposition 9.7 Let $\mathcal{F}_{k\delta-} = \sigma(\mathcal{F}_s,\ 0 \leq s < k\delta)$ be the σ-algebra of events up to time $k\delta$ (the time of the kth-branching) and $\rho_{k\delta-}^n = \lim_{t \nearrow k\delta} \rho_t^n$. For $t \in [i\delta, (i+1)\delta)$, we have†

$$\rho_t^n(\varphi) = \pi_0^n(\varphi) + \bar{M}_i^{n,\varphi} + \sum_{k=1}^i (\rho_{k\delta-}^n(\varphi) - \rho_{(k-1)\delta}^n(\varphi))$$
$$+ (\rho_t^n(\varphi) - \rho_{i\delta}^n(\varphi)),$$

where $\bar{M}^{n,\varphi} = \{\bar{M}_k^{n,\varphi},\ k > 0\}$ is the martingale

$$\bar{M}_i^{n,\varphi} = \sum_{k=1}^i \left(\rho_{k\delta}^n(\varphi) - \rho_{k\delta-}^n(\varphi)\right)$$
$$= \frac{1}{n} \sum_{k=1}^i \xi_{i\delta}^n \sum_{j'=1}^n (o_{j'}^n(i\delta) - n\bar{a}_{j'}^{n,i\delta}) \varphi(v_{j'}^n(i\delta)), \quad \text{for } i \geq 0.$$

Next, by Itô's formula, from (9.4) and (9.5), we get that

$$da_j^n(t)\varphi\left(v_j^n(t)\right) = a_j^n(t) A\varphi(v_j^n(t))\,dt$$
$$+ a_j^n(t)((\nabla\varphi)^\top \sigma)(v_j^n(t))\,dV_t$$
$$+ a_j^n(t)\varphi(v_j^n(t)) h(v_j^n(t))^\top dY_t$$

for $\varphi \in C_b^2(\mathbb{R}^d)$. Hence for $t \in [k\delta, (k+1)\delta)$, for $k = 0, 1, \ldots, i$, we have

$$\rho_t^n(\varphi) - \rho_{k\delta}^n(\varphi) = \int_{k\delta}^t \xi_{k\delta}^n \sum_{j=1}^n da_j^n(s)\varphi(v_j^n(s))$$
$$= \int_{k\delta}^t \rho_s^n(A\varphi)\,ds$$
$$+ \frac{1}{n} \sum_{j=1}^n \int_{k\delta}^t \xi_{k\delta}^n a_j^n(s)((\nabla\varphi)^\top \sigma)(v_j^n(s))\,dV_s^j$$
$$+ \sum_{r=1}^m \int_{k\delta}^t \rho_s^n(h_r\varphi)\,dY_s^r.$$

---

† We use the standard convention $\sum_{k=1}^0 = 0$.

Similarly

$$\rho_{k\delta-}^n(\varphi) - \rho_{(k-1)\delta}^n(\varphi) = \int_{(k-1)\delta}^{k\delta} \rho_s^n(A\varphi)\,ds$$
$$+ \frac{1}{n}\sum_{j=1}^n \int_{(k-1)\delta}^{k\delta} \xi_{k\delta}^n a_j^n(s)((\nabla\varphi)^\top \sigma)(v_j^n(s))\,dV_s^j$$
$$+ \sum_{r=1}^m \int_{(k-1)\delta}^{k\delta} \rho_s^n(h_r\varphi)\,dY_s^r.$$

9.16 Following Lemma 3.29, the process $t \mapsto \rho_t(\mathbf{1})$ has the explicit representation (3.55). That is,

$$\rho_t(\mathbf{1}) = \exp\left(\int_0^t \pi_s(h^\top)\,dY_s - \frac{1}{2}\int_0^t \pi_s(h^\top)\pi_s(h)\,ds\right).$$

As in Exercise 9.10 with $p = -2$, for $t \in [0, T]$,

$$\rho_t(\mathbf{1})^{-2} \leq \exp(3mt\|h\|_\infty^2)\,M_t,$$

where $M = \{M_t,\ t \in [0, T]\}$ is the exponential martingale defined as

$$M_t \triangleq \exp\left(-2\int_0^t \pi_s(h^\top)\,dY_s - 2\int_0^t \pi_s(h^\top)\pi_s(h)\,ds\right).$$

Using an argument similar to that used in the solution of Exercise 3.10 based on the Gronwall inequality and the Burkholder–Davis–Gundy inequality (see Theorem B.36 in the appendix), one shows that

$$\tilde{\mathbb{E}}\left[\sup_{t\in[0,T]} M_t\right] < \infty;$$

hence the claim.

10
Particle Filters in Discrete Time

The purpose of this chapter is to present a rigorous mathematical treatment of the convergence of particle filters in the (simpler) framework where both the signal X and the observation Y are discrete time processes. This restriction means that this chapter does not use stochastic calculus. The chapter is organized as follows. In the following section we describe the discrete time framework. In Section 10.2 we deduce the recurrence formula for the conditional distribution of the signal in discrete time. In Section 10.3 we deduce necessary and sufficient conditions for sequences of (random) measures to converge to the conditional distribution of the signal. In Section 10.4 we describe a generic class of particle filters which are shown to converge in the following section.

10.1 The Framework

Let the signal $X = \{X_t,\ t \in \mathbb{N}\}$ be a stochastic process defined on the probability space $(\Omega, \mathcal{F}, \mathbb{P})$ with values in $\mathbb{R}^d$. Let $\mathcal{F}_t^X$ be the filtration generated by the process; that is,

$$\mathcal{F}_t^X \triangleq \sigma(X_s,\ s \in [0,t]).$$

We assume that X is a Markov chain. That is, for all $t \in \mathbb{N}$ and $A \in \mathcal{B}(\mathbb{R}^d)$,

$$\mathbb{P}\left(X_{t+1} \in A \mid \mathcal{F}_t^X\right) = \mathbb{P}\left(X_{t+1} \in A \mid X_t\right). \tag{10.1}$$

The transition kernel of the Markov chain X is the function $K_t(\cdot,\cdot)$ defined on $\mathbb{R}^d \times \mathcal{B}(\mathbb{R}^d)$ such that, for all $t \in \mathbb{N}$ and $x \in \mathbb{R}^d$,

$$K_t(x,A) = \mathbb{P}(X_{t+1} \in A \mid X_t = x). \tag{10.2}$$

The transition kernel K_t is required to have the following properties.

i. $K_t(x,\cdot)$ is a probability measure on $(\mathbb{R}^d, \mathcal{B}(\mathbb{R}^d))$, for all $t \in \mathbb{N}$ and $x \in \mathbb{R}^d$.

A. Bain, D. Crisan, *Fundamentals of Stochastic Filtering*,
DOI 10.1007/978-0-387-76896-0_10, © Springer Science+Business Media, LLC 2009

ii. $K_t(\cdot, A) \in B(\mathbb{R}^d)$, for all $t \in \mathbb{N}$ and $A \in \mathcal{B}(\mathbb{R}^d)$.

The distribution of X is uniquely determined by its initial distribution and its transition kernel (see Theorem A.11 for details of how a stochastic process may be constructed from its transition kernels). Let us denote by q_t the distribution of the random variable X_t,

$$q_t(A) \triangleq \mathbb{P}(X_t \in A).$$

Then, from (10.2), it follows that q_t satisfies the recurrence formula

$$q_{t+1} = K_t q_t, \quad t \geq 0,$$

where $K_t q_t$ is the measure defined by

$$(K_t q_t)(A) \triangleq \int_{\mathbb{R}^d} K_t(x, A) q_t(\mathrm{d}x). \tag{10.3}$$

Hence, by induction it follows that

$$q_t = K_{t-1} \ldots K_1 K_0 q_0, \quad t > 0.$$

Exercise 10.1. For arbitrary $\varphi \in B(\mathbb{R}^d)$ and $t \geq 0$, define $K_t \varphi$ as

$$K_t \varphi(x) = \int_{\mathbb{R}^d} \varphi(y) K_t(x, \mathrm{d}y).$$

i. Prove that $K_t \varphi \in B(\mathbb{R}^d)$ for any $t \geq 0$.
ii. Prove that $K_t q_t$ is a probability measure for any $t \geq 0$.
iii. Prove that, for any $\varphi \in B(\mathbb{R}^d)$ and $t > 0$, we have

$$K_t q_t(\varphi) = q_t(K_t \varphi),$$

hence in general

$$q_t(\varphi) = q_0(\varphi_t), \quad t > 0,$$

where $\varphi_t = K_0 K_1 \ldots K_{t-1} \varphi \in B(\mathbb{R}^d)$.

Let the observation process $Y = \{Y_t, \; t \in \mathbb{N}\}$ be an $\mathbb{R}^m$-valued stochastic process defined as follows

$$Y_t \triangleq h(t, X_t) + W_t, \quad t > 0, \tag{10.4}$$

and $Y_0 = 0$. In (10.4), $h : \mathbb{N} \times \mathbb{R}^d \to \mathbb{R}^m$ is a Borel-measurable function and for all $t \in \mathbb{N}$, $W_t : \Omega \to \mathbb{R}^m$ are mutually independent random vectors with laws absolutely continuous with respect to the Lebesgue measure λ on $\mathbb{R}^m$. We denote by $g(t, \cdot)$ the density of W_t with respect to λ and we further assume that $g(t, \cdot) \in B(\mathbb{R}^d)$ and is a strictly positive function.

The filtering problem consists of computing the conditional distribution of the signal given the σ-algebra generated by the observation process from time

0 up to the current time i.e. computing the (random) probability measure π_t, where

$$\pi_t(A) \triangleq \mathbb{P}(X_t \in A \mid \sigma(Y_{0:t})), \qquad (10.5)$$
$$\pi_t f = \mathbb{E}\left[f(X_t) \mid \sigma(Y_{0:t})\right]$$

for all $A \in \mathcal{B}(\mathbb{R}^d)$ and $f \in B(\mathbb{R}^d)$, where $Y_{0:t}$ is the random vector $Y_{0:t} \triangleq (Y_0, Y_1, \ldots, Y_t)$.[†] For arbitrary $y_{0:t} \triangleq (y_0, y_1, \ldots, y_t) \in (\mathbb{R}^m)^{t+1}$, let $\pi_t^{y_{0:t}}$ be the (non-random) probability measure defined as

$$\pi_t^{y_{0:t}}(A) \triangleq \mathbb{P}\left(X_t \in A \mid Y_{0:t} = y_{0:t}\right), \qquad (10.6)$$
$$\pi_t^{y_{0:t}} f = \mathbb{E}\left[f(X_t) \mid Y_{0:t} = y_{0:t}\right]$$

for all $A \in \mathcal{B}(\mathbb{R}^d)$ and $f \in B(\mathbb{R}^d)$. Then $\pi_t = \pi_t^{Y_{0:t}}$. While π_t is a random probability measure, $\pi_t^{y_{0:t}}$ is a deterministic probability measure. We also introduce p_t and $p_t^{y_{0:t-1}}$, $t > 0$ the predicted conditional probability measures defined by

$$p_t^{y_{0:t-1}}(A) \triangleq \mathbb{P}\left(X_t \in A \mid Y_{0:t-1} = y_{0:t-1}\right),$$
$$p_t^{y_{0:t-1}} f = \mathbb{E}\left[f(X_t) \mid Y_{0:t-1} = y_{0:t-1}\right].$$

Again $p_t = p_t^{Y_{0:t-1}}$.

In the statistics and engineering literature the probability q_t is commonly called the *prior distribution* of the signal X_t, whilst π_t is called the (Bayesian) *posterior distribution*.

10.2 The Recurrence Formula for π_t

The following lemma gives the density of the random vector $Y_{s:t} = (Y_s, \ldots, Y_t)$ for arbitrary $s, t \in \mathbb{N}$, $s \le t$.

Lemma 10.2. *Let $\mathbb{P}_{Y_{s:t}} \in \mathcal{P}((\mathbb{R}^m)^{t-s+1})$ be the probability distribution of $Y_{s:t}$ and λ be the Lebesgue measure on $((\mathbb{R}^m)^{t-s+1}, \mathcal{B}((\mathbb{R}^m)^{t-s+1}))$. Then, for all $0 < s \le t < \infty$, $\mathbb{P}_{Y_{s:t}}$ is absolutely continuous with respect to λ and its Radon–Nikodym derivative is*

$$\frac{d\mathbb{P}_{Y_{s:t}}}{d\lambda}(y_{s:t}) = \Upsilon(y_{s:t}) \triangleq \int_{(\mathbb{R}^d)^{t-s+1}} \prod_{i=s}^{t} g_i(y_i - h(i, x_i)) \mathbb{P}_{X_{s:t}}(dx_{s:t}),$$

where $\mathbb{P}_{X_{s:t}} \in \mathcal{P}((\mathbb{R}^d)^{t-s+1})$ is the probability distribution of the random vector $X_{s:t} = (X_s, \ldots, X_t)$.

[†] $\{Y_{0:t}, t \in \mathbb{N}\}$ is the path process associated with the observation process $Y = \{Y_t, t \in \mathbb{N}\}$. That is, $\{Y_{0:t}, t \in \mathbb{N}\}$ records the entire history of Y up to time t, not just its current value.

Proof. Let $C_{s:t} = C_s \times \cdots \times C_t$, where C_r are arbitrary Borel sets, $C_r \in \mathcal{B}(\mathbb{R}^m)$ for all $s \le r \le t$. We need to prove that

$$\mathbb{P}_{Y_{s:t}}(C_{s:t}) = \mathbb{P}\left(\{Y_{s:t} \in C_{s:t}\}\right) = \int_{C_{s:t}} \Upsilon(y_{s:t}) dy_s \ldots dy_t. \tag{10.7}$$

Using the properties of the conditional probability,

$$\mathbb{P}\left(Y_{s:t} \in C_{s:t}\right) = \int_{(\mathbb{R}^d)^{t-s+1}} \mathbb{P}\left(Y_{s:t} \in C_{s:t} \mid X_{s:t} = x_{s:t}\right) \mathbb{P}_{X_{s:t}}\left(dx_{s:t}\right). \tag{10.8}$$

Since $(X_s, \ldots, X_t)$ is independent of $(W_s, \ldots, W_t)$, from (10.4) it follows that

$$\mathbb{P}\left(Y_{s:t} \in C_{s:t} \mid X_{s:t} = x_{s:t}\right) = \mathbb{E}\left[\prod_{i=s}^{t} 1_{C_i}\left(h(i, X_i) + W_i\right) \mid X_{s,t} = x_{s:t}\right]$$

$$= \mathbb{E}\left[\prod_{i=s}^{t} 1_{C_i}\left(h(i, x_i) + W_i\right)\right],$$

thus by the mutual independence of $W_s, \ldots, W_t$,

$$\mathbb{P}\left(Y_{s:t} \in C_{s:t} \mid X_{s:t} = x_{s:t}\right) = \prod_{i=s}^{t} \mathbb{E}\left[1_{C_i}\left(h(i, x_i) + W_i\right)\right]$$

$$= \prod_{i=s}^{t} \int_{C_i} g_i(y_i - h(i, x_i)) \, dy_i. \tag{10.9}$$

By combining (10.8) and (10.9) and applying Fubini's theorem, we obtain (10.7). □

Remark 10.3. A special case of (10.9) gives that

$$\mathbb{P}\left(Y_t \in dy_t \mid X_t = x_t\right) = g_t(y_t - h(t, x_t)) \, dy_t,$$

which explains why the function $g_t^{y_t} : \mathbb{R}^d \to \mathbb{R}$ defined by

$$g_t^{y_t}(x) = g_t(y_t - h(t, x)), \quad x \in \mathbb{R}^d \tag{10.10}$$

is commonly referred to as the *likelihood* function.

Since g_i for $i = s, \ldots, t$ are strictly positive, the density of the random vector $(Y_s, \ldots, Y_t)$ is also strictly positive. This condition can be relaxed (i.e. g_i required to be non-negative), however, the relaxation requires a more involved theoretical treatment of the particle filter.

The recurrence formula for π_t involves two operations defined on $\mathcal{P}(\mathbb{R}^d)$: a transformation via the transition kernel K_t and a *projective* product associated with the likelihood function $g_t^{y_t}$ defined as follows.

10.2 The Recurrence Formula for π_t

Definition 10.4. *Let $p \in \mathcal{P}(\mathbb{R}^d)$ be a probability measure, and let $\varphi \in B(\mathbb{R}^d)$ be a non-negative function such that $p(\varphi) > 0$. The projective product $\varphi * p$ is the (set) function $\varphi * p \colon \mathcal{B}(\mathbb{R}^d) \to \mathbb{R}$ defined by*

$$\varphi * p(A) \triangleq \frac{\int_A \varphi(x) p(\mathrm{d}x)}{p(\varphi)}$$

for any $A \in \mathcal{B}(\mathbb{R}^d)$.

In the above definition, recall that

$$p(\varphi) = \int_{\mathbb{R}^d} \varphi(x) p(\mathrm{d}x).$$

Exercise 10.5. Prove that $\varphi * p$ is a probability measure on $\mathcal{B}(\mathbb{R}^d)$.

The projective product $\varphi * p$ is a probability measure which is absolutely continuous with respect to p, whose Radon–Nikodym derivative with respect to p is proportional to φ, viz:

$$\frac{\mathrm{d}(\varphi * p)}{\mathrm{d}p} = c\varphi,$$

where c is the normalizing constant, $c = 1/p(\varphi)$.

The following result gives the recurrence formula for the conditional probability of the signal. The prior and the posterior distributions coincide at time 0, $\pi_0 = q_0$, since $Y_0 = 0$ (i.e. no observations are available at time 0).

Proposition 10.6. *For any fixed path $(y_0, y_1, \ldots, y_t, \ldots)$ the sequence of (non-random) probability measures $(\pi_t^{y_{0:t}})_{t \geq 0}$ satisfies the following recurrence relation*

$$\pi_t^{y_{0:t}} = g_t^{y_t} * K_{t-1} \pi_{t-1}^{y_{0:t-1}}, \quad t > 0. \tag{10.11}$$

The recurrence formula (10.11) holds $\mathbb{P}_{Y_{0:t}}$-almost surely.[†] Equivalently, the conditional distribution of the signal satisfies the following recurrence relation

$$\pi_t = g_t^{Y_t} * K_{t-1} \pi_{t-1}, \quad t > 0, \tag{10.12}$$

and the recurrence is satisfied $\mathbb{P}$-almost surely.

Proof. For all $f \in B(\mathbb{R}^d)$, using the Markov property of X and the definition of the transition kernel K,

$$\mathbb{E}\left[f(X_t) \mid \mathcal{F}_{t-1}^X\right] = \mathbb{E}\left[f(X_t) \mid X_{t-1}\right] = K_{t-1} f(X_{t-1}).$$

[†] Equivalently, formula (10.11) holds true λ-almost surely where λ is the Lebesgue measure on $(\mathbb{R}^m)^{t+1}$.

Since $W_{0:t-1}$ is independent of $X_{0:t}$, from property (f) of conditional expectation,†

$$\mathbb{E}\left[f(X_t) \mid \mathcal{F}_{t-1}^X \vee \sigma(W_{0:t-1})\right] = \mathbb{E}\left[f(X_t) \mid \mathcal{F}_{t-1}^X\right],$$

hence, using property (d) of conditional expectation

$$\begin{aligned}p_t f &= \mathbb{E}\left[f(X_t) \mid Y_{0:t-1}\right] \\ &= \mathbb{E}\left[\mathbb{E}\left[f(X_t) \mid \mathcal{F}_{t-1}^X \vee \sigma(W_{0:t-1})\right] \mid \sigma(Y_{0:t-1})\right] \\ &= \mathbb{E}\left[K_{t-1}f(X_{t-1}) \mid \sigma(Y_{0:t-1})\right] \\ &= \pi_{t-1}(K_{t-1}f),\end{aligned}$$

which implies that $p_t = K_{t-1}\pi_{t-1}$ (as in Exercise 10.1 part (iii) or equivalently $p_t^{y_{0:t-1}} = K_{t-1}\pi_{t-1}^{y_{0:t-1}}$.

Next we prove that $\pi_t^{y_{0:t}} = g_t^{y_t} * p_t^{y_{0:t-1}}$. Let $C_{0:t} = C_0 \times \cdots \times C_t$ where $C_r \in \mathcal{B}(\mathbb{R}^m)$ for $r = 0, 1, \ldots, t$. We need to prove that for any $A \in \mathcal{B}(\mathbb{R}^d)$,

$$\int_{C_{0:t}} \pi_t^{y_{0:t}}(A)\, \mathbb{P}_{Y_{0:t}}(dy_{0:t}) = \int_{C_{0:t}} g_t^{y_t} * p_t^{y_{0:t-1}}(A)\, \mathbb{P}_{Y_{0:t}}(dy_{0:t}). \quad (10.13)$$

By (A.2), the left-hand side of (10.13) is equal to $\mathbb{P}(\{X_t \in A\} \cap \{Y_{0:t} \in C_{0:t}\})$. Since $\sigma(X_{0:t}, W_{0:t-1}) \supset \sigma(X_t, Y_{0:t-1})$, from property (f) of conditional expectation

$$\mathbb{P}(Y_t \in C_t \mid X_t, Y_{0:t-1}) = \mathbb{E}(\mathbb{P}(Y_t \in C_t \mid X_{0:t}, W_{0:t-1}) \mid X_t, Y_{0:t-1}) \quad (10.14)$$

and using property (d) of conditional expectations and (10.9)

$$\begin{aligned}\mathbb{P}(Y_t \in C_t \mid X_{0:t}, W_{0:t-1}) &= \mathbb{P}(Y_t \in C_t \mid X_{0:t}) \\ &= \mathbb{P}(Y_{0:t} \in (\mathbb{R}^m)^t \times C_t \mid X_{0:t}) \\ &= \int_{C_t} g_t(y_t - h(t, X_t))\, dy_t. \end{aligned} \quad (10.15)$$

From (10.14) and (10.15),

$$\begin{aligned}\mathbb{P}(Y_t \in C_t \mid X_t, Y_{0:t-1}) &= \mathbb{E}(\mathbb{P}(Y_t \in C_t \mid X_t, W_{0:t-1}) \mid X_t, Y_{0:t-1}) \\ &= \int_{C_t} g_t(y_t - h(t, X_t))\, dy_t.\end{aligned}$$

This gives us

$$\mathbb{P}(Y_t \in C_t \mid X_t = x_t, Y_{0:t-1} = y_{0:t-1}) = \int_{C_t} g_t^{y_t}(x_t)\, dy_t, \quad (10.16)$$

where g^{y_t} is defined in (10.10); hence

---

† See Section A.2 for a list of the properties of conditional expectation.

10.2 The Recurrence Formula for π_t

$$\mathbb{P}_{Y_{0:t}}(C_{0:t}) = \mathbb{P}\left(\{Y_t \in C_t\} \cap \{X_t \in \mathbb{R}^d\} \cap \{Y_{0:t-1} \in C_{0:t-1}\}\right)$$

$$= \int_{\mathbb{R}^d \times C_{0:t-1}} \mathbb{P}\left(Y_t \in C_t \mid X_t = x_t, Y_{0:t-1} = y_{0:t-1}\right)$$

$$\mathbb{P}_{X_t, Y_{0:t-1}}(\mathrm{d}x_t, \mathrm{d}y_{0:t-1})$$

$$= \int_{\mathbb{R}^d \times C_{0:t-1}} \int_{C_t} g_t^{y_t}(x_t)\, \mathrm{d}y_t\, p_t^{y_{0:t-1}}(\mathrm{d}x_t) \mathbb{P}_{Y_{0:t-1}}(\mathrm{d}y_{0:t-1})$$

$$= \int_{C_{0:t}} \int_{\mathbb{R}^d} g_t^{y_t}(x_t) p_t^{y_{0:t-1}}(\mathrm{d}x_t)\, \mathbb{P}_{Y_{0:t-1}}(\mathrm{d}y_{0:t-1})\, \mathrm{d}y_t. \tag{10.17}$$

In (10.17), we used the identity

$$\mathbb{P}_{X_t, Y_{0:t-1}}(\mathrm{d}x_t, \mathrm{d}y_{0:t-1}) = p_t^{y_{0:t-1}}(\mathrm{d}x_t) \mathbb{P}_{Y_{0:t-1}}(\mathrm{d}y_{0:t-1}), \tag{10.18}$$

which is again a consequence of the vector-valued equivalent of (A.2), since for all $A \in \mathcal{B}(\mathbb{R}^d)$, we have

$$\mathbb{P}\left((X_t, Y_{0:t-1}) \in A \times C_{0:t-1}\right)$$

$$= \int_{C_{0:t-1}} \mathbb{P}\left(X_t \in A \mid Y_{0:t-1} = y_{0:t-1}\right) \mathbb{P}_{Y_{0:t-1}}(\mathrm{d}y_{0:t-1})$$

$$= \int_{A \times C_{0:t-1}} p_t^{y_{0:t-1}}(\mathrm{d}x_t) \mathbb{P}_{Y_{0:t-1}}(\mathrm{d}y_{0:t-1}).$$

From (10.17)

$$\mathbb{P}_{Y_{0:t}}(\mathrm{d}y_{0:t}) = p_t^{y_{0:t-1}}\left(g_t^{y_t}\right)\, \mathrm{d}y_t \mathbb{P}_{Y_{0:t-1}}(\mathrm{d}y_{0:t-1}).$$

Hence the second term in (10.13) is equal to

$$\int_{C_{0:t}} g_t^{y_t} * p_t^{y_{0:t-1}}(A) \mathbb{P}_{Y_{0:t}}(\mathrm{d}y_{0:t})$$

$$= \int_{C_{0:t}} \frac{\int_A g_t^{y_t}(x_t) p_t^{y_{0:t-1}}(\mathrm{d}x_t)}{p_t^{y_{0:t-1}}(g_t^{y_t})} \mathbb{P}_{Y_{0:t}}(\mathrm{d}y_{0:t})$$

$$= \int_{C_{0:t}} \int_A g_t^{y_t}(x_t) p_t^{y_{0:t-1}}(\mathrm{d}x_t)\, \mathrm{d}y_t \mathbb{P}_{Y_{0:t-1}}(\mathrm{d}y_{0:t-1}).$$

Finally, using (10.16) and (10.18),

$$\int_{C_{0:t}} g_t^{y_t} * p_t^{y_{0:t-1}}(A) \mathbb{P}_{Y_{0:t}}(\mathrm{d}y_{0:t})$$

$$= \int_{A \times C_{0:t-1}} \left(\int_{C_t} g_t^{y_t}(x_t) \mathrm{d}y_t\right) p_t^{y_{0:t-1}}(\mathrm{d}x_t) \mathbb{P}_{Y_{0:t-1}}(\mathrm{d}y_{0:t-1})$$

$$= \int_{A \times C_{0:t-1}} \mathbb{P}\left(Y_t \in C_t \mid X_t = x_t, Y_{0:t-1} = y_{0:t-1}\right)$$

$$\times \mathbb{P}_{X_t, Y_{0:t-1}}(\mathrm{d}x_t, \mathrm{d}y_{0:t-1})$$

$$= \mathbb{P}\left(\{X_t \in A\} \cap \{Y_{0:t} \in C_{0:t}\}\right).$$

From the earlier discussion this is sufficient to establish the result. □

As it can be seen from its proof, the recurrence formula (10.12) can be rewritten in the following expanded way,

$$\pi_{t-1} \mapsto p_t = K_{t-1}\pi_{t-1} \mapsto \pi_t = g_t^{Y_t} * p_t, \quad t > 0. \quad (10.19)$$

The first step is called the *prediction* step: it occurs at time t before the arrival of the new observation Y_t. The second step is the *updating* step as it takes into account the new observation Y_t. A similar expansion holds true for the recurrence formula (10.11); that is,

$$\pi_{t-1}^{y_{0:t-1}} \mapsto p_t^{y_{0:t-1}} = K_{t-1}\pi_{t-1}^{y_{0:t-1}} \mapsto \pi_t^{y_{0:t}} = g_t^{y_t} * p_t^{y_{0:t-1}}, \quad t > 0. \quad (10.20)$$

The simplicity of the recurrence formulae (10.19) and (10.20) is misleading. A closed formula for the posterior distribution exists only in exceptional cases (the linear/Gaussian filter). The main difficulty resides in the updating step: the projective product is a non-linear transformation involving the computation of the normalising constant $p_t(g_t^{Y_t})$ or $p_t^{y_{0:t-1}}(g_t^{y_t})$ which requires an integration over a (possibly) high-dimensional space. In Section 10.4 we present a generic class of particle filters which can be used to approximate numerically the posterior distribution. Before that we state and prove necessary and sufficient criteria for sequences of approximations to converge to the posterior distribution.

10.3 Convergence of Approximations to π_t

We have two sets of criteria: for the case when the observation is a priori fixed to a particular outcome, that is, say

$$Y_0 = y_0, Y_1 = y_1, \ldots$$

and for the case when the observation remains random. The first case is the simpler of the two, since the measures to be approximated are not random.

10.3.1 The Fixed Observation Case

We look first at the case when the observation process has an arbitrary, but fixed, value $y_{0:T}$, where T is a finite time horizon. We assume that the recurrence formula (10.20) for $\pi_t^{y_{0:t}}$ – the conditional distribution of the signal given the event $\{Y_{0:t} = y_{0:t}\}$ – holds true for the particular observation path $y_{0:t}$ for all $0 \le t \le T$ (remember that (10.20) is valid $\mathbb{P}_{Y_{0:t}}$-almost surely). As stated above, (10.20) requires the computation of the predicted conditional probability measure $p_t^{y_{0:t-1}}$:

$$\pi_{t-1}^{y_{0:t-1}} \longrightarrow p_t^{y_{0:t-1}} \longrightarrow \pi_t^{y_{0:t}}.$$

10.3 Convergence of Approximations to π_t

Therefore it is natural to study algorithms which provide recursive approximations for $\pi_t^{y_{0:t}}$ using intermediate approximations for $p_t^{y_{0:t-1}}$. Denote by $(\pi_t^n)_{n=1}^\infty$ the approximating sequence for $\pi_t^{y_{0:t}}$ and $(p_t^n)_{n=1}^\infty$ the approximating sequence for $p_t^{y_{0:t-1}}$. Is is assumed that the following three conditions are satisfied.

- π_t^n and p_t^n are random measures, not necessarily probability measures.
- $p_t^n \neq 0$, $\pi_t^n \neq 0$ (i.e. no approximation should be trivial).
- $p_t^n g_t^{y_t} > 0$ for all $n > 0$, $0 \leq t \leq T$.

Let $\bar{\pi}_t^n$ be defined as a (random) probability measure absolutely continuous with respect to p_t^n for $t \in \mathbb{N}$ and $n \geq 1$ such that

$$\bar{\pi}_t^n = g_t^{y_t} * p_t^n; \tag{10.21}$$

thus

$$\bar{\pi}_t^n f = \frac{p_t^n(f g^{y_t})}{p_t^n g^{y_t}}. \tag{10.22}$$

The following theorems give necessary and sufficient conditions for the convergence of p_t^n to $p_t^{y_{0:t-1}}$ and π_t^n to $\pi_t^{y_{0:t}}$. In order to simplify notation, for the remainder of this subsection, dependence on $y_{0:t}$ is suppressed and $\pi_t^{y_{0:t}}$ is denoted by π_t, $p_t^{y_{0:t-1}}$ by p_t and $g_t^{y_t}$ by g_t. It is important to remember that the observation process is a given fixed path $y_{0:T}$.

Theorem 10.7. *For all $f \in B(\mathbb{R}^d)$ and all $t \in [0, T]$ the limits*

a0. $\lim_{n \to \infty} \mathbb{E}[|\pi_t^n f - \pi_t f|] = 0$,
b0. $\lim_{n \to \infty} \mathbb{E}[|p_t^n f - p_t f|] = 0$,

hold if and only if for all $f \in B(\mathbb{R}^d)$ and all $t \in [0, T]$ we have

a1. $\lim_{n \to \infty} \mathbb{E}[|\pi_0^n f - \pi_0 f|] = 0$,
b1. $\lim_{n \to \infty} \mathbb{E}[|p_t^n f - K_{t-1} \pi_{t-1}^n f|] = \lim_{n \to \infty} \mathbb{E}[|\pi_t^n f - \bar{\pi}_t^n f|] = 0$.

Proof. The necessity of conditions (a0) and (b0) is proved by induction. The limit (a0) follows in the starting case of $t = 0$ from (a1). We need to show that if π_{t-1}^n converges in expectation to π_{t-1} and p_t^n converges in expectation to p_t then π_t^n converges in expectation to π_t. Since $p_t = K_{t-1} \pi_{t-1}$, for all $f \in B(\mathbb{R}^d)$, by the triangle inequality

$$|p_t^n f - p_t f| \leq |p_t^n f - K_{t-1} \pi_{t-1}^n f| + |K_{t-1} \pi_{t-1}^n f - K_{t-1} \pi_{t-1} f|. \tag{10.23}$$

The expected value of the first term on the right-hand side of (10.23) converges to zero from (b1). Also using Exercise 10.1, $K_{t-1} f \in B(\mathbb{R}^d)$ and $K_{t-1} \pi_{t-1}^n f = \pi_{t-1}^n(K_{t-1} f)$ and $K_{t-1} \pi_{t-1} f = \pi_{t-1}(K_{t-1} f)$ hence

$$\lim_{n \to \infty} \mathbb{E}[|K_{t-1} \pi_{t-1}^n f - K_{t-1} \pi_{t-1} f|] = 0.$$

By taking expectation of both sides of (10.23),

266   10 Particle Filters in Discrete Time

$$\lim_{n\to\infty} \mathbb{E}\left[|p_t^n f - p_t f|\right] = 0, \tag{10.24}$$

which establishes condition (a0). From (10.22)

$$\bar{\pi}_t^n f - \pi_t f = \frac{p_t^n(fg_t)}{p_t^n g_t} - \frac{p_t(fg_t)}{p_t g_t}$$

$$= -\frac{p_t^n(fg_t)}{p_t^n g_t} \frac{1}{p_t g_t}(p_t^n g_t - p_t g_t) + \left(\frac{p_t^n(fg_t)}{p_t g_t} - \frac{p_t(fg_t)}{p_t g_t}\right),$$

and as $|p_t^n(fg_t)| \leq \|f\|_\infty p_t^n g_t$,

$$|\bar{\pi}_t^n f - \pi_t f| \leq \frac{\|f\|_\infty}{p_t g_t}|p_t^n g_t - p_t g_t| + \frac{1}{p_t g_t}|p_t^n(fg_t) - p_t(fg_t)|. \tag{10.25}$$

Therefore

$$\mathbb{E}\left[|\bar{\pi}_t^n f - \pi_t f|\right] \leq \frac{\|f\|_\infty}{p_t g_t}\mathbb{E}\left[|p_t^n g_t - p_t g_t|\right]$$

$$+ \frac{1}{p_t g_t}\mathbb{E}\left[|p_t^n(fg_t) - p_t(fg_t)|\right]. \tag{10.26}$$

From (10.24) both terms on the right-hand side of (10.26) converge to zero. Finally,

$$|\pi_t^n f - \pi_t f| \leq |\pi_t^n f - \bar{\pi}_t^n f| + |\bar{\pi}_t^n f - \pi_t f|. \tag{10.27}$$

As the expected value of the first term on the right-hand side of (10.27) converges to zero using (b1) and the expected value of the second term converges to zero using (10.26), $\lim_{n\to\infty}\mathbb{E}\left[|\pi_t^n f - \pi_t f|\right] = 0$.

For the sufficiency part, assume that conditions (a0) and (b0) hold. Thus for all $t \geq 0$ and for all $f \in B(\mathbb{R}^d)$,

$$\lim_{n\to\infty}\mathbb{E}\left[|\pi_t^n f - \pi_t f|\right] = \lim_{n\to\infty}\mathbb{E}\left[|p_t^n f - p_t f|\right] = 0.$$

Clearly condition (a1) follows as a special case of (a0) with $t = 0$. Since $p_t = K_{t-1}\pi_{t-1}$, we have for all $f \in B(\mathbb{R}^d)$,

$$\mathbb{E}\left[|p_t^n f - K_{t-1}\pi_{t-1}^n f|\right] \leq \mathbb{E}\left[|p_t^n f - p_t f|\right]$$
$$+ \mathbb{E}\left[|\pi_{t-1}(K_{t-1}f) - \pi_{t-1}^n(K_{t-1}f)|\right], \tag{10.28}$$

which implies the first limit in (b1). From (10.26),

$$\lim_{n\to\infty}\mathbb{E}\left[|\pi_t f - \bar{\pi}_t^n f|\right] = 0$$

and by the triangle inequality

$$\mathbb{E}\left[|\pi_t^n f - \bar{\pi}_t^n f|\right] \leq \mathbb{E}\left[|\pi_t^n f - \pi_t f|\right] + \mathbb{E}\left[|\pi_t f - \bar{\pi}_t^n f|\right] \tag{10.29}$$

from which the second limit in (b1) follows.   □

10.3 Convergence of Approximations to π_t

Thus conditions (a1) and (b1) imply that p_t^n converges in expectation to p_t and π_t^n converges in expectation to π_t (see Section A.10 for the definition of convergence in expectation). The convergence in expectation of p_t^n and of π_t^n holds if and only if conditions (a1) and (b1) are satisfied for all $f \in C_b(\mathbb{R}^d)$ (not necessarily for all $f \in B(\mathbb{R}^d)$) provided additional constraints are imposed on the transition kernel of the signal and of the likelihood functions; see Corollary 10.10 below.

Definition 10.8. *The transition kernel K_t is said to satisfy the* Feller property *if $K_t f \in C_b(\mathbb{R}^d)$ for all $f \in C_b(\mathbb{R}^d)$.*

Exercise 10.9. Let $\{V_t\}_{t=1}^\infty$ be a sequence of independent one-dimensional standard normal random variables.

i. Let $X = \{X_t,\ t \in \mathbb{N}\}$ be given by the following recursive formula
$$X_{t+1} = a(X_t) + V_t,$$
where $a \colon \mathbb{R} \to \mathbb{R}$ is a continuous function. Show that the corresponding transition kernel for X satisfies the Feller property.
ii. Let $X = \{X_t,\ t \in \mathbb{N}\}$ be given by the following recursive formula
$$X_{t+1} = X_t + \operatorname{sgn}(X_t) + V_t.$$
Then show that the corresponding transition kernel for X does not satisfy the Feller property.

The following result gives equivalent conditions for the convergence in expectation.

Corollary 10.10. *Assume that the transition kernel for X is Feller and that the likelihood functions g_t are all continuous. Then the sequences p_t^n, π_t^n converge in expectation to p_t and π_t for all $t \in [0,T]$ if and only if conditions (a1) and (b1) are satisfied for all $f \in C_b(\mathbb{R}^d)$ and all $t \in [0,T]$.*

Proof. The proof is a straightforward modification of the proof of Theorem 10.7. The Feller property is used in the convergence to zero of the second term on the right-hand side of (10.23):
$$\lim_{n \to \infty} \mathbb{E}\left[\left|K_{t-1}\pi_{t-1}^n f - K_{t-1}\pi_{t-1} f\right|\right]$$
$$= \lim_{n \to \infty} \mathbb{E}\left[\left|\pi_{t-1}^n (K_{t-1} f) - \pi_{t-1} (K_{t-1} f)\right|\right] = 0.$$

That is, only if $K_{t-1} f$ is continuous, we can conclude that the limit above is zero. The continuity of g_t is used to conclude that both terms on the right-hand side of (10.26) converge to zero. □

Following Remark A.38 in the appendix, if there exists a positive constant $p > 1$ such that

$$\mathbb{E}\left[|\pi_t^n f - \pi_t f|^{2p}\right] \leq \frac{c_f}{n^p}, \qquad (10.30)$$

where c_f is a positive constant depending on the test function f, but independent of n, then, for any $\varepsilon \in (0, 1/2 - 1/(2p))$ there exists a positive random variable $c_{f,\varepsilon}$ almost surely finite such that

$$|\pi_t^n f - \pi_t f| \leq \frac{c_{f,\varepsilon}}{n^\varepsilon}.$$

In particular $\pi_t^n f$ converges to $\pi_t f$ almost surely. Moreover if (10.30) holds for any $f \in \mathcal{M}$ where $\mathcal{M}$ is a countable convergence determining set (as defined in Section A.10), then, almost surely, π_t^n converges to π_t in the weak topology. This means that there exists a set $\bar{\Omega} \in \mathcal{F}$ such that $\mathbb{P}(\bar{\Omega}) = 1$ and for any $\omega \in \bar{\Omega}$ the corresponding sequence of probability measures $\pi_t^{n,\omega}$ satisfies

$$\lim_{n \to \infty} \pi_t^{n,\omega}(f) = \pi_t(f),$$

for any $f \in C_b(\mathbb{R}^d)$. This cannot be extended to the convergence for any $f \in B(\mathbb{R}^d)$ (i.e. to the stronger, so-called convergence in total variation, of $\pi_t^{n,\omega}$ to π_t).

Exercise 10.11. Let μ be the uniform measure on the interval $[0, 1]$ and $(\mu_n)_{n \geq 1}$ be the sequence of probability measures

$$\mu_n = \frac{1}{n} \sum_{i=1}^{n} \delta_{i/n}.$$

i. Show that $(\mu_n)_{n \geq 1}$ converges to μ in the weak topology.
ii. Let $f = 1_{\mathbb{Q} \cap [0,1]} \in B(\mathbb{R}^d)$ be the indicator set of all the rational numbers in $[0, 1]$. Show that $\mu_n(f) \not\to \mu(f)$, hence μ_n does not converge to μ in total variation.

Having rates of convergence for the higher moments of the error terms $\pi_t^n f - \pi_t f$ as in (10.30) is therefore very useful as they imply the almost sure convergence of the approximations in the weak topology with no additional assumptions required on the transition kernels of the signal and the likelihood function. However, if we wish a result in the same vein as that of Theorem 10.7, the same assumptions as in Corollary 10.10 must be imposed. The following theorem gives us the corresponding criterion for the almost sure convergence of p_t^n to p_t and π_t^n to π_t in the weak topology. The theorem makes use of the metric $d_{\mathcal{M}}$ as defined in Section A.10 which generates the weak topology on $\mathcal{M}_F(\mathbb{R}^d)$. The choice of the metric is not important; any metric which generates the weak topology may be used.

Theorem 10.12. *Assume that the transition kernel for X is Feller and that the likelihood functions g_t are all continuous for all $t \in [0,T]$. Then the sequence p_t^n converges almost surely to p_t and π_t^n converges almost surely to π_t for all $t \in [0,T]$ if and only if the following two conditions are satisfied for all $t \in [0,T]$*

a2. $\lim_{n\to\infty} \pi_0^n = \pi_0$, $\mathbb{P}$-a.s.
b2. $\lim_{n\to\infty} d_{\mathcal{M}}\left(p_t^n, \pi_{t-1}^n K_{t-1}\right) = \lim_{n\to\infty} d_{\mathcal{M}}\left(\pi_t^n, \bar{\pi}_t^n\right) = 0$, $\mathbb{P}$-a.s.

Proof. The sufficiency of the conditions (a2) and (b2) is proved as above by induction using inequalities (10.23), (10.25) and (10.27). It remains to prove that (a2) and (b2) are necessary. Assume that for all $t \geq 0$ p_t^n converges almost surely to p_t and π_t^n converges almost surely to π_t This implies that $\pi_{t-1}^n K_{t-1}$ converges almost surely to p_t (which is equal to $\pi_{t-1} K_{t-1}$) and using (10.25), that $\bar{\pi}_t^n$ converges almost surely to π_t.

Hence, almost surely $\lim_{n\to\infty} d_{\mathcal{M}}(p_t^n, p_t) = 0$, $\lim_{n\to\infty} d_{\mathcal{M}}(\pi_t^n, \pi_t) = 0$, $\lim_{n\to\infty} d_{\mathcal{M}}(\pi_{t-1}^n K_{t-1}, p_t) = 0$ and $\lim_{n\to\infty} d_{\mathcal{M}}(\bar{\pi}_t^n, \pi_t) = 0$. Finally, using the triangle inequality

$$d_{\mathcal{M}}\left(p_t^n, \pi_{t-1}^n K_{t-1}\right) \leq d_{\mathcal{M}}\left(p_t^n, p_t\right) + d_{\mathcal{M}}\left(p_t, \pi_{t-1}^n K_{t-1}\right)$$

and

$$d_{\mathcal{M}}\left(\pi_t^n, \bar{\pi}_t^n\right) \leq d_{\mathcal{M}}\left(\pi_t^n, \pi_t\right) + d_{\mathcal{M}}\left(\pi_t, \bar{\pi}_t^n\right),$$

which imply (b2). □

Remark 10.13. Theorems 10.7 and 10.12 and Corollary 10.10 are very natural. They say that we obtain approximations of $p_t^{y_{0:t-1}}$ and $\pi_t^{y_{0:t}}$ for all $t \in [0,T]$ if and only if we start from an approximation of π_0 and then 'follow closely' the recurrence formula (10.20) for $p_t^{y_{0:t-1}}$ and $\pi_t^{y_{0:t}}$.

The natural question arises as to whether we can lift the results to the case when the observation process is random and not just a given fixed observation path.

10.3.2 The Random Observation Case

In the previous section both the converging sequences and the limiting measures depend on the fixed value of the observation. Let us look first at the convergence in mean. If for an arbitrary $f \in B(\mathbb{R}^d)$, the condition

$$\lim_{n\to\infty} \mathbb{E}\left[|\pi_t^{n,y_{0:t}} f - \pi_t^{y_{0:t}} f|\right] = 0,$$

holds for $\mathbb{P}_{Y_{0:t}}$-almost all values $y_{0:t}$ and there exists a $\mathbb{P}_{Y_{0:t}}$-integrable function $w(y_{0:t})$ such that, for all $n \geq 0$,

$$\mathbb{E}[|\pi_t^{n,y_{0:t}} f - \pi_t^{y_{0:t}} f|] \le w_f(y_{0:t}) \quad \mathbb{P}_{Y_{0:t}}\text{-a.s.,}^\dagger \tag{10.31}$$

then by the dominated convergence theorem,

$$\lim_{n\to\infty} \mathbb{E}\left[\left|\pi_t^{n,Y_{0:t}} f - \pi_t f\right|\right]$$
$$= \lim_{n\to\infty} \int_{(\mathbb{R}^m)^{t+1}} \mathbb{E}[|\pi_t^{n,y_{0:t}} f - \pi_t^{y_{0:t}} f|] \, \mathbb{P}_{Y_{0:t}}(dy_{0:t}) = 0.$$

Hence conditions (a1) and (b1) are also sufficient for convergence in the random observation case. In particular, if (a1) and (b1) are satisfied for any $f \in C_b(\mathbb{R}^d)$ and the two additional assumptions of Corollary 10.10 hold then $\pi_t^{n,Y_{0:t}}$ converges in expectation to π_t. Similar remarks apply to p_t. Also, the existence of rates of convergence for higher moments and appropriate integrability conditions can lead to the $\mathbb{P}$-almost sure convergence of $\pi_t^{n,Y_{0:t}}$ to π_t.

However, a necessary and sufficient condition can not be obtained in this manner, since $\lim_{n\to\infty} \mathbb{E}[|\pi_t^{n,Y_{0:t}} f - \pi_t f|] = 0$ does not imply

$$\lim_{n\to\infty} \mathbb{E}[|\pi_t^{n,y_{0:t}} f - \pi_t^{y_{0:t}} f|] = 0$$

for $\mathbb{P}_{Y_{0:t}}$-almost all values $y_{0:t}$.

The randomness of the approximating measures $p_t^{n,Y_{0:t-1}}$ and $\pi_t^{n,Y_{0:t}}$ now comes from two sources; one is the (random) observation Y and the other one is the actual construction of the approximations. In the case of particle approximations, randomness is introduced in the system during each of the propagation steps (see the next section for details). As the following convergence results show, the effect of the second source of randomness vanishes asymptotically (the approximating measures converge to p_t and π_t).

The following proposition is the equivalent of Theorem 10.7 for the random observation case. Here and throughout the remainder of the section the dependence on the process Y is suppressed from the notations $p_t^{n,Y_{0:t}}$, $\pi_t^{n,Y_{0:t}}$, $g_t^{Y_t}$, and so on.

Proposition 10.14. *Assume that for any $t \ge 0$, there exists a constant $c_t > 0$ such that $p_t g_t \ge c_t$. Then, for all $f \in B(\mathbb{R}^d)$ and all $t \ge 0$ the limits*

a0'. $\lim_{n\to\infty} \mathbb{E}[|\pi_t^n f - \pi_t f|] = 0$,
b0'. $\lim_{n\to\infty} \mathbb{E}[|p_t^n f - p_t f|] = 0$,

hold if and only if for all $f \in B(\mathbb{R}^d)$ and all $t \ge 0$

a1'. $\lim_{n\to\infty} \mathbb{E}[|\pi_0^n f - \pi_0 f|] = 0$,
b1'. $\lim_{n\to\infty} \mathbb{E}[|p_t^n f - K_{t-1}\pi_{t-1}^n f|] = \lim_{n\to\infty} \mathbb{E}[|\pi_t^n f - \bar{\pi}_t^n f|] = 0$.

---

† Condition (10.31) is trivially satisfied for approximations which are probability measures since in this case $w_f = 2\|f\|_\infty$ satisfies the condition.

Proof. The proof follows step by step that of Theorem 10.7. The only step that differs slightly is the proof of convergence to zero of $\mathbb{E}[|\bar{\pi}_t^n f - \pi_t f|]$. Using the equivalent of the inequality (10.25)

$$\mathbb{E}\left[|\bar{\pi}_t^n f - \pi_t f|\right] \leq \|f\|_\infty \mathbb{E}\left[\frac{1}{p_t g_t} |p_t^n g_t - p_t g_t|\right]$$
$$+ \mathbb{E}\left[\frac{1}{p_t g_t} |p_t^n(f g_t) - p_t(f g_t)|\right]. \tag{10.32}$$

Since $1/(p_t g_t)$ is now random it can not be taken outside the expectations as in (10.26). However, by using the assumption $p_t g_t \geq c_t$, we deduce that

$$\mathbb{E}\left[|\bar{\pi}_t^n f - \pi_t f|\right] \leq \frac{\|f\|_\infty}{c_t} \mathbb{E}\left[|p_t^n g_t - p_t g_t|\right] + \frac{1}{c_t} \mathbb{E}\left[|p_t^n(f g_t) - p_t(f g_t)|\right]$$

and hence the required convergence. □

The condition that $p_t g_t \geq c_t$ is difficult to check in practice. It is sometimes replaced by the condition that $\mathbb{E}[1/(p_t g_t)^2] < \infty$ together with the convergence to zero of the second moments of $p_t^n g_t - p_t g_t$ and $p_t^n(f g_t) - p_t(f g_t)$ (see the proof of convergence of the particle filter in continuous time described in the previous chapter).

As in the previous case, conditions (a1′) and (b1′) imply that p_t^n converges in expectation to p_t and π_t^n converges in expectation to π_t. A result analogous to Corollary 10.10 is true for the convergence in expectation of p_t^n and π_t^n, provided that the same additional constraints are imposed on the transition kernel of the signal and of the likelihood functions.

The existence of rates of convergence for the higher moments of the error terms $\pi_t^n f - \pi_t f$ as in (10.30) can be used to deduce the almost sure convergence of the approximations in the weak topology with no additional constraints imposed upon the transition kernel of the signal or the likelihood function. However, in order to prove a similar result to Theorem 10.7, the same assumptions as in Corollary 10.10 must be imposed. The following theorem gives us the corresponding criterion for the almost sure convergence of p_t^n to p_t and π_t^n to π_t in the weak topology. The result is true without the need to use the cumbersome assumption $p_t g_t \geq c_t$ for any $t \geq 0$. It makes use of the metric $d_\mathcal{M}$, defined in Section A.10, which generates the weak topology on $\mathcal{M}_F(\mathbb{R}^d)$. The choice of the metric is not important; any metric which generates the weak topology may be used.

Proposition 10.15. *Assume that the transition kernel for X is Feller and that the likelihood functions g_t are all continuous. Then the sequence p_t^n converges almost surely to p_t and π_t^n converges almost surely to π_t, for all $t \geq 0$ if and only if, for all $t \geq 0$,*

a2′. $\lim_{n \to \infty} \pi_0^n = \pi_0$, $\mathbb{P}$-a.s.
b2′. $\lim_{n \to \infty} d_\mathcal{M}\left(p_t^n, K_{t-1} \pi_{t-1}^n\right) = \lim_{n \to \infty} d_\mathcal{M}\left(\pi_t^n, \bar{\pi}_t^n\right) = 0.$

Proof. The proof is similar to that of Theorem 10.12, the only difference being the proof that $\lim_{n\to\infty} p_t^n = p_t$, $\mathbb{P}$-a.s. implies $\lim_{n\to\infty} \bar{\pi}_t^n = \pi_t$, $\mathbb{P}$-a.s. which is as follows. Let $\mathcal{M}$ be a convergence determining set of functions in $C_b(\mathbb{R}^d)$, for instance, the set used to construct the metric $d_{\mathcal{M}}$. Then almost surely

$$\lim_{n\to\infty} p_t^n g_t = p_t g_t \quad \text{and} \quad \lim_{n\to\infty} p_t^n(g_t f) = p_t(g_t f) \quad \text{for all } f \in \mathcal{M}.$$

Hence, again almost surely, we have

$$\lim_{n\to\infty} \bar{\pi}_t^n f = \lim_{n\to\infty} \frac{p_t^n(g_t f)}{p_t^n g_t}$$

$$= \frac{p_t(g_t f)}{p_t g_t}(\omega) = \pi_t f, \quad \forall f \in \mathcal{M}$$

which implies $\lim_{n\to\infty} \bar{\pi}_t^n = \pi_t$, $\mathbb{P}$-a.s. □

In the next section we present examples of approximations to the posterior distribution which satisfy the conditions of these results. The algorithms used to produce these approximations are called *particle filters* or *sequential Monte Carlo methods*.

10.4 Particle Filters in Discrete Time

The algorithms presented below involve the use of a system of n particles which evolve (mutate) according to the law of X. After each mutation the system is corrected: each particle is replaced by a random number of particles whose mean is proportional to the likelihood of the position of the particle. After imposing some weak restrictions on the offspring distribution of the particles, the empirical measure associated with the particle systems is proven to converge (as n tends to ∞) to the conditional distribution of the signal given the observation.

Denote by π_t^n the approximation to π_t and by p_t^n the approximation to p_t. The particle filter has the following description.

1. **Initialization** $[t = 0]$.
   For $i = 1, \ldots, n$, sample $x_0^{(i)}$ from π_0.
2. **Iteration** $[t-1 \text{ to } t]$.
   Let $x_{t-1}^{(i)}$, $i = 1, \ldots, n$ be the positions of the particles at time $t-1$.
   a) For $i = 1, \ldots, n$, sample $\bar{x}_t^{(i)}$ from $K_{t-1}(x_{t-1}^{(i)}, \cdot)$. Compute the (normalized) weight $w_t^{(i)} = g_t(\bar{x}_t^{(i)})/(\sum_{j=1}^n g_t(\bar{x}_t^{(j)}))$.
   b) Replace each particle by $\xi_t^{(i)}$ offspring such that $\sum_{i=1}^n \xi_t^{(i)} = n$. Denote the positions of the offspring particles by $x_t^{(i)}$, $i = 1, \ldots, n$.

It follows from the above that the particle filter starts from π_0^n: the empirical measure associated with a set of n random particles of mass $1/n$ whose positions $x_0^{(i)}$ for $i = 1, \ldots, n$ form a sample of size n from π_0,

$$\pi_0^n \triangleq \frac{1}{n} \sum_{i=1}^{n} \delta_{x_0^{(i)}}.$$

In general, define π_t^n to be

$$\pi_t^n \triangleq \frac{1}{n} \sum_{i=1}^{n} \delta_{x_t^{(i)}},$$

where $x_t^{(i)}$ for $i = 1, \ldots, n$ are the positions of the particles of mass $1/n$ obtained after the second step of the iteration. Let $\bar{\pi}_t^n$ be the weighted measure

$$\bar{\pi}_t^n \triangleq \sum_{i=1}^{n} w_t^{(i)} \delta_{\bar{x}_t^{(i)}}.$$

We introduce the following σ-algebras

$$\mathcal{F}_t = \sigma(x_s^{(i)}, \bar{x}_s^{(i)}, s \le t, \ i = 1, \ldots, n)$$
$$\bar{\mathcal{F}}_t = \sigma(x_s^{(i)}, \bar{x}_s^{(i)}, s < t, \ \bar{x}_t^{(i)}, \ i = 1, \ldots, n).$$

Obviously $\bar{\mathcal{F}}_t \subset \mathcal{F}_t$ and the (random) probability measures p_t^n and $\bar{\pi}_t^n$ are $\bar{\mathcal{F}}_t$-measurable whilst π_t^n is $\mathcal{F}_t$-measurable for any $t \ge 0$. The random variables $\bar{x}_t^{(i)}$ for $i = 1, \ldots, n$ are chosen to be mutually independent conditional upon $\mathcal{F}_{t-1}$.

The iteration uses π_{t-1}^n to obtain π_t^n, but not any of the previous approximations. Following part (a) of the iteration, each particle changes its position according to the transition kernel of the signal. Let p_t^n be the empirical distribution associated with the cloud of particles of mass $1/n$ after part (a) of the iteration

$$p_t^n = \frac{1}{n} \sum_{i=1}^{n} \delta_{\bar{x}_t^{(i)}}.$$

This step of the algorithm is known as the *importance sampling* step (popular in the statistics literature) or *mutation* step (inherited from the genetic algorithms literature).

Exercise 10.16. Prove that $\mathbb{E}\left[p_t^n \mid \mathcal{F}_{t-1}\right] = K_{t-1}^n \pi_{t-1}^n$.

Remark 10.17. An alternative way to obtain p_t^n from π_{t-1}^n is to sample n times from the measure $K_{t-1} \pi_{t-1}^n$ and define p_t^n to be the empirical measure associated with this sample.

We assume that the offspring vector $\xi_t = (\xi_t^{(i)})_{i=1}^n$ satisfies the following two conditions.

1. The conditional mean number of offspring is proportional to $w_t^{(i)}$. More precisely
$$\mathbb{E}\left[\xi_t^{(i)} \mid \bar{\mathcal{F}}_t\right] = nw_t^{(i)}. \tag{10.33}$$

2. Let A_t^n be the conditional covariance matrix of the random vector $\xi_t \triangleq (\xi_t^{(i)})_{i=1}^n$,
$$A_t^n \triangleq \mathbb{E}\left[(\xi_t - nw_t)^\top (\xi_t - nw_t) \mid \bar{\mathcal{F}}_t\right]$$
with entries
$$(A_t^n)_{ij} = \mathbb{E}\left[\left(\xi_t^{(i)} - nw_t^{(i)}\right)\left(\xi_t^{(j)} - nw_t^{(j)}\right) \mid \bar{\mathcal{F}}_t\right],$$
where $w_t \triangleq (w_t^{(i)})_{i=1}^n$ is the vector of weights. Then assume that there exists a constant c_t, such that
$$q^\top A_t^n q \leq nc_t \tag{10.34}$$
for any n-dimensional vector $q = (q^{(i)})_{i=1}^n \in \mathbb{R}^n$, such that $|q^{(i)}| \leq 1$ for $i = 1, \ldots, n$.

Exercise 10.18. Prove that the following identity holds
$$\pi_t^n = \frac{1}{n}\sum_{i=1}^n \xi_t^{(i)} \delta_{\bar{x}_t^{(i)}},$$
and that $\mathbb{E}[\pi_t^n \mid \bar{\mathcal{F}}_t] = \bar{\pi}_t^n$.

Step (b) of the iteration is called the *selection* step. The particles obtained after the first step of the recursion are multiplied or discarded according to the magnitude of the likelihood weights. In turn the likelihood weights are proportional to the likelihood of the new observation given the corresponding position of the particle (see Remark 10.3). Hence if $nw_t^{(i)}$ is small, fewer offspring are expected than if $nw_t^{(i)}$ is large. Since
$$nw_t^{(i)} = \frac{g_t\left(\bar{x}_t^{(i)}\right)}{\frac{1}{n}\sum_{j=1}^n g_t\left(\bar{x}_t^{(j)}\right)},$$

$nw_t^{(i)}$ is small when the corresponding value of the likelihood function $g_t(\bar{x}_t^{(i)})$ is smaller than the likelihood function averaged over the positions of all the particles. In conclusion, the effect of part (b) of the iteration is that it discards particles in unlikely positions and multiplies those in more likely ones. Following Exercise 10.18, this is done in an unbiased manner: the conditional expectation of the approximation after applying the step is equal to the weighted sample obtained after the first step of the recursion. That is, the average of the mass $\xi_t^{(i)}/n$ associated with particle i is equal to $w_t^{(i)}$, the weight of the particle before applying the step.

Exercise 10.19. Prove that, for all $f \in B(\mathbb{R}^d)$, we have

$$\mathbb{E}\left[(\pi_t^n f - \bar{\pi}_t^n f)^2\right] \leq \frac{c_t \|f\|_\infty^2}{n}.$$

Exercise 10.19 implies that the randomness introduced in part (b) of the iteration, as measured by the second moment of $\pi_t^n f - \bar{\pi}_t^n f$, tends to zero with rate given by $1/n$, where n is the number of particles in the system.

Lemma 10.20. *Condition (10.34) is equivalent to*

$$q^\top A_t^n q \leq n \bar{c}_t \tag{10.35}$$

for any n-dimensional vector $q = (q^{(i)})_{i=1}^n \in [0,1]^n$, where $\bar{c}_t$ is a fixed constant.

Proof. Obviously (10.34) implies (10.35), so we only need to show the reverse implication. Let $q \in \mathbb{R}^n$ be an arbitrary vector such that $q = (q^{(i)})_{i=1}^n$, $|q^{(i)}| \leq 1$, $i = 1, \ldots, n$. Let also

$$q_+^{(i)} \triangleq \max\left(q^{(i)}, 0\right), \quad q_-^{(i)} \triangleq \max\left(-q^{(i)}, 0\right), \quad 0 \leq q_+^{(i)}, q_-^{(i)} \leq 1$$

and $q_+ = (q_+^{(i)})_{i=1}^n$ and $q_- = (q_-^{(i)})_{i=1}^n$. Then $q = q_+ - q_-$. Define $\|\cdot\|_A$ to be the semi-norm associated with the matrix A; that is,

$$\|q\|_A \triangleq \sqrt{q^\top A q}.$$

If all the eigenvalues of A are strictly positive, then $\|\cdot\|_A$ is a genuine norm. Using the triangle inequality and (10.35),

$$\|q\|_{A_t^n} \leq \|q_+\|_{A_t^n} + \|q_-\|_{A_t^n} \leq 2\sqrt{n\bar{c}_t},$$

which implies that (10.34) holds with $c_t = 4\bar{c}_t$. $\square$

10.5 Offspring Distributions

In order to have a complete description of the particle filter we need to specify the offspring distribution. The most popular offspring distribution is the multinomial distribution

$$\xi_t = \text{Multinomial}\left(n, w_t^{(1)}, \ldots, w_t^{(n)}\right);$$

that is,

276   10 Particle Filters in Discrete Time

$$\mathbb{P}\left(\xi_t^{(i)} = n^{(i)}, i = 1, \ldots, n\right) = \frac{n!}{\prod_{i=1}^{n} n^{(i)}!} \prod_{i=1}^{n} \left(w_t^{(i)}\right)^{n^{(i)}}.$$

The multinomial distribution is the empirical distribution of an n-sample from the distribution $\bar{\pi}_t^n$. In other words, if we sample (with replacement) n times from the population of particles with positions $\bar{x}_t^{(i)}$, $i = 1, \ldots, n$ according to the probability distribution given by the corresponding weights $w_t^{(i)}$, $i = 1, \ldots, n$ and denote by $\xi_t^{(i)}$ the number of times that the particle with position $\bar{x}_t^{(i)}$ is chosen, then $\xi_t = (\xi_t^{(i)})_{i=1}^n$ has the above multinomial distribution.

Lemma 10.21. *If ξ_t has a multinomial distribution then it satisfies the unbiasedness condition; that is,*

$$\mathbb{E}\left[\xi_t^{(i)} \mid \bar{\mathcal{F}}_t\right] = n w_t^{(i)},$$

for any $i = 1, \ldots, n$. Also ξ_t satisfies condition (10.34).

Proof. The unbiasedness condition follows immediately from the properties of the multinomial distribution. Also

$$\mathbb{E}\left[\left(\xi_t^{(i)} - n w_t^{(i)}\right)^2 \mid \bar{\mathcal{F}}_t\right] = n w_t^{(i)} \left(1 - w_t^{(i)}\right)$$

$$\mathbb{E}\left[\left(\xi_t^{(i)} - n w_t^{(i)}\right)\left(\xi_t^{(j)} - n w_t^{(j)}\right) \mid \bar{\mathcal{F}}_t\right] = -n w_t^{(i)} w_t^{(j)}, \quad i \neq j.$$

Then for all $q = (q^{(i)})_{i=1}^n \in [-1, 1]^n$,

$$q^\top A_t^n q = \sum_{i=1}^{n} n w_t^{(i)} \left(1 - w_t^{(i)}\right) \left(q^{(i)}\right)^2 - 2 \sum_{1 \leq i < j \leq n} n w_t^{(i)} w_t^{(j)} q^{(i)} q^{(j)}$$

$$= n \sum_{i=1}^{n} w_t^{(i)} \left(q^{(i)}\right)^2 - n \left(\sum_{i=1}^{n} w_t^{(i)} q^{(i)}\right)^2$$

$$\leq n \sum_{i=1}^{n} w_t^{(i)},$$

and since $\sum_{i=1}^{n} w_t^{(i)} = 1$, (10.34) holds with $c_t = 1$. □

The particle filter with this choice of offspring distribution is called the *bootstrap filter* or the *sampling importance resampling* algorithm (SIR algorithm). It was introduced by Gordon, Salmond and Smith in [106] (see the last section for further historical remarks). Within the context of the bootstrap filter, the second step is called the *resampling* step.

The bootstrap filter is quick and easy to implement and amenable to parallelisation. This explains its great popularity among practitioners. However, it is suboptimal: the resampling step replaces the (normalised) weights $w_t^{(i)}$ by the random masses $\xi_t^{(i)}/n$, where $\xi_t^{(i)}$ is the number of offspring of the ith particle. Since ξ_t has a multinomial distribution, $\xi_t^{(i)}$ can take any value between 0 and n. That is, even when $w_t^{(i)}$ is high (the position of the ith particle is very likely), the ith particle may have very few offspring or even none at all (albeit with small probability).

If ξ_t is obtained by residual sampling, rather than by independent sampling with replacement, then the above disadvantage can be avoided. In this case

$$\xi_t = [nw_t] + \bar{\xi}_t. \tag{10.36}$$

In (10.36), $[nw_t]$ is the (row) vector of integer parts of the quantities $nw_t^{(i)}$. That is,

$$[nw_t] = \left(\left[nw_t^{(1)}\right], \ldots, \left[nw_t^{(n)}\right]\right),$$

and $\bar{\xi}_t$ has multinomial distribution

$$\bar{\xi}_t = \text{Multinomial}\left(\bar{n}, \bar{w}_t^{(1)}, \ldots, \bar{w}_t^{(n)}\right),$$

where the integer $\bar{n}$ is given by

$$\bar{n} \triangleq n - \sum_{i=1}^{n} \left[nw_t^{(i)}\right] = \sum_{i=1}^{n} \left\{nw_t^{(i)}\right\}$$

and the weights $\bar{w}_t^{(i)}$ are given by

$$\bar{w}_t^{(i)} \triangleq \frac{\left\{nw_t^{(i)}\right\}}{\sum_{i=1}^{n}\left\{nw_t^{(i)}\right\}}.$$

By using residual sampling to obtain ξ_t, we ensure that the original weights $w_t^{(i)}$ are replaced by a random weight which is at least $[nw_t^{(i)}]/n$. This is the closest integer multiple of $1/n$ lower than the actual weight $w_t^{(i)}$. In this way, eliminating particles with likely positions is no longer possible. As long as the corresponding weight is larger than $1/n$, the particle will have at least one offspring.

Lemma 10.22. *If ξ_t has distribution given by (10.36), it satisfies both the unbiasedness condition (10.33) and condition (10.34).*

Proof. The unbiasedness condition follows from the properties of the multinomial distribution:

$$\mathbb{E}\left[\xi_t^{(i)} \mid \bar{\mathcal{F}}_t\right] = \left[nw_t^{(i)}\right] + \mathbb{E}\left[\bar{\xi}_t^{(i)} \mid \bar{\mathcal{F}}_t\right]$$
$$= \left[nw_t^{(i)}\right] + \bar{n}\bar{w}_t^{(i)}$$
$$= \left[nw_t^{(i)}\right] + \left\{nw_t^{(i)}\right\} = nw_t^{(i)}.$$

Also
$$\mathbb{E}\left[\left(\xi_t^{(i)} - n w_t^{(i)}\right)^2 \mid \bar{\mathcal{F}}_t\right] = \mathbb{E}\left[\left(\bar{\xi}_t^{(i)} - \{n w_t^{(i)}\}\right)^2 \mid \bar{\mathcal{F}}_t\right]$$
$$= \bar{n} \bar{w}_t^{(i)} \left(1 - \bar{w}_t^{(i)}\right)$$

and
$$\mathbb{E}\left[\left(\xi_t^{(i)} - n w_t^{(i)}\right)\left(\xi_t^{(j)} - n w_t^{(j)}\right) \mid \bar{\mathcal{F}}_t\right] = -\bar{n} \bar{w}_t^{(i)} \bar{w}_t^{(j)}.$$

Then for all $q = (q^{(i)})_{i=1}^n \in [-1,1]^n$, we have

$$q^\top A_t^n q = \sum_{i=1}^n \bar{n} \bar{w}_t^{(i)} \left(1 - \bar{w}_t^{(i)}\right) \left(q^{(i)}\right)^2 - 2 \sum_{1 \le i < j \le n} \bar{n} \bar{w}_t^{(i)} \bar{w}_t^{(j)} q^{(i)} q^{(j)}$$
$$= \sum_{i=1}^n \bar{n} \bar{w}_t^{(i)} \left(q^{(i)}\right)^2 - \bar{n} \left(\sum_{i=1}^n \bar{w}_t^{(i)} q^{(i)}\right)^2$$
$$\le \sum_{i=1}^n \bar{n} \bar{w}_t^{(i)},$$

and since $\sum_{i=1}^n \bar{n} \bar{w}_t^{(i)} = \sum_{i=1}^n \{n w_t^{(i)}\} < n$, (10.34) holds with $c_t = 1$. □

Exercise 10.23. In addition to the bound on the second moment of $\pi_t^n f - \bar{\pi}_t^n f$ resulting by imposing the assumption (10.34) on the offspring distribution ξ_t (see Exercise 10.19), prove that if ξ_t has multinomial distribution or the distribution given by (10.36), then there exists a constant c such that, for all $f \in B(\mathbb{R}^d)$, we have

$$\mathbb{E}\left[(\pi_t^n f - \bar{\pi}_t^n f)^4 \mid \bar{\mathcal{F}}_t\right] \le \frac{c \|f\|_\infty^4}{n^2}.$$

The residual sampling distribution is still suboptimal; the correction step now replaces the weight $w_t^{(i)}$ by the deterministic mass $[n w_t^{(i)}]/n$ to which it adds a random mass given by $\bar{\xi}_t^{(i)}/n$, where $\bar{\xi}_t^{(i)}$ can take any value between 0 and $\bar{n}$. This creates a problem for particles with small weights. Even when $w_t^{(i)}$ is small (the position of the ith particle is very unlikely) it may have a large number of offspring: up to $\bar{n}$ offspring are possible (albeit with small probability). The multinomial distribution also suffers from this problem.

If ξ_t is obtained by using the branching algorithm described in Section 9.2.1, then both the above difficulties are eliminated. In this case, the number of offspring $\xi_t^{(i)}$ for each individual particle has the distribution

$$\xi_t^{(i)} = \begin{cases} \left[n w_t^{(i)}\right] & \text{with probability } 1 - \{n w_t^{(i)}\} \\ \left[n w_t^{(i)}\right] + 1 & \text{with probability } \{n w_t^{(i)}\}, \end{cases} \quad (10.37)$$

whilst $\sum_{i=1}^{n} \xi_t^{(i)}$ remains equal to n.

If the particle has a weight $w_t^{(i)} > 1/n$, then the particle will have offspring. Thus if the corresponding likelihood function $g_t(\bar{x}_t^{(i)})$ is larger than the likelihood averaged over all the existing particles $(1/n) \sum_{j=1}^{n} g_t(\bar{x}_t^{(j)})$, then the ith site is selected and the higher the weight $w_t^{(i)}$ the more offspring the ith particle will have. If $w_t^{(i)}$ is less than or equal to $1/n$, the particle will have at most one offspring. It will have no offspring with probability $1 - nw_t^{(i)}$, as in this case $nw_t^{(i)} = \{nw_t^{(i)}\}$. Hence, if $w_t^{(i)} \ll 1/n$, no mass is likely to be assigned to site i; the ith particle is very unlikely and it is eliminated from the sample.

The algorithm described in Section 9.2.1 belongs to a class of algorithms called tree-based branching algorithms. If ξ_t is obtained by using the branching algorithm described in Section 9.2.1, then it is optimal in the sense that, for any $i = 1, \ldots, n$, $\xi_t^{(i)}$ has the smallest possible variance amongst all integer-valued random variables with the given mean $nw_t^{(i)}$. Hence, the algorithm ensures that minimal randomness, as measured by the variance of the mass allocated to individual sites, is introduced to the system. The minimal variance property for the distribution produced by any tree-based branching algorithm holds true not only for individual sites but also for all groups of sites corresponding to a node of the building binary tree. A second optimality property of this distribution is that it has the minimal relative entropy with respect to the measure $\bar{\pi}_t$ which it replaces in the class of all empirical distributions of n particles of mass $1/n$. The interested reader should consult Crisan [60] for details of these properties. See also Künsch [169] for further results on the distribution produced by the branching algorithm.

Lemma 10.24. *If ξ_t is produced by the algorithm described in Section 9.2.1, it satisfies both unbiasedness condition (10.33) and condition (10.34).*

Proof. The unbiasedness condition immediately follows from (10.37)

$$\mathbb{E}\left[\xi_t^{(i)} \mid \bar{\mathcal{F}}_t\right] = \left[nw_t^{(i)}\right]\left(1 - \{nw_t^{(i)}\}\right) + \left(\left[nw_t^{(i)}\right] + 1\right)\{nw_t^{(i)}\} = nw_t^{(i)}.$$

Also

$$\mathbb{E}\left[\left(\xi_t^{(i)} - nw_t^{(i)}\right)^2 \mid \bar{\mathcal{F}}_t\right] = \{nw_t^{(i)}\}\left(1 - \{nw_t^{(i)}\}\right),$$

and from Proposition 9.3, part (e),

$$\mathbb{E}\left[\left(\xi_t^{(i)} - nw_t^{(i)}\right)\left(\xi_t^{(j)} - nw_t^{(j)}\right) \mid \bar{\mathcal{F}}_t\right] \leq 0.$$

Then for all $q = (q^{(i)})_{i=1}^{n} = [0,1]^n$, we have

$$q^\top A_t^n q \leq \sum_{i=1}^{n} \{nw_t^{(i)}\}\left(1 - \{nw_t^{(i)}\}\right),$$

and since $\{nw_t^{(i)}\}(1-\{nw_t^{(i)}\}) < \frac{1}{4}$, following Lemma 10.20, condition (10.34) holds with $c_t = \frac{1}{4}$. □

For further theoretical results related to the properties of the above offspring distributions, see Chopin [51] and Künsch [169].

There exists another algorithm that satisfies the same minimal variance property of the branching algorithm described above. It was introduced by Carpenter, Clifford and Fearnhead in the context of particle approximations (see [38]). The method had appeared earlier in the field of genetic algorithms and it is known under the name of *stochastic universal sampling* (see Baker [6] and Whitley [268]). However, the offspring distribution generated by this method does not satisfy condition (10.34) and the convergence of the particle filter with this method is still an open question.[†]

All offspring distributions presented above leave the total number of particles constant and satisfy (10.34). However, the condition that the total number of particles does not change is not essential.

One can choose the individual offspring numbers $\xi_t^{(i)}$ to be mutually independent given $\bar{\mathcal{F}}_t$. As alternatives for the distribution of the integer-valued random variables $\xi_t^{(i)}$ the following can be used.

1. $\xi_t^{(i)} = B(n, w_t^{(i)})$; that is, $\xi_t^{(i)}$ are binomially distributed with parameters $(n, w_t^{(i)})$.
2. $\xi_t^{(i)} = P(nw_t^{(i)})$; that is, $\xi_t^{(i)}$ are Poisson distributed with parameters $nw_t^{(i)}$.
3. $\xi_t^{(i)}$ are Bernoulli distributed with distribution given by (10.37).

Exercise 10.25. Show that if the individual offspring numbers $\xi_t^{(i)}$ are mutually independent given $\bar{\mathcal{F}}_t$ and have any of the three distributions described above, then ξ_t satisfies both the unbiasedness condition and condition (10.34).

The Bernoulli distribution is the optimal choice for independent offspring distributions. Since $\sum_{i=1}^n \xi_t^{(i)}$ is no longer equal to n, the approximating measure π_t^n is no longer a probability measure. However, following the unbiasedness condition (10.33) and condition (10.34), the total mass $\pi_t^n(\mathbf{1})$ of the approximating measure is a martingale which satisfies, for any $t \in [0,T]$,

$$\mathbb{E}\left[(\pi_t^n(\mathbf{1}) - 1)^2\right] \leq \frac{c}{n},$$

where $c = c(T)$ is a constant independent of n. This implies that for large n the mass oscillations become very small. Indeed, by Chebyshev's inequality

$$\mathbb{P}\left(|\pi_t^n(\mathbf{1}) - 1| \geq \varepsilon\right) \leq \frac{c}{n\varepsilon^2}.$$

[†] See Künsch [169] for some partial results.

Hence, having a non-constant number of particles does not necessarily lead to instability. The oscillations in the number of particles can in themselves constitute an indicator of the convergence of the algorithm. Such an offspring distribution with independent individual offspring numbers is easy to implement and saves computational effort. An algorithm with variable number of particles is presented in Crisan, Del Moral and Lyons [67]. Theorems 10.7 and 10.12 and all other results presented above can be used in order to prove the convergence of the algorithm in [67] and indeed any algorithm based on such offspring distributions.

10.6 Convergence of the Algorithm

First fix the observation process to an arbitrary value $y_{0:T}$, where T is a finite time horizon and we prove that the random measures resulting from the class of algorithm described above converge to $\pi_t^{y_{0:t}}$ and $p_t^{y_{0:t-1}}$ for all $0 \leq t \leq T$.

Exercise 10.26. Prove that π_0^n converges in expectation to π_0 and also $\lim_{n\to\infty} \pi_0^n = \pi_0$, $\mathbb{P}$-a.s.

Theorem 10.27. *Let $(p_t^n)_{n=1}^\infty$ and $(\pi_t^n)_{n=1}^\infty$ be the measure-valued sequences produced by the class of algorithms described above. Then, for all $0 \leq t \leq T$, we have*
$$\lim_{n\to\infty} \mathbb{E}\left[|\pi_t^n f - \pi_t f|\right] = \lim_{n\to\infty} \mathbb{E}\left[|p_t^n f - p_t f|\right] = 0,$$
for all $f \in B(\mathbb{R}^d)$. In particular, $(p_t^n)_{n=1}^\infty$ converges in expectation to $p_t^{y_{0:t-1}}$ and $(\pi_t^n)_{n=1}^\infty$ converges in expectation to $\pi_t^{y_{0:t}}$ for all $0 \leq t \leq T$.

Proof. We apply Theorem 10.7. Since (a1) holds as a consequence of Exercise 10.26, it is only necessary to verify condition (b1). From Exercise 10.16, $\mathbb{E}\left[p_t^n f \mid \mathcal{F}_t\right] = \pi_{t-1}^n(K_{t-1}f)$ and using the independence of the sample $\{\bar{x}_t^{(i)}\}_{i=1}^n$ conditional on $\mathcal{F}_{t-1}$,

$$\mathbb{E}\left[\left(p_t^n f - \pi_{t-1}^n(K_{t-1}f)\right)^2 \mid \mathcal{F}_{t-1}\right]$$
$$= \frac{1}{n^2}\mathbb{E}\left[\left(\sum_{i=1}^n f\left(\bar{x}_t^{(i)}\right) - K_{t-1}f\left(x_{t-1}^{(i)}\right)\right)^2 \mid \mathcal{F}_{t-1}\right]$$
$$= \frac{1}{n^2}\sum_{i=1}^n \mathbb{E}\left[\left(f\left(\bar{x}_t^{(i)}\right)\right)^2 \mid \mathcal{F}_{t-1}\right]$$
$$- \frac{1}{n^2}\sum_{i=1}^n \left(\mathbb{E}\left[K_{t-1}f\left(x_{t-1}^{(i)}\right) \mid \mathcal{F}_{t-1}\right]\right)^2$$
$$= \frac{1}{n}\pi_{t-1}^n\left(K_{t-1}f^2 - (K_{t-1}f)^2\right).$$

Therefore $\mathbb{E}[(p_t^n f - \pi_{t-1}^n K_{t-1} f)^2] \leq \|f\|_\infty^2 / n$ and the first limit in (b1) is satisfied. The second limit in (b1) follows from Exercise 10.19. □

Corollary 10.28. *For all $0 \leq t \leq T$, there exists a constant k_t such that*

$$\mathbb{E}\left[(\pi_t^n f - \pi_t f)^2\right] \leq \frac{k_t \|f\|_\infty^2}{n}, \tag{10.38}$$

for all $f \in B(\mathbb{R}^d)$.

Proof. We proceed by induction. Since $\{x_0^{(i)}, i = 1, \ldots, n\}$ is an n-independent sample from π_0,

$$\mathbb{E}\left[(\pi_0^n f - \pi_0 f)^2\right] \leq \frac{\|f\|_\infty^2}{n},$$

hence by Jensen's inequality (10.38) is true for $t = 0$ with $k_0 = 1$. Now assume that (10.38) holds at time $t - 1$. Then

$$\mathbb{E}\left[(\pi_{t-1}^n (K_{t-1} f) - \pi_{t-1}(K_{t-1} f))^2\right] \leq \frac{k_{t-1} \|K_{t-1} f\|_\infty^2}{n} \leq \frac{k_{t-1} \|f\|_\infty^2}{n}. \tag{10.39}$$

Also from the proof of Theorem 10.27,

$$\mathbb{E}\left[(p_t^n f - \pi_{t-1}^n K_{t-1} f)^2\right] \leq \frac{\|f\|_\infty^2}{n}. \tag{10.40}$$

By using inequality (10.23) and the triangle inequality for the L_2-norm,

$$\mathbb{E}\left[(p_t^n f - p_t f)^2\right] \leq \frac{\hat{k}_t \|f\|_\infty^2}{n}, \tag{10.41}$$

where $\hat{k}_t = (\sqrt{k_{t-1}} + 1)^2$. In turn, (10.41) and (10.25) imply that

$$\mathbb{E}\left[(\bar{\pi}_t^n f - \pi_t f)^2\right] \leq \frac{\bar{k}_t \|f\|_\infty^2}{n}, \tag{10.42}$$

where $\bar{k}_t = 4\hat{k}_t \|g_t\|_\infty^2 / (p_t g_t)^2$. From Exercise 10.19.

$$\mathbb{E}\left[(\pi_t^n f - \bar{\pi}_t^n f)^2\right] \leq \frac{c_t \|f\|_\infty^2}{n}, \tag{10.43}$$

where c_t is the constant appearing in (10.34). Finally from (10.42), (10.43) and the triangle inequality (10.27), (10.38) holds with $k_t = (\sqrt{c_t} + \sqrt{\bar{k}_t})^2$. This completes the induction step. □

Condition (10.34) is essential in establishing the above rate of convergence. A more general condition than (10.34) is possible, for example, that there exists $\alpha > 0$ such that

$$q^\top A_t^n q \leq n^\alpha c_t \tag{10.44}$$

for any $q \in [-1, 1]^n$. In this case, inequality (10.43) would become

$$\mathbb{E}[(\pi_t^n f - \bar{\pi}_t^n f)^2] \leq \frac{c_t \|f\|_\infty^2}{n^{2-\alpha}}.$$

Hence the overall rate of convergence would take the form

$$\mathbb{E}\left[(\pi_t^n f - \pi_t f)^2\right] \leq \frac{k_t \|f\|_\infty^2}{n^{\max(2-\alpha,1)}}$$

for all $f \in B(\mathbb{R}^d)$. Hence if $\alpha > 1$ we will see a deterioration in the overall rate of convergence. On the other hand, if $\alpha < 1$ no improvement in the rate of convergence is obtained as the error in all the other steps of the particle filter remains of order $1/n$. So $\alpha = 1$ is the most suitable choice for condition (10.34).

Theorem 10.29. *If the offspring distribution is multinomial or is given by (10.36), then for all $0 \leq t \leq T$,*

$$\lim_{n \to \infty} p_t^n = p_t^{y_{0:t-1}} \quad \text{and} \quad \lim_{n \to \infty} \pi_t^n = \pi_t^{y_{0:t}} \quad \mathbb{P}\text{-a.s.}$$

Proof. We apply Theorem 10.12. Since condition (a2) holds as a consequence of Exercise 10.26, it is only necessary to verify condition (b2). Let $\mathcal{M} \subset C_b(\mathbb{R}^d)$ be a countable, convergence determining set of functions (see Section A.10 for details). Following Exercise 10.16, for any $f \in \mathcal{M}$,

$$\mathbb{E}\left[f\left(\bar{x}_t^{(i)}\right) \mid \mathcal{F}_{t-1}\right] = K_{t-1} f\left(x_{t-1}^{(i)}\right)$$

and using the independence of the sample $\{\bar{x}_t^{(i)}\}_{i=1}^n$ conditional on $\mathcal{F}_{t-1}$,

$$\mathbb{E}\left[\left(p_t^n f - K_{t-1} \pi_{t-1}^n f\right)^4 \mid \mathcal{F}_{t-1}\right]$$

$$= \mathbb{E}\left[\left(\frac{1}{n} \sum_{i=1}^n \left(f\left(\bar{x}_t^{(i)}\right) - K_{t-1} f\left(x_{t-1}^{(i)}\right)\right)\right)^4 \mid \mathcal{F}_{t-1}\right]$$

$$= \frac{1}{n^4} \sum_{i=1}^n \mathbb{E}\left[\left(f\left(\bar{x}_t^{(i)}\right) - K_{t-1} f\left(x_{t-1}^{(i)}\right)\right)^4 \mid \mathcal{F}_{t-1}\right]$$

$$+ \frac{6}{n^4} \sum_{1 \leq i < j \leq n} \mathbb{E}\left[\left(f\left(\bar{x}_t^{(i)}\right) - K_{t-1} f\left(x_{t-1}^{(i)}\right)\right)^2\right.$$

$$\left.\left(f\left(\bar{x}_t^{(j)}\right) - K_{t-1} f\left(x_{t-1}^{(j)}\right)\right)^2 \mid \mathcal{F}_{t-1}\right]. \quad (10.45)$$

Observe that since $\|K_{t-1} f\|_\infty \leq \|f\|_\infty$,

$$\mathbb{E}\left[\left(f\left(\bar{x}_t^{(i)}\right) - K_{t-1} f\left(x_{t-1}^{(i)}\right)\right)^4 \mid \mathcal{F}_{t-1}\right] \leq 16 \|f\|_\infty^4$$

and

$$\mathbb{E}\left[\left(f\left(\bar{x}_t^{(i)}\right) - K_{t-1}f\left(x_{t-1}^{(i)}\right)\right)^2 \left(f\left(\bar{x}_t^{(j)}\right) - K_{t-1}f\left(x_{t-1}^{(j)}\right)\right)^2 \Big| \mathcal{F}_{t-1}\right]$$
$$\leq 16\|f\|_\infty^4.$$

Hence by taking the expectation of both terms in (10.45)

$$\mathbb{E}\left[(p_t^n f - K_{t-1}\pi_{t-1}^n f)^4\right] \leq \frac{16\|f\|_\infty^4}{n^3} + \frac{6}{n^4}\frac{n(n-1)}{2}16\|f\|_\infty^4$$
$$\leq \frac{48\|f\|_\infty^4}{n^2}. \qquad (10.46)$$

From (10.46), following Remark A.38 in the appendix, for any $\varepsilon \in (0, \frac{1}{4})$ there exists a positive random variable $c_{f,\varepsilon}$ which is almost surely finite such that

$$\left|p_t^n f - K_{t-1}\pi_{t-1}^n f\right| \leq \frac{c_{f,\varepsilon}}{n^\varepsilon}.$$

In particular $|p_t^n f - K_{t-1}\pi_{t-1}^n f|$ converges to zero, $\mathbb{P}$-a.s., for any $f \in \mathcal{M}$. Therefore $\lim_{n\to\infty} d_\mathcal{M}\left(p_t^n, K_{t-1}\pi_{t-1}^n\right) = 0$ which is the first limit in (b2). Similarly, following Exercise 10.23, one proves that, for all $f \in \mathcal{M}$,

$$\mathbb{E}\left[(\pi_t^n f - \bar{\pi}_t^n f)^4\right] \leq \frac{c\|f\|_\infty^4}{n^2} \qquad (10.47)$$

which implies that $\lim_{n\to\infty} d_\mathcal{M}(\pi_t^n, \bar{\pi}_t^n) = 0$, hence also the second limit in b2. holds. □

We now consider the case where the observation process is no longer a particular fixed outcome, but is random. With similar arguments one uses Propositions 10.14 and 10.15 to prove the following.

Corollary 10.30. *Assume that for all $t \geq 0$, there exists a constant $c_t > 0$ such that $p_t g_t \geq c_t$. Then we have*

$$\lim_{n\to\infty} \mathbb{E}\left[\left|\pi_t^{n,Y_{0:t}}f - \pi_t f\right|\right] = \lim_{n\to\infty} \mathbb{E}\left[\left|p_t^{n,Y_{0:t-1}}f - p_t f\right|\right] = 0$$

for all $f \in B(\mathbb{R}^d)$ and all $t \geq 0$. In particular, $(p_t^{n,Y_{0:t-1}})_{n=1}^\infty$ converges in expectation to $p_t^{y_{0:t-1}}$ and $(\pi_t^{n,Y_{0:t}})_{n=1}^\infty$ converges in expectation to $\pi_t^{y_{0:t}}$ for all $t \geq 0$.

Corollary 10.31. *If the offspring distribution is multinomial or is given by (10.36), then*

$$\lim_{n\to\infty} p_t^{n,Y_{0:t-1}} = p_t \quad \text{and} \quad \lim_{n\to\infty} \pi_t^{n,Y_{0:t}} = \pi_t \qquad \mathbb{P}\text{-a.s.}$$

for all $t \geq 0$.

10.7 Final Discussion

The results presented in Section 10.3 provide efficient techniques for proving convergence of particle algorithms. The necessary and sufficient conditions ((a0), (b0)), ((a1), (b1)) and ((a2), (b2)) are natural and easy to verify as it can be seen in the proofs of Theorems 10.27 and 10.29.

The necessary and sufficient conditions can be applied when the algorithms studied provide both π_t^n (the approximation to π_t) and also p_t^n (the intermediate approximation to p_t). Algorithms are possible where π_t^n is obtained from π_{t-1}^n without using the approximation for p_t. In other words one can perform the mutation step using a different transition from that of the signal. In the statistics literature, the transition kernel K_t is usually called the *importance distribution*. Should a kernel (or importance distribution) $\bar{K}_t$ be used which is different from that of the signal K_t, the form of the weights appearing in the selection step of the particle filter must be changed. The results presented in Section 10.3 then apply for p_t now given by $\bar{K}_{t-1}\pi_{t-1}$ and the weighted measure $\bar{\pi}_t^n$ defined in (10.21) given by

$$\bar{\pi}_t^n = \sum_{i=1}^n \bar{w}_t^{(i)} \delta_{\bar{x}_t^{(i)}},$$

where $\bar{w}_t^{(i)}$ are the new weights. See Doucet et al. [83] and Pitt and Shephard [244] and the references contained therein which describe the use of such importance distributions.

As already pointed out, the randomness introduced in the system at each selection step must be kept to a minimum as it affects the rate of convergence of the algorithm. Therefore one should not apply the selection step after every new observation arrives. Assume that the information received from the observation is 'bad' (i.e. the signal-to-noise ratio is small). Because of this, the likelihood function is close to being constant and the corresponding weights are all (roughly) equal; $\bar{w}_t^{(i)} \simeq 1/n$. In other words, the observation is uninformative; it cannot distinguish between different sites and all particles are equally likely. In this case no selection procedure needs to be performed. The observation is stored in the weights of the approximation $\bar{\pi}_t^n$ and carried forward to the next step. If a correction procedure is nevertheless performed and ξ_t has a minimal variance distribution, all particles will have a single offspring 'most of the time'. In other words the system remains largely unchanged with high probability. However, with small probability, the ith particle might have no offspring (if $\bar{w}_t^{(i)} < 1/n$) or two offspring (if $\bar{w}_t^{(i)} > 1/n$). Hence randomness still enters the system and this can affect the convergence rates (see Crisan and Lyons [66] for a related result in the continuous time framework). If ξ_t does not have a minimal variance distribution, the amount of randomness is even higher. It remains an open question as to when and how often one should use the selection procedure.

The first paper on the sequential Monte Carlo methods was that of Handschin and Mayne [120] which appeared in 1969. Unfortunately, Handschin and Mayne's paper appeared at a time when the lack of computing power meant that it could not be implemented; thus their ideas were overlooked. In the late 1980s, the advances in computer hardware rekindled interest in obtaining sequential Monte Carlo methods for approximating the posterior distribution. The first paper describing numerical integration for Bayesian filtering was published by Kitagawa [150] in 1987. The area developed rapidly following the publication of the bootstrap filter by Gordon, Salmond and Smith [106] in 1993. The development of the bootstrap filter was inspired by the earlier work of Rubin [251] on the SIR algorithm from 1987. The use of the algorithm has spread very quickly among engineers and computer scientists. An important example is the work of Isard and Blake in computer vision (see [132, 133, 134]).

The first convergence results on particle filters in discrete time were published by Del Moral in 1996 (see [214, 215]). Together with Rigal and Salut, he produced several earlier LAAS-CNRS reports which were originally classified ([219, 221, 220]) which contain the description of the bootstrap filter. The condition (10.34) was introduced by Crisan, Del Moral and Lyons in 1999 (see [67]). The tree-based branching algorithm appeared in Crisan and Lyons [66].

In the last ten years we have witnessed a rapid development of the theory of particle filters in discrete time. The discrete time framework has been extensively studied and a multitude of convergence and stability results have been proved. A comprehensive account of these developments in the wider context of approximations of Feynman–Kac formulae can be found in Del Moral [216] and the references therein.

10.8 Solutions to Exercises

10.1

i. For $\varphi = I_A$ where A is an arbitrary Borel set, $K_t\varphi \in B(\mathbb{R}^d)$ by property (ii) of the transition kernel. By linearity, the same is true for φ being a simple function, that is, a linear combination of indicator functions. Consider next an arbitrary $\varphi \in B(\mathbb{R}^d)$. Then there exists a sequence of simple functions $(\varphi_n)_{n\geq 0}$ uniformly bounded which converges to φ. Then by the dominated convergence theorem $K_t(\varphi_n)(x)$ converges to $K_t(\varphi)(x)$ for any $x \in \mathbb{R}^d$. Hence $K_t(\varphi)(x)$ is Borel-measurable. The boundedness results from the fact that

$$|K_t\varphi(x)| = \left|\int_{\mathbb{R}^d} \varphi(y) K_t(x, \mathrm{d}y)\right| \leq \|\varphi\|_\infty \int_{\mathbb{R}^d} K_t(x, \mathrm{d}y) = \|\varphi\|_\infty$$

for any $x \in \mathbb{R}^d$; hence $\|K_t\varphi\|_\infty \leq \|\varphi\|_\infty$.

ii. Let $A_i \in \mathcal{B}(\mathbb{R}^d)$ be a sequence of disjoint sets for $i = 1, 2, \ldots$, then using property (i) of K_t,

$$K_t q_t(\cup_{i=1}^{\infty} A_i) = \int_{\mathbb{R}^d} K_t(x, \cup_{i=1}^{\infty} A_i) q_t(\mathrm{d}x)$$

$$= \int_{\mathbb{R}^d} \lim_{N \to \infty} \sum_{i=1}^{N} K_t(x, A_i) q_t(\mathrm{d}x)$$

$$= \lim_{N \to \infty} \sum_{i=1}^{N} \int_{\mathbb{R}^d} K_t(x, A_i) q_t(\mathrm{d}x) = \sum_{i=1}^{\infty} (K_t q_t)(A_i),$$

where the bounded convergence theorem was used to interchange the limit and the integral (using $K_t(x, \Omega) = 1$ as the bound). Consequently $K_t q_t$ is countably additive and hence a measure. To check that it is a probability measure

$$K_t q_t(\Omega) = \int_{\mathbb{R}^d} K_t(x, \Omega) q_t(\mathrm{d}x) = \int_{\mathbb{R}^d} q_t(\mathrm{d}x) = 1.$$

iii.

$$(K_t q_t)(\varphi) = \int_{y \in \mathbb{R}^d} \varphi(y) \int_{x \in \mathbb{R}^d} K_t(x, \mathrm{d}y) q_t(\mathrm{d}x).$$

By Fubini's theorem, which is applicable since φ is bounded and as a consequence of (ii) $K_t q_t$ is a probability measure, which implies that $K_t q_t(|\varphi|) \leq \|\varphi\|_\infty < \infty$),

$$(K_t q_t)(\varphi) = \int_{x \in \mathbb{R}^d} q_t(\mathrm{d}x) \int_{y \in \mathbb{R}^d} \varphi(y) K_t(x, \mathrm{d}y) = q_t(K_t \varphi)$$

and the general case follows by induction.

10.5 Finite additivity is trivial from the linearity of the integral, and countable additivity follows from the bounded convergence theorem, since φ is bounded. Thus $\varphi * p$ is a measure. It is clear that $\varphi * p(\Omega) = p(\varphi)/p(\varphi) = 1$, so it is a probability measure.

10.9

i. For arbitrary $\varphi \in C_b(\mathbb{R}^d)$. It is clear that

$$K_t \varphi(x) = \int_{\mathbb{R}^d} \varphi(y) \frac{1}{\sqrt{2\pi}} \exp\left(-\frac{(y - a(x))^2}{2}\right)$$

$$= \int_{\mathbb{R}^d} \varphi(y + a(x)) \frac{1}{\sqrt{2\pi}} \exp\left(-\frac{y^2}{2}\right).$$

Then by the dominated convergence theorem using the continuity of a,

$$\lim_{x \to x_0} K_t(\varphi_n)(x) = K_t(\varphi_n)(x_0)$$

for arbitrary $x_0 \in \mathbb{R}^d$, hence the continuity of $K_t \varphi$.

ii. Choose a strictly increasing $\varphi \in C_b(\mathbb{R}^d)$. Then, as above

$$\lim_{x \uparrow 0} K_t(\varphi_n)(x) = \int_{\mathbb{R}^d} \varphi(y-1) \frac{1}{\sqrt{2\pi}} \exp\left(-\frac{y^2}{2}\right)$$
$$< \int_{\mathbb{R}^d} \varphi(y+1) \frac{1}{\sqrt{2\pi}} \exp\left(-\frac{y^2}{2}\right) = \lim_{x \downarrow 0} K_t(\varphi_n)(x).$$

10.11

i. Let $f \in C_b(\mathbb{R})$; then

$$\mu_n f = \frac{1}{n} \sum_{i=1}^{n} f(i/n)$$

and

$$\mu f = \int_0^1 f(x) \, dx.$$

As $f \in C_b(\mathbb{R})$, it is Riemann integrable. Therefore the Riemann approximation $\mu_n f \to \mu f$ as $n \to \infty$.

ii. If $f = 1_{\mathbb{Q} \cap [0,1]}$ then $\mu_n f = 1$, yet $\mu f = 0$. Hence $\mu_n f \not\to \mu f$ as $n \to \infty$.

10.16 Following part (a) of the iteration we get, for arbitrary $f \in B(\mathbb{R}^d)$, that

$$\mathbb{E}\left[f\left(\bar{x}_t^{(i)}\right) \mid \mathcal{F}_{t-1}\right] = K_{t-1}f(x_{t-1}^{(i)}) \qquad \forall i = 1, \ldots, n;$$

hence

$$\mathbb{E}[p_t^n f \mid \mathcal{F}_t] = \frac{1}{n} \sum_{i=1}^{n} \mathbb{E}\left[f\left(\bar{x}_t^{(i)}\right) \mid \mathcal{F}_{t-1}\right] = \frac{1}{n} \sum_{i=1}^{n} K_{t-1}f\left(x_{t-1}^{(i)}\right)$$
$$= K_{t-1}\pi_{t-1}^n(f) = \pi_{t-1}(K_{t-1}f).$$

10.18 The first assertion follows trivially from part (b) of the iteration. Next observe that, for all $f \in B(\mathbb{R}^d)$,

$$\mathbb{E}\left[\pi_t^n f \mid \bar{\mathcal{F}}_t\right] = \frac{1}{n} \sum_{i=1}^{n} f(\bar{x}_t^{(i)}) \mathbb{E}\left[\xi_t^{(i)} \mid \bar{\mathcal{F}}_t\right] = \bar{\pi}_t^n f,$$

since $\mathbb{E}[\xi_t^{(i)} \mid \bar{\mathcal{F}}_t] = nw_t^{(i)}$ for any $i = 1, \ldots, n$.

10.19

$$\mathbb{E}\left[(\pi_t^n f - \bar{\pi}_t^n f)^2 \mid \bar{\mathcal{F}}_t\right] = \frac{1}{n^2} \sum_{i,j=1}^{n} f\left(\bar{x}_t^{(i)}\right) f\left(\bar{x}_t^{(j)}\right) (A_t^n)_{ij}. \qquad (10.48)$$

By applying (10.34) with $q = (q^{(i)})_{i=1}^{d}$, where $q^{(i)} = f(\bar{x}_t^{(i)})/\|f\|_\infty$, we get that

$$\sum_{i,j=1}^{n} \frac{f\left(\bar{x}_{t}^{(i)}\right)}{\|f\|_{\infty}} (A_{t}^{n})_{ij} \frac{f\left(\bar{x}_{t}^{(j)}\right)}{\|f\|_{\infty}} \leq nc_{t}. \tag{10.49}$$

The exercise now follows from (10.48) and (10.49).

10.23 The multinomial distribution is the empirical distribution of an n-sample from the distribution $\bar{\pi}_{t}^{n}$. Hence, in this case, π_{t}^{n} has the representation

$$\pi_{t}^{n} = \frac{1}{n} \sum_{1}^{n} \delta_{\zeta_{t}^{(i)}},$$

where $\zeta_{t}^{(i)}$ are random variables mutually independent given $\bar{\mathcal{F}}_{t}$ such that $\mathbb{E}[f(\zeta_{t}^{(i)}) \mid \bar{\mathcal{F}}_{t}] = \bar{\pi}_{t}^{n} f$ for any $f \in B(\mathbb{R}^{d})$. Using the independence of the sample $\{\zeta_{t}^{(i)}\}_{i=1}^{n}$ conditional on $\bar{\mathcal{F}}_{t}$,

$$\mathbb{E}[(\pi_{t}^{n} f - \bar{\pi}_{t}^{n} f)^{4} \mid \bar{\mathcal{F}}_{t}] = \mathbb{E}\left[\frac{1}{n}\left(\sum_{i=1}^{n}(f(\zeta_{t}^{(i)}) - \bar{\pi}_{t}^{n} f)\right)^{4} \middle| \bar{\mathcal{F}}_{t}\right]$$

$$= \frac{1}{n^{4}} \sum_{i=1}^{n} \mathbb{E}[(f(\zeta_{t}^{(i)}) - \bar{\pi}_{t}^{n} f)^{4} \mid \bar{\mathcal{F}}_{t}]$$

$$+ \frac{6}{n^{4}} \sum_{1 \leq i < j \leq n} \mathbb{E}[(f(\zeta_{t}^{(i)}) - \bar{\pi}_{t}^{n} f)^{2} (f(\zeta_{t}^{(j)}) - \bar{\pi}_{t}^{n} f)^{2} \mid \bar{\mathcal{F}}_{t}].$$

Observe that since $|\bar{\pi}_{t}^{n} f| \leq \|f\|_{\infty}$,

$$\mathbb{E}[(f(\zeta_{t}^{(i)}) - \bar{\pi}_{t}^{n} f)^{4} \mid \bar{\mathcal{F}}_{t}] \leq 16\|f\|_{\infty}^{4}$$

and

$$\mathbb{E}[(f(\zeta_{t}^{(i)}) - \bar{\pi}_{t}^{n} f)^{2}(f(\zeta_{t}^{(j)}) - \bar{\pi}_{t}^{n} f)^{2} \mid \bar{\mathcal{F}}_{t}] \leq 16\|f\|_{\infty}^{4}.$$

Hence

$$\mathbb{E}[(\pi_{t}^{n} f - \bar{\pi}_{t}^{n} f)^{4} \mid \bar{\mathcal{F}}_{t}] \leq \frac{16\|f\|_{\infty}^{4}}{n^{3}} + \frac{48\|f\|_{\infty}^{4}(n-1)}{n^{3}} \leq \frac{48\|f\|_{\infty}^{4}}{n^{2}}.$$

The bound for the case when the offspring distribution is given by (10.36) is proved in a similar manner.

10.25 π_{0}^{n} is the empirical measure associated with a set of n random particles of mass $1/n$ whose positions $x_{0}^{(i)}$ for $i = 1, \ldots, n$ form a sample of size n from π_{0}. Hence, in particular, $\mathbb{E}[f(x_{0}^{(i)})] = \pi_{0}^{n} f$ for any $f \in B(\mathbb{R}^{d})$ and by a similar argument to that in Exercise 10.23,

$$\mathbb{E}[(\pi_{0}^{n} f - \pi_{0} f)^{4} \mid \bar{\mathcal{F}}_{t}] \leq \frac{48\|f\|_{\infty}^{4}}{n^{2}},$$

which implies the convergence in expectation by Jensen's inequality and the almost sure convergence follows from Remark A.38 in the appendix.

10.26 Immediate from the computation of the first and second moments of the binomial, Poisson and Bernoulli distributions and the fact that, due to the independence of the random variables $\xi_t^{(i)}$, the conditional covariance matrix A_t^n is diagonal,

$$(A_t^n)_{ij} = \mathbb{E}\left[\left(\xi_t^{(i)} - nw_t^{(i)}\right)\left(\xi_t^{(j)} - nw_t^{(j)}\right) \mid \bar{\mathcal{F}}_t\right] = 0, \quad i \neq j,$$

where $w_t \triangleq (w_t^{(i)})_{i=1}^n$ is the vector of weights. Hence

$$q^\top A_t^n q = \sum_{i=1}^n (q^{(i)})^2 \mathbb{E}\left[\left(\xi_t^{(i)} - nw_t^{(i)}\right)^2 \mid \bar{\mathcal{F}}_t\right]$$

for any n-dimensional vector $q = \left(q^{(i)}\right)_{i=1}^n \in [-1,1]^n$.

Part III

Appendices

A
Measure Theory

A.1 Monotone Class Theorem

Let $\mathbb{S}$ be a set. A family $\mathcal{C}$ of subsets of $\mathbb{S}$ is called a π-system if it is closed under finite intersection. That is, for any $A, B \in \mathcal{C}$ we have that $A \cap B \in \mathcal{C}$.

Theorem A.1. *Let $\mathcal{H}$ be a vector space of bounded functions from $\mathbb{S}$ into $\mathbb{R}$ containing the constant function $\mathbf{1}$. Assume that $\mathcal{H}$ has the property that for any sequence $(f_n)_{n \geq 1}$ of non-negative functions in $\mathcal{H}$ such that $f_n \nearrow f$ where f is a bounded function on $\mathbb{S}$, then $f \in \mathcal{H}$. Also assume that $\mathcal{H}$ contains the indicator function of every set in some π-system $\mathcal{C}$. Then $\mathcal{H}$ contains every bounded $\sigma(\mathcal{C})$-measurable function of $\mathbb{S}$.*

For a proof of Theorem A.1 and other related results see Williams [272] or Rogers and Williams [248].

A.2 Conditional Expectation

Let $(\Omega, \mathcal{F}, \mathbb{P})$ be a probability space and $\mathcal{G} \subset \mathcal{F}$ be a sub-σ-algebra of $\mathcal{F}$. The *conditional expectation* of an integrable $\mathcal{F}$-measurable random variable ξ given $\mathcal{G}$ is defined as the integrable $\mathcal{G}$-measurable random variable, denoted by $\mathbb{E}[\xi \mid \mathcal{G}]$, with the property that

$$\int_A \xi \, d\mathbb{P} = \int_A \mathbb{E}[\xi \mid \mathcal{G}] \, d\mathbb{P}, \quad \text{for all } A \in \mathcal{G}. \tag{A.1}$$

Then $\mathbb{E}[\xi \mid \mathcal{G}]$ exists and is almost surely unique (for a proof of this result see for example Williams [272]). By this we mean that if $\bar{\xi}$ is another $\mathcal{G}$-measurable integrable random variable such that

$$\int_A \bar{\xi} \, d\mathbb{P} = \int_A \mathbb{E}[\xi \mid \mathcal{G}] \, d\mathbb{P}, \quad \text{for all } A \in \mathcal{G},$$

then $\mathbb{E}[\xi \mid \mathcal{G}] = \bar{\xi}$, $\mathbb{P}$-a.s.

The following are some of the important properties of the conditional expectation which are used throughout the text.

a. If $\alpha_1, \alpha_2 \in \mathbb{R}$ and ξ_1, ξ_2 are $\mathcal{F}$-measurable, then

$$\mathbb{E}[\alpha_1 \xi_1 + \alpha_2 \xi_2 \mid \mathcal{G}] = \alpha_1 \mathbb{E}[\xi_1 \mid \mathcal{G}] + \alpha_2 \mathbb{E}[\xi_2 \mid \mathcal{G}], \qquad \mathbb{P}\text{-a.s.}$$

b. If $\xi \geq 0$, then $\mathbb{E}[\xi \mid \mathcal{G}] \geq 0$, $\mathbb{P}$-a.s.
c. If $0 \leq \xi_n \nearrow \xi$, then $\mathbb{E}[\xi_n \mid \mathcal{G}] \nearrow \mathbb{E}[\xi \mid \mathcal{G}]$, $\mathbb{P}$-a.s.
d. If $\mathcal{H}$ is a sub-σ-algebra of $\mathcal{G}$, then $\mathbb{E}\left[\mathbb{E}[\xi \mid \mathcal{G}] \mid \mathcal{H}\right] = \mathbb{E}[\xi \mid \mathcal{H}]$, $\mathbb{P}$-a.s.
e. If ξ is $\mathcal{G}$-measurable, then $\mathbb{E}[\xi \eta \mid \mathcal{G}] = \xi \mathbb{E}[\eta \mid \mathcal{G}]$, $\mathbb{P}$-a.s.
f. If $\mathcal{H}$ is independent of $\sigma(\sigma(\xi), \mathcal{G})$, then

$$\mathbb{E}[\xi \mid \sigma(\mathcal{G}, \mathcal{H})] = \mathbb{E}[\xi \mid \mathcal{G}], \qquad \mathbb{P}\text{-a.s.}$$

The *conditional probability* of a set $A \in \mathcal{F}$ with respect to the σ-algebra $\mathcal{G}$ is the random variable denoted by $\mathbb{P}(A \mid \mathcal{G})$ defined as $\mathbb{P}(A \mid \mathcal{G}) \triangleq \mathbb{E}[I_A \mid \mathcal{G}]$, where I_A is the indicator function of the set A. From (A.1),

$$\mathbb{P}(A \cap B) = \int_B \mathbb{P}(A \mid \mathcal{G}) \, \mathrm{d}\mathbb{P}, \qquad \text{for all } B \in \mathcal{G}. \tag{A.2}$$

This definition of conditional probability has the shortcoming that the conditional probability $\mathbb{P}(A \mid \mathcal{G})$ is only defined outside of a null set which depends upon the set A. As there may be an uncountable number of possible choices for A, $\mathbb{P}(\cdot \mid \mathcal{G})$ may not be a probability measure.

Under certain conditions regular conditional probabilities as in Definition 2.28 exist. Regular conditional distributions (following the nomenclature in Breiman [23] whose proof we follow) exist under much less restrictive conditions.

Definition A.2. *Let $(\Omega, \mathcal{F}, \mathbb{P})$ be a probability space, $(E, \mathcal{E})$ be a measurable space, $X : \Omega \to E$ be an $\mathcal{F}/\mathcal{E}$-measurable random element and $\mathcal{G}$ a sub-σ-algebra of $\mathcal{F}$. A function $\mathbb{Q}(\omega, B)$ defined for all $\omega \in \Omega$ and $B \in \mathcal{E}$ is called a regular conditional distribution of X with respect to $\mathcal{G}$ if*

(a) For each $B \in \mathcal{E}$, the map $\mathbb{Q}(\cdot, B)$ is $\mathcal{G}$-measurable.
(b) For each $\omega \in \Omega$, $\mathbb{Q}(\omega, \cdot)$ is a probability measure on $(E, \mathcal{E})$.
(c) For any $B \in \mathcal{E}$,

$$\mathbb{Q}(\cdot, B) = \mathbb{P}(X \in B \mid \mathcal{G}) \qquad \mathbb{P}\text{-a.s.} \tag{A.3}$$

Theorem A.3. *If the space $(E, \mathcal{E})$ in which X takes values is a Borel space, that is, if there exists a function $\varphi : E \to \mathbb{R}$ such that φ is $\mathcal{E}$-measurable and φ^{-1} is $\mathcal{B}(\mathbb{R})$-measurable, then the regular conditional distribution of the variable X conditional upon $\mathcal{G}$ in the sense of Definition A.2 exists.*

Proof. Consider the case when $(E, \mathcal{E}) = (\mathbb{R}, \mathcal{B}(\mathbb{R}))$. First we construct a regular version of the distribution function $\mathbb{P}(X < x \mid \mathcal{G})$. Define a countable family of random variables by selecting versions $Q_q(\omega) = \mathbb{P}(X < q \mid \mathcal{G})(\omega)$. For each $q \in \mathbb{Q}$, define for $r, q \in \mathbb{Q}$,

$$M_{r,q} \triangleq \{\omega : Q_r < Q_q\}$$

and then define the set on which monotonicity of the distribution function fails

$$M \triangleq \bigcup_{\substack{r > q \\ r, q \in \mathbb{Q}}} M_{r,q}.$$

It is clear from property (b) of the conditional expectation that $\mathbb{P}(M) = 0$. Similarly define for $q \in \mathbb{Q}$,

$$N_q \triangleq \left\{ \omega : \lim_{r \uparrow q} Q_r \neq Q_q \right\}$$

and

$$N \triangleq \bigcup_{q \in \mathbb{Q}} N_q;$$

by property (c) of conditional expectation it follows that $\mathbb{P}(N_q) = 0$, so $\mathbb{P}(N) = 0$. Finally define

$$L_\infty \triangleq \left\{ \omega : \lim_{\substack{q \to \infty \\ q \in \mathbb{Q}}} Q_q \neq 1 \right\} \quad \text{and} \quad L_{-\infty} \triangleq \left\{ \omega : \lim_{\substack{q \to -\infty \\ q \in \mathbb{Q}}} Q_q \neq 0 \right\},$$

and again $\mathbb{P}(L_\infty) = \mathbb{P}(L_{-\infty}) = 0$.

Define

$$F(x \mid \mathcal{G}) \triangleq \begin{cases} \lim_{\substack{r \uparrow x \\ r \in \mathbb{Q}}} Q_r & \text{if } \omega \notin M \cup N \cup L_\infty \cup L_{-\infty} \\ \Phi(x) & \text{otherwise,} \end{cases}$$

where $\Phi(x)$ is the distribution function of the normal $N(0, 1)$ distribution (its choice is arbitrary). It follows using property (c) of conditional expectation applied to the functions $f_{r_i} = 1_{(-\infty, r_i)}$ with $r_i \in \mathbb{Q}$ a sequence such that $r_i \uparrow x$ that $F(x \mid \mathcal{G})$ satisfies all the properties of a distribution function and is a version of $\mathbb{P}(X < x \mid \mathcal{G})$.

This distribution function can be extended to define a measure $\mathbb{Q}(\cdot \mid \mathcal{G})$. Let $\mathcal{H}$ be the class of $B \in \mathcal{B}(\mathbb{R})$ such that $\mathbb{Q}(B \mid \mathcal{G})$ is a version of $\mathbb{P}(X \in B \mid \mathcal{G})$. It is clear that $\mathcal{H}$ contains all finite disjoint unions of intervals of the form $[a, b)$ for $a, b \in \mathbb{R}$ so by the monotone class theorem A.1 the result follows.

In the general case, $Y = \varphi(X)$ is a real-valued random variable and so has regular conditional distribution such that for $B \in \mathcal{B}(\mathbb{R})$, $\hat{\mathbb{Q}}(B \mid \mathcal{G}) = \mathbb{P}(Y \in B \mid \mathcal{G})$; thus define

$$\mathbb{Q}(B \mid \mathcal{G}) \triangleq \hat{\mathbb{Q}}(\varphi(B) \mid \mathcal{G}),$$

and since φ^{-1} is measurable it follows that $\mathbb{Q}$ has the required properties. $\square$

Lemma A.4. *If X is as in the statement of Theorem A.3 and ψ is a $\mathcal{E}$-measurable function such that $\mathbb{E}[|\psi(X)|] < \infty$ then if $\mathbb{Q}(\cdot \mid \mathcal{G})$ is a regular conditional distribution for X given $\mathcal{G}$ it follows that*

$$\mathbb{E}[\psi(X) \mid \mathcal{G}] = \int_E \psi(x) \mathbb{Q}(\mathrm{d}x \mid \mathcal{G}).$$

Proof. If $A \in \mathcal{B}$ then it is clear that the result follows from (A.3). By linearity this extends to simple functions, by monotone convergence to non-negative functions, and in general write $\psi = \psi^+ - \psi^-$. □

A.3 Topological Results

Definition A.5. *A metric space (E, d) is said to be separable if it has a countable dense set. That is, for any $x \in E$, given $\varepsilon > 0$ we can find y in this countable set such that $d(x, y) < \varepsilon$.*

Lemma A.6. *Let (X, ρ) be a separable metric space. Then X is homeomorphic to a subspace of $[0, 1]^\mathbb{N}$, the space of sequences of real numbers in $[0, 1]$ with the topology of co-ordinatewise convergence.*

Proof. Define a bounded version of the metric $\hat{\rho} \triangleq \rho/(1 + \rho)$; it is easily checked that this is a metric on X, and the space $(X, \hat{\rho})$ is also separable. Clearly the metric satisfies the bounds $0 \leq \hat{\rho} \leq 1$. As a consequence of separability we can choose a countable set $x_1, x_2, \ldots$ which is dense in $(X, \hat{\rho})$.

Define $J = [0, 1]^\mathbb{N}$ and endow this space with the metric d which generated the topology of co-ordinatewise convergence. Define $\alpha : X \to J$,

$$\alpha : x \mapsto (\hat{\rho}(x, x_1), \hat{\rho}(x, x_2), \ldots).$$

Suppose $x^{(n)} \to x$ in X; then by continuity of $\hat{\rho}$ it is immediate that $\hat{\rho}(x^{(n)}, x_k) \to \hat{\rho}(x, x_k)$ for each $k \in \mathbb{N}$ and thus $\alpha(x^{(n)}) \to \alpha(x)$.

Conversely if $\alpha(x^{(n)}) \to \alpha(x)$ then this implies that $\hat{\rho}(x^{(n)}, x_k) \to \hat{\rho}(x, x_k)$ for each k. Then by the triangle inequality

$$\hat{\rho}(x^{(n)}, x) \leq \hat{\rho}(x^{(n)}, x_k) + \hat{\rho}(x_k, x)$$

and since $\hat{\rho}(x^{(n)}, x_k) \to \hat{\rho}(x, x_k)$ it is immediate that

$$\limsup_{n \to \infty} \hat{\rho}(x^{(n)}, x) \leq 2\hat{\rho}(x_k, x) \quad \forall k.$$

As this holds for all $k \in \mathbb{N}$ and the x_ks are dense in X we may pick a sequence $x_{m_k} \to x$ whence $\hat{\rho}(x^{(n)}, x) \to 0$ as $n \to \infty$. Hence α is a homeomorphism $X \to J$. □

The following is a standard result and the proof is based on that in Rogers and Williams [248] who reference Bourbaki [22] Chapter IX, Section 6, No 1.

A.3 Topological Results 297

Theorem A.7. *A complete separable metric space X is homeomorphic to a Borel subset of a compact metric space.*

Proof. By Lemma A.6 there is a homeomorphism $\alpha : X \to J$. Let d denote the metric giving the topology of co-ordinatewise convergence on J. We must now consider $\alpha(X)$ and show that it is a countable intersection of open sets in J and hence belongs to the σ-algebra of open sets, the Borel σ-algebra.

For $\varepsilon > 0$ and $x \in X$ we can find $\delta(\varepsilon)$ such that for any $y \in X$, $d(\alpha(x), \alpha(y)) < \delta$ implies that $\hat{\rho}(x,y) < \varepsilon$. For $n \in \mathbb{N}$ set $\varepsilon = 1/(2n)$ and then consider the ball $B(\alpha(x), \delta(\varepsilon) \wedge \varepsilon)$. It is immediate that the d-diameter of this ball is at most $1/n$. But also, as a consequence of the choice of δ, the image under α^{-1} of the intersection of this ball with X has $\hat{\rho}$-diameter at most $1/n$.

Let $\overline{\alpha(X)}$ be the closure of $\alpha(X)$ under the metric d in J. Define a set $U_n \subseteq \overline{\alpha(X)}$ to be the set of $x \in \overline{\alpha(X)}$ such that there exists an open ball $N_{x,n}$ about x of d-diameter less than $1/n$, with $\hat{\rho}$-diameter of the image under α^{-1} of the intersection of $\alpha(X)$ and this ball less than $1/n$. By the argument of the previous paragraph we see that if $x \in \alpha(X)$ we can always find such a ball; hence $\alpha(X) \subseteq U_n$.

For $x \in \bigcap_n U_n$ choose $x_n \in \alpha(X) \cap \bigcap_{k \leq n} N_{x,k}$. By construction $d(x, x_k) \leq 1/n$, thus $x_n \to x$ as $n \to \infty$ under the d metric on J. However, for $r \geq n$ both points x_r and x_n are in $N_{x,n}$ thus $\hat{\rho}(\alpha^{-1}(x_r), \alpha^{-1}(x_n)) \leq 1/n$, so $(\alpha^{-1}(x_r))_{r \geq 1}$ is a Cauchy sequence in $(X, \hat{\rho})$. But this space is complete so there exists $y \in X$ such that $\alpha^{-1}(x_n) \to y$. As α is a homeomorphism this implies that $d(x_n, \alpha(y)) \to 0$. Hence by uniqueness of limits $x = \alpha(y)$ and thus it is immediate that $x \in \alpha(X)$. Therefore $\bigcap_n U_n \subseteq \alpha(X)$; since $\alpha(X) \subseteq U_n$ it follows immediately that

$$\alpha(X) = \bigcap_n U_n. \tag{A.4}$$

It is now necessary to show that U_n is relatively open in $\overline{\alpha(X)}$. From the definition of U_n, for any $x \in U_n$ we can find $N_{x,n}$ with diameter properties as above which is a subset of J containing x. For any arbitrary $z \in \overline{\alpha(X)}$, by (A.4) there exists $x \in U_n$ such that $z \in N_{x,n}$; then by choosing $N_{z,n} = N_{x,n}$ it is clear that $z \in U_n$. Therefore $N_{x,n} \cap \overline{\alpha(X)} \subseteq U_n$ from which we conclude that U_n is relatively open in $\overline{\alpha(X)}$. Therefore we can write $U_n = \overline{\alpha(X)} \cap V_n$ where V_n is open in J

$$\alpha(X) = \bigcap_n U_n = \overline{\alpha(X)} \cap \left(\bigcap_n V_n \right), \tag{A.5}$$

where V_n are open subsets of J. It only remains to show that $\overline{\alpha(X)}$ can be expressed as a countable intersection of open sets; this is easily done since

$$\overline{\alpha(X)} = \bigcap_n \{x \in J : d(x, \alpha(X)) < 1/n\},$$

therefore it follows that $\overline{\alpha(X)}$ is a countable intersection of open sets in J. Together with (A.5) it follows that $\alpha(X)$ is a countable intersection of open sets. □

Theorem A.8. *Any compact metric space X is separable.*

Proof. Consider the open cover of X which is the uncountable union of all balls of radius $1/n$ centred on each point in X. As X is compact there exists a finite subcover. Let $x_1^n, \ldots, x_{N_n}^n$ be the centres of the balls in one such finite subcover. By a diagonal argument we can construct a countable set which is the union of all these centres for all $n \in \mathbb{N}$. This set is clearly dense in X and countable, so X is separable. □

Theorem A.9. *If E is a compact metric space then the set of continuous real-valued functions defined on E is separable.*

Proof. By Theorem A.8, the space E is separable. Let $x_1, x_2, \ldots$ be a countable dense subset of E. Define $h_0(x) = 1$, and $h_n(x) = d(x, x_n)$, for $n \geq 1$. Now define an algebra of polynomials in these h_ns with coefficients in the rationals

$$A = \left\{ x \mapsto \sum q_{k_0,\ldots,k_r}^{n_0,\ldots,n_r} h_{k_0}^{n_0}(x) \ldots h_{k_r}^{n_r}(x) : q_{k_0,\ldots,k_r}^{n_0,\ldots,n_r} \in \mathbb{Q} \right\}.$$

The closure of A is an algebra containing constant functions and it is clear that it separates points in E, therefore by the Stone–Weierstrass theorem, it follows that A is dense in $C(E)$. □

Corollary A.10. *If E is a compact metric space then there exists a countable set $f_1, f_2, \ldots$ which is dense in $C(E)$.*

Proof. By Theorem A.8 E is separable, so by Theorem A.9 the space $C(E)$ is separable and hence has a dense countable subset. □

A.4 Tulcea's Theorem

Tulcea's theorem (see Tulcea [265]) is frequently stated in the form for product spaces and their σ-algebras (for a very elegant proof in this vein see Ethier and Kurtz [95, Appendix 9]) and this form is sufficient to establish the existence of stochastic processes. We give the theorem in a more general form where the measures are defined on the same space X, but defined on an increasing family of σ-algebras $\mathcal{B}_n$ as this makes the important condition on the atoms of the σ-algebras clear. The approach taken here is based on that in Stroock and Varadhan [261].

Define the *atom* $A(x)$ of the Borel σ-algebra $\mathcal{B}$ on the space X, for $x \in X$ by

$$A(x) \triangleq \bigcap \{B : B \in \mathcal{B},\ x \in B\}, \qquad (A.6)$$

that is, $A(x)$ is the smallest element of $\mathcal{B}$ which contains x.

A.4 Tulcea's Theorem

Theorem A.11. *Let $(X, \mathcal{B})$ be a measurable space and let $\mathcal{B}_n$ be an increasing family of sub-σ-algebras of $\mathcal{B}$ such that $\mathcal{B} = \sigma(\bigcup_{n=1}^{\infty} \mathcal{B}_n)$. Suppose that these σ-algebras satisfy the following constraint. If A_n is a sequence of atoms such that $A_n \in \mathcal{B}_n$ and $A_1 \supseteq A_2 \supseteq \cdots$ then $\bigcap_{n=0}^{\infty} A_n \neq \emptyset$.*

Let $\mathbb{P}_0$ be a probability measure defined on $\mathcal{B}_0$ and let π_n be a family of probability kernels, where $\pi_n(x, \cdot)$ is a measure on $(X, \mathcal{B}_n)$ and the mapping $x \mapsto \pi_n(x, \cdot)$ is $\mathcal{B}_{n-1}$-measurable. Such a probability kernel allows us to define inductively a family of probability measures on $(X, \mathcal{B}_n)$ via

$$\mathbb{P}_n(A) \triangleq \int_X \pi_n(x, A) \mathbb{P}_{n-1}(dx), \tag{A.7}$$

with the starting point for the induction being given by the probability measure $\mathbb{P}_0$.

Suppose that the kernels $\pi_n(x, \cdot)$ satisfy the compatibility condition that for $x \notin N_n$, where N_n is a $\mathbb{P}_n$-null set, the kernel $\pi_{n+1}(x, \cdot)$ is supported on $A_n(x)$ (i.e. if $B \in \mathcal{B}_{n+1}$ and $B \cap A_n(x) = \emptyset$ then $\pi_{n+1}(x, B) = 0$). That is, starting from a point x, the transition measure only contains with positive probability transitions to points y such that x and y belong to the same atom of $\mathcal{B}_n$.

Then there exists a unique probability measure $\mathbb{P}$ defined on $\mathcal{B}$ such that $\mathbb{P}|_{\mathcal{B}_n} = \mathbb{P}_n$ for all $n \in \mathbb{N}$.

Proof. It is elementary to see that $\mathbb{P}_n$ as defined in (A.7) is a probability measure on $\mathcal{B}_n$ and that $\mathbb{P}_{n+1}$ agrees with $\mathbb{P}_n$ on $\mathcal{B}_n$. We can then define a set function $\mathbb{P}$ on $\bigcup \mathcal{B}_n$ by setting $\mathbb{P}(B_n) = \mathbb{P}_n(B_n)$ for $B_n \in \mathcal{B}_n$.

From the definition (A.7), for $B \in \mathcal{B}_n$ we have defined $\mathbb{P}_n$ inductively via the transition functions

$$\mathbb{P}_n(B_n) = \int_X \cdots \int_X \pi_n(q_{n-1}, B) \pi_{n-1}(q_{n-2}, dq_{n-1}) \cdots \pi_1(q_0, dq_1) \mathbb{P}_0(dq_0). \tag{A.8}$$

To simplify the notation define $\pi^{m,n}$ such that $\pi^{m,n}(x, \cdot)$ is a measure on $\mathcal{M}(X, \mathcal{B}_n)$ as follows.

If $m \geq n \geq 0$ and $B \in \mathcal{B}_n$, then define $\pi^{m,n}(x, B) = 1_B(x)$ which is clearly $\mathcal{B}_n$-measurable and hence as $\mathcal{B}_m \supseteq \mathcal{B}_n$, $x \mapsto \pi^{m,n}(x, B)$ is also $\mathcal{B}_m$-measurable. If $m < n$ define $\pi^{m,n}$ inductively using the transition kernel π_n,

$$\pi^{m,n}(x, B) \triangleq \int_X \pi_n(y_{n-1}, B) \pi^{m,n-1}(x, dy_{n-1}). \tag{A.9}$$

It is clear that in both cases $x \mapsto \pi^{m,n}(x, \cdot)$ is $\mathcal{B}_m$-measurable. Thus $\pi^{m,n}$ can be viewed as a transition kernel from $(X, \mathcal{B}_m)$ to $(X, \mathcal{B}_n)$. From these definitions, for $m < n$

$$\pi^{m,n}(x, B) = \int_X \cdots \int_X \pi_n(y_{n-1}, B) \cdots \pi_{m+1}(y_m, dy_{m+1}) \pi^{m,m}(x, dy_m)$$

$$= \int_X \cdots \int_X \pi_n(y_{n-1}, B) \cdots \pi_{m+2}(y_{m+1}, dy_{m+2}) \pi_{m+1}(x, dy_{m+1}).$$

It therefore follows from the above with $m = 0$ and (A.8) that for $B \in \mathcal{B}_n$,

$$\mathbb{P}(B_n) = \mathbb{P}_n(B_n) = \int_X \pi^{0,n}(y_0, B)\mathbb{P}_0(dy_0). \qquad (A.10)$$

We must show that $\mathbb{P}$ is a probability measure on $\bigcup_{n=0}^{\infty} \mathcal{B}_n$, as then the Carathéodory extension theorem[†] establishes the existence of an extension to a probability measure on $(X, \sigma(\bigcup_{n=0}^{\infty} \mathcal{B}))$. The only non-trivial condition which must be verified for $\mathbb{P}$ to be a measure is countable additivity.

A necessary and sufficient condition for countable additivity of $\mathbb{P}$ is that if $B_n \in \bigcup_n \mathcal{B}_n$, are such that $B_1 \supseteq B_2 \supseteq \cdots$ and $\bigcap_n B_n = \emptyset$ then $\mathbb{P}(B_n) \to 0$ as $n \to \infty$ (the proof can be found in many books on measure theory, see for example page 200 of Williams [272]). It is clear that the non-trivial cases are covered by considering $B_n \in \mathcal{B}_n$ for each $n \in \mathbb{N}$.

We argue by contradiction; suppose that $\mathbb{P}(B_n) \geq \varepsilon > 0$ for all $n \in \mathbb{N}$. We must exhibit a point of $\bigcap_n B_n$; as we started with the assumption that this intersection was empty, this is the desired contradiction.

Define
$$F_n^0 \triangleq \{x \in X : \pi^{0,n}(x, B_n) \geq \varepsilon/2\}. \qquad (A.11)$$

Since $x \mapsto \pi^{0,n}(x, \cdot)$ is $\mathcal{B}_0$-measurable, it follows that $F_0^n \in \mathcal{B}_0$. Then from (A.10) it is clear that
$$\mathbb{P}(B_n) \leq \mathbb{P}_0(F_n^0) + \varepsilon/2.$$

As by assumption $\mathbb{P}(B_n) \geq \varepsilon$ for all $n \in \mathbb{N}$, we conclude that $\mathbb{P}_0(F_n^0) \geq \varepsilon/2$ for all $n \in \mathbb{N}$.

Suppose that $x \in F_{n+1}^0$; then $\pi^{0,n+1}(x, B_{n+1}) \geq \varepsilon/2$. But $B_{n+1} \subseteq B_n$, so $\pi^{0,n+1}(x, B_n) \geq \varepsilon/2$. From (A.9) it follows that

$$\pi^{0,n+1}(x, B_n) = \int_X \pi_{n+1}(y_n, B_n)\pi^{0,n}(x, dy_n),$$

for $y \notin N_n$, the probability measure $\pi_{n+1}(y, \cdot)$ is supported on $A_n(y)$. As $B_n \in \mathcal{B}_n$, from the definition of an atom, it follows that $y \in B_n$ if and only if $A_n(y) \subseteq B_n$, thus $\pi_{n+1}(y, B_n) = 1_{B_n}(y)$ for $y \notin N_n$. So on integration we obtain that $\pi^{0,n}(x, B_n) = \pi^{0,n+1}(x, B_n) \geq \varepsilon/2$. Thus $x \in F_0^n$. So we have shown that $F_0^{n+1} \subseteq F_0^n$.

Since $\mathbb{P}_0(F_0^n) \geq \varepsilon/2$ for all n and the F_n form a non-increasing sequence, it is then immediate that $\mathbb{P}_0(\bigcap_{n=0}^{\infty} F_0^n) \geq \varepsilon/2$, whence we can find $x_0 \notin N_0$ such that $\pi^{0,n}(x_0, B_n) \geq \varepsilon/2$ for all $n \in \mathbb{N}$.

Now we proceed inductively; suppose that we have found $x_0, x_1, \ldots x_{m-1}$ such that $x_0 \notin N_0$ and $x_i \in A_{i-1}(x_{i-1}) \setminus N_i$ for $i = 1, \ldots, m-1$, and

---

[†] Carathéodory extension theorem: Let $\mathbb{S}$ be a set, $\mathcal{S}_0$ be an algebra of subsets of $\mathbb{S}$ and $\mathcal{S} = \sigma(\mathcal{S}_0)$. Let μ_0 be a countably additive map $\mu_0 : \mathcal{S}_0 \to [0, \infty]$; then there exists a measure μ on $(\mathbb{S}, \mathcal{S})$ such that $\mu = \mu_0$ on $\mathcal{S}_0$. Furthermore if $\mu_0(\mathbb{S}) < \infty$, then this extension is unique. For a proof of the theorem see, for example, Williams [272] or Rogers and Williams [248].

$\pi^{i,n}(x_i, B_n) \geq \varepsilon/2^{i+1}$ for all $n \in \mathbb{N}$ for each $i = 0, \ldots, m-1$. We have already established the result for the case $m = 0$. Now define

$$F_n^m \triangleq \{x \in X : \pi^{m,n}(x, B_n) \geq \varepsilon/2^{m+1}\};$$

from the integral representation for $\pi^{m,n}$,

$$\pi^{m-1,n}(x, B_n) = \int_X \pi^{m,n}(y_m, B_n) \pi_m(x, \mathrm{d}y_m),$$

it follows by an argument analogous to that for F_n^0, that

$$\varepsilon/2^m \leq \pi^{m-1,n}(x_{m-1}, B_n) \leq \varepsilon/2^{m+1} + \pi_m(x_{m-1}, F_n^m),$$

where the inequality on the left hand side follows from the inductive hypothesis. As in the case for $m = 0$, we can deduce that $F_{n+1}^m \subseteq F_n^m$. Thus

$$\pi_m\left(x_{m-1}, \bigcap_{n=0}^{\infty} F_n^m\right) \geq \varepsilon/2^{m+1}, \qquad (A.12)$$

which implies that we can choose $x_m \in \bigcap_{n=0}^{\infty} F_n^m$, such that $\pi^{m,n}(x_m, B_n) > \varepsilon/2^{m+1}$ for all $n \in \mathbb{N}$, and from (A.12) as the set of suitable x_m has strictly positive probability, it cannot be empty, and we can choose an x_m not in the $\mathbb{P}_m$-null set N_m. Therefore, this choice can be made such that $x_m \in A_{m-1}(x_{m-1}) \setminus N_m$. This establishes the required inductive step.

Now consider the case of $\pi^{n,n}(x_n, B_n)$; we see from the definition that this is just $1_{B_n}(x_n)$, but by choice of the x_ns, $\pi^{n,n}(x_n, B_n) > 0$. Consequently as $x_n \notin N_n$, by the support property of the transition kernels, it follows that $A_n(x_n) \subseteq B_n$ for each n. Thus $\bigcap A_n(x_n) \subset \bigcap B_n$ and if we define $K_n \triangleq \bigcap_{i=0}^n A_i(x_i)$ it follows that $x_n \in K_n$ and K_n is a descending sequence; by the σ-algebra property it is clear that $K_n \in \mathcal{B}_n$, and since $A_n(x_n)$ is an atom in $\mathcal{B}_n$ it follows that $K_n = A_n(x_n)$. We thus have a decreasing sequence of atoms; by the initial assumption, such an intersection is non-empty, that is, $\bigcap A_n(x_n) \neq \emptyset$ which implies that $\bigcap B_n \neq \emptyset$, but this is a contradiction, since we assumed that this intersection was empty. Therefore $\mathbb{P}$ is countably additive and the existence of an extension follows from the theorem of Carathéodory. $\square$

A.4.1 The Daniell–Kolmogorov–Tulcea Theorem

The Daniell–Kolmogorov–Tulcea theorem gives conditions under which the law of a stochastic process can be extended from its finite-dimensional distributions to its full (infinite-dimensional) law.

The original form of this result due to Daniell and Kolmogorov (see Doob [81] or Rogers and Williams [248, section II.30]) requires topological conditions on the space X; the space X needs to be Borel, that is, homeomorphic to a

Borel set in some space, which is the case if X is a complete separable metric space as a consequence of Theorem A.7.

It is possible to take an alternative probabilistic approach using Tulcea's theorem. In this approach the finite-dimensional distributions are related to each other through the use of regular conditional probabilities as transition kernels; while this does not explicitly use topological conditions, such conditions may be required to establish the existence of these regular conditional probabilities (as was seen in Exercise 2.29 regular conditional probabilities are guaranteed to exist if X is a complete separable metric space).

We use the notation X^I for the I-fold product space generated by X, that is, $X^I = \prod_{i \in I} X_i$ where X_is are identical copies of X, and let $\mathcal{B}^I$ denote the product σ-algebra on X^I; that is, $\mathcal{B}^I = \prod_{i \in I} \mathcal{B}_i$ where $\mathcal{B}_i$ are copies of $\mathcal{B}$. If U and V are finite subsets of the index set I, let π_U^V denote the restriction map from X^V to X^U.

Theorem A.12. *Let X be a complete separable metric space. Let μ_U be a family of probability measures on $(X^U, \mathcal{B}^U)$, for U any finite subset of I. Suppose that these measures satisfy the compatibility condition for $U \subset V$*

$$\mu_U = \mu_V \circ \pi_U^V.$$

Then there exists a unique probability measure on $(X^I, \mathcal{B}^I)$ such that $\mu_U = \mu \circ \pi_U^I$ for any U a finite subset of I.

Proof. Let $\mathrm{Fin}(I)$ denote the set of all finite subsets of I. It is immediate from the compatibility condition that we can find a finitely additive μ_0 which is a probability measure on $(X^I, \bigcup_{F \in \mathrm{Fin}(I)} (\pi_F^I)^{-1}(\mathcal{B}_F))$, such that for $U \in \mathrm{Fin}(I)$, $\mu_U = (\pi_U^I)^{-1} \circ \mu_0$. If we can show that μ_0 is countably additive, then the Carathéodory extension theorem implies that μ_0 can be extended to a measure μ on $(X^I, \mathcal{B}^I)$.

We cannot directly use Tulcea's theorem to construct the extension measure; however we can use it to show that μ_0 is countably additive. Suppose A_n is a non-increasing family of sets $A_n \in \bigcup_{F \in \mathrm{Fin}(I)} (\pi_F^I)^{-1}(\mathcal{B}_F)$ such that $A_n \downarrow \emptyset$; we must show that $\mu_0(A_n) \to 0$.

Given the A_is, we can find finite subsets F_i of I such that $A_i \in (\pi_{F_i}^I)^{-1} \mathcal{B}_{F_i}$ for each i. Without loss of generality we can choose this sequence so that $F_0 \subset F_1 \subset F_2 \subset \cdots$. Define $\mathcal{F}_n \triangleq (\pi_{F_n}^I)^{-1}(\mathcal{B}_{F_n}) \subset \mathcal{B}_I$. As a consequence of the product space structure, these σ-algebras satisfy the condition that the intersection of a decreasing family of atoms $Z_n \in \mathcal{F}_n$ is non-empty.

For $q \in X^I$ and $B \in \mathcal{F}_n$, let

$$\pi_n(q, B) \triangleq \left(\mu_{F_n} \left| \left(\pi_{F_{n-1}}^{F_n} \right)^{-1} (\mathcal{B}_{F_{n-1}}) \right) \left(\pi_{F_n}^I(q), \left(\pi_{F_n}^I \right)^{-1}(B) \right) \right.,$$

where $(\mu_{F_n} \mid \mathcal{G})(\omega, \cdot)$ for $\mathcal{G} \subset \mathcal{B}_{F_n}$ is the regular conditional probability distribution of μ_{F_n} given $\mathcal{G}$. We refer to the properties of regular conditional probability distribution using the nomenclature of Definition 2.28. This π_n is a probability kernel from $(X^I, \mathcal{F}_{n-1})$ to $(X^I, \mathcal{F}_n)$, i.e. $\pi^n(q, \cdot)$ is a measure on $(X^I, \mathcal{F}_n)$ and the map $q \mapsto \pi_n(q, \cdot)$ is $\mathcal{F}_{n-1}$-measurable (which follows from property (b) of regular conditional distribution). In order to apply Tulcea's theorem we must verify that the compatibility condition is satisfied i.e. $\pi_n(q, \cdot)$ is supported for a.e. q on the atom in $\mathcal{F}_{n-1}$ containing q which is denoted $A_{n-1}(q)$. This is readily established by computing $\pi(q, (A_{n-1}(q))^c)$ and using property (c) of regular conditional distribution and the fact that $q \notin (A_{n-1}(q))^c$. Thus we can apply Tulcea's theorem to find a unique probability measure μ on $(X^I, \sigma(\bigcup_{n=0}^{\infty} \mathcal{F}_n))$ such that μ is equal to μ_0 on $\bigcup_{n=0}^{\infty} \mathcal{F}_n$. Hence as $A_n \in \mathcal{F}_n$, it follows that $\mu(A_n) = \mu_0(A_n)$ for each n and therefore since μ is countably additive $\mu_0(A_n) \downarrow 0$ which establishes the required countable additivity of μ_0. $\square$

A.5 Càdlàg Paths

A càdlàg (*continue à droite, limite à gauche*) path is one which is right continuous with left limits; that is, x_t has càdlàg paths if for all $t \in [0, \infty)$, the limit x_{t-} exists and $x_t = x_{t+}$. Such paths are sometimes described as RCLL (right continuous with left limits). The space of càdlàg functions from $[0, \infty)$ to E is conventionally denoted $D_E[0, \infty)$.

Useful references for this material are Billingsley [19, Chapter 3], Ethier and Kurtz [95, Sections 3.5–3.9], and Whitt [269, Chapter 12].

A.5.1 Discontinuities of Càdlàg Paths

Clearly càdlàg paths can only have left discontinuities, i.e. points t where $x_t \neq x_{t-}$.

Lemma A.13. *For any $\varepsilon > 0$, a càdlàg path taking values in a metric space (E, d) has at most a finite number of discontinuities of size in the metric d greater than ε; that is, the set*

$$D = \{t \in [0, T] : d(x_t, x_{t-}) > \varepsilon\}$$

contains at most a finite number of points.

Proof. Let τ be the supremum of $t \in [0, T]$ such that $[0, t)$ can be finitely subdivided $0 < t_0 < t_1 < \cdots < t_k = t$ with the subdivision having the property that for $i = 0, \ldots, k-1$, $\sup_{s, r \in [t_i, t_{i+1})} d(x_s, x_r) < \varepsilon$. As right limits exist at 0 it is clear that $\tau > 0$ and since a left limit exists at $\tau-$ it is clear that the interval $[0, \tau)$ can be thus subdivided. Right continuity implies that there exists $\delta > 0$ such that for $0 \leq t' - t < \delta$, then $d(x_{t'}, x_t) < \varepsilon$; consequently the result holds for $[0, t')$, which contradicts the fact that τ is the supremum unless $\tau = T$, consequently $\tau = T$. Therefore $[0, T)$ can be so subdivided:

jumps of size greater than ε can only occur at the t_is, of which there are a finite number and thus there must be at most a finite number of such jumps. □

Lemma A.14. *Let X be a càdlàg stochastic process taking values in a metric space (E, d); then*

$$\{t \in [0, \infty) : \mathbb{P}(X_{t-} \neq X_t) > 0\}$$

contains at most countably many points.

Proof. For $\varepsilon > 0$, define

$$J_t(\varepsilon) \triangleq \{\omega : d(X_t(\omega), X_{t-}(\omega)) > \varepsilon\}$$

Fix ε, then for any $T > 0$, $\delta > 0$ we show that there are at most a finite number of points $t \in [0, T]$ such that $\mathbb{P}(J_t(\varepsilon)) > \delta$. Suppose this is false, and an infinite sequence t_i of disjoint times $t_i \in [0, T]$ exists. Then by Fatou's lemma

$$\mathbb{P}\left(\liminf_{i \to \infty} (J_{t_i}(\varepsilon))^c\right) \leq \liminf_{i \to \infty} \mathbb{P}((J_{t_i}(\varepsilon))^c)$$

thus

$$\mathbb{P}\left(\limsup_{i \to \infty} J_{t_i}(\varepsilon)\right) \geq \limsup_{i \to \infty} \mathbb{P}(J_{t_i}(\varepsilon)) > \delta,$$

so the event that $J_t(\varepsilon)$ occurs for an infinite number of the t_is has strictly positive probability and is hence non empty. This implies that there is a càdlàg path with an infinite number of jumps in $[0, T]$ of size greater than ε, which contradicts the conclusion of Lemma A.13. Taking the union over a countable sequence $\delta_n \downarrow 0$, it then follows that $\mathbb{P}(J_t(\varepsilon)) > 0$ for at most a countable set of $t \in [0, T]$.

Clearly $\mathbb{P}(J_t(\varepsilon)) \to \mathbb{P}(X_t \neq X_{t-})$ as $\varepsilon \to 0$, thus the set $\{t \in [0, T] : \mathbb{P}(X_t \neq X_{t-}) > 0\}$ contains at most a countable number of points. By taking the countable union over $T \in \mathbb{N}$, it follows that $\{t \in [0, \infty) : \mathbb{P}(X_t \neq X_{t-}) > 0\}$ is at most countable. □

A.5.2 Skorohod Topology

Consider the sequence of functions $x^n(t) = 1_{\{t \geq 1/n\}}$, and the function $x(t) = 1_{\{t > 0\}}$ which are all elements of $D_E[0, \infty)$. In the uniform topology which we used on $C_E[0, \infty)$, as $n \to \infty$ the sequence x_n does not converge to x; yet considered as càdlàg paths it appears natural that x^n should converge to x since the location of the unit jump of x_n converges to the location of the unit jump of x. A different topology is required. The Skorohod topology is the most frequently used topology on the space $D_E[0, \infty)$ which resolves this problem. Let $\lambda : [0, \infty) \to [0, \infty)$, and define

$$\gamma(\lambda) \triangleq \operatorname*{esssup}_{t \geq 0} |\log \lambda'(t)|$$
$$= \sup_{s > t \geq 0} \left| \log \frac{\lambda(s) - \lambda(t)}{s - t} \right|.$$

Let Λ be the subspace of Lipschitz continuous increasing functions from $[0, \infty) \to [0, \infty)$ such that $\lambda(0) = 0$, $\lim_{t \to \infty} \lambda(t) = \infty$ and $\gamma(\lambda) < \infty$.

The Skorohod topology is most readily defined in terms of a metric which induces the topology. For $x, y \in D_E[0, \infty)$ define a metric $d_{D_E}(x, y)$ by

$$d_{D_E}(x, y) = \inf_{\lambda \in \Lambda} \left[\gamma(\lambda) \vee \int_0^\infty e^{-u} d(x, y, \lambda, u) \, du \right],$$

where

$$d(x, y, \lambda, u) = \sup_{t \geq 0} d(x(t \wedge u), y(\lambda(t) \wedge u)).$$

It is of course necessary to verify that this satisfies the definition of a metric. This is straightforward, but details may be found in Ethier and Kurtz [95, Chapter 3, pages 117-118]. For the functions x_n and x in the example, it is clear that $d_{D_\mathbb{R}}(x_n, x) \to 0$ as $n \to \infty$. While there are other simpler topologies which have this property, the following proposition is the main reason why the Skorohod topology is the preferred choice of topology on D_E.

Proposition A.15. *If the metric space (E, d) is complete and separable, then $(D_E[0, \infty), d_{D_E})$ is also complete and separable.*

Proof. The following proof follows Ethier and Kurtz [95]. As E is separable, it has a countable dense set. Let $\{x_n\}_{n \geq 1}$ be such a set. Given n, $0 = t_0 < t_1 < \cdots < t_n$ where $t_j \in \mathbb{Q}^+$ and $i_j \in \mathbb{N}$ for $j = 0, \ldots, n$ define the piecewise constant function

$$x(t) = \begin{cases} x_{i_k} & t_k \leq t < t_{k+1} \\ x_{i_n} & t \geq t_n. \end{cases}$$

The set of all such functions forms a dense subset of $D_E[0, \infty)$, therefore the space is separable.

To show that the space is complete, suppose that $\{y_n\}_{n \geq 1}$ is a Cauchy sequence in $(D_E[0, \infty), d_{D_E})$, which implies that there exists an increasing sequence of numbers N_k such that for $n, m \geq N_k$,

$$d_{D_E}(y_n, y_m) \leq 2^{-k-1} e^{-k}.$$

Set $v_k = y_{N_k}$; then $d_{D_E}(v_k, v_{k+1}) \leq 2^{-k-1} e^{-k}$. Thus there exists λ_k such that

$$\int_0^\infty e^{-u} d(v_k, v_{k+1}, \lambda_k, u) \, du < 2^{-k} e^{-k}.$$

As $d(x, y, \lambda, u)$ is monotonic increasing in u, it follows that for any $v \geq 0$,

$$\int_0^\infty e^{-u} d(x,y,\lambda,u)\,du \geq d(x,y,\lambda,v) \int_v^\infty e^{-u}\,du = e^{-v} d(x,y,\lambda,v).$$

Therefore it is possible to find $\lambda_k \in \Lambda$ and $u_k > k$ such that

$$\max(\gamma(\lambda_k), d(v_k, v_{k+1}, \lambda_k, u_k)) \leq 2^{-k}. \tag{A.13}$$

Then form the limit of the composition of the functions λ_i

$$\mu_k \triangleq \lim_{n \to \infty} \lambda_{k+n} \circ \cdots \lambda_{k+1} \circ \lambda_k.$$

It then follows that

$$\gamma(\mu_k) \leq \sum_{i=k}^\infty \gamma(\lambda_i) \leq \sum_{i=k}^\infty 2^{-i} = 2^{-k+1} < \infty;$$

thus $\mu_k \in \Lambda$. Using the bound (A.13) it follows that for $k \in \mathbb{N}$,

$$\sup_{t \geq 0} d\left(v_k(\mu_k^{-1}(t) \wedge u_k), v_{k+1}(\mu_{k+1}^{-1}(t) \wedge u_k)\right)$$
$$= \sup_{t \geq 0} d\left(v_k(\mu_k^{-1}(t) \wedge u_k), v_{k+1}(\lambda_k \circ \mu_k^{-1}(t) \wedge u_k)\right)$$
$$= \sup_{t \geq 0} d\left(v_k(t \wedge u_k), v_{k+1}(\lambda_k(t) \wedge u_k)\right)$$
$$= d\left(v_k, v_{k+1}, \lambda_k, u_k\right)$$
$$\leq 2^{-k}.$$

Since (E, d) is complete, it now follows that $z_k = v_k \circ \mu_k^{-1}$ converges uniformly on compact sets of t to some limit, which we denote z. As each z_k has càdlàg paths, it follows that the limit also has càdlàg paths and thus belongs to $D_E[0, \infty)$. It only remains to show that v_k converges to z in the Skorohod topology. This follows since, $\gamma(\mu_k^{-1}) \to 0$ as $k \to \infty$ and for fixed $T > 0$,

$$\lim_{k \to \infty} \sup_{0 \leq t \leq T} d\left(v_k \circ \mu_k^{-1}(t), z(t)\right) = 0.$$

□

A.6 Stopping Times

In this section, the notation $\mathcal{F}_t^o$ is used to emphasise that this filtration has not been augmented.

Definition A.16. *A random variable T taking values in $[0, \infty)$ is said to be an $\mathcal{F}_t^o$-stopping time, if for all $t \geq 0$, the event $\{T \leq t\} \in \mathcal{F}_t^o$.*

The subject of stopping times is too large to cover in any detail here. For more details see Rogers and Williams [248], or Dellacherie and Meyer [77, Section IV.3].

Lemma A.17. *A random variable T taking values in $[0, \infty)$ is an $\mathcal{F}^o_{t+}$-stopping time if and only if $\{T < t\} \in \mathcal{F}^o_t$ for all $t \geq 0$.*

Proof. If $\{T < t\} \in \mathcal{F}^o_t$ for all $t \geq 0$ then since

$$\{T \leq t\} = \bigcap_{\varepsilon > 0} \{T < t + \varepsilon\},$$

it follows that $\{T \leq t\} \in \mathcal{F}^o_{t+\varepsilon}$ for any $t \geq 0$ and $\varepsilon > 0$, thus $\{T \leq t\} \in \mathcal{F}^o_{t+}$. Thus T is an $\mathcal{F}^o_{t+}$-stopping time.

Conversely if T is an $\mathcal{F}^o_{t+}$-stopping time then since

$$\{T < t\} = \bigcup_{n=1}^{\infty} \{T \leq t - 1/n\}$$

and each $\{T \leq t - 1/n\} \in \mathcal{F}^o_{(t-1/n)+} \subseteq \mathcal{F}^o_t$, therefore $\{T < t\} \in \mathcal{F}^o_t$. □

Lemma A.18. *Let T^n be a sequence of $\mathcal{F}^o_t$-stopping times. Then $T = \inf_n T^n$ is an $\mathcal{F}^o_{t+}$-stopping time.*

Proof. Write the event $\{\inf_n T^n < t\}$ as

$$\left\{\inf_n T^n < t\right\} = \bigcap_n \{T^n < t\}.$$

By Lemma A.17 each term in this intersection belongs to $\mathcal{F}^o_{t+}$, therefore so does the intersection which again by Lemma A.17 implies that $\inf_n T^n$ is a $\mathcal{F}^o_{t+}$-stopping time. □

Lemma A.19. *Let X be a real-valued, continuous, adapted process and $a \in \mathbb{R}$. Define $T_a \triangleq \inf\{t \geq 0 : X_t \geq a\}$. Then T_a is a $\mathcal{F}_t$-stopping time*

Proof. The set $\{\omega : X_q(\omega) \geq a\}$ is $\mathcal{F}_q$-measurable for any $q \in \mathbb{Q}^+$ as X is $\mathcal{F}_t$-adapted. Then using the path continuity of X,

$$\{T_a \leq t\} = \left\{\omega : \inf_{0 \leq s \leq t} X_s(\omega) \geq a\right\} = \bigcup_{q \in \mathbb{Q}^+ : 0 \leq q \leq t} \{\omega : X_q(\omega) \geq a\}.$$

Thus $\{T_a \leq t\}$ may be written as a countable union of $\mathcal{F}_t$-measurable sets and so is itself $\mathcal{F}_t$-measurable. Hence T_a is a $\mathcal{F}_t$-stopping time. □

Theorem A.20 (Début Theorem). *Let X be a process defined in some topological space (S) (with its associated Borel σ-algebra $\mathcal{B}(\mathbb{S}))$. Assume that X is progressively measurable relative to a filtration $\mathcal{F}^o_t$. Then for $A \in \mathcal{B}(\mathbb{S})$, the mapping $D_A = \inf\{t \geq 0; X_t \in A\}$ defines an $\mathcal{F}_t$-stopping time, where $\mathcal{F}_t$ is the augmentation of $\mathcal{F}^o_t$.*

For a proof see Theorem IV.50 of Dellacherie and Meyer [77]. See also Rogers and Williams [248, 249] for related results.

We require a technical result regarding the augmentation aspect of the usual conditions which is used during the innovations derivation of the filtering equations.

Lemma A.21. *Let $\mathcal{G}_t$ be the filtration $\mathcal{F}_t^o \vee \mathcal{N}$ where $\mathcal{N}$ contains all the $\mathbb{P}$-null sets. If T is a $\mathcal{G}_t$-stopping time, then there exists a $\mathcal{F}_{t+}^o$-stopping time T' such that $T = T'$ $\mathbb{P}$-a.s. In addition if $L \in \mathcal{G}_T$ then there exists $M \in \mathcal{F}_{T+}^o$ such that $L = M$ $\mathbb{P}$-a.s.*

Proof. Consider a stopping time of the form $T = a 1_A + \infty 1_{A^c}$ where $a \in \mathbb{R}^+$ and $A \in \mathcal{G}_a$; in this case let B be an element of $\mathcal{F}_a^o$ such that the symmetric difference $A \triangle B$ is a $\mathbb{P}$-null set and define $T' = a 1_B + \infty 1_{B^c}$. For a general $\mathcal{G}_t$-stopping time T use a dyadic approximation. Let

$$S^{(n)} \triangleq \sum_{k=0}^{\infty} k 2^{-n} 1_{\{(k-1)2^{-n} \leq T < k 2^{-n}\}}.$$

Clearly $S^{(n)}$ is $\mathcal{G}_T$-measurable and by construction $S^{(n)} \geq T$. Thus S^n is a $\mathcal{G}_t$-stopping time. But the stopping time $S^{(n)}$ takes values in a countable set, so

$$S^{(n)} = \inf_k \left\{ k 2^{-n} 1_{A_k} + \infty I_{A_k^c} \right\},$$

where $A_k \triangleq \{S^{(n)} = k 2^{-n}\}$. The result has already been proved for stopping times of the form of those inside the infimum. As $T = \lim_n S^{(n)} = \inf_n S^{(n)}$, consequently the result holds for all $\mathcal{G}_t$-stopping times. As a consequence of this limiting operation $\mathcal{F}_{t+}^o$ appears instead of $\mathcal{F}_t^o$.

To prove the second assertion, let $L \in \mathcal{G}_T$. By the first part since $L \in \mathcal{G}_\infty$ there exists $L' \in \mathcal{F}_\infty^o$ such that $L = L'$ $\mathbb{P}$-a.s. Let $V = T 1_L + \infty 1_{L^c}$ a.s. Using the first part again, $\mathcal{F}_{t+}^o$-stopping times V' and T' can be constructed such that $V = V'$ a.s. and $T = T'$ a.s. Define $M \triangleq \{L' \cap \{T' = \infty\}\} \cup \{V' = T' < \infty\}$. Clearly M is $\mathcal{F}_{T+}^o$-measurable and it follows that $L = M$ $\mathbb{P}$-a.s. □

The following lemma is trivial, but worth stating to avoid confusion in the more complex proof which follows.

Lemma A.22. *Let $\mathcal{X}_t^o$ be the unaugmented σ-algebra generated by a process X. Then for T an $\mathcal{X}_t^o$-stopping time, if $T(\omega) \leq t$ and $X_s(\omega) = X_s(\omega')$ for $s \leq t$ then $T(\omega') \leq t$.*

Proof. As T is a stopping time, $\{T \leq t\} \in \mathcal{X}_t^o = \sigma(X_s : 0 \leq s \leq t)$ from which the result follows. □

Corollary A.23. *Let $\mathcal{X}_t^o$ be the unaugmented σ-algebra generated by a process X. Then for T an $\mathcal{X}_t^o$-stopping time, if $T(\omega) \leq t$ and $X_s(\omega) = X_s(\omega')$ for $s \leq t$ then $T(\omega') = T(\omega)$.*

Proof. Apply Lemma A.22 with $t = T(\omega)$ to conclude $T(\omega') \leq T(\omega)$. By symmetry, $T(\omega) \leq T(\omega')$ from which the result follows. □

Lemma A.24. *Let $\mathcal{X}_t^o$ be the unaugmented σ-algebra generated by a process X. Then for T a $\mathcal{X}_t^o$-stopping time, for all $t \geq 0$,*

$$\mathcal{X}_{t \wedge T}^o = \sigma\left\{X_{s \wedge T} : 0 \leq s \leq t\right\}.$$

Proof. Since $T \wedge t$ is also a $\mathcal{X}_t^o$-stopping time, it suffices to show

$$\mathcal{X}_T^o = \sigma\left\{X_{s \wedge T} : s \geq 0\right\}.$$

The definition of the σ-algebra associated with a stopping time is that

$$\mathcal{X}_T^o \triangleq \left\{B \in \mathcal{X}_\infty^o : B \cap \{T \leq s\} \in \mathcal{X}_s^o \text{ for all } s \geq 0\right\}.$$

If $A \in \mathcal{F}_T^o$ then it follows from this definition that

$$T_A = \begin{cases} T & \text{if } \omega \in A, \\ +\infty & \text{otherwise,} \end{cases}$$

defines a $\mathcal{X}_t^o$-stopping time. Conversely if for some set A, the time T_A defined as above is a stopping time it follows that $A \in \mathcal{X}_T^o$. Therefore we will have established the result if we can show that $A \in \sigma\{X_{s \wedge T} : s \geq 0\}$ is a necessary and sufficient condition for T_A to be a stopping time.

For the first implication, assume that T_A is a $\mathcal{X}_t^o$-stopping time. It is necessary to show that $A \in \sigma\{X_{s \wedge T} : s \geq 0\}$. Suppose that $\omega, \omega' \in \Omega$ are such that $X_s(\omega) = X_s(\omega')$ for $s \leq T(\omega)$. We will establish that $A \in \sigma\{X_{s \wedge T} : s \geq 0\}$ if we show $\omega \in A$ implies that $\omega' \in A$.

If $T(\omega) = \infty$ then it is immediate that the trajectories $X_s(\omega)$ and $X_s(\omega')$ are identical and hence $\omega' \in A$. Therefore consider $T(\omega) < \infty$; if $\omega \in A$ then $T_A(\omega) = T(\omega)$ and since it was assumed that T_A is a $\mathcal{X}_t^o$-stopping time the fact that $X_s(\omega)$ and $X_s(\omega')$ agree for $s \leq T_A(\omega)$ implies by Corollary A.23 that $T_A(\omega') = T_A(\omega) = T(\omega) < \infty$ and from $T_A(\omega') < \infty$ it follows that $\omega' \in A$.

We must now prove the opposite implication; that is, given that T is a stopping time and $A \in \sigma\{X_{s \wedge T} : s \geq 0\}$, we must show that T_A is a stopping time.

Given arbitrary $t \geq 0$, if $T_A(\omega) \leq t$ and $X_s(\omega) = X_s(\omega')$ for $s \leq t$ it follows that $\omega \in A$ (since $T_A(\omega) < \infty$). If $T(\omega) \leq t$ and $X_s(\omega) = X_s(\omega')$ for $s \leq t$, since T is a stopping time it follows from Corollary A.23 that $T(\omega) = T(\omega')$. Since we assumed $A \in \sigma\{X_{s \wedge T} : s \geq 0\}$ it follows that $\omega' \in A$ from which we deduce $T_A(\omega) = T(\omega) = T(\omega') = T_A(\omega')$ whence

$$\{T_A(\omega) \leq t, X_s(\omega) = X_s(\omega') \quad \text{for all } s \leq t\} \Rightarrow T_A(\omega') \leq t,$$

which implies that $\{T_A(\omega) \leq t\} \in \mathcal{X}_t^o$ and hence that T_A is a $\mathcal{X}_t^o$-stopping time. □

For many arguments it is required that the augmentation of the filtration generated by a process be right continuous. While left continuity of sample paths does imply left continuity of the filtration, right continuity (or even continuity) of the sample paths does not imply that the augmentation of the generated filtration is right continuous. This can be seen by considering the event that a process has a local maximum at time t which may be in $\mathcal{X}_{t+}$ but not $\mathcal{X}_t$ (see the solution to Problem 7.1 (iii) in Chapter 2 of Karatzas and Shreve [149]). The following proposition gives an important class of process for which the right continuity does hold.

Proposition A.25. *If X is a d-dimensional strong Markov process, then the augmentation of the filtration generated by X is right continuous.*

Proof. Denote by $\mathcal{X}^o$ the (unaugmented) filtration generated by the process X. If $0 \leq t_0 < t_1 < \cdots < t_n \leq s < t_{n+1} \cdots < t_m$, then by application of the strong Markov property to the trivial $\mathcal{X}_{t+}^o$-stopping time s,

$$\mathbb{P}(X_{t_0} \in \varGamma_0, \ldots X_{t_m} \in \varGamma_m \mid \mathcal{F}_{s+})$$
$$= 1_{\{X_{t_0} \in \varGamma_0, \ldots, X_{t_n} \in \varGamma_n\}} \mathbb{P}(X_{t_{n+1}} \in \varGamma_{n+1}, \ldots, X_m \in \varGamma_m \mid X_s).$$

The right-hand side in this equation is clearly $\mathcal{X}_s$-measurable and it is $\mathbb{P}$-a.s. equal to $\mathbb{P}(X_{t_0} \in \varGamma_0, \ldots X_{t_m} \in \varGamma_m \mid \mathcal{F}_{s+})$. As this holds for all cylinder sets, it follows that for all $F \in \mathcal{X}_\infty^o$ there exists a $\mathcal{X}_s^o$-measurable random variable which is $\mathbb{P}$-a.s. equal to $\mathbb{P}(F \mid \mathcal{X}_{s+}^o)$.

Suppose that $F \in \mathcal{X}_{s+}^o \subseteq \mathcal{X}_\infty^o$; then clearly $\mathbb{P}(F \mid \mathcal{X}_{s+}^o) = 1_F$. As above there exists a $\mathcal{X}_s^o$-measurable random variable $\hat{1}_F$ such that $\hat{1}_F = 1_F$ a.s. Define the event $G \triangleq \{\omega : \hat{1}_F(\omega) = 1\}$, then $G \in \mathcal{X}_s^o$ and the events G and F differ by at most a null set (i.e. the symmetric difference $G \triangle F$ is null). Therefore $F \in \mathcal{X}_s$, which establishes that $\mathcal{X}_{s+}^o \subseteq \mathcal{X}_s$ for all $s \geq 0$.

It is clear that $\mathcal{X}_s \subseteq \mathcal{X}_{s+}$. Now prove the converse implication. Suppose that $F \in \mathcal{X}_{s+}$, which implies that for all n, $F \in \mathcal{X}_{s+1/n}$. Therefore there exists $G_n \in \mathcal{X}_{s+1/n}^o$ such that F and G_n differ by a null set. Define $G \triangleq \bigcap_{m=1}^\infty \bigcup_{n=m}^\infty G_n$. Then clearly $G \in \mathcal{X}_{s+}^o \subseteq \mathcal{X}_s$ (by the result just proved). To show that $F \in \mathcal{X}_s$, it suffices to show that this G differs by at most a null set from F. Consider

$$G \setminus F \subseteq \left(\bigcup_{n=1}^\infty G_n \right) \setminus F = \bigcup_{n=1}^\infty (G_n \setminus F),$$

where the right-hand side is a countable union of null sets; thus $G \setminus F$ is null. Secondly

$$F \setminus G = F \cap \left(\bigcap_{m=1}^\infty \bigcup_{n=m}^\infty G_n \right)^c = F \cap \left(\bigcup_{m=1}^\infty \bigcap_{n=m}^\infty G_n^c \right)$$
$$= \bigcup_{m=1}^\infty \left(F \cap \left(\bigcap_{n=m}^\infty G_n^c \right) \right) \subseteq \bigcup_{m=1}^\infty F \cap G_m^c = \bigcup_{m=1}^\infty (F \setminus G_m),$$

and again the right-hand side is a countable union of null sets, thus $F \setminus G$ is null. Therefore $F \in \mathcal{X}_s$, which implies that $\mathcal{X}_{s+} \subseteq \mathcal{X}_s$; hence $\mathcal{X}_s = \mathcal{X}_{s+}$. □

A.7 The Optional Projection

Proof of Theorem 2.7

Proof. The proof uses a monotone class argument (Theorem A.1). Let $\mathcal{H}$ be the class of bounded measurable processes for which an optional projection exists. The class of functions $1_{[s,t)} 1_F$, where $s < t$ and $F \in \mathcal{F}$ can readily be seen to form a π-system which generates the measurable processes. Define Z to be a càdlàg version of the martingale $t \mapsto \mathbb{E}(1_F \mid \mathcal{F}_t)$ (which necessarily exists since we have assumed that the usual conditions hold); then we may set

$$^o\left(1_{[s,t)} 1_F\right)(r, \omega) = 1_{[s,t)}(r) Z_r(\omega).$$

It is necessary to check that the defining condition (2.8) is satisfied. Let T be a stopping time. Then from Doob's optional sampling theorem (which is applicable in this case without restrictions on T, because the martingale Z is bounded and hence uniformly integrable) that

$$\mathbb{E}[1_F \mid \mathcal{F}_T] = \mathbb{E}[Z_\infty \mid \mathcal{F}_T] = Z_T$$

whence

$$\mathbb{E}[1_F 1_{\{T<\infty\}} \mid \mathcal{F}_T] = Z_T 1_{\{T<\infty\}} \quad \mathbb{P}\text{-a.s.}$$

To apply the Monotone class theorem A.1 it is necessary to check that if X_n is a bounded monotone sequence in $\mathcal{H}$ with limit X then the optional projections ^oX_n converge to the optional projection of X. Consider

$$Y \triangleq \liminf_{n \to \infty} {}^oX_n 1_{\{|\liminf_{n \to \infty} {}^oX_n| < \infty\}}.$$

We must check that Y is the optional projection of X. Thanks to property (c) of conditional expectation the condition (2.8) is immediate. Consequently $\mathcal{H}$ is a monotone class and thus by the monotone class theorem A.1 the optional projection exists for any bounded $\mathcal{B} \times \mathcal{F}$-measurable process. To extend to the unbounded non-negative case consider $X \wedge n$ and pass to the limit.

In order to verify that the projection is unique up to indistinguishability, consider two candidates for the optional projection Y and Z. For any stopping time T from (2.8) it follows that

$$Y_T 1_{\{T<\infty\}} = Z_T 1_{\{T<\infty\}}, \quad \mathbb{P}\text{-a.s.} \tag{A.14}$$

Define $F \triangleq \{(t, \omega) : Z_t(\omega) \neq Y_t(\omega)\}$. Since both Z and Y are optional processes the set F is an optional subset of $[0, \infty) \times \Omega$. Write $\pi : [0, \infty) \times \Omega \to \Omega$ for the canonical projection map $\pi : (t, \omega) \mapsto \omega$. Now argue by contradiction. Suppose that Z and Y are not indistinguishable; this implies that $\mathbb{P}(\pi(F)) > 0$. By the optional section theorem (see Dellacherie and Meyer [77, IV.84]) it follows that given $\varepsilon > 0$ there exists a stopping time U such that when $U(\omega) < \infty$, $(U(\omega), \omega) \in F$ and $\mathbb{P}(U < \infty) \geq \mathbb{P}(\pi(F)) - \varepsilon$. As it has been assumed that $\mathbb{P}(\pi(F)) > 0$, by choosing ε sufficiently small, $\mathbb{P}(U < \infty) > 0$. It follows that on some set of non-zero probability $1_{\{U<\infty\}} Y_U \neq 1_{\{U<\infty\}} Z_U$. But from (A.14) this may only hold on a null set, which is a contradiction. Therefore $\mathbb{P}(\pi(F)) = 0$ and it follows that Z and Y are indistinguishable. □

Lemma A.26. *If almost surely $X_t \geq 0$ for all $t \geq 0$ then $°X_t \geq 0$ for all $t \geq 0$ almost surely.*

Proof. Use the monotone class argument (Theorem A.1) in the proof of the existence of the optional projection, noting that if $F \in \mathcal{F}$ then the càdlàg version of $\mathbb{E}[1_F \mid \mathcal{F}_t]$ is non-negative a.s. Alternatively use the optional section theorem as in the proof of uniqueness. □

A.7.1 Path Regularity

Introduce the following notation for a one-sided limit, in this case the right limit
$$\limsup_{s \downarrow \downarrow t} x_s \triangleq \limsup_{s \to t: s > t} x_s = \inf_{v > t} \sup_{t < u \leq v} x_u,$$
a similar notation with $s \uparrow \uparrow t$ being used for the left limit.

The following lemma is required to establish right continuity. It can be applied to the optional projection since being optional it must also be progressively measurable.

Lemma A.27. *Let X be a progressively measurable stochastic process taking values in $\mathbb{R}$; then $\liminf_{s \downarrow \downarrow t} X_s$ and $\limsup_{s \downarrow \downarrow t} X_s$ are progressively measurable.*

Proof. It is sufficient to consider the case of lim sup. Let $b \in \mathbb{R}$ be such that $b > 0$, then define
$$X_t^n \triangleq \begin{cases} \sup_{kb2^{-n} \leq s < (k+1)b2^{-n}} X_s & \text{if } b(k-1)2^{-n} \leq t < bk2^{-n}, k < 2^n, \\ \limsup_{s \downarrow \downarrow b} X_s & \text{if } b(1 - 2^{-n}) \leq t \leq b. \end{cases}$$

For every $t \leq b$, the supremum in the above definition is $\mathcal{F}_b$-measurable since X is progressively measurable; thus the random variable X_t^n is $\mathcal{F}_b$-measurable. For every $\omega \in \Omega$, $X_t^n(\omega)$ has trajectories which are right continuous for $t \in [0, b]$. Therefore X^n is $\mathcal{B}([0, b]) \otimes \mathcal{F}_b$-measurable and is thus progressively measurable. On $[0, b]$ it is clear that $\limsup_{n \to \infty} X_t^n = \limsup_{s \downarrow \downarrow t} X_s$, hence $\limsup_{s \downarrow \downarrow t} X_s$ is progressively measurable. □

A.7 The Optional Projection

In a similar vein, the following lemma is required in order to establish the existence of left limits. For the left limits the result is stronger and the lim inf and lim sup are previsible and thus also progressively measurable.

Lemma A.28. *Let X be a progressively measurable stochastic process taking values in $\mathbb{R}$; then $\liminf_{s\uparrow\uparrow t} X_s$ and $\limsup_{s\uparrow\uparrow t} X_s$ are previsible.*

Proof. It suffices to consider $\limsup_{s\uparrow\uparrow t} X_t$. Define

$$X_t^n \triangleq \sum_{k>0} 1_{\{k2^{-n} < t \leq (k+1)2^{-n}\}} \sup_{(k-1)2^{-n} < s \leq k2^{-n}} X_s,$$

from this definition it is clear that X_t^n is previsible as it is a sum of left continuous, adapted, processes. But as $\limsup_{n\to\infty} X_t^n = \limsup_{s\uparrow\uparrow t} X_s$, it follows that $\limsup_{s\uparrow\uparrow t} X_s$ is previsible. □

Proof of Theorem 2.9

Proof. First observe that if Y_t is bounded then ${}^o Y_t$ must also be bounded. There are three things which must be established; first, the existence of right limits; second, right continuity; and third the existence of left limits. Because of the difference between Lemmas A.27 and A.28 the cases of left and right limits are not identical. The first part of the proof establishes the existence of right limits. It is sufficient to show that

$$\mathbb{P}\left(\liminf_{s\downarrow\downarrow t} {}^o Y_s < \limsup_{s\downarrow\downarrow t} {}^o Y_s \text{ for some } t \in [0, \infty)\right) = 0. \qquad (A.15)$$

The following steps are familiar from the proof of Doob's martingale regularization theorem which is used to guarantee the existence of càdlàg modifications of martingales. If the right limit does not exist at $t \in [0,\infty)$, that is, if $\liminf_{s\downarrow\downarrow t} {}^o Y_s < \limsup_{s\downarrow\downarrow t} {}^o Y_s$, then rationals a,b can be found such that $\liminf_{s\downarrow\downarrow t} {}^o Y_s < a < b < \limsup_{s\downarrow\downarrow t} {}^o Y_s$. The event that the right limit does not exist has thus been decomposed into a countable union over the rationals:

$$\left\{\omega : \liminf_{s\downarrow\downarrow t} {}^o Y_s(\omega) < \limsup_{s\downarrow\downarrow t} {}^o Y_s(\omega) \text{ for some } t \in [0,\infty)\right\} =$$
$$\bigcup_{a,b\in\mathbb{Q}} \left\{\omega : \liminf_{s\downarrow\downarrow t} {}^o Y_s(\omega) < a < b < \limsup_{s\downarrow\downarrow t} {}^o Y_s(\omega) \text{ for some } t \in [0,\infty)\right\}.$$

The lim sup and lim inf processes are progressively measurable by Lemma A.27, therefore for rationals $a < b$, the set

$$E_{a,b} \triangleq \left\{(t,\omega) : \liminf_{s\downarrow\downarrow t} {}^o Y_s < a < b < \limsup_{s\downarrow\downarrow t} {}^o Y_s\right\},$$

is progressively measurable.

Now argue by contradiction; suppose that (A.15) is not true. Then from the decomposition into a countable union, it follows that we can find $a, b \in \mathbb{Q}$ such that $a < b$ and

$$0 < \mathbb{P}\left(\liminf_{s \downarrow \downarrow t} {}^\circ Y_s < a < b < \limsup_{s \downarrow \downarrow t} {}^\circ Y_s \text{ for some } t \in [0, \infty)\right) = \mathbb{P}(\pi(E_{a,b})),$$

where the projection π is defined for $A \subset [0, \infty) \times \Omega$, by $\pi(A) = \{\omega : (\omega, t) \in A\}$. Define

$$S_{a,b} \triangleq \inf\{t \geq 0 : (t, \omega) \in E_{a,b}\},$$

which is the début of a progressively measurable set, and thus by the Début theorem (Theorem A.20 applied to the progressive process $1_{E_{a,b}}(t, \omega)$) is a stopping time (and hence optional). For a given ω, this stopping time $S_{a,b}(\omega)$ is the first time where $\liminf_{s \downarrow \downarrow t} {}^\circ Y_s$ and $\limsup_{s \downarrow \downarrow t} {}^\circ Y_s$ straddle the interval $[a, b]$ and thus the right limit fails to exist at this point.

If $\omega \in \pi(E_{a,b})$ then there exists $t \in [0, \infty)$ such that $(t, \omega) \in E_{a,b}$ and this implies $t \geq S(\omega)$, whence $S(\omega) < \infty$. Thus, if $\mathbb{P}(\pi(E_{a,b})) > 0$ then this implies $\mathbb{P}(S_{a,b} < \infty) > 0$. Thus a consequence of the assumption that (A.15) is false is that we can find $a, b \in \mathbb{Q}$, with $a < b$ such that $\mathbb{P}(S_{a,b} < \infty) > 0$. This will lead to a contradiction. For the remainder of the argument we can keep a and b fixed and consequently we write S in place of $S_{a,b}$.

Define

$$A_0 \triangleq \{(t, \omega) : S(\omega) < t < S(\omega) + 1, {}^\circ Y_t(\omega) < a\};$$

it then follows that the projection $\pi(A_0) = \{S < \infty\}$. Thus by the optional section theorem, since A_0 is optional (S is a stopping time and ${}^\circ Y_t$ is a priori optional), we can find a stopping time S_0 such that on $\{S_0 < \infty\}$, $(S_0(\omega), \omega) \in A_0$ and

$$\mathbb{P}(S_0 < \infty) > (1 - 1/2)\mathbb{P}(S < \infty).$$

Define

$$A_1 \triangleq \{(t, \omega) : S(\omega) < t < (S(\omega) + 1/2) \wedge S_0(\omega), {}^\circ Y_t(\omega) > b\}$$

and again by the optional section theorem we can find a stopping time S_1 such that on $\{S_1 < \infty\}$, $(S_1(\omega), \omega) \in A_1$ and

$$\mathbb{P}(S_1 < \infty) > (1 - 1/2^2)\mathbb{P}(S < \infty).$$

We can carry on this construction inductively defining

$$A_{2k} \triangleq \left\{(t, \omega) : S(\omega) < t < (S(\omega) + 2^{-2k}) \wedge S_{2k-1}(\omega), {}^\circ Y_t(\omega) < a\right\},$$

and

$$A_{2k+1} \triangleq \left\{(t, \omega) : S(\omega) < t < (S(\omega) + 2^{-(2k+1)}) \wedge S_{2k}(\omega), {}^\circ Y_t(\omega) > b\right\}.$$

A.7 The Optional Projection

We can construct stopping times using the optional section theorem such that for each i, on $\{S_i < \infty\}$, $(S_i(\omega), \omega) \in A_i$, and such that

$$\mathbb{P}(S_i < \infty) > \left(1 - 2^{-(i+1)}\right) \mathbb{P}(S < \infty).$$

On the event $\{S_i < \infty\}$ it is clear that $S_i < S_{i-1}$ and $S_i < S + 2^{-i}$. Also if $S = \infty$ it follows that $S_i = \infty$ for all i, thus $S_i < \infty$ implies $S < \infty$, so $\mathbb{P}(S_i < \infty, S < \infty) = \mathbb{P}(S_i < \infty)$, whence

$$\begin{aligned}\mathbb{P}(S_i = \infty, S < \infty) &= \mathbb{P}(S < \infty) - \mathbb{P}(S_i < \infty, S < \infty) \\ &= \mathbb{P}(S < \infty) - \mathbb{P}(S_i < \infty) \le \mathbb{P}(S < \infty)/2^{i+1}.\end{aligned}$$

Thus

$$\sum_{i=0}^{\infty} \mathbb{P}(S_i = \infty, S < \infty) \le \mathbb{P}(S < \infty) \le 1 < \infty,$$

so by the first Borel–Cantelli lemma the probability that infinitely many of the events $\{S_i = \infty, S < \infty\}$ occur is zero. In other words for $\omega \in \{S < \infty\}$, we can find an index $i_0(\omega)$ such that for $i \ge i_0$, the sequence S_i converges in a decreasing fashion to S and ${}^oY_{S_i} < a$ for even i, and ${}^oY_{S_i} > b$ for odd i.

Define $R_i = \sup_{j \ge i} S_j$, which is a monotonically decreasing sequence. Almost surely, $R_i = S_i$ for i sufficiently large, therefore $\lim_{i \to \infty} R_i = S$ a.s. and on the event $\{S < \infty\}$, for i sufficiently large, ${}^oY_{R_i} < a$ for i even, and ${}^oY_{R_i} > b$ for i odd. Set $T_i = R_i \wedge N$. On $\{S < N\}$, for j sufficiently large $S_j < N$, hence using the boundedness of ${}^oY$ to enable interchange of limit and expectation

$$\limsup_{i \to \infty} \mathbb{E}\left[{}^oY_{T_{2i}}\right] \le a\mathbb{P}(S < N) + \mathbb{E}\left[{}^oY_N 1_{\{S \ge N\}}\right],$$

$$\liminf_{i \to \infty} \mathbb{E}\left[{}^oY_{T_{2i+1}}\right] \ge b\mathbb{P}(S < N) + \mathbb{E}\left[{}^oY_N 1_{\{S \ge N\}}\right].$$

But since T_i is bounded by N, from the definition of the optional projection (2.8) it is clear that

$$\mathbb{E}[{}^oY_{T_i}] = \mathbb{E}\left[\mathbb{E}\left[Y_{T_i} 1_{T_i < \infty} \mid \mathcal{F}_{T_i}\right]\right] = \mathbb{E}[Y_{T_i}]. \tag{A.16}$$

Thus, since Y has right limits, by an application of the bounded convergence theorem $\mathbb{E}[Y_{T_i}] \to \mathbb{E}[Y_T]$, and so as $i \to \infty$

$$\mathbb{E}[{}^oY_{T_i}] \to \mathbb{E}[{}^oY_T]. \tag{A.17}$$

Thus

$$\limsup_{i \to \infty} \mathbb{E}[{}^oY_{T_i}] = \limsup_{i \to \infty} \mathbb{E}[{}^oY_{T_{2i}}] \text{ and } \liminf_{i \to \infty} \mathbb{E}[{}^oY_{T_i}] = \liminf_{i \to \infty} \mathbb{E}[{}^oY_{T_{2i+1}}],$$

so, if $\mathbb{P}(S < N) > 0$ we see that since $a < b$, $\limsup_{i \to \infty} \mathbb{E}[{}^oY_{T_i}] < \liminf_{i \to \infty} \mathbb{E}[{}^oY_{T_i}]$, which is a contradiction therefore $\mathbb{P}(S < N) = 0$. As

N was chosen arbitrarily, this implies that $\mathbb{P}(S = \infty) = 1$ which is a contradiction, since we assumed $\mathbb{P}(S < \infty) > 0$. Thus a.s., right limits of ${}^oY_t$ exist.

Now we must show that ${}^oY_t$ is right continuous. Let ${}^oY_{t+}$ be the process of right limits. As this process is adapted and right continuous, it follows that it is optional. Consider for $\varepsilon > 0$, the set

$$A_\varepsilon \triangleq \{(t, \omega) : {}^oY_t(\omega) \geq {}^oY_{t+}(\omega) + \varepsilon\}.$$

Suppose that $\mathbb{P}(\pi(A_\varepsilon)) > 0$, from which we deduce a contradiction. By the optional section theorem, for $\delta > 0$, we can find a stopping time S such that on $S < \infty$, $(S(\omega), \omega) \in A_\varepsilon$, and $\mathbb{P}(S < \infty) = \mathbb{P}(\pi(A_\varepsilon)) - \delta$. We may choose δ such that $\mathbb{P}(S < \infty) > 0$. Let $S_n = S + 1/n$, and bound these times by some N, which is chosen sufficiently large that $\mathbb{P}(S < N) > 0$. Thus set $T_n \triangleq S_n \wedge N$ and $T \triangleq S \wedge N$. Hence by bounded convergence

$$\lim_{n \to \infty} \mathbb{E}[{}^oY_{T_n}] = \mathbb{E}[{}^oY_N \mathbf{1}_{S \geq N}] + \mathbb{E}[{}^oY_{T+} \mathbf{1}_{S < N}], \tag{A.18}$$

but

$$\mathbb{E}[{}^oY_T] = \mathbb{E}[{}^oY_N \mathbf{1}_{S \geq N}] + \mathbb{E}[{}^oY_T \mathbf{1}_{S < N}]. \tag{A.19}$$

As the right-hand sides of (A.18) and (A.19) are not equal we conclude that $\lim_{n \to \infty} \mathbb{E}({}^oY_{T_n}) \neq \mathbb{E}({}^oY_T)$, which contradicts (A.17). Therefore $\mathbb{P}(\pi(A_\varepsilon)) = 0$. The same argument can be applied to

$$B_\varepsilon = \{(t, \omega) : {}^oY_t(\omega) \leq {}^oY_{t+}(\omega) - \varepsilon\},$$

which allows us to conclude that $\mathbb{P}(\pi(B_\varepsilon)) = 0$; hence

$$\mathbb{P}\left({}^oY_t = {}^oY_{t+}, \ \forall t \in [0, \infty)\right) = 1,$$

and thus, up to indistinguishability, the process ${}^oY_t$ is right continuous.

The existence of left limits is approached in a similar fashion; by Lemma A.28, the processes $\liminf_{s \uparrow \uparrow t} {}^oY_s$ and $\limsup_{s \uparrow \uparrow t} {}^oY_s$ are previsible and hence optional. For $a, b \in \mathbb{Q}$ we define

$$F_{a,b} \triangleq \left\{(t, \omega) : \liminf_{s \uparrow \uparrow t} {}^oY_s(\omega) < a < b < \limsup_{s \uparrow \uparrow t} {}^oY_s(\omega)\right\}.$$

We assume $\mathbb{P}(\pi(F_{a,b})) > 0$ and deduce a contradiction. Since $F_{a,b}$ is optional, we may apply the optional section theorem to find an optional time T such that on $\{T < \infty\}$, the point $(T(\omega), \omega) \in F_{a,b}$ and with $\mathbb{P}(T < \infty) > \varepsilon$. Define

$$C_0 \triangleq \{(t, \omega) : t < T(\omega), {}^oY_t < a\},$$

which is itself optional; thus another application of the optional section theorem constructs a stopping time R_0 such that on $\{R_0 < \infty\}$ $(R(\omega), \omega) \in C_0$ and since $R_0 < T$ it is clear that $\mathbb{P}(R_0 < \infty) > \varepsilon$.

Then define

$$C_1 \triangleq \{(t,\omega) : R_0(\omega) < t < T(\omega), {}^oY_t > b\},$$

which is optional and by the optional section theorem we can find a stopping time R_1 such that on $R_1(\omega) < \infty$, $(R_1(\omega), \omega) \in C_1$ and again $R_1 < T$ implies that $\mathbb{P}(R_1 < \infty) > \varepsilon$. Proceed inductively.

We have constructed an increasing sequence of optional times R_k such that on the event $\{T < \infty\}$, $Y_{R_k} < a$ for even k, and ${}^oY_{R_k} > b$ for odd k. Define $L_k = R_k \wedge N$ for some N; then this is an increasing sequence of bounded stopping times and clearly on $\{T < N\}$ the limit $\lim_n \mathbb{E}[{}^oY_{L_n}]$ does not exist. But since L_n is bounded, from (A.16) it follows that this limit must exist a.s.; hence $\mathbb{P}(T < N) = 0$, which as N was arbitrary implies $\mathbb{P}(T < \infty) = 0$, which is a contradiction. □

The results used in the above proof are due to Doob and can be found in a very clear paper [82] which is discussed further in Benveniste [16]. These papers work in the context of separable processes, which are processes whose graph is the closure of the graph of the process with time restricted to some countable set D. That is, for every $t \in [0, \infty)$ there exists a sequence $t_i \in D$ such that $t_i \to t$ and $x_{t_i} \to x_t$. In these papers 'rules of play' disallow the use of the optional section theorem except when unavoidable and the above results are proved without its use. These results can be extended (with the addition of extra conditions) to *optionally separable* processes, which are similarly defined, but the set D consists of a countable collection of stopping times and by an application of the optional section theorem it can be shown that every optional process is optionally separable. The direct approach via the optional section theorems is used in Dellacherie and Meyer [79].

A.8 The Previsible Projection

The optional projection (called the *projection bien measurable* in some early articles) has been discussed extensively and is the projection which is of importance in the theory of filtering; a closely related concept is the previsible (or predictable) projection. Some of the very early theoretical papers make use of this projection. By convention we take $\mathcal{F}_{0-} = \mathcal{F}_0$.

Theorem A.29. *Let X be a bounded measurable process; then there exists an optional process ${}^oX$ called the* previsible projection *of X such that for every previsible stopping time T,*

$$ {}^pX_T 1_{\{T<\infty\}} = \mathbb{E}\left[X_T 1_{\{T<\infty\}} \mid \mathcal{F}_{T-}\right]. \tag{A.20}$$

This process is unique up to indistinguishability, i.e. any processes which satisfy these conditions will be indistinguishable.

Proof. As in the proof of Theorem 2.7, let F be a measurable set, and define Z_t to be a càdlàg version of the martingale $\mathbb{E}[1_F \mid \mathcal{F}_t]$. Then we define the previsible projection of $1_{(s,t]}1_F$ by

$$^p\left(1_{(s,t]}1_F\right)(r,\omega) = 1_{(s,t]}(r)Z_{r-}(\omega).$$

We must verify that this satisfies (A.20); let T be a previsible stopping time. Then we can find a sequence T_n of stopping times such that $T_n \leq T_{n+1} < T$ for all n. By Doob's optional sampling theorem applied to the uniformly integrable martingale Z;

$$\mathbb{E}[1_F \mid \mathcal{F}_{T_n}] = \mathbb{E}[Z_\infty \mid \mathcal{F}_{T_n}] = Z_{T_n},$$

now pass to the limit as $n \to \infty$, using the martingale convergence theorem (see Theorem B.1), and we get

$$Z_{T-} = \mathbb{E}\left[Z_\infty \mid \vee_{n=1}^\infty \mathcal{F}_{T_n}\right]$$

and from the definition of the σ-algebra of $T-$ it follows that

$$Z_{T-} = \mathbb{E}[Z_\infty \mid \mathcal{F}_{T-}].$$

To complete the proof, apply the monotone class theorem A.1 as in the proof for the optional projection and use the same optional section theorem argument for uniqueness. □

The previsible and optional projection are actually very similar, as the following theorem illustrates.

Theorem A.30. *Let X be a bounded measurable process; then the set*

$$\{(t,\omega) : {}^oX_t(\omega) \neq {}^pX_t(\omega)\}$$

is a countable union of graphs of stopping times.

Proof. Again we use the monotone class argument. Consider the process $1_{[s,t)(r)}1_F$, from (2.8) and (A.20) the set of points of difference is

$$\{(t,\omega) : Z_t(\omega) \neq Z_{t-}(\omega)\}$$

and since Z is a càdlàg process we can define a sequence T_n of stopping times corresponding to the nth discontinuity of Z, and by Lemma A.13 there are at most countably many such discontinuities, therefore the points of difference are contained in the countable union of the graphs of these T_ns. □

A.9 The Optional Projection Without the Usual Conditions

The proof of the optional projection theorem in Section A.7 depends crucially on the usual conditions to construct a càdlàg version of a martingale, both the augmentation by null sets and the right continuity of the filtration being used. The result can be proved on the uncompleted σ-algebra by making suitable modifications to the process constructed by Theorem 2.7. These results were first established in Dellacherie and Meyer [78] and take their definitive form in [77], the latter approach being followed here. The proofs in this section are of a more advanced nature and make use of a number of non-trivial results about σ-algebras of stopping times which are not proved here. These results and their proofs can be found in, for example, Rogers and Williams [249]. As usual let $\mathcal{F}_t^o$ denote the unaugmented σ-algebra corresponding to $\mathcal{F}_t$.

Lemma A.31. *Let $L \subset \mathbb{R}^+ \times \Omega$ be such that*

$$L = \bigcup_n \{(S_n(\omega), \omega) : \omega \in \Omega\},$$

where the S_n are positive $\mathcal{F}_t^o$-stopping times. We can find disjoint $\mathcal{F}_t^o$-stopping times T_n to replace the S_n such that

$$L = \bigcup_n \{(T_n(\omega), \omega) : \omega \in \Omega\}.$$

Proof. Define $T_1 = S_1$ and define

$$A_n \triangleq \{\omega \in \Omega : S_1 \neq S_n,\ S_2 \neq S_n, \ldots, S_{n-1} \neq S_n\}.$$

Then it is clear that $A_n \in \mathcal{F}_{S_n}^o$. From the definition of this σ-algebra, if we define

$$T_n \triangleq S_n 1_{A_n} + \infty 1_{A_n^c},$$

then this T_n is a stopping time. It is clear that this process may be continued inductively. The disjointness of the T_ns follows by construction. □

Given this lemma the following result is useful when modifying a process as it allows us to break down the 'bad' set A of points in a useful fashion.

Lemma A.32. *Let A be a subset of $\mathbb{R}^+ \times \Omega$ contained in a countable union of graphs of positive random variables then $A = K \cup L$ where K and L are disjoint measurable sets such that K is contained in a disjoint union of graphs of optional times and L intersects the graph of any optional time on an evanescent set.*[†]

[†] A set $A \subset [0, \infty) \times \Omega$ is evanescent if the projection $\pi(A) = \{\omega : \exists t \in [0, \infty) \text{ such that } (\omega, t) \in A\}$ is contained in a $\mathbb{P}$-null set. Two indistinguishable processes differ on an evanescent set.

Proof. Let $\mathcal{V}$ denote the set of all optional times. For Z a positive random variable define $V(Z) = \cup_{T \in \mathcal{V}} \{\omega : Z(\omega) = T(\omega)\}$; consequently there is a useful decomposition $Z = Z' \wedge Z''$, where

$$Z' = Z 1_{V(Z)} + \infty 1_{V(Z)^c}$$
$$Z'' = Z 1_{V(Z)^c} + \infty 1_{V(Z)}.$$

From the definition of $V(Z)$ the set $\{(Z'(\omega), \omega) : \omega \in \Omega\}$ is contained in the graph of a countable number of optional times and if T is an optional time then $\mathbb{P}(Z'' = T < \infty) = 0$. Let the covering of A by a countable family of graphs of random variables be written

$$A \subseteq \bigcup_{n=1}^{\infty} \{(Z_n(\omega), \omega) : \omega \in \Omega\}$$

and form a decomposition of each random variable $Z_n = Z'_n \wedge Z''_n$ as above. Clearly $\bigcup_{n=1}^{\infty} \{(Z'_n(\omega), \omega) : \omega \in \Omega\}$ is also covered by a countable union of graphs of optional times and by Lemma A.31 we can find a sequence of disjoint optional times T_n such that

$$\bigcup_{n=1}^{\infty} \{(Z'_n(\omega), \omega) : \omega \in \Omega\} \subseteq \bigcup_{n=1}^{\infty} \{(T_n(\omega), \omega) : \omega \in \Omega\}.$$

Define

$$K = A \cap \bigcup_{n=1}^{\infty} \{(Z'_n(\omega), \omega) : \omega \in \Omega\} = A \cap \bigcup_n \{(T_n(\omega), \omega) : \omega \in \Omega\}$$

$$L = A \cap \bigcup_{n=1}^{\infty} \{(Z''(\omega), \omega) : \omega \in \Omega\} = A \setminus \bigcup_n \{(T_n(\omega), \omega) : \omega \in \Omega\}.$$

Clearly $A = K \cup L$, hence this is a decomposition of A which has the required properties. □

Lemma A.33. *For every $\mathcal{F}_t$-optional process X_t there is an indistinguishable $\mathcal{F}^o_{t+}$-optional process.*

Proof. Let T be an $\mathcal{F}_t$-stopping time. Consider the process $X_t = 1_{[0,T)}$, which is càdlàg and $\mathcal{F}_t$-adapted, and hence $\mathcal{F}_t$-optional. By Lemma A.21 there exists an $\mathcal{F}^o_{t+}$-stopping time T' such that $T = T'$ a.s. If we define $X'_t = 1_{[0,T')}$, then since this process is càdlàg and $\mathcal{F}^o_{t+}$-adapted, it is clearly an $\mathcal{F}^o_{t+}$-optional process.

$$\mathbb{P}(\omega : X'_t(\omega) = X_t(\omega) \; \forall t) = \mathbb{P}(T = T') = 1,$$

which implies that the processes X' and X are indistinguishable.

We extend from processes of the form $1_{[0,T)}$ to the whole of $\mathcal{O}$ using the monotone class framework (Theorem A.1) to extend to bounded optional processes, and use truncation to extended to the unbounded case. □

A.9 The Optional Projection Without the Usual Conditions

Lemma A.34. *For every $\mathcal{F}_t$-previsible process, there is an indistinguishable $\mathcal{F}_t^o$-previsible process.*

Proof. We first show that if T is $\mathcal{F}_t$-previsible; then there exists T' which is $\mathcal{F}_t^o$-previsible, such that $T = T'$ a.s. As $\{T = 0\} \in \mathcal{F}_{0-}$, we need only consider the case where $T > 0$. Let T_n be a sequence of $\mathcal{F}_t$-stopping times announcing[†] T. By Lemma A.33 it is clear that we can find R_n an $\mathcal{F}_{t+}^o$-stopping time such that $R_n = T_n$ a.s. Define $L_n \triangleq \max_{i=1,\ldots,n} R_n$; clearly this is an increasing sequence of stopping times. Let this sequence have limit L.

Define $A_n \triangleq \{L_n = 0\} \cup \{L_n < L\}$ and define

$$M_n = \begin{cases} L_n \wedge n & \text{if } \omega \in A_n \\ +\infty & \text{otherwise.} \end{cases}$$

Since the sets A_n are decreasing, the stopping times M_n form an increasing sequence and the sequence M_n announces everywhere its limit T'. This limit is strictly positive. Because T' is announced, T' is an $\mathcal{F}_t^o$-previsible time and $T = T'$ a.s. Finish the proof by a monotone class argument as in Lemma A.33. □

The main result of this section is the following extension of the optional projection theorem which does not require the imposition of the usual conditions.

Theorem A.35. *Given a stochastic process X, we can construct an $\mathcal{F}_t^o$ optional process Z_t such that for every stopping time T,*

$$Z_T 1_{\{T<\infty\}} = \mathbb{E}\left[Z_T 1_{\{T<\infty\}} \mid \mathcal{F}_T\right], \tag{A.21}$$

and this process is unique up to indistinguishability.

Proof. By the optional projection theorem 2.7 we can construct an $\mathcal{F}_t$-optional process $\bar{Z}_t$ which satisfies (A.21). By Lemma A.33 we can find a process Z_t which is indistinguishable from $\bar{Z}_t$ but which is $\mathcal{F}_{t+}^o$-optional. In general this process Z will not be $\mathcal{F}_t^o$-optional. We must therefore modify it.

Similarly using Theorem A.29, we can construct an $\mathcal{F}_t$-previsible process Y_t, and using Lemma A.34, we can find an $\mathcal{F}_t^o$-previsible process $\bar{Y}_t$ which is indistinguishable from the process Y_t.

Let $H = \{(t, \omega) : Y_t(\omega) \neq Z_t(\omega)\}$; then it follows by Theorem A.30, that this graph of differences is contained within a countable disjoint union of graphs of random variables. Thus by Lemma A.32 we may write $H = K \cup L$

[†] A stopping time T is called *announceable* if there exists an *announcing* sequence $(T_n)_{n \geq 1}$ for T. This means that for any $n \geq 1$ and $\omega \in \Omega$, $T_n(\omega) \leq T_{n+1}(\omega) < T(\omega)$ and $T_n(\omega) \nearrow T(\omega)$. A stopping time T is announceable if and only if it is previsible. For details see Rogers and Williams [248].

such that for T any $\mathcal{F}_t^o$-stopping time $\mathbb{P}(\omega : (T(\omega), \omega) \in L) = 0$ and there exists a sequence of $\mathcal{F}_t^o$-stopping times T_n such that

$$K \subset \bigcup_n \{(T_n(\omega), \omega) : \omega \in \Omega\}.$$

For each n let Z_n be a version of $\mathbb{E}[X_{T_n} 1_{\{T_n < \infty\}} \mid \mathcal{F}_{T_n}^o]$; then we can define

$$Z_t(\omega) \triangleq \begin{cases} Y_t(\omega) & \text{if } (t, \omega) \notin \bigcup_n \{(T_n(\omega), \omega) : \omega \in \Omega\} \\ Z_n(\omega) & \text{if } (t, \omega) \in \{(T_n(\omega), \omega) : \omega \in \Omega\}. \end{cases} \quad (A.22)$$

It is immediate that this Z_t is $\mathcal{F}_t^o$-optional. Let us now show that it satisfies (A.21). Let T be an $\mathcal{F}_t^o$-optional time and let $A \in \mathcal{F}_T^o$. Set $A_n = A \cap \{T = T_n\}$; thus $A \in \mathcal{F}_{T_n}^o$. Let $B = A \setminus \bigcup_n A_n$ and thus $B \in \mathcal{F}_T^o$.

From the definition (A.22),

$$Z_T 1_{A_n} 1_{T < \infty} = Z_n 1_{A_n} 1_{T_n < \infty} = 1_{A_n} \mathbb{E}[1_{T_n < \infty} X_{T_n} \mid \mathcal{F}_{T_n}^o]$$
$$= \mathbb{E}[X_{T_n} 1_{A_n} 1_{T_n < \infty} \mid \mathcal{F}_{T_n}^o] = \mathbb{E}[X_T 1_{A_n} 1_{T < \infty} \mid \mathcal{F}_T^o].$$

Consequently

$$\mathbb{E}[1_{A_n} Z_T 1_{T < \infty}] = \mathbb{E}[1_{A_n} \mathbb{E}[1_{T < \infty} X_T \mid \mathcal{F}_T^o]] = \mathbb{E}[1_{A_n} X_T 1_{T < \infty}].$$

So on A_n the conditions are satisfied. Now consider B, on which a.s. $T \neq T_n$ for all n; hence $(T(\omega), \omega) \notin L$. Since $\mathbb{P}((T(\omega), \omega) \in K) = 0$, it follows that a.s. $(T(\omega), \omega) \notin H$. Recalling the definition of H this implies that $Y_t(\omega) = \zeta_t(\omega)$ a.s.; from the Definition A.22 on B, $Z_T = Y_T$, thus

$$\mathbb{E}[1_B Z_T] = \mathbb{E}[1_B \zeta_T] = \mathbb{E}[1_B \mathbb{E}[X_T \mid \mathcal{F}_{T+}^o]] = \mathbb{E}[1_B X_T].$$

Thus on A_n for each n and on B the process Z is an optional projection of X. The uniqueness argument using the optional section theorem is exactly analogous to that used in the proof of Theorem 2.7. $\square$

A.10 Convergence of Measure-valued Random Variables

Let $(\Omega, \mathcal{F}, \mathbb{P})$ be a probability space and let $(\mu^n)_{n=1}^\infty$ be a sequence of random measures, $\mu^n : \Omega \to \mathcal{M}(\mathbb{S})$ and $\mu : \Omega \to \mathcal{M}(\mathbb{S})$ be another measure-valued random variable. In the following we define two types of convergence for sequences of measure-valued random variables:

1. $\lim_{n \to \infty} \mathbb{E}[|\mu^n f - \mu f|] = 0$ for all $f \in C_b(\mathbb{S})$.
2. $\lim_{n \to \infty} \mu^n = \mu$, $\mathbb{P}$-a.s.

We call the first type of convergence *convergence in expectation*. If there exists an integrable random variable $w : \Omega \to \mathbb{R}$ such that $\mu^n(1) \leq w$ for all n, then $\lim_{n \to \infty} \mu^n = \mu$, $\mathbb{P}$-a.s., implies that μ^n converged to μ in expectation by the dominated convergence theorem. The extra condition is satisfied if $(\mu^n)_{n=1}^\infty$ is a sequence of random probability measures, since in this case, $\mu^n(1) = 1$ for all n. We also have the following.

A.10 Convergence of Measure-valued Random Variables

Remark A.36. If μ^n converges in expectation to μ, then there exist sequences $n(m)$ such that $\lim_{m\to\infty} \mu^{n(m)} = \mu$, $\mathbb{P}$-a.s.

Proof. Since $\mathcal{M}(\mathbb{S})$ is isomorphic to $(0,\infty) \times \mathcal{P}(\mathbb{S})$, with the isomorphism being given by

$$\nu \in \mathcal{M}(\mathbb{S}) \mapsto (\nu(1), \nu/\nu(1)) \in (0,\infty) \times \mathcal{P}(\mathbb{S}),$$

it follows from Theorem 2.18 that there exists a countable convergence determining set of functions[†]

$$\mathcal{M} \triangleq \{\varphi_0, \varphi_1, \varphi_2, \ldots\}, \tag{A.23}$$

where φ_0 is the constant function equal to 1 everywhere and $\varphi_i \in C_b(\mathbb{S})$ for any $i > 0$. Since

$$\lim_{n \to \infty} \mathbb{E}\left[|\mu^n f - \mu f|\right] = 0$$

for all $f \in \{\varphi_0, \varphi_1, \varphi_2, \ldots\}$ and the set $\{\varphi_0, \varphi_1, \varphi_2, \ldots\}$ is countable, one can find a subsequence $n(m)$ such that, with probability one, $\lim_{m\to\infty} \mu^{n(m)} \varphi_i = \mu\varphi_i$ for all $i \geq 0$, hence the claim. $\square$

If a suitable bound on the rate of convergence for $\mathbb{E}\left[|\mu^n f - \mu f|\right]$ is known, then the sequence $n(m)$ can be specified explicitly. For instance we have the following.

Remark A.37. Assume that there exists a countable convergence determining set $\mathcal{M}$ such that, for any $f \in \mathcal{M}$,

$$\mathbb{E}\left[|\mu^n f - \mu f|\right] \leq \frac{c_f}{\sqrt{n}},$$

where c_f is a positive constant independent of n, then $\lim_{m\to\infty} \mu^{m^3} = \mu$, $\mathbb{P}$-a.s.

Proof. By Fatou's lemma

$$\mathbb{E}\left[\sum_{m=1}^{\infty} \left|\mu^{m^3} f - \mu f\right|\right] \leq \lim_{n\to\infty} \sum_{m=1}^{n} \mathbb{E}\left[\left|\mu^{m^3} f - \mu f\right|\right]$$

$$\leq c_f \sum_{m=1}^{\infty} \frac{1}{m^{3/2}} < \infty.$$

Hence

$$\sum_{m=1}^{\infty} \left|\mu^{m^3} f - \mu f\right| < \infty \qquad \mathbb{P}\text{-a.s.},$$

[†] Recall that $\mathcal{M}$ is a convergence determining set if, for any sequence of finite measures ν_n, $n = 1, 2, \ldots$ and ν being another finite measure for which $\lim_{n\to\infty} \nu_n f = \nu f$ for all $f \in \mathcal{M}$, it follows $\lim_{n\to\infty} \nu_n = \nu$.

therefore
$$\lim_{m\to\infty} \mu^{m^3} f = \mu f \quad \text{for any } f \in \mathcal{M}.$$

Since $\mathcal{M}$ is countable and convergence determining, it also follows that $\lim_{m\to\infty} \mu^{m^3} = \mu$, $\mathbb{P}$-a.s. $\square$

Let $d : \mathcal{P}(\mathbb{S}) \times \mathcal{P}(\mathbb{S}) \to [0, \infty)$ be the metric defined in Theorem 2.19; that is, for $\mu, \nu \in \mathcal{P}(\mathbb{S})$,
$$d(\mu, \nu) = \sum_{i=1}^{\infty} \frac{|\mu\varphi_i - \nu\varphi_i|}{2^i},$$

where $\varphi_1, \varphi_2, \ldots$ are elements of $C_b(\mathbb{S})$ such that $\|\varphi_i\|_\infty = 1$ and let $\varphi_0 = \mathbf{1}$. We can extend d to a metric on $\mathcal{M}(\mathbb{S})$ as follows.

$$d_\mathcal{M} : \mathcal{M}(\mathbb{S}) \times \mathcal{M}(\mathbb{S}) \to [0, \infty), \qquad d(\mu, \nu) \triangleq \sum_{i=0}^{\infty} \frac{1}{2^i} |\mu\varphi_i - \nu\varphi_i|. \quad (A.24)$$

The careful reader should check that $d_\mathcal{M}$ is a metric and that indeed $d_\mathcal{M}$ induces the weak topology on $\mathcal{M}(\mathbb{S})$. Using $d_\mathcal{M}$, the almost sure convergence 2. is equivalent to

2′. $\lim_{n\to\infty} d_\mathcal{M}(\mu^n, \mu) = 0$, $\quad \mathbb{P}$-a.s.

If there exists an integrable random variable $w \colon \Omega \to \mathbb{R}$ such that $\mu^n(\mathbf{1}) \le w$ for all n, then similarly, (1) implies

1′. $\lim_{n\to\infty} \mathbb{E}\left[d_\mathcal{M}(\mu^n, \mu)\right] = 0$.

However, a stronger condition (such as tightness) must be imposed in order for condition (1) to be equivalent to condition (1′).

It is usually the case that convergence in expectation is easier to establish than almost sure convergence. However, if we have control on the higher moments of the error variables $\mu^n f - \mu f$ then we can deduce the almost sure convergence of μ^n to μ. The following remark shows how this can be achieved and is used repeatedly in Chapters 8, 9 and 10.

Remark A.38. i. Assume that there exists a positive constant $p > 1$ and a countable convergence determining set $\mathcal{M}$ such that, for any $f \in \mathcal{M}$, we have
$$\mathbb{E}\left[|\mu^n f - \mu f|^{2p}\right] \le \frac{c_f}{n^p},$$
where c_f is a positive constant independent of n. Then, for any $\varepsilon \in (0, 1/2 - 1/(2p))$ there exists a positive random variable $c_{f,\varepsilon}$ almost surely finite such that
$$|\mu^n f - \mu f| \le \frac{c_{f,\varepsilon}}{n^\varepsilon}.$$

In particular, $\lim_{n\to\infty} \mu^n = \mu$, $\mathbb{P}$-a.s.

ii. Similarly, assume that there exists a positive constant $p > 1$ and a countable convergence determining set $\mathcal{M}$ such that

$$\mathbb{E}\left[d_{\mathcal{M}}(\mu^n, \mu)^{2p}\right] \leq \frac{c}{n^p},$$

where $d_{\mathcal{M}}$ is the metric defined in (A.24) and c is a positive constant independent of n. Then, for any $\varepsilon \in (0, 1/2 - 1/(2p))$ there exists a positive random variable c_ε almost surely finite such that

$$|\mu^n f - \mu f| \leq \frac{c_\varepsilon}{n^\varepsilon}, \quad \mathbb{P}\text{-a.s.}$$

In particular, $\lim_{n \to \infty} \mu^n = \mu$, $\mathbb{P}$-a.s.

Proof. As in the proof of Remark A.37,

$$\mathbb{E}\left[\sum_{n=1}^\infty n^{2\varepsilon p} |\mu^n f - \mu f|^{2p}\right] \leq c_f \sum_{m=1}^\infty \frac{1}{n^{p-2\varepsilon p}} < \infty,$$

since $p - 2\varepsilon p > 1$. Let $c_{f,\varepsilon}$ be the random variable

$$c_{f,\varepsilon} = \left(\sum_{n=1}^\infty n^{2\varepsilon p} |\mu^n f - \mu f|^{2p}\right)^{1/2p}.$$

As $(c_{f,\varepsilon})^{2p}$ is integrable, $c_{f,\varepsilon}$ is almost surely finite and

$$n^\varepsilon |\mu^n f - \mu f| \leq c_{f,\varepsilon}.$$

Therefore $\lim_{n \to \infty} \mu^n f = \mu f$ for any $f \in \mathcal{M}$. Again, since $\mathcal{M}$ is countable and convergence determining, it also follows that $\lim_{n \to \infty} \mu^n = \mu$, $\mathbb{P}$-a.s. Part (ii) of the remark follows in a similar manner. □

A.11 Gronwall's Lemma

An important and frequently used result in the theory of stochastic differential equations is Gronwall's lemma.

Lemma A.39 (Gronwall). *Let x, y and z be measurable non-negative functions on the real numbers. If y is bounded and z is integrable on $[0, T]$ for some $T \geq 0$, and for all $0 \leq t \leq T$,*

$$x_t \leq z_t + \int_0^t x_s y_s \, \mathrm{d}s, \tag{A.25}$$

then for all $0 \leq t \leq T$,

$$x_t \leq z_t + \int_0^t z_s y_s \exp\left(\int_s^t y_r \, \mathrm{d}r\right) \mathrm{d}s.$$

Proof. Multiplying both sides of the inequality (A.25) by $y_t \exp\left(-\int_0^t y_s\,ds\right)$ yields

$$x_t y_t \exp\left(-\int_0^t y_s\,ds\right) - \left(\int_0^t x_s y_s\,ds\right) y_t \exp\left(-\int_0^t y_s\,ds\right)$$
$$\leq z_t y_t \exp\left(-\int_0^t y_s\,ds\right).$$

The left-hand side can be written as the derivative of a product,

$$\frac{d}{dt}\left[\left(\int_0^t x_s y_s\,ds\right) \exp\left(-\int_0^t y_s\,ds\right)\right] \leq z_t y_t \exp\left(-\int_0^t y_s\,ds\right),$$

which can be integrated to give

$$\left(\int_0^t x_s y_s\,ds\right) \exp\left(-\int_0^t y_s\,ds\right) \leq \int_0^t z_s y_s \exp\left(-\int_0^s y_r\,dr\right) ds,$$

or equivalently

$$\int_0^t x_s y_s\,ds \leq \int_0^t z_s y_s \exp\left(\int_s^t y_r\,dr\right) ds.$$

Combining this with the original equation (A.25) gives the desired result. □

Corollary A.40. *If x is a real-valued function such that for all $t \geq 0$,*

$$x_t \leq A + B\int_0^t x_s\,ds,$$

then for all $t \geq 0$,

$$x_t \leq Ae^{Bt}.$$

Proof. We have for $t \geq 0$,

$$x_t \leq A + \int_0^t ABe^{B(t-s)}\,ds$$
$$\leq A + ABe^{Bt}(e^{-tB} - 1)/(-B) = Ae^{Bt}.$$

□

A.12 Explicit Construction of the Underlying Sample Space for the Stochastic Filtering Problem

Let $(\mathbb{S}, d)$ be a complete separable metric space (a Polish space) and Ω^1 be the space of $\mathbb{S}$-valued continuous functions defined on $[0, \infty)$, endowed with the topology of uniform convergence on compact intervals and with the Borel σ-algebra associated denoted with $\mathcal{F}^1$,

$$\Omega^1 = C([0,\infty),\mathbb{S}), \qquad \mathcal{F}^1 = \mathcal{B}(\Omega^1). \tag{A.26}$$

Let X be an $\mathbb{S}$-valued process defined on this space; $X_t(\omega^1) = \omega^1(t)$, $\omega^1 \in \Omega^1$. We observe that X_t is measurable with respect to the σ-algebra $\mathcal{F}^1$ and consider the filtration associated with the process X,

$$\mathcal{F}_t^1 = \sigma(X_s,\ s \in [0,t]). \tag{A.27}$$

Let $A : C_b(\mathbb{S}) \to C_b(\mathbb{S})$ be an unbounded operator with domain $\mathcal{D}(A)$ with $\mathbf{1} \in \mathcal{D}(A)$ and $A\mathbf{1} = 0$ and let $\mathbb{P}^1$ be a probability measure which is a solution of the martingale problem associated with the infinitesimal generator A and the initial distribution $\pi_0 \in \mathcal{P}(\mathbb{S})$, i.e., under $\mathbb{P}^1$, the distribution of X_0 is π_0 and

$$M_t^\varphi = \varphi(X_t) - \varphi(X_0) - \int_0^t A\varphi(X_s)\,ds, \qquad \mathcal{F}_t^1, \qquad 0 \le t < \infty, \tag{A.28}$$

is a martingale for any $\varphi \in \mathcal{D}(A)$. Let also Ω^2 be defined similarly to Ω^1, but with $\mathbb{S} = \mathbb{R}^m$. Hence

$$\Omega^2 = C([0,\infty),\mathbb{R}^m), \qquad \mathcal{F}^2 = \mathcal{B}(\Omega^2). \tag{A.29}$$

We consider also V to be the canonical process in Ω^2, (i.e. $V_t(\omega^2) = \omega^2(t)$, $\omega^2 \in \Omega^2$) and $\mathbb{P}^2$ to be a probability measure such that V is an m-dimensional standard Brownian motion on $(\Omega^2, \mathcal{F}^2)$ with respect to it. We consider now the following.

$$\Omega \triangleq \Omega^1 \times \Omega^2,$$
$$\mathcal{F}' \triangleq \mathcal{F}^1 \otimes \mathcal{F}^2,$$
$$\mathbb{P} \triangleq \mathbb{P}^1 \otimes \mathbb{P}^2,$$
$$\mathcal{N} \triangleq \{B \subset \Omega : B \subset A,\ A \in \mathcal{F},\ \mathbb{P}(A) = 0\}$$
$$\mathcal{F} \triangleq \mathcal{F}' \vee \mathcal{N}.$$

So $(\Omega, \mathcal{F}, \mathbb{P})$ is a complete probability space and, under $\mathbb{P}$, X and V are two independent processes. They can be viewed as processes on the product space $(\Omega, \mathcal{F}, \mathbb{P})$ in the usual way: as projections onto their original spaces of definition. If W is the canonical process on Ω, then

$$W(t) = \omega(t) = (\omega^1(t), \omega^2(t))$$
$$X = p^1(\omega) \quad \text{where } p^1 : \Omega \to \Omega^1,\ p^1(\omega) = \omega^1$$
$$V = p^2(\omega) \quad \text{where } p^2 : \Omega \to \Omega^2,\ p^2(\omega) = \omega^2.$$

M_t^φ is also a martingale with respect to the larger filtration $\mathcal{F}_t$, where

$$\mathcal{F}_t = \sigma(X_s, V_s,\ s \in [0,t]) \vee \mathcal{N}.$$

Let $h : \mathbb{S} \to \mathbb{R}^m$ be a Borel-measurable function with the property that

$$\mathbb{P}\left(\int_0^T \|h(X_s)\|\,\mathrm{d}s < \infty\right) = 1 \quad \text{for all } T > 0,$$

Finally let Y be the following stochastic process (usually called the *observation process*)

$$Y_t = \int_0^t h(s, X_s)\,\mathrm{d}s + V_t, \quad t \geq 0.$$

B

Stochastic Analysis

B.1 Martingale Theory in Continuous Time

The subject of martingale theory is too large to cover in an appendix such as this. There are many useful references, for example, Rogers and Williams [248] or Doob [81].

Theorem B.1. *If $M = \{M_t, t \geq 0\}$ is a right continuous martingale bounded in L^p for $p \geq 1$, that is, $\sup_{t \geq 0} \mathbb{E}[|M_t|^p] < \infty$, then there exists an L^p-integrable random variable M_∞ such that $M_t \to M_\infty$ almost surely as $t \to \infty$. Furthermore,*

1. *If M is bounded in L^p for $p > 1$, then $M_t \to M_\infty$ in L^p as $t \to \infty$.*
2. *If M is bounded in L^1 and $\{M_t, t \geq 0\}$ is uniformly integrable then $M_t \to M_\infty$ in L^1 as $t \to \infty$.*

If either condition (1) or (2) holds then the extended process $\{M_t, t \in [0, \infty]\}$ is a martingale.

For a proof see Theorem 1.5 of Chung and Williams [53].

The following lemma provides a very useful test for identifying martingales.

Lemma B.2. *Let $M = \{M_t, t \geq 0\}$ be a càdlàg adapted process such that for each bounded stopping time T, $\mathbb{E}[|M_T|] < \infty$ and $\mathbb{E}[M_T] = \mathbb{E}[M_0]$ then M is a martingale.*

Proof. For $s < t$ and $A \in \mathcal{F}_s$ define

$$T(\omega) \triangleq \begin{cases} s & \text{if } \omega \in A, \\ t & \text{if } \omega \in A^c. \end{cases}$$

Then T is a stopping time and

$$\mathbb{E}[M_0] = \mathbb{E}[M_T] = \mathbb{E}[M_s 1_A] + \mathbb{E}[M_t 1_{A^c}],$$

and trivially for the stopping time t,

$$\mathbb{E}[M_0] = \mathbb{E}[M_t] = \mathbb{E}[M_t 1_A] + \mathbb{E}[M_t 1_{A^c}],$$

so $\mathbb{E}[M_t 1_A] = \mathbb{E}[M_s 1_A]$ which implies that $M_s = \mathbb{E}[M_t \mid \mathcal{F}_s]$ a.s. which together with the integrability condition implies M is a martingale. □

By a straightforward change to this proof the following corollary may be established.

Corollary B.3. *Let $\{M_t,\ t \geq 0\}$ be a càdlàg adapted process such that for each stopping time (potentially infinite) T, $\mathbb{E}[|M_T|] < \infty$ and $\mathbb{E}[M_T] = \mathbb{E}[M_0]$ then M is a uniformly integrable martingale.*

Definition B.4. *Let M be a stochastic process. If M_0 is $\mathcal{F}_0$-measurable and there exists an increasing sequence T_n of stopping times such that $T_n \to \infty$ a.s. and such that*

$$M_n^T = \{M_{t \wedge T_n} - M_0,\ t \geq 0\}$$

is a $\mathcal{F}_t$-adapted martingale for each $n \in \mathbb{N}$, then M is called a local martingale *and the sequence T_n is called a* reducing sequence *for the local martingale M.*

The initial condition M_0 is treated separately to avoid imposing integrability conditions on M_0.

B.2 Itô Integral

The stochastic integrals which arise in this book are the integrals of stochastic processes with respect to continuous local martingales. The following section contains a very brief overview of the construction of the Itô integral in this context and the necessary conditions on the integrands for the integral to be well defined.

The results are presented starting from the previsible integrands, since in the general theory of stochastic integration these form the natural class of integrators. The results then extend in the case of continuous martingale integrators to integrands in the class of progressively measurable processes and if the quadratic variation of the continuous martingale is absolutely continuous with respect to Lebesgue measure (as for example in the case of integrals with respect to Brownian motion) then this extends further to all adapted, jointly measurable processes. It is also possible to construct directly the stochastic integral with a continuous martingale integrator on the space of progressively measurable processes (this approach is followed in e.g. Ethier and Kurtz [95]).

There are numerous references which describe the material in this section in much greater detail; examples include Chung and Williams [53], Karatzas and Shreve [149], Protter [247] and Dellacherie and Meyer [79].

Definition B.5. *The previsible (predictable) σ-algebra denoted $\mathcal{P}$ is the σ-algebra of subsets of $[0, \infty) \times \Omega$ generated by left continuous processes valued in $\mathbb{R}$; that is, it is the smallest σ-algebra with respect to which all left continuous processes are measurable. A process is said to be* previsible *if it is $\mathcal{P}$-measurable.*

Lemma B.6. *Let $\mathcal{A}$ be the ring[†] of subsets of $[0, \infty) \times \Omega$ generated by the sets of the form $\{(s,t] \times A\}$ where $A \in \mathcal{F}_s$ and $0 \le s < t$ and the sets $\{0\} \times A$ for $A \in \mathcal{F}_0$. Then $\sigma(\mathcal{A}) = \mathcal{P}$.*

Proof. It suffices to show that any adapted left continuous process (as a generator of $\mathcal{P}$) can be approximated by finite linear combinations of indicator functions of elements of $\mathcal{A}$. Let H be a bounded adapted left continuous process; define

$$H_t = \lim_{k \to \infty} \lim_{n \to \infty} \sum_{i=2}^{nk} H_{(i-1)/n} 1_{((i-1)/n, i/n]}(t).$$

As H_t is adapted it follows that $H_{(i-1)/n} \in \mathcal{F}_{(i-1)/n}$, thus each term in the sum is $\mathcal{A}$-measurable, and therefore by linearity so is the whole sum. □

Definition B.7. *Define the vector space of elementary function $\mathcal{E}$ to be the space of finite linear combinations of indicator functions of elements of $\mathcal{A}$.*

Definition B.8. *For the indicator function $X = 1_{\{(s,t] \times A\}}$ for $A \in \mathcal{F}_s$, which an element of $\mathcal{E}$, we can define the stochastic integral*

$$\int_0^\infty X_r \, dM_r \triangleq 1_A (M_t - M_s).$$

For $X = 1_{\{0\} \times A}$ where $A \in \mathcal{F}_0$, define the integral to be identically zero. This definition can be extended by linearity to the space of elementary functions $\mathcal{E}$. Further define the integral between 0 and t by

$$\int_0^t X_r \, dM_r \triangleq \int_0^\infty 1_{[0,t]}(r) X_r \, dM_r.$$

Lemma B.9. *If M is a martingale and $X \in \mathcal{E}$ then $\int_0^t X_r \, dM_r$ is a $\mathcal{F}_t$-adapted martingale.*

Proof. Consider $X_t = 1_A 1_{(r,s]}(t)$ where $A \in \mathcal{F}_r$. From Definition B.8,

$$\int_0^t X_p \, dM_p = \int_0^\infty 1_{[0,t]}(p) X_p \, dM_p = 1_A (M_{s \wedge t} - M_{r \wedge t}),$$

and hence as M is a martingale and $A \in \mathcal{F}_r$, then by considering separately the cases $0 \le p \le r$, $r < p \le s$ and $p > s$ it follows that

---

[†] A ring is a class of subsets closed under finite unions and set differences $A \setminus B$ and which contains the empty set.

$$\mathbb{E}\left[\int_0^t X_r \, dM_r \,\bigg|\, \mathcal{F}_p\right] = \mathbb{E}\left[1_A(M_{s\wedge t} - M_{r\wedge t}) \mid \mathcal{F}_p\right]$$
$$= 1_A \mathbb{E}(M_{s\wedge p} - M_{r\wedge p}) = \int_0^p X_s \, dM_s.$$

By linearity, this result extends to $X \in \mathcal{E}$. □

B.2.1 Quadratic Variation

The total variation is the variation which is used in the construction of the usual Lebesgue–Stieltjes integral. This cannot be used to define a non-trivial stochastic integral, as any continuous local martingale of finite variation is indistinguishable from zero.

Definition B.10. *The quadratic variation process*[†] $\langle M \rangle_t$ *of a continuous square integrable martingale M is a continuous increasing process A_t starting from zero such that $M_t^2 - A_t$ is a martingale.*

Theorem B.11. *If M is a continuous square integrable martingale then the quadratic variation process $\langle M \rangle_t$ exists and is unique.*

The following proof is based on Theorem 4.3 of Chung and Williams [53] who attribute the argument to M. J. Sharpe.

Proof. Without loss of generality consider a martingale starting from zero. The result is first proved for a martingale which is bounded by C. For given $n \in \mathbb{N}$, define $t_j^n \triangleq j2^{-n}$ and $\hat{t}_j^n \triangleq t \wedge t_j^n$ for $j \in \mathbb{N}$ and

$$S_t^n \triangleq \sum_{j=0}^{\infty} \left(M_{\hat{t}_{j+1}^n} - M_{\hat{t}_j^n}\right)^2.$$

By rearrangement of terms in the summations

$$M_t^2 = \sum_{k=0}^{\infty} \left(M_{\hat{t}_{k+1}^n}^2 - M_{\hat{t}_k^n}^2\right)$$
$$= 2\sum_{k=0}^{\infty} M_{\hat{t}_k^n}\left(M_{\hat{t}_{k+1}^n} - M_{\hat{t}_k^n}\right) + \sum_{k=0}^{\infty}\left(M_{\hat{t}_{k+1}^n} - M_{\hat{t}_k^n}\right)^2.$$

Therefore

$$S_t^n = M_t^2 - 2\sum_{k=0}^{\infty} M_{\hat{t}_k^n}\left(M_{\hat{t}_{k+1}^n} - M_{\hat{t}_k^n}\right). \tag{B.1}$$

[†] Technically, if we were to study discontinuous processes what is being constructed here should be denoted $[M]_t$. The process $\langle M \rangle_t$, when it exists, is the dual previsible projection of $[M]_t$. In the continuous case, the two processes coincide, and historical precedent makes $\langle M \rangle_t$ the more common notation.

For fixed n and t the summation in (B.1) contains a finite number of non zero terms each of which is a continuous martingale. It therefore follows that the $S_t^n - M_t^2$ is a continuous martingale for each n.

It is now necessary to show that as $n \to \infty$, for fixed t, the sequence $\{S_t^n, n \in \mathbb{N}\}$ is a Cauchy sequence and therefore converges in L^2. If we consider fixed $m < n$ and for notational convenience write t_j for t_j^n, then it is possible to relate the points on the two dyadic meshes by setting $t_j' = 2^{-m}[t_j 2^m]$ and $\hat{t}_j \triangleq t \wedge t_j'$; that is, t_j' is the closest point on the coarser mesh to the left of t_j. It follows from (B.1) that

$$S_t^n - S_t^m = -2 \sum_{j=0}^{[2^n t]} \left(M_{\hat{t}_j} - M_{\hat{t}_j'}\right)\left(M_{\hat{t}_{j+1}} - M_{\hat{t}_j}\right). \tag{B.2}$$

Define $Z_j \triangleq M_{\hat{t}_j} - M_{\hat{t}_j'}$; as $t_j' \leq t_j$ it follows that Z_j is $\mathcal{F}_{\hat{t}_j}$-measurable. For $j < k$ since $Z_j(M_{\hat{t}_{j+1}} - M_{\hat{t}_j})Z_k$ is $\mathcal{F}_{\hat{t}_k}$-measurable it follows that

$$\mathbb{E}\left[Z_j\left(M_{\hat{t}_{j+1}} - M_{\hat{t}_j}\right)Z_k\left(M_{\hat{t}_{k+1}} - M_{\hat{t}_k}\right)\right] = 0. \tag{B.3}$$

Hence using (B.3) and the Cauchy–Schwartz inequality

$$\mathbb{E}\left[(S_t^n - S_t^m)^2\right] = 4\mathbb{E}\left[\sum_{j=0}^{[2^n t]} Z_j^2 \left(M_{\hat{t}_{j+1}} - M_{\hat{t}_j}\right)^2\right]$$

$$\leq 4\mathbb{E}\left[\sup_{\substack{0 \leq r \leq s \leq t \\ s-r < 2^{-m}}} (M_r - M_s)^2 \sum_{j=0}^{[2^n t]} \left(M_{\hat{t}_{j+1}} - M_{\hat{t}_j}\right)^2\right]$$

$$\leq 4 \sqrt{\mathbb{E}\left[\left(\sup_{\substack{0 \leq r \leq s \leq t \\ s-r < 2^{-m}}} (M_r - M_s)^2\right)^2\right]}$$

$$\times \sqrt{\mathbb{E}\left[\left(\sum_{j=0}^{[2^n t]} \left(M_{\hat{t}_{j+1}} - M_{\hat{t}_j}\right)^2\right)^2\right]}.$$

The first term tends to zero using the fact that M being continuous is uniformly continuous on the bounded time interval $[0, t]$. It remains to show that the second term is bounded. Write $a_j \triangleq (M_{\hat{t}_{j+1}} - M_{\hat{t}_j})^2$, for $j \in \mathbb{N}$; then

$$\mathbb{E}\left[\left(\sum_{j=0}^{[2^n t]}\left(M_{\hat{t}_{j+1}}-M_{\hat{t}_j}\right)^2\right)^2\right] = \mathbb{E}\left[\left(\sum_{j=0}^{[2^n t]} a_j\right)^2\right]$$

$$= \mathbb{E}\left[\sum_{j=0}^{[2^n t]} a_j^2 + 2\sum_{j=0}^{[2^n t]} a_j \sum_{k=j+1}^{[2^n t]} a_k\right]$$

$$= \mathbb{E}\left[\sum_{j=0}^{[2^n t]} a_j^2\right] + 2\mathbb{E}\left[\sum_{j=0}^{[2^n t]} a_j \mathbb{E}\left[\sum_{k=j+1}^{[2^n t]} a_k \,\Big|\, \mathcal{F}_{\hat{t}_{j+1}}\right]\right].$$

It is clear that since the a_js are non-negative and M is bounded by C that

$$\sum_{j=0}^{[2^n t]} a_j^2 \le \max_{l=0,\ldots,[2^n t]} a_l \sum_{j=0}^{[2^n t]} a_j \le 4C^2 \sum_{j=0}^{[2^n t]} a_j$$

and

$$\mathbb{E}\left[\sum_{k=j+1}^{[2^n t]} a_k \,\Big|\, \mathcal{F}_{\hat{t}_{j+1}}\right] = \sum_{k=j+1}^{\infty} \mathbb{E}\left[\left(M_{\hat{t}_{k+1}}-M_{\hat{t}_k}\right)^2 \,\Big|\, \mathcal{F}_{\hat{t}_{j+1}}\right]$$

$$= \sum_{k=j+1}^{\infty} \mathbb{E}\left[M_{\hat{t}_{k+1}}^2 - M_{\hat{t}_k}^2 \,\Big|\, \mathcal{F}_{\hat{t}_{j+1}}\right]$$

$$= \mathbb{E}\left[M_t^2 - M_{\hat{t}_{j+1}}^2 \,\Big|\, \mathcal{F}_{\hat{t}_{j+1}}\right]$$

$$\le C^2.$$

From these two bounds

$$\mathbb{E}\left[\left(\sum_{j=0}^{[2^n t]}\left(M_{\hat{t}_{j+1}}-M_{\hat{t}_j}\right)^2\right)^2\right] \le (4C^2 + 2C^2)\mathbb{E}\left[\sum_{j=0}^{[2^n t]} a_j\right]$$

$$= 6C^2 \mathbb{E}\left[M_t^2\right] < \infty.$$

As this bound holds uniformly in n, m, as n and $m \to \infty$ it follows that $S_t^n - S_t^m \to 0$ in the L^2 sense and hence the sequence $\{S_t^n,\ n \in \mathbb{N}\}$ converges in L^2 to a limit which we denote S_t. As the martingale property is preserved by L^2 limits, it follows that $\{M_t^2 - S_t,\ t \ge 0\}$ is a martingale.

It is necessary to show that S_t is increasing. Let $s < t$,

$$S_t - S_s = \lim_{n \to \infty}\left(S_t^n - S_s^n\right) \text{ in } L^2.$$

Then writing $k \triangleq \inf\{j : \hat{t}_j > s\}$,

$$S_t^n - S_s^n = \sum_{t_j > s}\left(M_{\hat{t}_{j+1}}-M_{\hat{t}_j}\right)^2 + \left(M_{\hat{t}_k}-M_{\hat{t}_{k-1}}\right)^2 - \left(M_s - M_{\hat{t}_{k-1}}\right)^2.$$

Clearly

$$\left|\left(M_{\hat{t}_k} - M_{\hat{t}_{k-1}}\right)^2 - \left(M_s - M_{\hat{t}_{k-1}}\right)^2\right| \leq 2 \sup_{\substack{0 \leq r \leq s \leq t \\ s-r < 2^{-m}}} (M_r - M_s)^2,$$

where the bound on the right-hand side tends to zero in L^2 as $n \to \infty$. Therefore in L^2

$$S_t - S_s = \lim_{n \to \infty} \sum_{t_j > s} \left(M_{\hat{t}_{j+1}} - M_{\hat{t}_j}\right)^2$$

and hence $S_t - S_s \geq 0$ almost surely, so the process S is a.s. increasing.

It remains to show that a version of S_t can be chosen which is almost surely continuous. By Doob's L^2-inequality applied to the martingale (B.2) it follows that

$$\mathbb{E}\left[\sup_{t \leq a} |S_t^n - S_t^m|^2\right] \leq 4 \mathbb{E}\left[(S_a^n - S_a^m)^2\right];$$

thus a suitable subsequence n_k can be chosen such that $S_t^{n_k}$ converges a.s. uniformly on compact time intervals to a limit S which from the continuity of M must be continuous a.s.

Uniqueness follows from the result that a continuous local martingale of finite variation is everywhere zero. Suppose the process A in the above definition were not unique. That is, suppose that also for some B_t continuous increasing from zero, $M_t^2 - B_t$ is a martingale. Then as $M_t^2 - A_t$ is also a martingale, by subtracting these two equations we get that $A_t - B_t$ is a martingale, null at zero. It clearly must have finite variation, and hence be zero.

To extend to the general case where the martingale M is not bounded use a sequence of stopping times

$$T_n \triangleq \inf\{t \geq 0 : |M_t| > n\};$$

then $\{M_t^{T_n}, t \geq 0\}$ is a bounded martingale to which the proof can be applied to construct $\langle M^{T_n} \rangle$. By uniqueness it follows that $\langle M^{T_n} \rangle$ and $\langle M^{T_{n+1}} \rangle$ agree on $[0, T_n]$ so a process $\langle M \rangle$ may be defined. □

Definition B.12. *Define a measure on $([0,\infty) \times \Omega, \mathcal{P})$ in terms of the quadratic variation of M via*

$$\mu_M(A) \triangleq \mathbb{E}\left[\int_0^\infty 1_A(s,\omega) \, d\langle M\rangle_s\right]. \tag{B.4}$$

In terms of this measure we can define an associated norm on a $\mathcal{P}$-measurable process X via

$$\|X\|_M \triangleq \int_{[0,\infty) \times \Omega} X^2 \, d\mu_M. \tag{B.5}$$

This norm can be written using (B.4) and (B.5) more simply as

$$\|X\|_M = \mathbb{E}\left[\int_0^\infty X_s^2 \, d\langle M\rangle_s\right].$$

Definition B.13. *Define* $\mathcal{L}_\mathcal{P}^2 \triangleq \{X \in \mathcal{P} : \|X\|_M < \infty\}$.

This space $\mathcal{L}_\mathcal{P}^2$ with associated norm $\|\cdot\|_M$ is a Banach space. Denote by $L_\mathcal{P}^2$ the space of equivalence classes of elements of $\mathcal{L}_\mathcal{P}^2$, where we consider the equivalence class of an element X to be all those elements $Y \in \mathcal{L}_\mathcal{P}^2$ which satisfy $\|X - Y\|_M = 0$.

Lemma B.14. *The space of bounded elements of $\mathcal{E}$, which we denote $\bar{\mathcal{E}}$ is dense in the subspace of bounded functions in $\mathcal{L}_\mathcal{P}^2$.*

Proof. This is a classical monotone class theorem proof which explains the requirement to work within spaces of bounded functions. Define

$$\mathcal{C} = \left\{H \in \mathcal{P} : H \text{ is bounded}, \forall \varepsilon > 0 \ \exists J \in \bar{\mathcal{E}} : \|H - J\|_M < \varepsilon\right\}.$$

It is clear that $\bar{\mathcal{E}} \subset \mathcal{C}$. Thus it also follows that the constant function one is included in $\mathcal{C}$. The fact that $\mathcal{C}$ is a vector space is immediate. It remains to verify that if $H_n \uparrow H$ where $H_n \in \mathcal{C}$ with H bounded that this implies $H \in \mathcal{C}$.

Fix $\varepsilon > 0$. By the bounded convergence theorem for Stieltjes integrals, it follows that $\|H_n - H\|_M \to 0$ as $n \to \infty$; thus we can find N such that for $n \geq N$, $\|H_n - H\|_M < \varepsilon/2$. As $H_N \in \mathcal{C}$, it follows that there exists $J \in \bar{\mathcal{E}}$ such that $\|J - H_N\|_M < \varepsilon/2$. Thus by the triangle inequality $\|H - J\|_M \leq \|H - H_N\|_M + \|H_N - J\|_M < \varepsilon$. Hence by the monotone class theorem $\sigma(\bar{\mathcal{E}}) \subset \mathcal{C}$. □

Lemma B.15. *For $X \in \mathcal{E}$ it follows that*

$$\mathbb{E}\left[\left(\int_0^\infty X_r \, dM_r\right)^2\right] = \|X\|_M.$$

Proof. Consider $X = 1_{(s,t] \times A}$ where $A \in \mathcal{F}_s$ and $s < t$. Then

$$\mathbb{E}\left[\left(\int_0^\infty X_t \, dM_r\right)^2\right] = \mathbb{E}\left[\left(\int_0^\infty 1_{(s,t]}(r)1_A \, dM_r\right)^2\right]$$
$$= \mathbb{E}\left[(M_t - M_s)^2 1_A\right]$$
$$= \mathbb{E}\left[1_A\left(M_t^2 - 2M_t M_s + M_s^2\right)\right]$$
$$= \mathbb{E}\left[1_A\left(M_t^2 + M_s^2\right)\right] - 2\mathbb{E}\left[1_A \mathbb{E}\left[M_t M_s \mid \mathcal{F}_s\right]\right]$$
$$= \mathbb{E}\left[1_A\left(M_t^2 - M_s^2\right)\right].$$

Then from the definition of μ_M it follows that

$$\mu_M((s,t] \times A) = \mathbb{E}[1_A(\langle M\rangle_t - \langle M\rangle_s)].$$

We know $M_t^2 - \langle M \rangle_t$ is a local martingale, so it follows that

$$\mu_M((s,t] \times A) = \mathbb{E}\left[\left(\int_0^\infty 1_{(s,t]}(r)1_A \, dM_r\right)^2\right]$$

and by linearity this extends to functions in $\mathcal{E}$. □

As a consequence of Lemma B.14 it follows that given any bounded $X \in \mathcal{L}_\mathcal{P}^2$ we can construct an approximating sequence $X^n \in \mathcal{E}$ such that $\|X^n - X\|_M \to 0$ as $n \to \infty$. Using Lemma B.15 it follows that $\int_0^\infty X_s^n \, dM_s$ is a Cauchy sequence in the L^2 sense; thus we can make the following definition.

Definition B.16. *For $X \in \mathcal{L}_\mathcal{P}^2$ we may define the Itô integral in the L^2 sense through the isometry*

$$\mathbb{E}\left[\left(\int_0^\infty X_r \, dM_r\right)^2\right] = \|X\|_M. \tag{B.6}$$

We must check that this extension of the stochastic integral is well defined. That is, consider another approximating sequence $Y_n \to X$; we must show that this converges to the same limit as the sequence X_n considered previously, but this is immediate from the isometry.

Remark B.17. From the above definition of the stochastic integral in an L^2 sense as a limit of approximations $\int_0^\infty X_r^n \, dM_r$, it follows that since convergence in L^2 implies convergence in probability we can also define the extension of the stochastic integral as a limit in probability. By a standard result, there exists a subsequence n_k such that $\int_0^\infty X_r^{n_k} \, dM_r$ converges a.s. as $k \to \infty$. It might appear that this would lead to a pathwise extension (i.e. a definition for each ω). However, this a.s. limit is not well defined: different choices of approximating sequence can give rise to limits which differ on (potentially different) null sets. As there are an uncountable number of possible approximating sequences the union of these null sets may not be null and thus the limit not well defined.

The following theorem finds numerous applications throughout the book, usually to show that the expectation of a particular stochastic integral term is 0.

Theorem B.18. *If $X \in \mathcal{L}_\mathcal{P}^2$ and M is a square integrable martingale then $\int_0^t X_s \, dM_s$ is a martingale.*

Proof. Let $X^n \in \mathcal{E}$ be sequence converging to X in the $\|\cdot\|_M$ norm; then by Lemma B.9 each $\int_0^t X_s^n \, dM_s$ is a martingale. By the Itô isometry $\int_0^t X_s^n \, dM_s$ converges to $\int_0^t X_s \, dM_s$ in L^2 and the martingale property is preserved by L^2 limits. □

B.2.2 Continuous Integrator

The foregoing arguments cannot be used to extend the definition of the stochastic integral to integrands outside of the class of previsible processes. For example, the previsible processes do not form a dense set in the space of progressively measurable processes so approximation arguments can not be used to extend the definition to progressively measurable integrands. The approach taken here is based on Chung and Williams [53].

Let $\tilde{\mu}_M$ be a measure on $[0, \infty) \times \Omega$ which is an extension of μ_M (that is μ_M and $\tilde{\mu}_M$ agree on $\mathcal{P}$ and $\tilde{\mu}_M$ is defined on a larger σ-algebra than $\mathcal{P}$).

Given a process X which is $\mathcal{B} \times \mathcal{F}$-measurable, if there is a previsible process Z such that

$$\int_{[0,\infty) \times \Omega} (X - Z)^2 \, \mathrm{d}\tilde{\mu}_M = 0, \tag{B.7}$$

which, by the usual Lebesgue argument, is equivalent to

$$\tilde{\mu}_M((t, \omega) : X_t(\omega) \neq Z_t(\omega)) = 0,$$

then we may define $\int_0^\infty X_s \, \mathrm{d}M_s \triangleq \int_0^\infty Z_s \, \mathrm{d}M_s$. In general we cannot hope to find such a Z for all $\mathcal{B} \times \mathcal{F}$-measurable X. However, in the case where the integrator M is continuous we can find such a previsible Z for all progressively measurable X.

Let $\tilde{\mathcal{N}}$ be the set of $\tilde{\mu}_M$ null sets and define $\tilde{\mathcal{P}} = \mathcal{P} \vee \tilde{\mathcal{N}}$; then it follows that for X a $\tilde{\mathcal{P}}$-measurable process, we can find a process Z in $\mathcal{P}$ such that $\tilde{\mu}_M((t, \omega) : X_t(\omega) \neq Z_t(\omega)) = 0$. Hence (B.7) will hold and consequently we may define $\int_0^\infty X_s \, \mathrm{d}M_s \triangleq \int_0^\infty Z_s \, \mathrm{d}M_s$. The following theorem is an important application of this result.

Theorem B.19. *Let M be a continuous martingale. Then if X is progressively measurable we can define the integral of X with respect to M in the Itô sense through the extension of the isometry*

$$\mathbb{E}\left[\left(\int_0^\infty X_s \, \mathrm{d}M_s\right)^2\right] = \mathbb{E}\left[\int_0^\infty X_s^2 \, \mathrm{d}\tilde{\mu}_M\right].$$

Proof. From the foregoing remarks, it is clear that it is sufficient to show that every progressively measurable process X is $\tilde{\mathcal{P}}$-measurable. There are two approaches to establishing this: one is direct via the previsible projection and the other indirect via the optional projection. In either case, the result of Lemma B.21 is established, and the conclusion of the theorem follows. □

Optional Projection Route

We begin with a measurability result which we need in the proof of the main result in this section.

Lemma B.20. *If X is progressively measurable and T is a stopping time, then $X_T 1_{\{T<\infty\}}$ is $\mathcal{F}_T$-measurable.*

Proof. For fixed t the map $\omega \mapsto X(t,\omega)$ defined on $[0,t] \times \Omega$ is $\mathcal{B}[0,t] \otimes \mathcal{F}$-measurable. Since T is a stopping time $\omega \mapsto T(\omega) \wedge t$ is $\mathcal{F}_t$-measurable. By composition of functions[†] it follows that $\omega \mapsto X(T(\omega) \wedge t, \omega)$ is $\mathcal{F}_t$-measurable. Now define $Y = X_T 1_{\{T \le \infty\}}$; for any t it is clear $Y 1_{\{T \le t\}} = X_{T \wedge t} 1_{\{T \le t\}}$. Hence on $\{T \le t\}$ it follows that Y is $\mathcal{F}_t$-measurable, which by the definition of $\mathcal{F}_T$ implies that Y is $\mathcal{F}_T$-measurable. □

Lemma B.21. *The set of progressively measurable functions on $[0,\infty) \times \Omega$ is contained in $\tilde{\mathcal{P}}$.*

Proof. First we must show that all optional processes are $\tilde{\mathcal{P}}$-measurable. This is straightforward: if τ is a stopping time we must show that $1_{[0,\tau]}$ is $\tilde{\mathcal{P}}$-measurable. But $1_{[0,\tau)}$ is previsible and thus automatically $\tilde{\mathcal{P}}$-measurable, hence it is sufficient to establish that $[\tau] \triangleq \{(\tau(\omega),\omega) : \tau(\omega) < \infty, \omega \in \Omega\} \in \tilde{\mathcal{P}}$. But

$$\tilde{\mu}_M([\tau]) = \mathbb{E}\left[\int_0^\infty 1_{\{\tau(\omega)=s\}} \, d\langle M \rangle_s\right] = \mathbb{E}[\langle M \rangle_t - \langle M \rangle_{t-}] = 0;$$

the final equality follows from the fact that M_t is continuous.

Starting from a progressively measurable process X, by Theorem 2.7 we can construct its optional projection ${}^oX$. From (B.4),

$$\tilde{\mu}_M((t,\omega) : {}^oX_t(\omega) \ne X_t(\omega)) = \mathbb{E}\left[\int_0^\infty 1_{\{{}^oX_s(\omega) \ne X_s(\omega)\}} \, d\langle M \rangle_s\right].$$

Define
$$\tau_t = \inf\{s \ge 0 : \langle M \rangle_s > t\};$$

since the set $(t,\infty) \times \Omega$ is progressively measurable, and $\langle M \rangle_t$ is continuous and hence progressively measurable, it follows that τ_t is a stopping time by the Début theorem (Theorem A.20). Hence,

$$\tilde{\mu}_M((t,\omega) : {}^oX_t(\omega) \ne X_t(\omega)) = \mathbb{E}\left[\int_0^\infty 1_{\{{}^oX_s(\omega) \ne X_s(\omega)\}} \, d\langle M \rangle_s\right]$$

$$= \mathbb{E}\left[\int_0^{\langle M \rangle_\infty} 1_{\{{}^oX_{\tau_s}(\omega) \ne X_{\tau_s}(\omega)\}} \, ds\right]$$

$$= \mathbb{E}\left[\int_0^\infty 1_{\{\tau_s < \infty\}} 1_{\{{}^oX_{\tau_s}(\omega) \ne X_{\tau_s}(\omega)\}} \, ds\right].$$

[†] It is important to realise that this argument depends fundamentally on the progressive measurability of X, it is in fact the same argument which is used (e.g. in Rogers and Williams [248, Lemma II.73.11]) to show that for progressively measurable X, X_T is $\mathcal{F}_T$-measurable for T an $\mathcal{F}_t$-stopping time.

Thus using Fubini's theorem

$$\tilde{\mu}_M((t,\omega) : {}^oX_t(\omega) \neq X_t(\omega)) = \mathbb{E}\left[\int_0^\infty 1_{\{\tau_s < \infty\}} 1_{\{{}^oX_{\tau_s} \neq X_{\tau_s}\}} \, ds\right]$$

$$= \int_0^\infty \mathbb{P}(\tau_s < \infty, \; {}^oX_{\tau_s} \neq X_{\tau_s}) \, ds.$$

From Lemma B.20 it follows that for any stopping time τ, $X_\tau 1_{\{\tau < \infty\}}$ is $\mathcal{F}_\tau$-measurable; thus from the definition of optional projection

$${}^oX_\tau 1_{\{\tau < \infty\}} = \mathbb{E}[X_\tau 1_{\{\tau < \infty\}} \mid \mathcal{F}_\tau]$$

$$= X_\tau 1_{\{\tau < \infty\}} \qquad \mathbb{P}\text{-a.s.}$$

Hence $\tilde{\mu}_M((t,\omega) : {}^oX_t(\omega) \neq X_t(\omega)) = 0$. But we have shown that the optional processes are $\tilde{\mathcal{P}}$-measurable, and ${}^oX$ is an optional process; thus from the definition of $\tilde{\mathcal{P}}$ there exists a previsible process Z such that $\tilde{\mu}_M((t,\omega) : Z_t(\omega) \neq {}^oX_t(\omega)) = 0$ hence using these two results $\tilde{\mu}_M((t,\omega) : Z_t(\omega) \neq X_t(\omega)) = 0$ which implies that X is $\tilde{\mathcal{P}}$-measurable. □

Previsible Projection Route

While the previous approach shows that the progressively measurable processes can be viewed as the class of integrands, the argument is not constructive. By considering the previsible projection we can provide a constructive argument. In brief, if X is progressively measurable and M is a continuous martingale then

$$\int_0^\infty X_s \, dM_s = \int_0^\infty {}^pX_s \, dM_s,$$

where ${}^pX$, the previsible projection of X, is a previsible process and the integral on the right-hand side is to be understood in the sense of Definition B.16.

Lemma B.22. *If X is progressively measurable and T is a previsible time, then $X_T 1_{\{T < \infty\}}$ is $\mathcal{F}_{T-}$-measurable.*

Proof. If T is a previsible time then there exists an announcing sequence $T^n \uparrow T$ such that T_n is a stopping time. By Lemma B.20 it follows for each n that $X_{T_n} 1_{\{T_n < \infty\}}$ is $\mathcal{F}_{T_n}$-measurable. Recall that

$$\mathcal{F}_{T-} = \bigvee_n \mathcal{F}_{T_n},$$

so if we define random variables $Y^n \triangleq X_{T_n} 1_{\{T_n < \infty\}}$ and

$$Y \triangleq \liminf_{n \to \infty} Y^n,$$

then it follows that Y is $\mathcal{F}_{T-}$-measurable. □

From the Début theorem,
$$\tau_t \triangleq \inf\{s \geq 0 : \langle M \rangle_s > t\}$$
is a $\mathcal{F}_t$-stopping time. Therefore $\tau_{t-1/n}$ is an increasing sequence of stopping times and their limit is
$$\hat{\tau}_t \triangleq \inf\{s \geq 0 : \langle M \rangle_s \geq t\}$$
therefore $\hat{\tau}_t$ is a previsible time. We can now complete the proof of Lemma B.21 using the definition of the previsible projection.

Proof. Starting from a progressively measurable process X by Theorem A.29 we can construct its previsible projection ${}^pX$, from (B.4),
$$\tilde{\mu}_M({}^pX_t(\omega) \neq X_t(\omega)) = \mathbb{E}\left[\int_0^\infty 1_{\{(s,\omega){}^pX_s(\omega) \neq X_s(\omega)\}}\, d\langle M \rangle_s\right].$$

Using the previsible time $\hat{\tau}_t$,
$$\tilde{\mu}_M({}^pX_t(\omega) \neq X_t(\omega)) = \mathbb{E}\left[\int_0^\infty 1_{\{{}^pX_s(\omega) \neq X_s(\omega)\}}\, d\langle M \rangle_s\right]$$
$$= \mathbb{E}\left[\int_0^{\langle M \rangle_\infty} 1_{\{{}^pX_{\hat{\tau}_s}(\omega) \neq X_{\hat{\tau}_s}(\omega)\}}\, ds\right]$$
$$= \mathbb{E}\left[\int_0^\infty 1_{\{\hat{\tau}_s < \infty\}} 1_{\{{}^pX_{\hat{\tau}_s}(\omega) \neq X_{\hat{\tau}_s}(\omega)\}}\, ds\right].$$

Thus using Fubini's theorem
$$\tilde{\mu}_M((t,\omega) : {}^pX_t(\omega) \neq X_t(\omega)) = \int_0^\infty \mathbb{P}(\hat{\tau}_s < \infty,\ {}^pX_{\hat{\tau}_s} \neq X_{\hat{\tau}_s})\, ds.$$

From Lemma B.22 it follows that for any previsible time $\hat{\tau}$, $X_{\hat{\tau}} 1_{\{\hat{\tau} < \infty\}}$ is $\mathcal{F}_{\hat{\tau}-}$-measurable; thus from the definition of previsible projection
$$ {}^pX_\tau 1_{\{\hat{\tau}<\infty\}} = \mathbb{E}[X_{\hat{\tau}} 1_{\{\hat{\tau}<\infty\}} \mid \mathcal{F}_{\hat{\tau}-}]$$
$$= X_{\hat{\tau}} 1_{\{\hat{\tau}<\infty\}} \quad \mathbb{P}\text{-a.s.}$$

Hence $\tilde{\mu}_M((t,\omega) : {}^pX_t(\omega) \neq X_t(\omega)) = 0$. Therefore X is $\tilde{\mathbb{P}}$-measurable. We also see that the previsible process Z in (B.7) is just the previsible projection of X. □

B.2.3 Integration by Parts Formula

The stochastic form of the integration parts formula leads to Itô's formula which is the most important result for practical computations.

Lemma B.23. *Let M be a continuous martingale. Then*

$$\langle M \rangle_t = M_t^2 - M_0^2 - 2 \int_0^t M_s \, dM_s.$$

Proof. Following the argument and notation of the proof of Theorem B.11 define X^n by

$$X_s^n(\omega) \triangleq \sum_{j=0}^{\infty} M_{t_j}(\omega) 1_{(t_j, t_{j+1}]}(s);$$

while X^n is defined in terms of an infinite number of non-zero terms, it is clear that $1_{[0,t]}(s) X_s^n \in \mathcal{E}$. Therefore using the definition B.8,

$$S_t^n = \sum_{j=0}^{\infty} \left(M_{\hat{t}_{j+1}}^2 - M_{\hat{t}_j}^2 - 2 M_{\hat{t}_j} \left(M_{\hat{t}_{j+1}} - M_{\hat{t}_j} \right) \right)$$

$$= M_t^2 - M_0^2 - \int_0^{\infty} 1_{[0,t]}(s) X_s^n \, dM_s.$$

As the process M is continuous, it is clear that for fixed ω, $X^n(\omega) \to M(\omega)$ uniformly on compact subsets of time and therefore by bounded convergence, $\|X^n 1_{[0,t]} - M 1_{[0,t]}\|_M$ tends to zero. Thus by the Itô isometry (B.6) the result follows. $\square$

Lemma B.24. *Let M and N be square integrable martingales; then*

$$M_t N_t = M_0 N_0 + \int_0^t M_s \, dN_s + \int_0^t N_s \, dM_s + \langle M, N \rangle_t.$$

Proof. Apply the polarization identity

$$\langle M, N \rangle_t = (\langle M + N \rangle_t - \langle M - N \rangle_t)/4$$

to the result of Lemma B.23, to give

$$\langle M, N \rangle_t = (1/4) \bigg((M_t + N_t)^2 - (M_0 + N_0)^2 - 2 \int_0^t (M_s + N_s) \, dM_s$$

$$- 2 \int_0^t (M_s + N_s) \, dN_s - (M_t - N_t)^2 - (M_0 - N_0)^2$$

$$- 2 \int_0^t (M_s - N_s) \, dM_s + 2 \int_0^t (M_s - N_s) \, dN_s \bigg)$$

$$= M_t N_t - M_0 N_0 - \int_0^t N_s \, dM_s - \int_0^t M_s \, dN_s.$$

$\square$

B.2.4 Itô's Formula

Theorem B.25. *If X is an $\mathbb{R}^d$-valued semimartingale and $f \in C^2(\mathbb{R}^d)$ then*

$$f(X_t) = f(X_0) + \sum_{i=1}^{d} \int_0^t \frac{\partial}{\partial x^i} f(X_s) \, \mathrm{d}X_s^i + \frac{1}{2} \sum_{i,j=1}^{d} \int_0^t \frac{\partial^2}{\partial x^i \partial x^j} f(X_s) \, \mathrm{d}\langle X^i, X^j \rangle_s.$$

The continuity condition on f in the statement of Itô's lemma is important; if it does not hold then the local time of X must be considered (see for example Chapter 7 of Chung and Williams [53] or Section IV. 43 of Rogers and Williams [249]).

Proof. We sketch a proof for $d = 1$. The finite variation case is the standard fundamental theorem of calculus for Stieltjes integration. Consider the case of M a martingale.

The proof is carried out by showing it holds for $f(x) = x^k$ for all k; by linearity it then holds for all polynomials and by a standard approximation argument for all $f \in C^2(\mathbb{R})$. To establish the result for polynomials proceed by induction. Suppose it holds for functions f and g; then by Lemma B.24,

$$\begin{aligned} \mathrm{d}(f(M_t)g(M_t)) &= f(M_t)\,\mathrm{d}g(M_t) + g(M_t)\,\mathrm{d}f(M_t) + \mathrm{d}\langle f(M_t), g(M_t)\rangle_t \\ &= f(M_t)(g'(M_t)\,\mathrm{d}M_t + \tfrac{1}{2}g''(M_t)\,\mathrm{d}\langle M\rangle_t) \\ &\quad + g(M_t)(f'(M_t)\,\mathrm{d}M_t + \tfrac{1}{2}f''(M_t)\,\mathrm{d}\langle M\rangle_t) \\ &\quad + g'(M_t)f'(M_t)\,\mathrm{d}\langle M\rangle_t. \end{aligned}$$

Since the result clearly holds for $f(x) = x$, it follows that it holds for all polynomials. The extension to $C^2(\mathbb{R})$ functions follows from a standard approximation argument (see e.g. Rogers and Williams [249] for details). $\square$

B.2.5 Localization

The integral may be extended to a larger class of integrands by the procedure of localization. Let H be a progressively measurable process. Define a non-decreasing sequence of stopping times

$$T_n \triangleq \inf_{t \geq 0} \left\{ \int_0^t H_s^2 \, \mathrm{d}\langle M \rangle_s > n \right\}; \tag{B.8}$$

then it is clear that the process $H_t^{T_n} \triangleq H_{t \wedge T_n}$ is in the space $\mathcal{L}_\mathcal{P}$. Thus the stochastic integral $\int_0^\infty H_s^{T_n} \, \mathrm{d}M_s$ is defined in the Itô sense of Definition B.16.

Theorem B.26. *If for all $t \geq 0$,*

$$\mathbb{P}\left(\int_0^t H_s^2 \, \mathrm{d}\langle M \rangle_s < \infty \right) = 1, \tag{B.9}$$

then we may define the stochastic integral

$$\int_0^\infty H_s \, \mathrm{d}M_s \triangleq \lim_{n \to \infty} \int_0^\infty H_s^{T_n} \, \mathrm{d}M_s.$$

Proof. Under condition (B.9) the sequence of stopping times T_n defined in (B.8) tends to infinity $\mathbb{P}$-a.s. It is straightforward to verify that this is well defined; that is, different choices of sequence T_n tending to infinity give rise to the same limit. □

This general definition of integral is then a *local martingale*. We can similarly extend to integrators M which are local martingales by using the minimum of a reducing sequence R_n for the local martingale M and the sequence T_n above.

B.3 Stochastic Calculus

A very useful result can be proved using the Itô calculus about the characterisation of Brownian motion, due to Lévy.

Theorem B.27. *Let $\{B^i\}_{t \geq 0}$ be continuous local martingales starting from zero for $i = 1, \ldots, n$. Then $B_t = (B_t^1, \ldots, B_t^n)$ is a Brownian motion with respect to $(\Omega, \mathcal{F}, \mathbb{P})$ adapted to the filtration $\mathcal{F}_t$, if and only if*

$$\langle B^i, B^j \rangle_t = \delta_{ij} t \qquad \forall i, j \in \{1, \ldots, n\}.$$

Proof. In these circumstances it follows that the statement B_t is a Brownian motion is by definition equivalent to stating that $B_t - B_s$ is independent of $\mathcal{F}_s$ and is distributed normally with mean zero and covariance matrix $(t-s)\mathbb{I}$.

Clearly if B_t is a Brownian motion then the covariation result follows trivially from the definitions. To establish the converse, we assume $\langle B^i, B^j \rangle_t = \delta_{ij} t$ for $i, j \in \{1, \ldots, n\}$ and prove that B_t is a Brownian motion.

Observe that for fixed $\theta \in \mathbb{R}^n$ we can define M_t^θ by

$$M_t^\theta = f(B_t, t) \triangleq \exp\left(i\theta^\top B_t + \frac{1}{2}\|\theta\|^2 t\right).$$

By application of Itô's formula to f we obtain (in differential form using the Einstein summation convention)

$$\begin{aligned}
\mathrm{d}\left(f(B_t, t)\right) &= \frac{\partial f}{\partial x^j}(B_t, t) \, \mathrm{d}B_t^j + \frac{\partial f}{\partial t}(B_t, t) \, \mathrm{d}t + \frac{1}{2}\frac{\partial^2 f}{\partial x^j \partial x^k}(B_t, t) \, \mathrm{d}\langle B^j, B^k \rangle_t \\
&= i\theta_j f(B_t, t) \, \mathrm{d}B_t^j + \frac{1}{2}\|\theta\|^2 f(B_t, t) \, \mathrm{d}t - \frac{1}{2}\theta_j \theta_k \delta_{jk} f(B_t, t) \, \mathrm{d}t \\
&= i\theta_j f(B_t, t) \, \mathrm{d}B_t^j.
\end{aligned}$$

Hence
$$M_t^\theta = 1 + \int_0^t \mathrm{d}(f(B_t, t)),$$
and is a sum of stochastic integrals with respect to continuous local martingales and is hence itself a continuous local martingale. But for each t, using $|\cdot|$ to denote the complex modulus
$$|M_t^\theta| = \exp\left(\frac{1}{2}\|\theta\|^2 t\right) < \infty.$$
Hence for any fixed time t_0, $M_t^{t_0}$ satisfies
$$|M_t^{t_0}| \le |M_\infty^{t_0}| < \infty,$$
and so is a bounded local martingale. Hence $\{M_t^{t_0}, t \ge 0\}$ is a genuine martingale. Thus for $0 \le s < t$ we have
$$\mathbb{E}\left[\exp(i\theta^\top (B_t - B_s)) \mid \mathcal{F}_s\right] = \exp\left(-\frac{1}{2}(t-s)\|\theta\|^2\right) \quad \text{a.s.}$$

However, this is the characteristic function of a multivariate normal random variable distributed as $N(O, (t-s)\mathbb{I})$. Thus by the Lévy characteristic function theorem $B_t - B_s$ is an $N(O, (t-s)\mathbb{I})$ random variable. □

B.3.1 Girsanov's Theorem

Girsanov's theorem for the change of drift underlies many important results. The result has an important converse but this is not used here.

Theorem B.28. *Let M be a continuous martingale, and let Z be the associated exponential martingale*
$$Z_t = \exp\left(M_t - \tfrac{1}{2}\langle M\rangle_t\right). \tag{B.10}$$
If Z is a uniformly integrable martingale, then a new measure $\mathbb{Q}$, equivalent to $\mathbb{P}$, may be defined by
$$\frac{\mathrm{d}\mathbb{Q}}{\mathrm{d}\mathbb{P}} \triangleq Z_\infty.$$
Furthermore, if X is a continuous $\mathbb{P}$ local martingale then $X_t - \langle X, M\rangle_t$ is a $\mathbb{Q}$-local martingale.

Proof. Since Z is a uniformly integrable martingale it follows from Theorem B.1 (martingale convergence) that $Z_t = \mathbb{E}[Z_\infty \mid \mathcal{F}_t]$. Hence $\mathbb{Q}$ constructed thus is a probability measure which is equivalent to $\mathbb{P}$. Now consider X, a $\mathbb{P}$-local martingale. Define a sequence of stopping times which tend to infinity via
$$T_n \triangleq \inf\{t \ge 0 : |X_t| \ge n \text{ or } |\langle X, M\rangle_t| \ge n\}.$$

Consider the process Y defined via

$$Y \triangleq X_t^{T_n} - \langle X^{T_n}, M \rangle_t.$$

By Itô's formula applied to (B.10), $\mathrm{d}Z_t = Z_t \mathrm{d}M_t$; a second application of Itô's formula yields

$$\begin{aligned}\mathrm{d}(Z_t Y_t) &= 1_{t \leq T_n} \left(Z_t \mathrm{d}Y_t + Y_t \mathrm{d}Z_t + \langle Z, Y \rangle_t \right) \\ &= 1_{t \leq T_n} \left(Z_t (\mathrm{d}X_t - \mathrm{d}\langle X, M \rangle_t) + Y_t Z_t \mathrm{d}M_t + \langle Z, Y \rangle_t \right) \\ &= 1_{t \leq T_n} \left(Z_t (\mathrm{d}X_t - \mathrm{d}\langle X, M \rangle_t) \right. \\ &\quad + (X_t - \langle X, M \rangle_t) Z_t \mathrm{d}M_t + Z_t \mathrm{d}\langle X, M \rangle_t) \\ &= 1_{t \leq T_n} \left((X_t - \langle X, M \rangle_t) Z_t \mathrm{d}M_t + Z_t \mathrm{d}X_t \right),\end{aligned}$$

where the result $\langle Z, Y \rangle_t = Z_t \langle X, M \rangle_t$ follows from the Kunita–Watanabe identity; hence ZY is a $\mathbb{P}$-local martingale. But Z is uniformly integrable and Y is bounded (by construction of the stopping time T_n), hence ZY is a genuine $\mathbb{P}$-martingale. Hence for $s < t$ and $A \in \mathcal{F}_s$, we have

$$\mathbb{E}_{\mathbb{Q}} \left[(Y_t - Y_s) 1_A \right] = \mathbb{E} \left[Z_\infty (Y_t - Y_s) 1_A \right] = \mathbb{E} \left[(Z_t Y_t - Z_s Y_s) 1_A \right] = 0;$$

hence Y is a $\mathbb{Q}$-martingale. Thus $X_t - \langle X, M \rangle_t$ is a $\mathbb{Q}$-local martingale, since T_n is a reducing sequence such that $(X - \langle X, M \rangle)^{T_n}$ is a $\mathbb{Q}$-martingale, and $T_n \uparrow \infty$ as $n \to \infty$. □

Corollary B.29. *Let W_t be a $\mathbb{P}$-Brownian motion and define $\mathbb{Q}$ as in Theorem B.28; then $\tilde{W}_t = W_t - \langle W, M \rangle_t$ is a $\mathbb{Q}$-Brownian motion.*

Proof. Since W is a Brownian motion it follows that $\langle W, W \rangle_t = t$ for all $t \geq 0$. Since $\tilde{W}_t$ is continuous and $\langle \tilde{W}, \tilde{W} \rangle_t = \langle W, W \rangle_t = t$, it follows from Lévy's characterisation of Brownian motion (Theorem B.27) that $\tilde{W}$ is a $\mathbb{Q}$-Brownian motion. □

The form of Girsanov's theorem in Theorem B.28 is too restrictive for many applications of interest. In particular the requirement that the martingale Z be uniformly integrable and the implied equivalence of $\mathbb{P}$ and $\mathbb{Q}$ on $\mathcal{F}$ rules out even such simple applications as transforming $X_t = \mu t + W_t$ to remove the constant drift. In this case the martingale $Z_t = \exp(\mu W_t - \frac{1}{2}\mu^2 t)$ is clearly not uniformly integrable. If we consider $A \in \mathcal{F}_\infty$ defined by

$$A = \left\{ \lim_{t \to \infty} \frac{X_t - \mu t}{t} = 0 \right\}, \tag{B.11}$$

it is clear that $\mathbb{P}(A) = 1$, yet under a measure $\mathbb{Q}$ under which X has no drift $\mathbb{Q}(A) = 0$. Since equivalent measures have the same null sets it would follow that if this measure which killed the drift were equivalent to $\mathbb{P}$ then A should also be null, a contradiction. Hence on $\mathcal{F}$ the measures $\mathbb{P}$ and $\mathbb{Q}$ cannot be equivalent.

If we consider restricting the definition of the measure $\mathbb{Q}$ to $\mathcal{F}_t$ for finite t then the above problem is avoided. In the example given earlier under $\mathbb{Q}^t$ the process X restricted to $[0, t]$ is a Brownian motion with zero drift. This approach via a family of consistent measures is used in the change of measure approach to filtering, which is described in Chapter 3. Since we have just shown that there does not exist any measure equivalent to $\mathbb{P}$ under which X is a Brownian motion on $[0, \infty)$ it is clear that we cannot, in general, find a measure $\mathbb{Q}$ defined on $\mathcal{F}_\infty$ such that the restriction of $\mathbb{Q}$ to $\mathcal{F}_t$ is $\mathbb{Q}^t$.

Define a set function on $\bigcup_{0 \leq t < \infty} \mathcal{F}_t$ by

$$\mathbb{Q}(A) = \mathbb{Q}^t(A), \qquad \forall A \in \mathcal{F}_t, \ \forall t \geq 0. \tag{B.12}$$

If we have a finite set $A_1, \ldots, A_n$ of elements of $\bigcup_{0 \leq t < \infty} \mathcal{F}_t$, then we can find s such that $A_i \in \mathcal{F}_s$ for $i = 1, \ldots, n$ and since $\mathbb{Q}^s$ is a probability measure it follows that the set function $\mathbb{Q}$ is finitely additive. It is immediate that $\mathbb{Q}(\emptyset) = 0$ and $\mathbb{Q}(\Omega) = 1$.

It is not obvious whether $\mathbb{Q}$ is countably additive. If $\mathbb{Q}$ is countably additive, then Carathéodory's theorem allows us to extend the definition of $\mathbb{Q}$ to $\sigma\left(\bigcup_{0 \leq t < \infty} \mathcal{F}_t\right) = \mathcal{F}_\infty$. This can be resolved in special situations by using Tulcea's theorem. The σ-algebras $\mathcal{F}_t$ are all defined on the same space, so the atom condition of Tulcea's theorem is non-trivial (contrast with the case of the product spaces used in the proof of the Daniell–Kolmogorov–Tulcea theorem), which explains why this extension cannot be carried out in general. The following corollary gives an important example where an extension is possible.

Corollary B.30. *Let $\Omega = C([0, \infty), \mathbb{R}^d)$ and let X_t be the canonical process on Ω. Define $\mathcal{F}_t^o = \sigma(X_s : 0 \leq s \leq t)$. If*

$$Z_t = \exp\left(M_t - \tfrac{1}{2}\langle M \rangle_t\right)$$

is a $\mathcal{F}_{t+}^o$-adapted martingale then there exists a unique measure $\mathbb{Q}$ on $(\Omega, \mathcal{F}_\infty^o)$ such that

$$\left.\frac{d\mathbb{Q}}{d\mathbb{P}}\right|_{\mathcal{F}_{t+}^o} = Z_t, \ \forall t$$

and the process $X_t - \langle X, M \rangle_t$ is a $\mathbb{Q}$ local martingale with respect to $\{\mathcal{F}_{t+}^o\}_{t \geq 0}$.

Proof. We apply Theorem B.28 to the process Z^t, which is clearly a uniformly integrable martingale (since $Z_s^t = \mathbb{E}[Z_t^t \mid \mathcal{F}_{s+}^o]$). We may thus define a family $\mathbb{Q}^t$ of measures equivalent to $\mathbb{P}$ on $\mathcal{F}_{t+}^o$. It is clear that these measures are consistent; that is for $s \leq t$, $\mathbb{Q}^t$ restricted to $\mathcal{F}_{s+}^o$ is identical to $\mathbb{Q}^s$.

For any finite set of times $t_1 < t_2 < \cdots$ such that $t_k \to \infty$ as $k \to \infty$, since the sample space $\Omega = C([0, \infty), \mathbb{R}^d)$ is a complete separable metric space, regular conditional probabilities in the sense of Definition 2.28 exist as a consequence of Exercise 2.29, and we may denote them $\mathbb{Q}^{t_k}(\cdot \mid \mathcal{F}_{t_{k-1}+})$ for $k = 1, 2, \ldots$.

The sequence of σ-algebras $\mathcal{F}^o_{t_k+}$ is clearly increasing. If we consider a sequence A_k of atoms with each $A_k \in \mathcal{F}^o_{t_k+}$ such that $A_1 \supseteq A_2 \supset \cdot$, then using the fact that these are the unaugmented σ-algebras on the canonical sample space it follows that $\cap_{k=1}^\infty A_k \neq \emptyset$. Therefore, using these regular conditional probabilities as the transition kernels, we may now apply Tulcea's theorem A.11 to construct a measure $\mathbb{Q}$ on $\mathcal{F}^o_\infty$ which is consistent with $\mathbb{Q}^{t_k}$ on $\mathcal{F}^o_{t_k+}$ for each k. The consistency condition ensures that the measure $\mathbb{Q}$ thus obtained is independent of the choice of the times t_ks. □

Corollary B.31. *Let W_t be a $\mathbb{P}$-Brownian motion and define $\mathbb{Q}$ as in Corollary B.30; then $\tilde{W}_t = W_t - \langle W, M \rangle_t$ is a $\mathbb{Q}$-Brownian motion with respect to $\mathcal{F}^o_{t+}$.*

Proof. As for Corollary B.29. □

B.3.2 Martingale Representation Theorem

The following representation theorem has many uses. The proof given here only establishes the existence of the representation. The results of Clark allow an explicit form to be established (see Nualart [227, Proposition 1.3.14] for details, or Section IV.41 of Rogers and Williams [249] for an elementary account).

Theorem B.32. *Let B be an m-dimensional Brownian motion and let $\mathcal{F}_t$ be the right continuous enlargement of the σ-algebra generated by B augmented[†] with the null sets $\mathcal{N}$. Let $T > 0$ be a constant time. If X is a square integrable random variable measurable with respect to the σ-algebra $\mathcal{F}_T$ then there exists a previsible ν_s such that*

$$X = \mathbb{E}[X] + \int_0^T \nu_s^\top \, dB_s. \tag{B.13}$$

Proof. To establish the respresentation (B.13), without loss of generality we may consider the case $\mathbb{E}X = 0$ (in the general case apply the result to $X - \mathbb{E}X$). Define the space

$$L_T^2 = \left\{ H : H \text{ is } \mathcal{F}_t\text{-previsible and } \mathbb{E}\left[\int_0^T \|H_s\|^2 \, ds\right] < \infty \right\}.$$

Consider the stochastic integral map

$$J : L_T^2 \to L^2(\mathcal{F}_T),$$

defined by

[†] This condition is satisfied automatically if the filtration satisfies the usual conditions.

$$J(H) = \int_0^T H_s^\top \, dB_s.$$

As a consequence of the Itô isometry theorem, this map is an isometry. Hence the image V under J of the Hilbert space L_T^2 is complete and hence a closed subspace of $L_0^2(\mathcal{F}_T) = \{H \in L^2(\mathcal{F}_T) : \mathbb{E}H = 0\}$. The theorem is proved if we can establish that the image is equal to the whole space $L_0^2(\mathcal{F}_T)$ for the image is the space of random variables X which admit a representation of the form (B.13).

Consider the orthogonal complement of V in $L_0^2(\mathcal{F}_T)$. We aim to show that every element of this orthogonal complement is zero. Suppose that Z is in the orthogonal complement of $L_0^2(\mathcal{F}_T)$; thus

$$\mathbb{E}(ZX) = 0 \quad \text{for all } X \in L_0^2(\mathcal{F}_T). \tag{B.14}$$

Define $Z_t = \mathbb{E}[Z \mid \mathcal{Y}_t]$ which is an L^2-bounded martingale. We know that the σ-algebra $\mathcal{F}_0$ is trivial by the Blumental 0–1 law therefore

$$Z_0 = \mathbb{E}[Z \mid \mathcal{F}_0] = \mathbb{E}(Z) = 0 \quad \mathbb{P}\text{-a.s.}$$

Let $H \in L_T^2$ and $N_T \triangleq J(H)$ and define $N_t \triangleq \mathbb{E}[N_T \mid \mathcal{F}_t]$ for $0 \le t \le T$. It is clear that $N_T \in V$. Let S be a stopping time such that $S \le T$; then by optional sampling

$$N_S = \mathbb{E}[N_T \mid \mathcal{F}_S] = \mathbb{E}\left[\int_0^S H_s^\top \, dB_s + \int_S^T H_s^\top \, dB_s \,\middle|\, \mathcal{F}_S\right] = J(H\mathbf{1}_{[0,S]}),$$

so consequently $N_S \in V$. The orthogonality relation (B.14) then implies that $\mathbb{E}(ZN_S) = 0$. Thus using the properties of conditional expectation

$$0 = \mathbb{E}[ZN_S] = \mathbb{E}[\mathbb{E}[ZN_S \mid \mathcal{F}_S]] = \mathbb{E}[N_S \mathbb{E}[Z \mid \mathcal{F}_S]] = \mathbb{E}[Z_S N_S].$$

Since this holds for S a bounded stopping time, and Z_T and N_T are square integrable, it follows that $Z_t N_t$ is a uniformly integrable martingale and hence $\langle Z, N \rangle_t$ is a null process.

Let ε_t be an element of the set S_t defined in Lemma B.39 where the stochastic process Y is taken to be the Brownian motion B. Extending J in the obvious way to m-dimensional vector processes, we have that

$$\varepsilon_t = 1 + J(i\varepsilon r \mathbf{1}_{[0,t]})$$

for some $r \in L^\infty([0,t], \mathbb{R}^m)$. Using the above, $Z_t \varepsilon_t = Z_0 + Z_t J(i\varepsilon r \mathbf{1}_{[0,t]})$. Both $\{Z_t J(i\varepsilon r \mathbf{1}_{[0,t]}), \, t \ge 0\}$ and $\{Z_t, \, t \ge 0\}$ are martingales and $Z_0 = 0$; hence

$$\mathbb{E}[\varepsilon_t Z_t] = \mathbb{E}[Z_0] + \mathbb{E}\left[Z_t J\left(i\varepsilon r \mathbf{1}_{[0,t]}\right)\right]$$
$$= \mathbb{E}(Z_0) = 0.$$

Thus since this holds for all $\varepsilon_t \in S_t$ and the set S_t is total this implies that $Z_t = 0$ $\mathbb{P}$-a.s. $\square$

Remark B.33. For X a square integrable $\mathcal{F}_t$-adapted martingale this result can be applied to X_T, followed by conditioning and use of the martingale property to obtain for any $0 \leq t \leq T$,

$$X_t = \mathbb{E}[X_T \mid \mathcal{F}_t] = \mathbb{E}[X_T] + \int_0^{t \wedge T} \nu_s^\top \, dB_s$$

$$= \mathbb{E}(X_0) + \int_0^t \nu_s^\top \, dB_s.$$

As the choice of the constant time T was arbitrary, it is clear that this result holds for all $t \geq 0$.

B.3.3 Novikov's Condition

One of the most useful conditions for checking whether a local martingale of exponential form is a martingale is that due to Novikov.

Theorem B.34. *If $Z_t = \exp(M_t - \frac{1}{2}\langle M \rangle_t)$ for M a continuous local martingale, then a sufficient condition for Z to be a martingale is that*

$$\mathbb{E}\left[\exp(\tfrac{1}{2}\langle M \rangle_t)\right] < \infty, \qquad 0 \leq t < \infty.$$

Proof. Define the stopping time

$$S_b = \inf\{t \geq 0 : M_s - s = b\}$$

and note that $\mathbb{P}(S_b < \infty) = 1$. Then define

$$Y_t \triangleq \exp(M_t - \tfrac{1}{2}t); \tag{B.15}$$

it follows by the optional stopping theorem that $\mathbb{E}[\exp(M_{S_b} - \frac{1}{2}S_b)] = 1$, which implies $\mathbb{E}[\exp(\frac{1}{2}S_b)] = e^{-b}$. Consider

$$N_t \triangleq Y_{t \wedge S_b}, \qquad t \geq 0,$$

which is also a martingale. Since $\mathbb{P}(S_b < \infty) = 1$ it follows that

$$N_\infty = \lim_{s \to \infty} N_s = \exp(M_{S_b} - \tfrac{1}{2}S_b).$$

By Fatou's lemma N_s is a supermartingale with last element. But $\mathbb{E}(N_\infty) = 1 = \mathbb{E}(N_0)$ whence N is a uniformly integrable martingale. So by optional sampling for any stopping time R,

$$\mathbb{E}\left[\exp\left(M_{R \wedge S_b} - \tfrac{1}{2}(R \wedge S_b)\right)\right] = 1.$$

Fix $t \geq 0$ and set $R = \langle M \rangle_t$. It then follows for $b < 0$,

$$\mathbb{E}\left(1_{S_b < \langle M \rangle_t} \exp\left(b + \tfrac{1}{2}S_b\right)\right) + \mathbb{E}\left(1_{S_b \geq \langle M \rangle_t} \exp\left(M_t - \tfrac{1}{2}\langle M \rangle_t\right)\right) = 1.$$

The first expectation is bounded by $e^b \mathbb{E}\left(\frac{1}{2}\langle M \rangle_t\right)$, thus from the condition of the theorem it converges to zero as $b \to -\infty$. The second term converges to $\mathbb{E}(Z_t)$ as a consequence of monotone convergence. Thus $\mathbb{E}(Z_t) = 1$. □

B.3.4 Stochastic Fubini Theorem

The Fubini theorem of measure theory has a useful extension to stochastic integrals. The form stated here requires a boundedness assumption and as such is not the most general form possible, but is that which is most useful for applications. We assume that all the stochastic integrals are with respect to continuous semimartingales, because this is the framework considered here. To extend the result it is simply necessary to stipulate that a càdlàg version of the stochastic integrals be chosen. For a more general form see Protter [247, Theorem IV.46].

In this theorem we consider a family of processes parametrised by an index $a \in A$, and let μ be a finite measure on the space $(A, \mathcal{A})$; that is, $\mu(A) < \infty$.

Theorem B.35. *Let X be a semimartingale and μ a finite measure. Let $H_t^a = H(t, a, \omega)$ be a bounded $\mathcal{B}[0, t] \otimes \mathcal{A} \otimes \mathcal{P}$ measurable process and $Z_t^a \triangleq \int_0^t H_s^a \, \mathrm{d}X_s$. If we define $H_t \triangleq \int_A H_t^a \mu(\mathrm{d}a)$ then $Y_t = \int_A Z^a \, \mu(\mathrm{d}a)$ is the process given by the stochastic integral $\int_0^t H_s \, \mathrm{d}X_s$.*

Proof. By stopping we can reduce the case to that of $X \in L^2$. As a consequence of the usual Fubini theorem it suffices to consider X a martingale. The proof proceeds via a monotone class argument. Suppose $H(t, a, \omega) = K(t, \omega) f(a)$ for f bounded $\mathcal{A}$-measurable. Then it follows that

$$Z_t = f(a) \int_0^t K(s, \omega) \, \mathrm{d}X_s,$$

and hence

$$\int_A Z_t^a \, \mu(\mathrm{d}a) = \int_A f(a) \left(\int_0^t K(s, \omega) \, \mathrm{d}X_s \right) \mu(\mathrm{d}a)$$

$$= \int_0^t K(s, \omega) \, \mathrm{d}X_s \int_A f(a) \, \mu(\mathrm{d}a)$$

$$= \int_0^t \left(\int_A f(a) \mu(\mathrm{d}a) K(s, \omega) \right) \mathrm{d}X_s$$

$$= \int_0^t H_s \, \mathrm{d}X_s.$$

Thus we have established the result in this simple case and by linearity to the vector space of finite linear combinations of bounded functions of this form. It remains to show the monotone property; that is, suppose that the result holds for H_n and $H_n \to H$. We must show that the result holds for H.

Let $Z_{n,t}^a \triangleq \int_0^t H_n^a \, \mathrm{d}X_s$. We are interested in convergence uniformly in t; thus note that

$$\mathbb{E}\left[\sup_t \left| \int_A Z_{n,t}^a \, \mu(\mathrm{d}a) - \int_A Z_t^a \, \mu(\mathrm{d}a) \right| \right] \leq \mathbb{E}\left[\int_A \sup_t |Z_{n,t}^a - Z_t^a| \, \mu(\mathrm{d}a) \right].$$

We show that the right-hand side tends to zero as $n \to \infty$. By Jensen's inequality and Cauchy–Schwartz we can compute as follows,

$$\left(\mathbb{E}\left[\int_A \sup_t |Z^a_{n,t} - Z^a_t| \mu(da)\right]\right)^2 \leq \mathbb{E}\left[\left(\int_A \sup_t |Z^a_{n,t} - Z^a_t| \mu(da)\right)^2\right]$$

$$\leq \int_A \mu(da) \, \mathbb{E}\left[\int_A \sup_t |Z^a_{n,t} - Z^a_t|^2 \mu(da)\right].$$

Then an application of the non-stochastic version of Fubini's theorem followed by Doob's L^2-inequality implies that

$$\frac{1}{\mu(A)} \mathbb{E}\left[\left(\int_A \sup_t |Z^a_{n,t} - Z^a_t| \mu(da)\right)^2\right] \leq \int_A \mathbb{E}\left[\sup_{s \in [0,T]} |Z^a_{n,s} - Z^a_s|^2\right] \mu(da)$$

$$\leq 4 \int_A \mathbb{E}\left[(Z^a_{n,\infty} - Z^a_\infty)^2\right] \mu(da)$$

$$\leq 4 \int_A \mathbb{E}[\langle Z^a_n - Z^a \rangle_\infty] \mu(da).$$

Then by the Kunita–Watanabe identity

$$\frac{1}{\mu(A)} \mathbb{E}\left(\int_A \sup_t |Z^a_{n,t} - Z^a_t| \mu(da)\right)^2$$

$$\leq 4 \int_A \mathbb{E}\left(\int_0^\infty (H^a_{n,s} - H^a_s)^2 \, d\langle X \rangle_s\right) \mu(da).$$

Since H_n increases monotonically to a bounded process H it follows that H_n and H are uniformly bounded; we may apply the dominated convergence theorem to the double integral and expectation and thus the right-hand side converges to zero. Thus

$$\lim_{n \to \infty} \mathbb{E}\left[\sup_t \left| \int_A Z^a_{n,t} \mu(da) - \int_A Z^a_t \mu(da) \right|\right] = 0. \tag{B.16}$$

We may conclude from this that

$$\int_A \sup_t |Z^a_{n,t} - Z^a_t| \mu(da) < \infty \quad \text{a.s.}$$

as a consequence of which $\int_A |Z^a_t| \mu(da) < \infty$ for all t a.s., and thus the integral $\int_A Z^a_t \mu(da)$ is defined a.s. for all t. Defining $H_{n,t} \triangleq \int_A H^a_{n,t} \mu(da)$, we have from (B.16) that $\int_0^t H_{n,s} dX_s$ converges in probability uniformly in t to $\int_A Z^a_t \mu(da)$. Since a priori the result holds for H_n we have that

$$\int_0^t H_{n,s} \, dX_s = \int_A Z^a_{n,t} \mu(da),$$

and since by the stochastic form of the dominated convergence theorem $\int_0^t H_{n,s}\,\mathrm{d}X_s$ tends to $\int_0^t H_s\,\mathrm{d}X_s$ as $n\to\infty$ it follows that

$$\int_0^t H_s\,\mathrm{d}X_s = \int_A Z_t^a\,\mu(\mathrm{d}a).$$

□

B.3.5 Burkholder–Davis–Gundy Inequalities

Theorem B.36. *If $F:[0,\infty)\to[0,\infty)$ is a continuous increasing function such that $F(0)=0$, and for every $\alpha>1$*

$$K_F = \sup_{x\in[0,\infty)} \frac{F(\alpha x)}{F(x)} < \infty,$$

then there exist constants c_F and C_F such that for every continuous local martingale M,

$$c_F \mathbb{E}\left[F\left(\sqrt{\langle M\rangle_\infty}\right)\right] \leq \mathbb{E}\left[F\left(\sup_{t\geq 0}|M_t|\right)\right] \leq C_F \mathbb{E}\left[F\left(\sqrt{\langle M\rangle_\infty}\right)\right].$$

An example of a suitable function F which satisfies the conditions of the theorem is $F(x)=x^p$ for $p>0$.

Various proofs exist of this result. The proof given follows Burkholder's approach in Chapter II of [36]. The proof requires the following lemma.

Lemma B.37. *Let X and Y be nonnegative real-valued random variables. Let $\beta>1$, $\delta>0$, $\varepsilon>0$ be such that for all $\lambda>0$,*

$$\mathbb{P}(X>\beta\lambda, Y\leq\delta\lambda) \leq \varepsilon\mathbb{P}(X>\lambda). \tag{B.17}$$

Let γ and η be such that $F(\beta\lambda)\leq\gamma F(\lambda)$ and $F(\delta^{-1}\lambda)\leq\eta F(\lambda)$. If $\gamma\varepsilon<1$ then

$$\mathbb{E}\left[F(X)\right] \leq \frac{\gamma\eta}{1-\gamma\varepsilon}\mathbb{E}\left[F(Y)\right].$$

Proof. Assume without loss of generality that $\mathbb{E}[F(X)]<\infty$. It is clear from (B.17) that for $\lambda>0$,

$$\mathbb{P}(X>\beta\lambda) = \mathbb{P}(X>\beta\lambda, Y\leq\delta\lambda) + \mathbb{P}(X>\beta\lambda, Y>\delta\lambda)$$
$$\leq \varepsilon\mathbb{P}(X>\lambda) + \mathbb{P}(Y>\delta\lambda). \tag{B.18}$$

Since $F(0)=0$ by assumption, it follows that

$$F(x) = \int_0^x \mathrm{d}F(\lambda) = \int_0^\infty I_{\{\lambda<x\}}\,\mathrm{d}F(\lambda);$$

thus by Fubini's theorem

$$\mathbb{E}[F(X)] = \int_0^\infty \mathbb{P}(X > \lambda)\,dF(\lambda).$$

Thus using (B.18) it follows that

$$\mathbb{E}[F(X/\beta)] = \int_0^\infty \mathbb{P}(X > \beta\lambda)\,dF(\lambda)$$
$$\leq \varepsilon \int_0^\infty \mathbb{P}(X > \lambda)\,dF(\lambda) + \int_0^\infty \mathbb{P}(Y > \delta\lambda)\,dF(\lambda)$$
$$\leq \varepsilon \mathbb{E}[F(X)] + \mathbb{E}[Y/\delta];$$

from the conditions on η, and γ it then follows that

$$\mathbb{E}[F(X/\beta)] \leq \varepsilon\gamma \mathbb{E}[F(X/\beta)] + \eta \mathbb{E}[F(Y)].$$

Since we assumed $\mathbb{E}[F(X)] < \infty$, and $\varepsilon\gamma < 1$, it follows that

$$\mathbb{E}[F(X/\beta)] \leq \frac{\eta}{1 - \varepsilon\gamma} \mathbb{E}[F(Y)],$$

and the result follows using the condition on γ. □

We can now prove the Burkholder–Davis–Gundy inequality, by using the above lemma.

Proof. Let $\tau = \inf\{u : |M_u| > \lambda\}$ which is an $\mathcal{F}_t$-stopping time. Define $N_t \triangleq (M_{\tau+t} - M_\tau)^2 - (\langle M\rangle_{\tau+t} - \langle M\rangle_\tau)$, which is a continuous $\mathcal{F}_{\tau+t}$-adapted local martingale. Choose $\beta > 1, 0 < \delta < 1$. On the event defined by $\{\sup_{t\geq 0} |M_t| > \beta\lambda, \langle M\rangle_\infty \leq \delta^2\lambda^2\}$ the martingale N_t must hit the level $(\beta - 1)^2\lambda^2 - \delta^2\lambda^2$ before it hits $-\delta^2\lambda^2$.

From elementary use of the optional sampling theorem the probability of a martingale hitting a level b before a level a is given by $-a/(b-a)$; thus

$$\mathbb{P}\left(\sup_{t\geq 0} |M_t| > \beta\lambda, \langle M\rangle_\infty \leq \delta^2\lambda^2 \mid \mathcal{F}_\tau\right) \leq \delta^2/(\beta - 1)^2.$$

Hence as $\beta > 1$,

$$\mathbb{P}\left(\sup_{t\geq 0} |M_t| > \beta\lambda, \langle M\rangle_\infty \leq \delta^2\lambda^2\right)$$
$$= \mathbb{P}\left(\sup_{t\geq 0} |M_t| > \beta\lambda, \langle M\rangle_\infty \leq \delta^2\lambda^2, \tau < \infty\right)$$
$$= \mathbb{E}\left[\mathbb{P}\left(\sup_{t\geq 0} |M_t| > \beta\lambda, \langle M\rangle_\infty \leq \delta^2\lambda^2 \mid \mathcal{F}_\tau\right) 1_{\tau<\infty}\right]$$
$$\leq \delta^2 \mathbb{P}(\tau < \infty)/(\beta - 1)^2.$$

It is immediate that since $\beta > 1$, $F(\beta\lambda) < K_F F(\lambda)$ and similarly since $\delta < 1$, $F(\lambda/\delta) < K_F F(\lambda)$, so we may take $\gamma = \eta = K_F$. Now we can choose $0 < \delta < 1$

sufficiently small that $\varepsilon\gamma = \delta^2/(\beta-1)^2 < 1/K_F$. Therefore all the conditions of Lemma B.37 are satisfied whence

$$\mathbb{E}\left[F\left(\sup_{t\geq 0}|M_t|\right)\right] \leq C\mathbb{E}\left[F\left(\sqrt{\langle M\rangle_\infty}\right)\right]$$

and the opposite inequality can be established similarly. □

B.4 Stochastic Differential Equations

Theorem B.38. *Let $f : \mathbb{R}^d \to \mathbb{R}^d$ and $\sigma : \mathbb{R}^d \to \mathbb{R}^p$ be Lipschitz functions. That is, there exist positive constants K_f and K_σ such that*

$$\|f(x) - f(y)\| \leq K_f\|x-y\|, \qquad \|\sigma(x) - \sigma(y)\| \leq K_\sigma\|x-y\|,$$

for all $x, y \in \mathbb{R}^d$.

Given a probability space $(\Omega, \mathcal{F}, \mathbb{P})$ and a filtration $\{\mathcal{F}_t,\ t \geq 0\}$ which satisfies the usual conditions, let W be an $\mathcal{F}_t$-adapted Brownian motion and let ζ be an $\mathcal{F}_0$-adapted random variable. Then there exists a unique continuous adapted process $X = \{X_t,\ t \geq 0\}$ which is a strong solution of the SDE,

$$X_t = \zeta + \int_0^t f(X_s)\,\mathrm{d}s + \int_0^t \sigma(X_s)\,\mathrm{d}W_s.$$

The proof of this theorem can be found as Theorem 10.6 of Chung and Williams [53] and is similar to the proof of Theorem 2.9 of Chapter 5 in Karatzas and Shreve [149].

B.5 Total Sets in L^1

The use of the following density result in stochastic filtering originated in the work of Krylov and Rozovskii.

Lemma B.39. *On the filtered probability space $(\Omega, \mathcal{F}, \tilde{\mathbb{P}})$ let Y be a Brownian motion starting from zero adapted to the filtration $\mathcal{Y}_t$; then define the set*

$$S_t = \left\{\varepsilon_t = \exp\left(i\int_0^t r_s^\top\,\mathrm{d}Y_s + \frac{1}{2}\int_0^t \|r_s\|^2\,\mathrm{d}s\right) : r \in L^\infty\left([0,t], \mathbb{R}^m\right)\right\} \quad \text{(B.19)}$$

Then S_t is a total set in $L^1(\Omega, \mathcal{Y}_t, \tilde{\mathbb{P}})$. That is, if $a \in L^1(\Omega, \mathcal{Y}_t, \tilde{\mathbb{P}})$ and $\tilde{\mathbb{E}}[a\varepsilon_t] = 0$, for all $\varepsilon_t \in S_t$, then $a = 0$ $\tilde{\mathbb{P}}$-a.s. Furthermore each process ε in the set S_t satisfies an SDE of the form

$$\mathrm{d}\varepsilon_t = i\varepsilon_t r_t^\top\,\mathrm{d}Y_t,$$

for some $r \in L^\infty([0,t], \mathbb{R}^m)$.

Proof. We follow the proof in Bensoussan [13, page 83]. Define a set

$$S'_t = \left\{ \varepsilon_t = \exp\left(i \int_0^t r_s^\top \, dY_s\right) \, r \in L^\infty([0,t], \mathbb{R}^m) \right\}.$$

Let a be a fixed element of $L^1(\Omega, \mathcal{Y}_t, \tilde{\mathbb{P}})$ such that $\tilde{\mathbb{E}}[a\varepsilon_t] = 0$ for all $\varepsilon_t \in S'_t$. This can easily be seen to be equivalent to the statement that $\tilde{\mathbb{E}}[a\varepsilon_t] = 0$ for all $\varepsilon_t \in S_t$, which we assume. To establish the result, we assume that $\tilde{\mathbb{E}}[a\varepsilon_t] = 0$ for all $\varepsilon_t \in S'_t$, and show that a is zero a.s. Take $t_1, t_2, \ldots, t_p \in (0, t)$ with $t_1 < t_2 < \cdots < t_p$, then given $l_1, l_2, \ldots, l_n \in \mathbb{R}^m$, define

$$\mu_p \triangleq l_p, \qquad \mu_{p-1} \triangleq l_p + l_{p-1}, \qquad \ldots \qquad \mu_1 \triangleq l_p + \cdots + l_1.$$

Adopting the convention that $t_0 = 0$, define a function

$$r_t = \begin{cases} \mu_h & \text{for } t \in (t_{h-1}, t_h), h = 1, \ldots, p, \\ 0 & \text{for } t \in (t_p, T), \end{cases}$$

whence as $Y_{t_0} = Y_0 = 0$,

$$\sum_{h=1}^p l_h^\top Y_{t_h} = \sum_{h=1}^p \mu_h^\top (Y_{t_h} - Y_{t_{h-1}}) = \int_0^t r_s^\top \, dY_s.$$

Hence for $a \in L^1(\Omega, \mathcal{Y}_t, \tilde{\mathbb{P}})$

$$\tilde{\mathbb{E}}\left[a \exp\left(i \sum_{h=1}^p l_h^\top Y_{t_h}\right)\right] = \tilde{\mathbb{E}}\left[a \exp\left(i \int_0^t r_s^\top \, dY_s\right)\right] = 0,$$

where the second equality follows from the fact that we have assumed $\mathbb{E}[a\varepsilon_t] = 0$ for all $\varepsilon \in S'_t$. By linearity therefore,

$$\tilde{\mathbb{E}}\left[a \sum_{k=1}^K c_k \exp\left(i \sum_{h=1}^p l_{h,k}^\top Y_{t_h}\right)\right] = 0,$$

where this holds for all K and for all coefficients $c_1, \ldots, c_K \in \mathbb{C}$, and values $l_{h,k} \in \mathbb{R}$. Let $F(x_1, \ldots, x_p)$ be a continuous bounded complex-valued function defined on $(\mathbb{R}^m)^p$. By Weierstrass' approximation theorem, there exists a uniformly bounded sequence of functions of the form

$$P^{(n)}(x_1, \ldots, x_p) = \sum_{k=1}^{K^n} c_k^{(n)} \exp\left(i \sum_{h=1}^p (l_{h,k}^{(n)})^\top x_h\right)$$

such that

$$\lim_{n \to \infty} P^{(n)}(x_1, \ldots, x_p) = F(x_1, \ldots, x_p).$$

Hence we have $\tilde{\mathbb{E}}[aF(Y_{t_1}, \ldots, Y_{t_p})] = 0$ for every continuous bounded function F, and by a further approximation argument, we can take F to be a bounded function, measurable with respect to the σ-algebra $\sigma(Y_{t_1}, \ldots, Y_{t_p})$. Since $t_1, t_2, \ldots, t_p$ were chosen arbitrarily, we obtain that $\tilde{\mathbb{E}}[ab] = 0$, for b any bounded $\mathcal{Y}_t$-measurable function. In particular it gives $\tilde{\mathbb{E}}[a^2 \wedge m] = 0$ for arbitrary m; hence $a = 0$ $\tilde{\mathbb{P}}$-a.s. □

The following corollary enables us to use a smaller set of functions in the definition of the set S_t, in particular we can consider only bounded continuous functions with any number m of bounded continuous derivatives.

Corollary B.40. *Assume the same conditions as in Lemma B.39. Define the set*

$$S_t^p = \left\{ \varepsilon_t = \exp\left(i \int_0^t r_s^\top \, dY_s + \frac{1}{2} \int_0^t \|r_s\|^2 \, ds \right) : r \in C_b^p([0,t], \mathbb{R}^m) \right\} \quad (B.20)$$

where m is an arbitrary non-negative integer. Then S_t^m is a total set in $L^1(\Omega, \mathcal{Y}_t, \tilde{\mathbb{P}})$. That is, if $a \in L^1(\Omega, \mathcal{Y}_t, \tilde{\mathbb{P}})$ and $\tilde{\mathbb{E}}[a\varepsilon_t] = 0$, for all $\varepsilon_t \in S_t$, then $a = 0$ $\tilde{\mathbb{P}}$-a.s. Furthermore each process ε in the set S_t satisfies an SDE of the form

$$d\varepsilon_t = i\varepsilon_t r_t^\top \, dY_t,$$

for some $r \in L^\infty([0,t], \mathbb{R}^m)$.

Proof. Let us prove the corollary for the case $p = 0$, that is, for r a bounded continuous function. To do this, as a consequence of Lemma B.39, it suffices to show that if $a \in L^1(\Omega, \mathcal{Y}_t, \tilde{\mathbb{P}})$ and $\tilde{\mathbb{E}}[a\varepsilon_t] = 0$, for all $\varepsilon_t \in S_t^0$, then $\tilde{\mathbb{E}}[a\varepsilon_t] = 0$, for all $\varepsilon_t \in S_t$. Pick an arbitrary $\varepsilon_t \in S_t$,

$$\varepsilon_t = \exp\left(i \int_0^t r_s^\top \, dY_s + \frac{1}{2} \int_0^t \|r_s\|^2 \, ds \right), \quad r \in L^\infty([0,t], \mathbb{R}^m).$$

First let us note that by the fundamental theorem of calculus, as $r \in L^\infty([0,t], \mathbb{R}^m)$, the function $p \colon [0,t] \to \mathbb{R}^m$ defined as

$$p_s = \int_0^s r_u \, du$$

is continuous and differentiable almost everywhere. Moreover, for almost all $s \in [0, t]$

$$\frac{dp_s}{ds} = r_s.$$

Now let $r^n \in C_b^0([0,t], \mathbb{R}^m)$ be defined as

$$r_s^n \triangleq n \left(p_s - p_{0 \vee s - 1/n} \right), \quad s \in [0, t].$$

Then r^n is uniformly bounded by same bound as r and from the above, for almost all $s \in [0, t]$, $\lim_{n \to \infty} r_s^n = r_s$. By the bounded convergence theorem,

$$\lim_{n\to\infty} \int_0^t \|r_s^n\|^2 \, ds = \int_0^t \|r_s\|^2 \, ds$$

and also

$$\lim_{n\to\infty} \tilde{\mathbb{E}}\left[\left(\int_0^t r_s^\top \, dY_s - \int_0^t (r_s^n)^\top \, dY_s\right)^2\right] = 0.$$

Hence at least for a subsequence $(r^{n_k})_{n_k > 0}$, by the Itô isometry

$$\lim_{k\to\infty} \int_0^t (r_s^{n_k})^\top \, dY_s = \int_0^t r_s^\top \, dY_s, \qquad \tilde{\mathbb{P}}\text{-a.s.}$$

and hence, the uniformly bounded sequence

$$\varepsilon_t^k = \exp\left(i \int_0^t (r_s^{n_k})^\top \, dY_s + \frac{1}{2} \int_0^t \|r_s^{n_k}\|^2 \, ds\right)$$

converges, $\tilde{\mathbb{P}}$-almost surely to ε_t. Then, via another use of the dominated convergence theorem

$$\tilde{\mathbb{E}}[a\varepsilon_t] = \lim_{k\to\infty} \tilde{\mathbb{E}}[a\varepsilon_t^k] = 0,$$

since $\varepsilon_t^k \in S_t^0$ for all $k \geq 0$. This completes the proof of the corollary for $p = 0$. For higher values of p, one iterates the above procedure. □

B.6 Limits of Stochastic Integrals

The following proposition is used in the proof of the Zakai equation.

Proposition B.41. *Let $(\Omega, \mathcal{F}, \mathbb{P})$ be a probability space, $\{B_t, \mathcal{F}_t\}$ be a standard n-dimensional Brownian motion defined on this space and Ψ_n, Ψ be an $\mathcal{F}_t$-adapted process such that $\int_0^t \Psi_n^2 \, ds < \infty$, $\int_0^t \Psi^2 \, ds < \infty$, $\mathbb{P}$-a.s. and*

$$\lim_{n\to\infty} \int_0^t \|\Psi_n - \Psi\|^2 \, ds = 0$$

in probability; then

$$\lim_{n\to\infty} \sup_{t\in[0,T]} \left|\int_0^t (\Psi_n^\top - \Psi^\top) \, dB_s\right| = 0$$

in probability.

Proof. Given arbitrary $t, \varepsilon, \eta > 0$ we first prove that for an n-dimensional process φ,

$$\mathbb{P}\left(\sup_{0\leq s\leq t}\left|\int_0^s \varphi_r^\top \, dB_r\right| \geq \varepsilon\right) \leq \mathbb{P}\left(\int_0^t \|\varphi_s\|^2 \, ds > \eta\right) + \frac{4\eta}{\varepsilon^2}. \qquad \text{(B.21)}$$

To this end, define

$$\tau_\eta \triangleq \inf\left\{t : \int_0^t \|\varphi_s\|^2\,\mathrm{d}s > \eta\right\},$$

and a corresponding stopped version of φ,

$$\varphi_s^\eta \triangleq \varphi_s 1_{[0,\tau_\eta]}(s).$$

Then using these definitions

$$\mathbb{P}\left(\sup_{0\le s\le t}\left|\int_0^s \varphi_r^\top\,\mathrm{d}B_r\right| \ge \varepsilon\right) = \mathbb{P}\left(\tau_\eta < t;\ \sup_{0\le s\le t}\left|\int_0^s \varphi_r^\top\,\mathrm{d}B_r\right| \ge \varepsilon\right)$$

$$+ \mathbb{P}\left(\tau_\eta \ge t;\ \sup_{0\le s\le t}\left|\int_0^s \varphi_r^\top\,\mathrm{d}B_r\right| \ge \varepsilon\right)$$

$$\le \mathbb{P}(\tau_\eta < t) + \mathbb{P}\left(\sup_{0\le s\le t}\left|\int_0^s (\varphi_r^\eta)^\top\,\mathrm{d}B_r\right| \ge \varepsilon\right)$$

$$\le \mathbb{P}\left(\int_0^t \|\varphi_s\|^2\,\mathrm{d}s > \eta\right)$$

$$+ \mathbb{P}\left(\sup_{0\le s\le t}\left|\int_0^s (\varphi_r^\eta)^\top\,\mathrm{d}B_r\right| \ge \varepsilon\right).$$

By Chebychev's inequality and Doob's L^2-inequality the second term on the right-hand side can be bounded

$$\mathbb{P}\left(\sup_{0\le s\le t}\left|\int_0^s (\varphi_r^\eta)^\top\,\mathrm{d}B_r\right| \ge \varepsilon\right) \le \frac{1}{\varepsilon^2}\mathbb{E}\left[\left(\sup_{0\le s\le t}\left|\int_0^s (\varphi_r^\eta)^\top\,\mathrm{d}B_r\right|\right)^2\right]$$

$$\le \frac{4}{\varepsilon^2}\mathbb{E}\left[\left(\int_0^t (\varphi_r^\eta)^\top\,\mathrm{d}B_r\right)^2\right]$$

$$\le \frac{4}{\varepsilon^2}\mathbb{E}\left[\int_0^t \|\varphi_r^\eta\|^2\,\mathrm{d}r\right]$$

$$\le \frac{4\eta}{\varepsilon^2},$$

which establishes (B.21). Applying this result with fixed ε to $\varphi = \Psi_n - \Psi$ yields

$$\mathbb{P}\left(\sup_{t\in[0,T]}\left|\int_0^t (\Psi_n^\top - \Psi^\top)\,\mathrm{d}B_s\right| \ge \varepsilon\right) \le \mathbb{P}\left(\int_0^t \|\Psi_n - \Psi\|^2\,\mathrm{d}s > \eta\right) + \frac{4\eta}{\varepsilon^2}.$$

Given arbitrary $\delta > 0$, by choosing $\eta < \delta\varepsilon^2/8$ the second term on the right-hand side is then bounded by $\delta/2$ and with this η by the condition of the proposition there exists $N(\eta)$ such that for $n \ge N(\eta)$ the first term is bounded by $\delta/2$. Thus the right-hand side can be bounded by δ. □

B.7 An Exponential Functional of Brownian motion

In this section we deduce an explicit expression of a certain exponential functional of Brownian motion which is used in Chapter 6. Let $\{B_t,\ t \geq 0\}$ be a d-dimensional standard Brownian motion. Let $\beta \colon [0, t] \to \mathbb{R}^d$ be a bounded measurable function, Γ a $d \times d$ real matrix and $\delta \in \mathbb{R}^d$. In this section, we compute the following functional of B,

$$I_t^{\beta,\Gamma,\delta} = \mathbb{E}\left[\exp\left(\int_0^t B_s^\top \beta_s \, ds - \tfrac{1}{2}\int_0^t \|\Gamma B_s\|^2 \, ds\right) \Big| B_t = \delta\right]. \tag{B.22}$$

In (B.22) we use the standard notation

$$B_s^\top \beta_s = \sum_{i=1}^d B_s^i \beta_s^i, \qquad \|\Gamma B_s\|^2 = \sum_{i,j=1}^d \left(\Gamma^{ij} B_s^j\right)^2, \qquad s \geq 0.$$

To obtain a closed formula for (B.22), we use Lévy's diagonalisation procedure, a powerful tool for deriving explicit formulae. Other results and techniques of this kind can be found in Yor [280] and the references contained therein. The orthogonal decomposition of B_s with respect to B_t is

$$B_s = \frac{s}{t} B_t + \left(B_s - \frac{s}{t} B_t\right), \qquad s \in [0, t],$$

and using the Fourier decomposition of the Brownian motion (as in Wiener's construction of the Brownian motion)

$$B_s = \frac{s}{t} B_t + \sum_{k \geq 1} \sqrt{\frac{2}{t}} \frac{\sin(ks\pi/t)}{k\pi/t} \xi_k, \qquad s \in [0, t], \tag{B.23}$$

where $\{\xi_k;\ k \geq 1\}$ are standard normal random vectors with independent entries, which are also independent of B_t and the infinite sum has a subsequence of its partial sums which almost surely converges uniformly (see Itô and McKean [135, page 22]), we obtain the following.

Lemma B.42. *Let $\nu \in \mathbb{R}$ and $\mu_k \in \mathbb{R}^d$, $k \geq 1$ be the following constants*

$$\nu^{\beta,\Gamma,\delta}(t) \triangleq \exp\left(\frac{1}{t}\int_0^t s\delta^\top \beta_s \, ds - \frac{1}{6}\|\Gamma\delta\|^2\, t\right)$$

$$\mu_k^{\beta,\Gamma,\delta}(t) \triangleq \int_0^t \frac{\sin(ks\pi/t)}{k\pi/t} \beta_s \, ds + (-1)^k \frac{t^2}{k^2\pi^2} \Gamma^\top \Gamma \delta, \qquad k \geq 1.$$

Then

$$I_t^{\beta,\Gamma,\delta} = \nu^{\beta,\Gamma,\delta}(t)\, \mathbb{E}\left[\exp\left(\sum_{k \geq 1}\left(\sqrt{\frac{2}{t}}\xi_k^\top \mu_k^{\beta,\Gamma,\delta}(t) - \frac{t^2}{2k^2\pi^2}\|\Gamma\xi_k\|^2\right)\right)\right]. \tag{B.24}$$

Proof. We have from (B.23),

$$\int_0^t B_s^\top \beta_s \, ds = \frac{1}{t} \int_0^t s\delta^\top \beta_s \, ds + \sum_{k \geq 1} \sqrt{\frac{2}{t}} \int_0^t \frac{\sin(ks\pi/t)}{k\pi/t} \xi_k^\top \beta_s \, ds \quad \text{(B.25)}$$

and similarly

$$\int_0^t \|\Gamma B_s\|^2 \, ds = \frac{1}{3} \|\Gamma \delta\|^2 t - 2\sqrt{\frac{2}{t}} \sum_{k \geq 1} (-1)^k \frac{t^2}{k^2 \pi^2} \xi_k^\top \Gamma^\top \Gamma \delta$$

$$+ \int_0^t \left\| \Gamma \left(\sqrt{\frac{2}{t}} \sum_{k \geq 1} \frac{\sin(ks\pi/t)}{k\pi/t} \xi_k \right) \right\|^2 ds. \quad \text{(B.26)}$$

Next using the standard orthonormality results for Fourier series

$$\int_0^t \left(\sqrt{\frac{2}{t}} \sin\left(\frac{ks\pi}{t}\right) \right)^2 ds = 1, \quad \forall k \geq 1,$$

$$\int_0^t \sin\left(\frac{k_1 s\pi}{t}\right) \sin\left(\frac{k_2 s\pi}{t}\right) ds = 0, \quad \forall k_1, k_2 \geq 1, \, k_1 \neq k_2,$$

it follows that

$$\int_0^t \left\| \Gamma \left(\sqrt{\frac{2}{t}} \sum_{k \geq 1} \frac{\sin(ks\pi/t)}{k\pi/t} \xi_k \right) \right\|^2 ds = \sum_{k \geq 1} \|\Gamma \xi_k\|^2 \frac{t^2}{k^2 \pi^2}. \quad \text{(B.27)}$$

The identity (B.24) follows immediately from equations (B.25), (B.26) and (B.27). □

Let P be an orthogonal matrix ($PP^\top = P^\top P = \mathbb{I}$) and D be a diagonal matrix $D = \text{diag}(\gamma_1, \gamma_2, \ldots, \gamma_d)$ such that $\Gamma^\top \Gamma = P^\top D P$. Obviously $(\gamma_i)_{i=1}^d$ are the eigenvalues of the real symmetric matrix $\Gamma^\top \Gamma$.

Lemma B.43. *Let $a_{i,k}^{\beta,\Gamma,\delta}(t)$, for $i = 1, \ldots, d$ and $k \geq 1$ be the following constants*

$$a_{i,k}^{\beta,\Gamma,\delta}(t) = \sum_{j=1}^d P^{ij} \left(\mu_k^{\beta,\Gamma,\delta}(t) \right)^j.$$

Then

$$I_t^{\beta,\Gamma,\delta} = \nu^{\beta,\Gamma,\delta}(t) \prod_{i=1}^d \frac{1}{\sqrt{\prod_{k \geq 1} \left[\frac{\gamma_i t^2}{k^2 \pi^2} + 1 \right]}} \exp\left(\sum_{k \geq 1} \frac{a_{i,k}^{\beta,\Gamma,\delta}(t)^2}{\left(\frac{t^2 \gamma_i}{k^2 \pi^2} + 1 \right) t} \right). \quad \text{(B.28)}$$

Proof. Let $\{\bar{\xi}_k,\ k \geq 1\}$ be the independent identically distributed standard normal random vectors defined by $\bar{\xi}_k = P\xi_k$ for any $k \geq 1$. As a consequence of Lemma B.42 we obtain that

$$I_t^{\beta,\Gamma,\delta} = \nu^{\beta,\Gamma,\delta}(t)\mathbb{E}\left[\exp\left(\sum_{k \geq 1}\left(\sqrt{\frac{2}{t}}\bar{\xi}_k^\top P\mu_k^{\beta,\Gamma,\delta}(t) - \frac{t^2}{2k^2\pi^2}\bar{\xi}_k^\top D\bar{\xi}_k\right)\right)\right]. \quad \text{(B.29)}$$

Define the σ-algebras

$$\mathcal{G}_k \triangleq \sigma(\bar{\xi}_p,\ p \geq k) \quad \text{and} \quad \mathcal{G} \triangleq \bigcap_{k \geq 1} \mathcal{G}_k.$$

Now define

$$\zeta \triangleq \exp\left(\sum_{k \geq 1}\left(\sqrt{\frac{2}{t}}\bar{\xi}_k^\top P\mu_k^{\beta,\Gamma,\delta}(t) - \frac{t^2}{2k^2\pi^2}\bar{\xi}_k^\top D\bar{\xi}_k\right)\right);$$

using the independence of $\bar{\xi}_1,\ldots,\bar{\xi}_n,\ldots$ and Kolmogorov's 0–1 Law (see Williams [272, page 46]), we see that

$$\mathbb{E}[\zeta] = \mathbb{E}\left[\zeta\ \bigg|\ \bigcap_{k \geq 1} \mathcal{G}_k\right].$$

Since $\mathcal{G}_k$ is a decreasing sequence of σ-algebras, the Lévy downward theorem (see Williams [272, page 136]) implies that

$$\mathbb{E}\left[\zeta\ \bigg|\ \bigcap_{k \geq 1} \mathcal{G}_k\right] = \lim_{k \to \infty} \mathbb{E}[\zeta \mid \mathcal{G}_k].$$

Hence we determine first $\mathbb{E}[\zeta \mid \mathcal{G}_k]$ and then take the limit as $k \to \infty$ to obtain the expectation in (B.29). Hence

$$\mathbb{E}[\zeta] = \prod_{k \geq 1}\mathbb{E}\left[\exp\left(\left(\sqrt{\frac{2}{t}}\bar{\xi}_k^\top P\mu_k^{\beta,\Gamma,\delta}(t) - \frac{t^2}{2k^2\pi^2}\bar{\xi}_k^\top D\bar{\xi}_k\right)\right)\right]$$

$$= \prod_{k \geq 1}\prod_{i=1}^d \frac{1}{\sqrt{2\pi}}\int_{-\infty}^{\infty}\exp\left(\sqrt{\frac{2}{t}}a_{i,k}^{\beta,\Gamma,\delta}(t)x - \frac{\left(\frac{t^2\gamma_i}{k^2\pi^2}+1\right)x^2}{2}\right)dx,$$

and identity (B.28) follows immediately. $\square$

Proposition B.44. *Let $f^{\beta,\Gamma}(t)$ be the following constant*

B.7 An Exponential Functional of Brownian motion

$$f^{\beta,\Gamma}(t) \triangleq \int_0^t \int_0^t \sum_{i=1}^d \frac{\sinh((s-t)\sqrt{\gamma_i})\sinh(s'\sqrt{\gamma_i})}{2\sqrt{\gamma_i}\sinh(t\sqrt{\gamma_i})}$$

$$\times \sum_{j=1}^d P^{ij}\beta_s^j \sum_{j'=1}^d P^{ij'}\beta_{s'}^{j'} \, ds \, ds',$$

and $R^{t,\beta,\Gamma}(\delta)$ be the following second-order polynomial in δ

$$R^{t,\beta,\Gamma}(\delta) \triangleq \left(\int_0^t \sum_{i=1}^d \frac{\sinh(s\sqrt{\gamma_i})}{\gamma_i \sinh(t\sqrt{\gamma_i})} \sum_{j=1}^d P^{ij}\beta_s^j \, ds \right) \sum_{j'=1}^d P^{ij'} \left(\Gamma^\top \Gamma \delta \right)^{j'}$$

$$- \sum_{i=1}^d \frac{\coth(t\sqrt{\gamma_i})}{2\gamma_i \sqrt{\gamma_i}} \left(\sum_{j=1}^d P^{ij} \left(\Gamma^\top \Gamma \delta \right)^j \right)^2.$$

Then

$$I_t^{\beta,\Gamma,\delta} = \prod_{i=1}^d \sqrt{\frac{t\sqrt{\gamma_i}}{\sinh(t\sqrt{\gamma_i})}} \exp\left(f^{\beta,\Gamma}(t) + R^{t,\beta,\Gamma}(\delta) + \frac{\|\delta\|^2}{2t} \right). \quad (B.30)$$

Proof. Using the classical identity (B.35), the infinite product in the denominator of (B.28) is equal to $\sinh(t\sqrt{\gamma_i})/(t\sqrt{\gamma_i})$. Then we need to expand the argument of the exponential in (B.28). The following argument makes use of the identities (B.32)–(B.34). We have that

$$a_{i,k}^{\beta,\Gamma,\delta}(t) = \int_0^t \frac{\sin(ks\pi/t)}{k\pi/t} c_i^{\beta,\Gamma}(s) \, ds + (-1)^k \frac{t^2}{k^2\pi^2} c_i^{\Gamma,\delta}$$

and

$$a_{i,k}^{\beta,\Gamma,\delta}(t)^2 = \int_0^t \int_0^t \frac{\sin(ks\pi/t)}{k\pi/t} \frac{\sin(ks'\pi/t)}{k\pi/t} c_i^{\beta,\Gamma}(s) c_i^{\beta,\Gamma}(s') \, ds \, ds'$$

$$+ 2(-1)^k \frac{t^2}{k^2\pi^2} c_i^{\Gamma,\delta} \int_0^t \frac{\sin(ks\pi/t)}{k\pi/t} c_i^{\beta,\Gamma}(s) \, ds$$

$$+ \left(\frac{t^2}{k^2\pi^2} c_i^{\Gamma,\delta} \right)^2, \quad (B.31)$$

where $c_i^{\beta,\Gamma}(s) = \sum_{j=1}^d P^{ij}\beta_s^j$ and $c_i^{\Gamma,\delta} = \sum_{j=1}^d P^{ij}\left(\Gamma^\top \Gamma \delta\right)^j$. Next we sum up over k each of the three terms on the right-hand side of (B.31). For the first term we use

$$\sum_{k\geq 1} \frac{\sin(ks\pi/t)\sin(ks'\pi/t)}{(k\pi/t)^2 t\left(t^2\gamma_i/(k^2\pi^2)+1\right)}$$

$$= \frac{t}{2\pi^2} \sum_{k\geq 1} \frac{\cos(k(s-s')\pi/t) - \cos(k(s+s')\pi/t)}{t^2\gamma_i/\pi^2 + k^2}$$

$$= \frac{t}{2\pi^2} \left(\frac{\pi}{2t\sqrt{\gamma_i}/\pi}\right) \frac{\cosh\left((s-t-s')\sqrt{\gamma_i}\right) - \cosh\left((s-t+s')\sqrt{\gamma_i}\right)}{\sinh\left(t\sqrt{\gamma_i}\right)}$$

$$= \frac{\sinh((s-t)\sqrt{\gamma_i})\sinh(s'\sqrt{\gamma_i})}{2\sqrt{\gamma_i}\sinh\left(t\sqrt{\gamma_i}\right)};$$

hence

$$\sum_{k\geq 1} \frac{\int_0^t \int_0^t \frac{\sin(ks\pi/t)}{k\pi/t}\frac{\sin(ks'\pi/t)}{k\pi/t} c_i^{\beta,\Gamma}(s)c_i^{\beta,\Gamma}(s')\,ds\,ds'}{t\left(t^2\gamma_i/(k^2\pi^2)+1\right)}$$

$$= \int_0^t \int_0^t \frac{\sinh((s-t)\sqrt{\gamma_i})\sinh(s'\sqrt{\gamma_i})}{2\sqrt{\gamma_i}\sinh\left(t\sqrt{\gamma_i}\right)} c_i^{\beta,\Gamma}(s)c_i^{\beta,\Gamma}(s')\,ds\,ds'.$$

For the second term,

$$\sum_{k\geq 1} \frac{(-1)^k \frac{t^2}{k^2\pi^2}\frac{\sin(ks\pi/t)}{k\pi/t}}{t(t^2\gamma_i/(k^2\pi^2)+1)} = \frac{t^2}{\pi^3}\sum_{k\geq 1}\frac{(-1)^k \sin(ks\pi/t)}{k\left(t^2\gamma_i/\pi^2 + k^2\right)}$$

$$= \frac{t^2}{\pi^3}\left(\frac{\pi}{2t^2\gamma_i/\pi^2}\frac{\sinh(s\sqrt{\gamma_i})}{\sinh(t\sqrt{\gamma_i})} - \frac{s\pi}{2t^3\gamma_i/\pi^2}\right)$$

$$= \left(\frac{1}{2\gamma_i}\frac{\sinh(s\sqrt{\gamma_i})}{\sinh(t\sqrt{\gamma_i})} - \frac{s}{2t\gamma_i}\right);$$

hence

$$\sum_{i=1}^d \sum_{k\geq 1} \frac{2(-1)^k \frac{t^2}{k^2\pi^2} c_i^{\Gamma,\delta}\int_0^t \frac{\sin(ks\pi/t)}{k\pi/t} c_i^{\beta,\Gamma}(s)\,ds}{t(t^2\gamma_i/(k^2\pi^2)+1)}$$

$$= \int_0^t \sum_{i=1}^d \left(\frac{\sinh(s\sqrt{\gamma_i})}{\sinh(t\sqrt{\gamma_i})} - \frac{s}{t}\right) \frac{c_i^{\beta,\Gamma}(s)c_i^{\Gamma,\delta}}{\gamma_i}\,ds$$

$$= \int_0^t \sum_{i=1}^d \frac{\sinh(s\sqrt{\gamma_i})}{\sinh(t\sqrt{\gamma_i})} \frac{c_i^{\beta,\Gamma}(s)c_i^{\Gamma,\delta}}{\gamma_i}\,ds$$

$$- \frac{1}{t}\int_0^t s\delta^\top \beta_s\,ds,$$

since $\sum_{i=1}^d c_i^{\beta,\Gamma}(s)c_i^{\Gamma,\delta}/\gamma_i = \delta^\top \beta_s$. For the last term we get

B.7 An Exponential Functional of Brownian motion

$$\sum_{k\geq 1} \frac{\left(t^2 c_i^{\Gamma,\delta}/(k^2\pi^2)\right)^2}{t\left(t^2\gamma_i/(k^2\pi^2)+1\right)} = \frac{t}{\gamma_i}\left(c_i^{\Gamma,\delta}\right)^2 \sum_{k\geq 1}\left(\frac{1}{k^2\pi^2} - \frac{1}{t^2\gamma_i + k^2\pi^2}\right)$$

$$= \frac{t}{\gamma_i}\left(c_i^{\Gamma,\delta}\right)^2\left(\frac{1}{6} + \frac{1}{2t^2\gamma_i} - \frac{1}{2t\sqrt{\gamma_i}}\coth(t\sqrt{\gamma_i})\right);$$

then

$$\sum_{i=1}^{d}\sum_{k\geq 1} \frac{\left(t^2 c_i^{\Gamma,\delta}/(k^2\pi^2)\right)^2}{t(t^2\gamma_i/(k^2\pi^2)+1)}$$

$$= \frac{\|\Gamma\delta\|^2}{6} + \frac{\|\delta\|^2}{2t} - \sum_{i=1}^{d}\frac{\coth(t\sqrt{\gamma_i})}{2\gamma_i\sqrt{\gamma_i}}\left(\sum_{j=1}^{d} P^{ij}\left(\Gamma^\top\Gamma\delta\right)^j\right)^2,$$

since $\sum_{i=1}^{d}\left(c_i^{\Gamma,\delta}\right)^2/\gamma_i = \|\Gamma\delta\|^2$, $\sum_{i=1}^{d}\left(c_i^{\Gamma,\delta}\right)^2/\gamma_i^2 = \|\delta\|^2$. In the above we used the following classical identities.

$$\sum_{k\geq 1}\frac{\cos kr}{z^2+k^2} = \frac{\pi}{2z}\frac{e^{(r-\pi)z}+e^{-(r-\pi)z}}{e^{\pi z}-e^{-\pi z}} - \frac{1}{2z^2}, \qquad \forall r\in(0,2\pi), \quad \text{(B.32)}$$

$$\sum_{k\geq 1}(-1)^k\frac{\sin kr}{k\left(z^2+k^2\right)} = \frac{\pi}{2z^2}\frac{e^{rz}-e^{-rz}}{e^{\pi z}-e^{-\pi z}} - \frac{r}{2z^2}, \qquad \forall r\in(-\pi,\pi), \quad \text{(B.33)}$$

$$\sum_{k\geq 1}\frac{1}{z^2+k^2\pi^2} = \frac{1}{2z}\left(\coth z - \frac{1}{z}\right), \qquad \text{(B.34)}$$

$$\prod_{k\geq 1}\left[1+\frac{l^2}{k^2}\right] = \frac{\sinh(\pi l)}{\pi l}, \qquad \text{(B.35)}$$

and $\sum_{k\geq 1} 1/k^2 = \pi^2/6$ (for proofs of these identities see for example, Macrobert [201]). We finally find the closed formula for the Brownian functional (B.22). □

In the one-dimensional case Proposition B.44 takes the following simpler form. This is the form of the result which is used in Chapter 6 to derive the density of π_t for the Beneš filter.

Corollary B.45. *Let $\{B_t, t\geq 0\}$ be a standard Brownian motion, $\beta:[0,t]\to\mathbb{R}$ be a bounded measurable function, and $\Gamma\in\mathbb{R}$ be a positive constant. Then*

$$\mathbb{E}\left[\exp\left(\int_0^t B_s\beta_s\,\mathrm{d}s - \frac{1}{2}\int_0^t \Gamma^2 B_s^2\,\mathrm{d}s\right)\bigg| B_t = \delta\right]$$
$$= \bar{f}^{\beta,\Gamma}(t)\exp\left(\left(\int_0^t \frac{\sinh(s\Gamma)}{\sinh(t\Gamma)}\beta_s\,\mathrm{d}s\right)\delta - \frac{\Gamma\coth(t\Gamma)}{2}\delta^2 + \frac{\delta^2}{2t}\right), \quad \text{(B.36)}$$

where

$$\bar{f}^{\beta,\Gamma}(t) = \sqrt{\frac{t\Gamma}{\sinh(t\Gamma)}} \exp\left(\int_0^t \int_0^t \frac{\sinh((s-t)\Gamma)\sinh(s'\Gamma)}{2\Gamma\sinh(t\Gamma)} \beta_s \beta_{s'} \, ds \, ds'\right).$$

References

1. Robert A. Adams. *Sobolev Spaces*. Academic Press, Orlando, FL, 2nd edition, 2003.
2. Lakhdar Aggoun and Robert J. Elliott. *Measure Theory and Filtering*, volume 15 of *Cambridge Series in Statistical and Probabilistic Mathematics*. Cambridge University Press, Cambridge, UK, 2004.
3. Deborah F. Allinger and Sanjoy K. Mitter. New results on the innovations problem for nonlinear filtering. *Stochastics*, 4(4):339–348, 1980/81.
4. Rami Atar, Frederi Viens, and Ofer Zeitouni. Robustness of Zakai's equation via Feynman-Kac representations. In *Stochastic Analysis, Control, Optimization and Applications*, Systems Control Found. Appl., pages 339–352. Birkhäuser Boston, Boston, MA, 1999.
5. Rami Atar and Ofer Zeitouni. Exponential stability for nonlinear filtering. *Ann. Inst. H. Poincaré Probab. Statist.*, 33(6):697–725, 1997.
6. J. E. Baker. Reducing bias and inefficiency in the selection algorithm. In John J. Grefenstette, editor, *Proceedings of the Second International Conference on Genetic Algorithms and their Applications*, pages 14–21, Mahwah, NJ, 1987. Lawrence Erlbaum.
7. John. S. Baras, Gilmer L. Blankenship, and William E. Hopkins Jr. Existence, uniqueness, and asymptotic behaviour of solutions to a class of Zakai equations with unbounded coefficients. *IEEE Trans. Automatic Control*, AC-28(2):203–214, 1983.
8. Eduardo Bayro-Corrochano and Yiwen Zhang. The motor extended Kalman filter: A geometric approach for rigid motion estimation. *J. Math. Imaging Vision*, 13(3):205–228, 2000.
9. V. E. Beneš. Exact finite-dimensional filters for certain diffusions with nonlinear drift. *Stochastics*, 5(1-2):65–92, 1981.
10. V. E. Beneš. New exact nonlinear filters with large Lie algebras. *Systems Control Lett.*, 5(4):217–221, 1985.
11. V. E. Beneš. Nonexistence of strong nonanticipating solutions to stochastic DEs: implications for functional DEs, filtering and control. *Stochastic Process. Appl.*, 5(3):243–263, 1977.
12. A. Bensoussan. On some approximation techniques in nonlinear filtering. In *Stochastic Differential Systems, Stochastic Control Theory and Applications*

(Minneapolis, MN, 1986), volume 10 of *IMA Vol. Math. Appl.*, pages 17–31. Springer, New York, 1988.
13. A. Bensoussan. *Stochastic Control of Partially Observable Systems*. Cambridge University Press, Cambridge, UK, 1992.
14. A. Bensoussan, R. Glowinski, and A. Rascanu. Approximation of the Zakai equation by splitting up method. *SIAM J. Control Optim.*, 28:1420–1431, 1990.
15. Alain Bensoussan. Nonlinear filtering theory. In *Recent advances in stochastic calculus (College Park, MD, 1987)*, Progr. Automat. Info. Systems, pages 27–64. Springer, New York, 1990.
16. Albert Benveniste. Séparabilité optionnelle, d'après doob. In *Séminaire de Probabilitiés, X (Univ. Strasbourg), Années universitaire 1974/1975*, volume 511 of *Lecture Notes in Math.*, pages 521–531. Springer Verlag, Berlin, 1976.
17. A. T. Bharucha-Reid. Review of Stratonovich, Conditional markov processes. *Mathematical Reviews*, (MR0137157), 1963.
18. A. G. Bhatt, G. Kallianpur, and R. L. Karandikar. Uniqueness and robustness of solution of measure-valued equations of nonlinear filtering. *Ann. Probab.*, 23(4):1895–1938, 1995.
19. P. Billingsley. *Convergence of Probability Measures*. Wiley, New York, 1968.
20. Jean-Michel Bismut and Dominique Michel. Diffusions conditionnelles. II. Générateur conditionnel. Application au filtrage. *J. Funct. Anal.*, 45(2):274–292, 1982.
21. B. Z. Bobrovsky and M. Zakai. Asymptotic a priori estimates for the error in the nonlinear filtering problem. *IEEE Trans. Inform. Theory*, 28:371–376, 1982.
22. N. Bourbaki. *Eléments de Mathématique: Topologie Générale [French]*. Hermann, Paris, France, 1958.
23. Leo Breiman. *Probability*. Classics in Applied Mathematics. SIAM, Philadelphia, PA, 1992.
24. Damiano Brigo, Bernard Hanzon, and François Le Gland. A differential geometric approach to nonlinear filtering: the projection filter. *IEEE Trans. Automat. Control*, 43(2):247–252, 1998.
25. Damiano Brigo, Bernard Hanzon, and François Le Gland. Approximate nonlinear filtering by projection on exponential manifolds of densities. *Bernoulli*, 5(3):495–534, 1999.
26. R. W. Brockett. Nonlinear systems and nonlinear estimation theory. In *Stochastic Systems: The Mathematics of Filtering and Identification and Applications (Les Arcs, 1980)*, volume 78 of *NATO Adv. Study Inst. Ser. C: Math. Phys. Sci.*, pages 441–477, Dordrecht-Boston, 1981. Reidel.
27. R. W. Brockett. Nonlinear control theory and differential geometry. In Z. Ciesielski and C. Olech, editors, *Proceedings of the International Congress of Mathematicians*, pages 1357–1367, Warsaw, 1984. Polish Scientific.
28. R. W. Brockett and J. M. C. Clark. The geometry of the conditional density equation. analysis and optimisation of stochastic systems. In *Proceedings of the International Conference, University of Oxford, Oxford, 1978*, pages 299–309, London-New York, 1980. Academic Press.
29. R. S. Bucy. Optimum finite time filters for a special non-stationary class of inputs. Technical Report Internal Report B. B. D. 600, March 31, Johns Hopkins Applied Physics Laboratory, 1959.
30. R. S. Bucy. Nonlinear filtering. *IEEE Trans. Automatic Control*, AC-10:198, 1965.

31. R. S. Bucy and P. D. Joseph. *Filtering for Stochastic Processes with Applications to Guidance*. Chelsea, New York, second edition, 1987.
32. A. Budhiraja and G. Kallianpur. Approximations to the solution of the Zakai equation using multiple Wiener and Stratonovich integral expansions. *Stochastics Stochastics Rep.*, 56(3-4):271–315, 1996.
33. A. Budhiraja and G. Kallianpur. The Feynman-Stratonovich semigroup and Stratonovich integral expansions in nonlinear filtering. *Appl. Math. Optim.*, 35(1):91–116, 1997.
34. A. Budhiraja and D. Ocone. Exponential stability in discrete-time filtering for non-ergodic signals. *Stochastic Process. Appl.*, 82(2):245–257, 1999.
35. Amarjit Budhiraja and Harold J. Kushner. Approximation and limit results for nonlinear filters over an infinite time interval. II. Random sampling algorithms. *SIAM J. Control Optim.*, 38(6):1874–1908 (electronic), 2000.
36. D. L. Burkholder. Distribution function inequalities for martingales. *Ann. Prob.*, 1(1):19–42, 1973.
37. Z. Cai, F. Le Gland, and H. Zhang. An adaptive local grid refinement method for nonlinear filtering. Technical Report 2679, INRIA, 1995.
38. J. Carpenter, P. Clifford, and P. Fearnhead. An improved particle filter for non-linear problems. *IEE Proceedings – Radar, Sonar and Navigation*, 146:2–7, 1999.
39. J. R. Carpenter, P. Clifford, and P. Fearnhead. Sampling strategies for Monte Carlo filters for non-linear systems. *IEE Colloquium Digest*, 243:6/1–6/3, 1996.
40. M. Chaleyat-Maurel. Robustesse du filtre et calcul des variations stochastique. *J. Funct. Anal.*, 68(1):55–71, 1986.
41. M. Chaleyat-Maurel. Continuity in nonlinear filtering. Some different approaches. In *Stochastic Partial Differential Equations and Applications (Trento, 1985)*, volume 1236 of *Lecture Notes in Math.*, pages 25–39. Springer, Berlin, 1987.
42. M. Chaleyat-Maurel and D. Michel. Des résultats de non existence de filtre de dimension finie. *Stochastics*, 13(1-2):83–102, 1984.
43. M. Chaleyat-Maurel and D. Michel. Hypoellipticity theorems and conditional laws. *Z. Wahrsch. Verw. Gebiete*, 65(4):573–597, 1984.
44. M. Chaleyat-Maurel and D. Michel. The support of the law of a filter in C^∞ topology. In *Stochastic Differential Systems, Stochastic Control Theory and Applications (Minneapolis, MN, 1986)*, volume 10 of *IMA Vol. Math. Appl.*, pages 395–407. Springer, New York, 1988.
45. M. Chaleyat-Maurel and D. Michel. The support of the density of a filter in the uncorrelated case. In *Stochastic Partial Differential Equations and Applications, II (Trento, 1988)*, volume 1390 of *Lecture Notes in Math.*, pages 33–41. Springer, Berlin, 1989.
46. M. Chaleyat-Maurel and D. Michel. Support theorems in nonlinear filtering. In *New Trends in Nonlinear Control Theory (Nantes, 1988)*, volume 122 of *Lecture Notes in Control and Inform. Sci.*, pages 396–403. Springer, Berlin, 1989.
47. M. Chaleyat-Maurel and D. Michel. A Stroock Varadhan support theorem in nonlinear filtering theory. *Probab. Theory Related Fields*, 84(1):119–139, 1990.
48. J. Chen, S. S.-T. Yau, and C.-W. Leung. Finite-dimensional filters with nonlinear drift. IV. Classification of finite-dimensional estimation algebras of maximal rank with state-space dimension 3. *SIAM J. Control Optim.*, 34(1):179–198, 1996.

49. J. Chen, S. S.-T. Yau, and C.-W. Leung. Finite-dimensional filters with nonlinear drift. VIII. Classification of finite-dimensional estimation algebras of maximal rank with state-space dimension 4. *SIAM J. Control Optim.*, 35(4):1132–1141, 1997.
50. W. L. Chiou and S. S.-T. Yau. Finite-dimensional filters with nonlinear drift. II. Brockett's problem on classification of finite-dimensional estimation algebras. *SIAM J. Control Optim.*, 32(1):297–310, 1994.
51. N. Chopin. Central limit theorem for sequential Monte Carlo methods and its application to Bayesian inference. *Annals of Statistics*, 32(6):2385–2411, 2004.
52. P.-L. Chow, R. Khasminskii, and R. Liptser. Tracking of signal and its derivatives in Gaussian white noise. *Stochastic Process. Appl.*, 69(2):259–273, 1997.
53. K. L. Chung and R. J. Williams. *Introduction to Stochastic Integration*. Birkhäuser, Boston, second edition, 1990.
54. B. Cipra. Engineers look to Kalman filtering for guidance. *SIAM News*, 26(5), 1993.
55. J. M. C. Clark. Conditions for one to one correspondence between an observation process and its innovation. Technical report, Centre for Computing and Automation, Imperial College, London, 1969.
56. J. M. C. Clark. The design of robust approximations to the stochastic differential equations of nonlinear filtering. In J. K. Skwirzynski, editor, *Communication Systems and Random Process Theory*, volume 25 of *Proc. 2nd NATO Advanced Study Inst. Ser. E, Appl. Sci.*, pages 721–734. Sijthoff & Noordhoff, Alphen aan den Rijn, 1978.
57. J. M. C. Clark, D. L. Ocone, and C. Coumarbatch. Relative entropy and error bounds for filtering of Markov processes. *Math. Control Signals Systems*, 12(4):346–360, 1999.
58. M. Cohen de Lara. Finite-dimensional filters. II. Invariance group techniques. *SIAM J. Control Optim.*, 35(3):1002–1029, 1997.
59. M. Cohen de Lara. Finite-dimensional filters. part I: The Wei normal technique. Part II: Invariance group technique. *SIAM J. Control Optim.*, 35(3):980–1029, 1997.
60. D. Crisan. Exact rates of convergence for a branching particle approximation to the solution of the Zakai equation. *Ann. Probab.*, 31(2):693–718, 2003.
61. D. Crisan. Particle approximations for a class of stochastic partial differential equations. *Appl. Math. Optim.*, 54(3):293–314, 2006.
62. D. Crisan, P. Del Moral, and T. Lyons. Interacting particle systems approximations of the Kushner-Stratonovitch equation. *Adv. in Appl. Probab.*, 31(3):819–838, 1999.
63. D. Crisan, J. Gaines, and T. Lyons. Convergence of a branching particle method to the solution of the Zakai equation. *SIAM J. Appl. Math.*, 58(5):1568–1590, 1998.
64. D. Crisan and T. Lyons. Nonlinear filtering and measure-valued processes. *Probab. Theory Related Fields*, 109(2):217–244, 1997.
65. D. Crisan and T. Lyons. A particle approximation of the solution of the Kushner-Stratonovitch equation. *Probab. Theory Related Fields*, 115(4):549–578, 1999.
66. D. Crisan and T. Lyons. Minimal entropy approximations and optimal algorithms for the filtering problem. *Monte Carlo Methods and Applications*, 8(4):343–356, 2002.

67. D. Crisan, P. Del Moral, and T. Lyons. Discrete filtering using branching and interacting particle systems. *Markov Processes and Related Fields*, 5(3):293–318, 1999.
68. R. W. R. Darling. Geometrically intrinsic nonlinear recursive filters. Technical report, Berkeley Statistics Department, 1998. http://www.stat.berkeley.edu/~darling/GINRF.
69. F. E. Daum. New exact nonlinear filters. In J. C. Spall, editor, *Bayesian Analysis of Time Series and Dynamic Models*, pages 199–226, New York, 1988. Marcel Dekker.
70. F. E. Daum. New exact nonlinear filters: Theory and applications. *Proc. SPIE*, 2235:636–649, 1994.
71. M. H. A. Davis. *Linear Estimation and Stochastic Control*. Chapman and Hall Mathematics Series. Chapman and Hall, London, 1977.
72. M. H. A. Davis. On a multiplicative functional transformation arising in nonlinear filtering theory. *Z. Wahrsch. Verw. Gebiete*, 54(2):125–139, 1980.
73. M. H. A. Davis. New approach to filtering for nonlinear systems. *Proc. IEE-D*, 128(5):166–172, 1981.
74. M. H. A. Davis. Pathwise nonlinear filtering. In M. Hazewinkel and J. C. Willems, editors, *Stochastic Systems: The Mathematics of Filtering and Identification and Applications, Proc. NATO Advanced Study Inst. Ser. C 78*, pages 505–528, Dordrecht-Boston, 1981. Reidel.
75. M. H. A. Davis. A pathwise solution of the equations of nonlinear filtering. *Theory Probability Applications [trans. of Teor. Veroyatnost. i Primenen.]*, 27(1):167–175, 1982.
76. M. H. A. Davis and M. P. Spathopoulos. Pathwise nonlinear filtering for nondegenerate diffusions with noise correlation. *SIAM J. Control Optim.*, 25(2):260–278, 1987.
77. Claude Dellacherie and Paul-André Meyer. *Probabilités et potentiel. Chapitres I à IV. [French] [Probability and potential. Chapters I–IV]* . Hermann, Paris, 1975.
78. Claude Dellacherie and Paul-André Meyer. Un noveau théorème de projection et de section [French]. In *Séminaire de Probabilités, IX (Seconde Partie, Univ. Strasbourg, Années universitaires 1973/1974 et 1974/1975)*, pages 239–245. Springer Verlag, New York, 1975.
79. Claude Dellacherie and Paul-André Meyer. *Probabilités et potentiel. Chapitres V à VIII. [French] [Probability and potential. Chapters V–VIII] Théorie des martingales*. Hermann, Paris, 1980.
80. Giovanni B. Di Masi and Wolfgang J. Runggaldier. An adaptive linear approach to nonlinear filtering. In *Applications of Mathematics in Industry and Technology (Siena, 1988)*, pages 308–316. Teubner, Stuttgart, 1989.
81. J. L. Doob. *Stochastic Processes*. Wiley, New York, 1963.
82. J. L. Doob. Stochastic process measurability conditions. *Annales de l'institut Fourier*, 25(3–4):163–176, 1975.
83. Arnaud Doucet, Nando de Freitas, and Neil Gordon. *Sequential Monte Carlo Methods in Practice*. Stat. Eng. Inf. Sci. Springer, New York, 2001.
84. T. E. Duncan. Likelihood functions for stochastic signals in white noise. *Information and Control*, 16:303–310, 1970.
85. T. E. Duncan. On the absolute continuity of measures. *Ann. Math. Statist.*, 41:30–38, 1970.

86. T. E. Duncan. On the steady state filtering problem for linear pure delay time systems. In *Analysis and control of systems (IRIA Sem., Rocquencourt, 1979)*, pages 25–42. INRIA, Rocquencourt, 1980.
87. T. E. Duncan. Stochastic filtering in manifolds. In *Control Science and Technology for the Progress of Society, Vol. 1 (Kyoto, 1981)*, pages 553–556. IFAC, Luxembourg, 1982.
88. T. E. Duncan. Explicit solutions for an estimation problem in manifolds associated with Lie groups. In *Differential Geometry: The Interface Between Pure and Applied Mathematics (San Antonio, TX, 1986)*, volume 68 of *Contemp. Math.*, pages 99–109. Amer. Math. Soc., Providence, RI, 1987.
89. T. E. Duncan. An estimation problem in compact Lie groups. *Systems Control Lett.*, 10(4):257–263, 1988.
90. R. J Elliott and V. Krishnamurthy. Exact finite-dimensional filters for maximum likelihood parameter estimation of continuous-time linear Gaussian systems. *SIAM J. Control Optim.*, 35(6):1908–1923, 1997.
91. R. J Elliott and J. van der Hoek. A finite-dimensional filter for hybrid observations. *IEEE Trans. Automat. Control*, 43(5):736–739, 1998.
92. Robert J. Elliott and Michael Kohlmann. Robust filtering for correlated multidimensional observations. *Math. Z.*, 178(4):559–578, 1981.
93. Robert J. Elliott and Michael Kohlmann. The existence of smooth densities for the prediction filtering and smoothing problems. *Acta Appl. Math.*, 14(3):269–286, 1989.
94. Robert J. Elliott and John B. Moore. Zakai equations for Hilbert space valued processes. *Stochastic Anal. Appl.*, 16(4):597–605, 1998.
95. Stewart N. Ethier and Thomas G. Kurtz. *Markov Processes: Characterization and Convergence*. Wiley, New York, 1986.
96. Marco Ferrante and Wolfgang J. Runggaldier. On necessary conditions for the existence of finite-dimensional filters in discrete time. *Systems Control Lett.*, 14(1):63–69, 1990.
97. W. H. Fleming and E. Pardoux. Optimal control of partially observed diffusions. *SIAM J. Control Optim.*, 20(2):261–285, 1982.
98. Wendell H. Fleming and Sanjoy K. Mitter. Optimal control and nonlinear filtering for nondegenerate diffusion processes. *Stochastics*, 8(1):63–77, 1982/83.
99. Patrick Florchinger. Malliavin calculus with time dependent coefficients and application to nonlinear filtering. *Probab. Theory Related Fields*, 86(2):203–223, 1990.
100. Patrick Florchinger and François Le Gland. Time-discretization of the Zakai equation for diffusion processes observed in correlated noise. In *Analysis and Optimization of Systems (Antibes, 1990)*, volume 144 of *Lecture Notes in Control and Inform. Sci.*, pages 228–237. Springer, Berlin, 1990.
101. Patrick Florchinger and François Le Gland. Time-discretization of the Zakai equation for diffusion processes observed in correlated noise. *Stochastics Stochastics Rep.*, 35(4):233–256, 1991.
102. Avner Friedman. *Partial Differential Equations of Parabolic Type*. Prentice-Hall, Englewood Cliffs, NJ, 1964.
103. P. Frost and T. Kailath. An innovations approach to least-squares estimation. III. *IEEE Trans. Autom. Control*, AC-16:217–226, 1971.
104. M. Fujisaki, G. Kallianpur, and H. Kunita. Stochastic differential equations for the non linear filtering problem. *Osaka J. Math.*, 9:19–40, 1972.

References 373

105. R. K. Getoor. On the construction of kernels. In *Séminaire de Probabilités, IX (Seconde Partie, Univ. Strasbourg, Années universitaires 1973/1974 et 1974/1975)*, volume 465 of *Lecture Notes in Math.*, pages 443–463. Springer Verlag, Berlin, 1975.
106. N. J. Gordon, D. J. Salmond, and A. F. M. Smith. Novel approach to nonlinear/non-Gaussian Bayesian state estimation. *IEE Proceedings, Part F*, 140(2):107–113, 1993.
107. B. Grigelionis. The theory of nonlinear estimation and semimartingales. *Izv. Akad. Nauk UzSSR Ser. Fiz.-Mat. Nauk*, (3):17–22, 97, 1981.
108. B. Grigelionis. Stochastic nonlinear filtering equations and semimartingales. In *Nonlinear Filtering and Stochastic Control (Cortona, 1981)*, volume 972 of *Lecture Notes in Math.*, pages 63–99. Springer, Berlin, 1982.
109. B. Grigelionis and R. Mikulevičius. On weak convergence to random processes with boundary conditions. In *Nonlinear Filtering and Stochastic Control (Cortona, 1981)*, volume 972 of *Lecture Notes in Math.*, pages 260–275. Springer, Berlin, 1982.
110. B. Grigelionis and R. Mikulevičius. Stochastic evolution equations and densities of the conditional distributions. In *Theory and Application of Random Fields (Bangalore, 1982)*, volume 49 of *Lecture Notes in Control and Inform. Sci.*, pages 49–88. Springer, Berlin, 1983.
111. B. Grigelionis and R. Mikulyavichyus. Robustness in nonlinear filtering theory. *Litovsk. Mat. Sb.*, 22(4):37–45, 1982.
112. I. Gyöngy. The approximation of stochastic partial differential equations and applications in nonlinear filtering. *Comput. Math. Appl.*, 19(1):47–63, 1990.
113. I. Gyöngy and N. V. Krylov. Stochastic partial differential equations with unbounded coefficients and applications. II. *Stochastics Stochastics Rep.*, 32(3-4):165–180, 1990.
114. I. Gyöngy and N. V. Krylov. On stochastic partial differential equations with unbounded coefficients. In *Stochastic partial differential equations and applications (Trento, 1990)*, volume 268 of *Pitman Res. Notes Math. Ser.*, pages 191–203. Longman Sci. Tech., Harlow, 1992.
115. István Gyöngy. On stochastic partial differential equations. Results on approximations. In *Topics in Stochastic Systems: Modelling, Estimation and Adaptive Control*, volume 161 of *Lecture Notes in Control and Inform. Sci.*, pages 116–136. Springer, Berlin, 1991.
116. István Gyöngy. Filtering on manifolds. *Acta Appl. Math.*, 35(1-2):165–177, 1994. White noise models and stochastic systems (Enschede, 1992).
117. István Gyöngy. Stochastic partial differential equations on manifolds. II. Nonlinear filtering. *Potential Anal.*, 6(1):39–56, 1997.
118. István Gyöngy and Nicolai Krylov. On the rate of convergence of splitting-up approximations for SPDEs. In *Stochastic inequalities and applications*, volume 56 of *Progr. Probab.*, pages 301–321. Birkhäuser, 2003.
119. István Gyöngy and Nicolai Krylov. On the splitting-up method and stochastic partial differential equations. *Ann. Probab.*, 31(2):564–591, 2003.
120. J. E. Handschin and D. Q. Mayne. Monte Carlo techniques to estimate the conditional expectation in multi-stage non-linear filtering. *Internat. J. Control*, 1(9):547–559, 1969.
121. M. Hazewinkel, S. I. Marcus, and H. J. Sussmann. Nonexistence of finite-dimensional filters for conditional statistics of the cubic sensor problem. *Systems Control Lett.*, 3(6):331–340, 1983.

122. M. Hazewinkel, S. I. Marcus, and H. J. Sussmann. Nonexistence of finite-dimensional filters for conditional statistics of the cubic sensor problem. In *Filtering and Control of Random Processes (Paris, 1983)*, volume 61 of *Lecture Notes in Control and Inform. Sci.*, pages 76–103, Berlin, 1984. Springer.
123. Michiel Hazewinkel. Lie algebraic methods in filtering and identification. In *VIIth International Congress on Mathematical Physics (Marseille, 1986)*, pages 120–137. World Scientific, Singapore, 1987.
124. Michiel Hazewinkel. Lie algebraic method in filtering and identification. In *Stochastic Processes in Physics and Engineering (Bielefeld, 1986)*, volume 42 of *Math. Appl.*, pages 159–176. Reidel, Dordrecht, 1988.
125. Michiel Hazewinkel. Non-Gaussian linear filtering, identification of linear systems, and the symplectic group. In *Modeling and Control of Systems in Engineering, Quantum Mechanics, Economics and Biosciences (Sophia-Antipolis, 1988)*, volume 121 of *Lecture Notes in Control and Inform. Sci.*, pages 299–308. Springer, Berlin, 1989.
126. Michiel Hazewinkel. Non-Gaussian linear filtering, identification of linear systems, and the symplectic group. In *Signal Processing, Part II*, volume 23 of *IMA Vol. Math. Appl.*, pages 99–113. Springer, New York, 1990.
127. A. J. Heunis. Nonlinear filtering of rare events with large signal-to-noise ratio. *J. Appl. Probab.*, 24(4):929–948, 1987.
128. A. J. Heunis. On the stochastic differential equations of filtering theory. *Appl. Math. Comput.*, 37(3):185–218, 1990.
129. A. J. Heunis. On the stochastic differential equations of filtering theory. *Appl. Math. Comput.*, 39(3, suppl.):3s–36s, 1990.
130. Andrew Heunis. Rates of convergence for an adaptive filtering algorithm driven by stationary dependent data. *SIAM J. Control Optim.*, 32(1):116–139, 1994.
131. Guo-Qing Hu, Stephen S. T. Yau, and Wen-Lin Chiou. Finite-dimensional filters with nonlinear drift. XIII. Classification of finite-dimensional estimation algebras of maximal rank with state space dimension five. Loo-Keng Hua: a great mathematician of the twentieth century. *Asian J. Math.*, 4(4):905–931, 2000.
132. M. Isard and A. Blake. Visual tracking by stochastic propagation of conditional density. In *Proceedings of the 4th European Conference on Computer Vision*, pages 343–356, New York, 1996. Springer Verlag.
133. M. Isard and A. Blake. Condensation conditional density propagation for visual tracking. *Int. J. Computer Vision*, 1998.
134. M. Isard and A. Blake. A mixed-state condensation tracker with automatic model switching. In *Proceedings of the 6th International Conference on Computer Vision*, pages 107–112, 1998.
135. K. Ito and H. P. McKean. *Diffusion Processes and Their Sample Paths*. Academic Press, New York, 1965.
136. Matthew R. James and François Le Gland. Numerical approximation for nonlinear filtering and finite-time observers. In *Applied Stochastic Analysis (New Brunswick, NJ, 1991)*, volume 177 of *Lecture Notes in Control and Inform. Sci.*, pages 159–175. Springer, Berlin, 1992.
137. T. Kailath. An innovations approach to least-squares estimation. I. linear filtering in additive white noise. *IEEE Trans. Autom. Control*, AC-13:646–655, 1968.

138. G. Kallianpur. White noise theory of filtering—Some robustness and consistency results. In *Stochastic Differential Systems (Marseille-Luminy, 1984)*, volume 69 of *Lecture Notes in Control and Inform. Sci.*, pages 217–223. Springer, Berlin, 1985.
139. G. Kallianpur and R. L. Karandikar. The Markov property of the filter in the finitely additive white noise approach to nonlinear filtering. *Stochastics*, 13(3):177–198, 1984.
140. G. Kallianpur and R. L. Karandikar. Measure-valued equations for the optimum filter in finitely additive nonlinear filtering theory. *Z. Wahrsch. Verw. Gebiete*, 66(1):1–17, 1984.
141. G. Kallianpur and R. L. Karandikar. A finitely additive white noise approach to nonlinear filtering: A brief survey. In *Multivariate Analysis VI (Pittsburgh, PA, 1983)*, pages 335–344. North-Holland, Amsterdam, 1985.
142. G. Kallianpur and R. L. Karandikar. White noise calculus and nonlinear filtering theory. *Ann. Probab.*, 13(4):1033–1107, 1985.
143. G. Kallianpur and R. L. Karandikar. *White Noise Theory of Prediction, Filtering and Smoothing*, volume 3 of *Stochastics Monographs*. Gordon & Breach Science, New York, 1988.
144. G. Kallianpur and C. Striebel. Estimation of stochastic systems: Arbitrary system process with additive white noise observation errors. *Ann. Math. Statist.*, 39(3):785–801, 1968.
145. Gopinath Kallianpur. *Stochastic filtering theory*, volume 13 of *Applications of Mathematics*. Springer, New York, 1980.
146. R. E. Kalman. A new approach to linear filtering and prediction problems. *J. Basic Eng.*, 82:35–45, 1960.
147. R. E. Kalman and R. S. Bucy. New results in linear filtering and prediction theory. *Trans. ASME, Ser. D, J. Basic Eng.*, 83:95–108, 1961.
148. Jim Kao, Dawn Flicker, Kayo Ide, and Michael Ghil. Estimating model parameters for an impact-produced shock-wave simulation: Optimal use of partial data with the extended Kalman filter. *J. Comput. Phys.*, 214(2):725–737, 2006.
149. I. Karatzas and S. E. Shreve. *Brownian Motion and Stochastic Calculus.*, volume 113 of *Graduate Texts in Mathematics*. Springer, New York, second edition, 1991.
150. Genshiro Kitagawa. Non-Gaussian state-space modeling of nonstationary time series. with comments and a reply by the author. *J. Amer. Statist. Assoc.*, 82(400):1032–1063, 1987.
151. P. E. Kloeden and E. Platen. *The Numerical Solution of Stochastic Differential Equations*. Springer, New York, 1992.
152. A. N. Kolmogorov. Sur l'interpolation et extrapolation des suites stationnaires. *C. R. Acad. Sci.*, 208:2043, 1939.
153. A. N. Kolmogorov. Interpolation and extrapolation. *Bulletin de l-académie des sciences de U.S.S.R., Ser. Math.*, 5:3–14, 1941.
154. Hayri Körezlioğlu and Wolfgang J. Runggaldier. Filtering for nonlinear systems driven by nonwhite noises: An approximation scheme. *Stochastics Stochastics Rep.*, 44(1-2):65–102, 1993.
155. M. G. Krein. On a generalization of some investigations of G. Szegö, W. M. smirnov, and A. N. Kolmogorov. *Dokl. Adad. Nauk SSSR*, 46:91–94, 1945.
156. M. G. Krein. On a problem of extrapolation of A. N. Kolmogorov. *Dokl. Akad. Nauk SSSR*, 46:306–309, 1945.

157. N. V. Krylov. On L_p-theory of stochastic partial differential equations in the whole space. *SIAM J. Math. Anal.*, 27(2):313–340, 1996.
158. N. V. Krylov. An analytic approach to SPDEs. In *Stochastic Partial Differential Equations: Six Perspectives*, number 64 in Math. Surveys Monogr., pages 185–242. Amer. Math. Soc., Providence, RI, 1999.
159. N. V. Krylov and B. L. Rozovskiĭ. The Cauchy problem for linear stochastic partial differential equations. *Izv. Akad. Nauk SSSR Ser. Mat.*, 41(6):1329–1347, 1448, 1977.
160. N. V. Krylov and B. L. Rozovskii. Conditional distributions of diffusion processes. *Izv. Akad. Nauk SSSR Ser. Mat.*, 42(2):356–378,470, 1978.
161. N. V. Krylov and B. L. Rozovskiĭ. Characteristics of second-order degenerate parabolic Itô equations. *Trudy Sem. Petrovsk.*, (8):153–168, 1982.
162. N. V. Krylov and B. L. Rozovskiĭ. Stochastic partial differential equations and diffusion processes. *Uspekhi Mat. Nauk*, 37(6(228)):75–95, 1982.
163. N. V. Krylov and A. Zatezalo. A direct approach to deriving filtering equations for diffusion processes. *Appl. Math. Optim.*, 42(3):315–332, 2000.
164. H. Kunita. *Stochastic Flows and Stochastic Differential Equations*. Number 24 in Cambridge Studies in Advanced Mathematics. Cambridge University Press, Cambridge, UK, 1990.
165. Hiroshi Kunita. Cauchy problem for stochastic partial differential equations arising in nonlinear filtering theory. *Systems Control Lett.*, 1(1):37–41, 1981/82.
166. Hiroshi Kunita. Stochastic partial differential equations connected with nonlinear filtering. In *Nonlinear Filtering and Stochastic Control (Cortona, 1981)*, volume 972 of *Lecture Notes in Math.*, pages 100–169. Springer, Berlin, 1982.
167. Hiroshi Kunita. Ergodic properties of nonlinear filtering processes. In *Spatial Stochastic Processes*, volume 19 of *Progr. Probab.*, pages 233–256. Birkhäuser Boston, 1991.
168. Hiroshi Kunita. The stability and approximation problems in nonlinear filtering theory. In *Stochastic Analysis*, pages 311–330. Academic Press, Boston, 1991.
169. Hans R. Künsch. Recursive Monte Carlo filters: Algorithms and theoretical analysis. *Ann. Statist.*, 33(5):1983–2021, 2005.
170. T. G. Kurtz and D. L. Ocone. Unique characterization of conditional distributions in nonlinear filtering. *Ann. Probab.*, 16(1):80–107, 1988.
171. T. G. Kurtz and J. Xiong. Numerical solutions for a class of SPDEs with application to filtering. In *Stochastics in Finite and Infinite Dimensions*, Trends Math., pages 233–258. Birkhäuser Boston, 2001.
172. Thomas G. Kurtz. Martingale problems for conditional distributions of Markov processes. *Electron. J. Probab.*, 3:no. 9, 29 pp. (electronic), 1998.
173. Thomas G. Kurtz and Daniel Ocone. A martingale problem for conditional distributions and uniqueness for the nonlinear filtering equations. In *Stochastic Differential Systems (Marseille-Luminy, 1984)*, volume 69 of *Lecture Notes in Control and Inform. Sci.*, pages 224–234. Springer, Berlin, 1985.
174. Thomas G. Kurtz and Jie Xiong. Particle representations for a class of nonlinear SPDEs. *Stochastic Process. Appl.*, 83(1):103–126, 1999.
175. H. Kushner. On the differential equations satisfied by conditional densities of markov processes, with applications. *SIAM J. Control*, 2:106–119, 1964.
176. H. Kushner. Technical Report JA2123, M.I.T Lincoln Laboratory, March 1963.
177. H. J. Kushner. Approximations of nonlinear filters. *IEEE Trans. Automat. Control*, AC-12:546–556, 1967.

178. H. J. Kushner. Dynamical equations for optimal nonlinear filtering. *J. Differential Equations*, 3:179–190, 1967.
179. H. J. Kushner. A robust discrete state approximation to the optimal nonlinear filter for a diffusion. *Stochastics*, 3(2):75–83, 1979.
180. H. J. Kushner. Robustness and convergence of approximations to nonlinear filters for jump-diffusions. *Matemática Aplicada e Computacional*, 16(2):153–183, 1997.
181. H. J. Kushner and P. Dupuis. *Numerical Methods for Stochastic Control Problems in Continuous Time*. Number 24 in Applications of Mathematics. Springer, New York, 1992.
182. Harold J. Kushner. *Weak Convergence Methods and Singularly Perturbed Stochastic Control and Filtering Problems*, volume 3 of *Systems & Control: Foundations & Applications*. Birkhäuser Boston, 1990.
183. Harold J. Kushner and Amarjit S. Budhiraja. A nonlinear filtering algorithm based on an approximation of the conditional distribution. *IEEE Trans. Autom. Control*, 45(3):580–585, 2000.
184. Harold J. Kushner and Hai Huang. Approximate and limit results for nonlinear filters with wide bandwidth observation noise. *Stochastics*, 16(1-2):65–96, 1986.
185. S. Kusuoka and D. Stroock. The partial Malliavin calculus and its application to nonlinear filtering. *Stochastics*, 12(2):83–142, 1984.
186. F. Le Gland. Time discretization of nonlinear filtering equations. In *Proceedings of the 28th IEEE-CSS Conference Decision Control, Tampa, FL*, pages 2601–2606, 1989.
187. François Le Gland. Splitting-up approximation for SPDEs and SDEs with application to nonlinear filtering. In *Stochastic Partial Differential Equations and Their Applications (Charlotte, NC, 1991)*, volume 176 of *Lecture Notes in Control and Inform. Sci.*, pages 177–187. Springer, New York, 1992.
188. François Le Gland and Nadia Oudjane. Stability and uniform approximation of nonlinear filters using the Hilbert metric and application to particle filters. *Ann. Appl. Probab.*, 14(1):144–187, 2004.
189. J. Lévine. Finite-dimensional realizations of stochastic PDEs and application to filtering. *Stochastics Stochastics Rep.*, 37(1–2):75–103, 1991.
190. Robert Liptser and Ofer Zeitouni. Robust diffusion approximation for nonlinear filtering. *J. Math. Systems Estim. Control*, 8(1):22 pp. (electronic), 1998.
191. Robert S. Liptser and Wolfgang J. Runggaldier. On diffusion approximations for filtering. *Stochastic Process. Appl.*, 38(2):205–238, 1991.
192. Robert S. Liptser and Albert N. Shiryaev. *Statistics of Random Processes. I General Theory*, volume 5 of *Stochastic Modelling and Applied Probablility*. Springer, New York, second edition, 2001. Translated from the 1974 Russian original by A. B. Aries.
193. Robert S. Liptser and Albert N. Shiryaev. *Statistics of Random Processes. II Applications*, volume 6 of *Stochastic Modelling and Applied Probability*. Springer, New York, second edition, 2001. Translated from the 1974 Russian original by A. B. Aries.
194. S. Lototsky, C. Rao, and B. Rozovskii. Fast nonlinear filter for continuous-discrete time multiple models. In *Proceedings of the 35th IEEE Conference on Decision and Control, Kobe, Japan, 1996*, volume 4, pages 4060–4064, Madison, WI, 1997. Omnipress.

195. S. V. Lototsky. Optimal filtering of stochastic parabolic equations. In *Recent Developments in Stochastic Analysis and Related Topics*, pages 330–353. World Scientific, Hackensack, NJ, 2004.
196. S. V. Lototsky. Wiener chaos and nonlinear filtering. *Appl. Math. Optim.*, 54(3):265–291, 2006.
197. Sergey Lototsky, Remigijus Mikulevicius, and Boris L. Rozovskii. Nonlinear filtering revisited: A spectral approach. *SIAM J. Control Optim.*, 35(2):435–461, 1997.
198. Sergey Lototsky and Boris Rozovskii. Stochastic differential equations: A Wiener chaos approach. In *From Stochastic Calculus to Mathematical Finance*, pages 433–506. Springer, New York, 2006.
199. Sergey V. Lototsky. Nonlinear filtering of diffusion processes in correlated noise: analysis by separation of variables. *Appl. Math. Optim.*, 47(2):167–194, 2003.
200. Vladimir M. Lucic and Andrew J. Heunis. On uniqueness of solutions for the stochastic differential equations of nonlinear filtering. *Ann. Appl. Probab.*, 11(1):182–209, 2001.
201. T. M. Macrobert. *Functions of a Complex Variable*. St. Martin's Press, New York, 1954.
202. Michael Mangold, Markus Grotsch, Min Sheng, and Achim Kienle. State estimation of a molten carbonate fuel cell by an extended Kalman filter. In *Control and Observer Design for Nonlinear Finite and Infinite Dimensional Systems*, volume 322 of *Lecture Notes in Control and Inform. Sci.*, pages 93–109. Springer, New York, 2005.
203. S. J. Maybank. Path integrals and finite-dimensional filters. In *Stochastic Partial Differential Equations (Edinburgh, 1994)*, volume 216 of *London Math. Soc. Lecture Note Ser.*, pages 209–229, Cambridge, UK, 1995. Cambridge University Press.
204. Stephen Maybank. Finite-dimensional filters. *Phil. Trans. R. Soc. Lond. A*, 354(1710):1099–1123, 1996.
205. Paul-André Meyer. Sur un problème de filtration [French]. In *Séminaire de Probabilitiés, VII (Univ. Strasbourg), Années universitaire 1971/1972*, volume 321 of *Lecture Notes in Math.*, pages 223–247. Springer Verlag, Berlin, 1973.
206. Paul-André Meyer. La théorie de la prédiction de F. Knight [French]. In *Séminaire de Probabilitiés, X (Univ. Strasbourg), Années universitaire 1974/1975*, volume 511 of *Lecture Notes in Math.*, pages 86–103. Springer Verlag, Berlin, 1976.
207. Dominique Michel. Régularité des lois conditionnelles en théorie du filtrage non-linéaire et calcul des variations stochastique. *J. Funct. Anal.*, 41(1):8–36, 1981.
208. R. Mikulevicius and B. L. Rozovskii. Separation of observations and parameters in nonlinear filtering. In *Proceedings of the 32nd IEEE Conference on Decision and Control, Part 2, San Antonio*. IEEE Control Systems Society, 1993.
209. R. Mikulevicius and B. L. Rozovskii. Fourier-Hermite expansions for nonlinear filtering. *Teor. Veroyatnost. i Primenen.*, 44(3):675–680, 1999.
210. Sanjoy K. Mitter. Existence and nonexistence of finite-dimensional filters. *Rend. Sem. Mat. Univ. Politec. Torino*, Special Issue:173–188, 1982.
211. Sanjoy K. Mitter. Geometric theory of nonlinear filtering. In *Mathematical Tools and Models for Control, Systems Analysis and Signal Processing, Vol.*

References 379

3 (Toulouse/Paris, 1981/1982), Travaux Rech. Coop. Programme 567, pages 37–60. CNRS, Paris, 1983.
212. Sanjoy K. Mitter and Nigel J. Newton. A variational approach to nonlinear estimation. *SIAM J. Control Optim.*, 42(5):1813–1833 (electronic), 2003.
213. Sanjoy K. Mitter and Irvin C. Schick. Point estimation, stochastic approximation, and robust Kalman filtering. In *Systems, Models and Feedback: Theory and Applications (Capri, 1992)*, volume 12 of *Progr. Systems Control Theory*, pages 127–151. Birkhäuser Boston, 1992.
214. P. Del Moral. Non-linear filtering: Interacting particle solution. *Markov Processes Related Fields*, 2:555–580, 1996.
215. P. Del Moral. Non-linear filtering using random particles. *Theory Probability Applications*, 40(4):690–701, 1996.
216. P. Del Moral. *Feynman-Kac formulae. Genealogical and Interacting Particle Systems with Applications*. Springer, New York, 2004.
217. P. Del Moral and J. Jacod. The Monte-Carlo method for filtering with discrete-time observations: Central limit theorems. In *Numerical Methods and Stochastics (Toronto, ON, 1999)*, Fields Inst. Commun., 34, pages 29–53. Amer. Math. Soc., Providence, RI, 2002.
218. P. Del Moral and L. Miclo. Branching and interacting particle systems approximations of Feynman-Kac formulae with applications to non-linear filtering. In *Séminaire de Probabilités, XXXIV*, volume 1729 of *Lecture Notes in Math.*, pages 1–145. Springer, Berlin, 2000.
219. P. Del Moral, J. C. Noyer, G. Rigal, and G. Salut. Traitement particulaire du signal radar : détection, estimation et reconnaissance de cibles aériennes. Technical Report 92495, LAAS, Dcembre 1992.
220. P. Del Moral, G. Rigal, and G. Salut. Estimation et commande optimale non-linéaire : un cadre unifié pour la résolution particulaire. Technical Report 91137, LAAS, 1991.
221. P. Del Moral, G. Rigal, and G. Salut. Filtrage non-linéaire non-gaussien appliqué au recalage de plates-formes inertielles. Technical Report 92207, LAAS, Juin 1992.
222. R. E. Mortensen. Stochastic optimal control with noisy observations. *Internat. J. Control*, 1(4):455–464, 1966.
223. Christian Musso, Nadia Oudjane, and Francois Le Gland. Improving regularised particle filters. In *Sequential Monte Carlo Methods in Practice*, Stat. Eng. Inf. Sci., pages 247–271. Springer, New York, 2001.
224. David E. Newland. Harmonic wavelet analysis. *Proc. Roy. Soc. London Ser. A*, 443(1917):203–225, 1993.
225. Nigel J. Newton. Observation sampling and quantisation for continuous-time estimators. *Stochastic Process. Appl.*, 87(2):311–337, 2000.
226. Nigel J. Newton. Observations preprocessing and quantization for nonlinear filters. *SIAM J. Control Optim.*, 38(2):482–502 (electronic), 2000.
227. David Nualart. *The Malliavin Calculus and Related Topics*. Springer, New York, second edition, 2006.
228. D. L. Ocone. Asymptotic stability of Beneš filters. *Stochastic Anal. Appl.*, 17(6):1053–1074, 1999.
229. Daniel Ocone. Multiple integral expansions for nonlinear filtering. *Stochastics*, 10(1):1–30, 1983.

230. Daniel Ocone. Application of Wiener space analysis to nonlinear filtering. In *Theory and Applications of Nonlinear Control Systems (Stockholm, 1985)*, pages 387–400. North-Holland, Amsterdam, 1986.
231. Daniel Ocone. Stochastic calculus of variations for stochastic partial differential equations. *J. Funct. Anal.*, 79(2):288–331, 1988.
232. Daniel Ocone. Entropy inequalities and entropy dynamics in nonlinear filtering of diffusion processes. In *Stochastic Analysis, Control, Optimization and Applications*, Systems Control Found. Appl., pages 477–496. Birkhäuser Boston, 1999.
233. Daniel Ocone and Étienne Pardoux. A Lie algebraic criterion for nonexistence of finite-dimensionally computable filters. In *Stochastic Partial Differential Equations and Applications, II (Trento, 1988)*, volume 1390 of *Lecture Notes in Math.*, pages 197–204. Springer, Berlin, 1989.
234. O. A. Oleĭnik and E. V. Radkevič. *Second Order Equations with Nonnegative Characteristic Form*. Plenum Press, New York, 1973.
235. Levent Ozbek and Murat Efe. An adaptive extended Kalman filter with application to compartment models. *Comm. Statist. Simulation Comput.*, 33(1):145–158, 2004.
236. E. Pardoux. *Equations aux dérivées partielles stochastiques non linéarires monotones*. PhD thesis, Univ Paris XI, Orsay, 1975.
237. E. Pardoux. Stochastic partial differential equations and filtering of diffusion processes. *Stochastics*, 3(2):127–167, 1979.
238. E. Pardoux. Filtrage non linéaire et equations aux dérivées partielles stochastiques associées. In *Ecole d'Eté de Probabilités de Saint-Flour XIX – 1989*, volume 1464 of *Lecture Notes in Mathematics*, pages 67–163. Springer, 1991.
239. P. Parthasarathy. *Probability Measures on Metric Spaces*. Academic Press, New York, 1967.
240. J. Picard. Efficiency of the extended Kalman filter for nonlinear systems with small noise. *SIAM J. Appl. Math.*, 51(3):843–885, 1991.
241. Jean Picard. Approximation of nonlinear filtering problems and order of convergence. In *Filtering and Control of Random Processes (Paris, 1983)*, volume 61 of *Lecture Notes in Control and Inform. Sci.*, pages 219–236. Springer, Berlin, 1984.
242. Jean Picard. An estimate of the error in time discretization of nonlinear filtering problems. In *Theory and Applications of Nonlinear Control Systems (Stockholm, 1985)*, pages 401–412. North-Holland, Amsterdam, 1986.
243. Jean Picard. Nonlinear filtering of one-dimensional diffusions in the case of a high signal-to-noise ratio. *SIAM J. Appl. Math.*, 46(6):1098–1125, 1986.
244. Michael K. Pitt and Neil Shephard. Filtering via simulation: Auxiliary particle filters. *J. Amer. Statist. Assoc.*, 94(446):590–599, 1999.
245. M. Pontier, C. Stricker, and J. Szpirglas. Sur le théorème de representation par raport a l'innovation [French]. In *Séminaire de Probabilités, XX (Univ. Strasbourg, Années universitaires 1984/1985)*, volume 1204 of *Lecture Notes in Math.*, pages 34–39. Springer Verlag, Berlin, 1986.
246. Yu. V. Prokhorov. Convergence of random processes and limit theorems in probability theory. *Theory Probability Applications [Teor. Veroyatnost. i Primenen.]*, 1(2):157–214, 1956.
247. P. Protter. *Stochastic Integration and Differential Equations*. Springer, Berlin, second edition, 2003.

248. L. C. G. Rogers and D. Williams. *Diffusions, Markov Processes and Martingales: Volume I Foundations.* Cambridge University Press, Cambridge, UK, second edition, 2000.
249. L. C. G. Rogers and D. Williams. *Diffusions, Markov Processes and Martingales: Volume II Itô Calculus.* Cambridge University Press, Cambridge, UK, second edition, 2000.
250. B. L. Rozovskii. *Stochastic Evolution Systems.* Kluwer, Dordrecht, 1990.
251. D. B. Rubin. A noniterative sampling/importance resampling alternative to the data augmentation algorithm for creating a few imputations when the fraction of missing information is modest: The SIR algorithm (discussion of Tanner and Wong). *J. Amer. Statist. Assoc.*, 82:543–546, 1987.
252. Laurent Saloff-Coste. *Aspects of Sobolev-Type Inequalities*, volume 289 of *London Mathematical Society Lecture Note Series.* Cambridge University Press, Cambridge, UK, 2002.
253. G. C. Schmidt. Designing nonlinear filters based on Daum's theory. *J. of Guidance, Control Dynamics*, 16(2):371–376, 1993.
254. Carla A.I. Schwartz and Bradley W. Dickinson. Characterizing finite-dimensional filters for the linear innovations of continuous-time random processes. *IEEE Trans. Autom. Control*, 30(3):312–315, 1985.
255. A. N. Shiryaev. Some new results in the theory of controlled random processes [Russian]. In *Transactions of the Fourth Prague Conference on Information Theory, Statistical Decision Functions, Random Processes (Prague, 1965)*, pages 131–203. Academia Prague, 1967.
256. Elias M. Stein. *Singular Integrals and Differentiability Properties of Functions.* Number 30 in Princeton Mathematical Series. Princeton University Press, Princeton, NJ, 1970.
257. R. L. Stratonovich. On the theory of optimal non-linear filtration of random functions. *Teor. Veroyatnost. i Primenen.*, 4:223–225, 1959.
258. R. L. Stratonovich. Application of the theory of Markov processes for optimum filtration of signals. *Radio Eng. Electron. Phys*, 1:1–19, 1960.
259. R. L. Stratonovich. Conditional Markov processes. *Theory Probability Applications [translation of Teor. Verojatnost. i Primenen.]*, 5(2):156–178, 1960.
260. R. L. Stratonovich. *Conditional Markov Processes and Their Application to the Theory of Optimal Control*, volume 7 of *Modern Analytic and Computational Methods in Science and Mathematics.* Elsevier, New York, 1968. Translated from the Russian by R. N. and N. B. McDonough for Scripta Technica.
261. D. W. Stroock and S. R. S. Varadhan. *Multidimensional Diffusion Processes.* Springer, New York, 1979.
262. Daniel W. Stroock. *Probability Theory, An Analytic View.* Cambridge University Press, Cambridge, UK, 1993.
263. M. Sun and R. Glowinski. Pathwise approximation and simulation for the Zakai filtering equation through operator splitting. *Calcolo*, 30(3):219–239 (1994), 1993.
264. J. Szpirglas. Sur l'équivalence d'équations différentielles stochastiques à valeurs mesures intervenant dans le filtrage Markovien non linéaire [French]. *Ann. Inst. H. Poincaré Sect. B (N.S.)*, 14(1):33–59, 1978.
265. I. Tulcea. Measures dans les espaces produits [French]. *Atti. Accad. Naz. Lincei Rend. Cl. Sci. Fis. Math. Nat.*, 8(7):208–211, 1949.

266. A. S. Üstünel. Some comments on the filtering of diffusions and the Malliavin calculus. In *Stochastic analysis and related topics (Silivri, 1986)*, volume 1316 of *Lecture Notes in Math.*, pages 247–266. Springer, Berlin, 1988.
267. A. Yu. Veretennikov. On backward filtering equations for SDE systems (direct approach). In *Stochastic Partial Differential equations (Edinburgh, 1994)*, volume 216 of *London Math. Soc. Lecture Note Ser.*, pages 304–311, Cambridge, UK, 1995. Cambridge Univ. Press.
268. D. Whitley. A genetic algorithm tutorial. *Statist. Comput.*, 4:65–85, 1994.
269. Ward Whitt. *Stochastic Process Limits. An Introduction to Stochastic-Process Limits and Their Application to Queues*. Springer, New York, 2002.
270. N. Wiener. *Extrapolation, Interpolation, and Smoothing of Stationary Time Series: With Engineering Applications*. MIT Press, Cambridge, MA, 1949.
271. N. Wiener. *I Am a Mathematician*. Doubleday, Garden City, NY; Victor Gollancz, London, 1956.
272. D. Williams. *Probability with Martingales*. Cambridge University Press, Cambridge, UK, 1991.
273. W. M. Wonham. Some applications of stochastic differential equations to optimal nonlinear filtering. *J. Soc. Indust. Appl. Math. Ser. A Control*, 2:347–369, 1965.
274. Xi Wu, Stephen S.-T. Yau, and Guo-Qing Hu. Finite-dimensional filters with nonlinear drift. XII. Linear and constant structure of Wong-matrix. In *Stochastic Theory and Control (Lawrence, KS, 2001)*, volume 280 of *Lecture Notes in Control and Inform. Sci.*, pages 507–518, Berlin, 2002. Springer.
275. T. Yamada and S. Watanabe. On the uniqueness of solutions of stochastic differential equations. *J. Math. Kyoto Univ.*, 11:151–167, 1971.
276. Shing-Tung Yau and Stephen S. T. Yau. Finite-dimensional filters with nonlinear drift. XI. Explicit solution of the generalized Kolmogorov equation in Brockett-Mitter program. *Adv. Math.*, 140(2):156–189, 1998.
277. Stephen S.-T. Yau. Finite-dimensional filters with nonlinear drift. I. A class of filters including both Kalman-Bucy filters and Benes filters. *J. Math. Systems Estim. Control*, 4(2):181–203, 1994.
278. Stephen S.-T. Yau and Guo-Qing Hu. Finite-dimensional filters with nonlinear drift. X. Explicit solution of DMZ equation. *IEEE Trans. Autom. Control*, 46(1):142–148, 2001.
279. Marc Yor. Sur les théories du filtrage et de la prédiction [French]. In *Séminaire de Probabilitiés, XI (Univ. Strasbourg), Années universitaire 1975/1976*, volume 581 of *Lecture Notes in Math.*, pages 257–297. Springer Verlag, Berlin, 1977.
280. Marc Yor. *Some Aspects of Brownian Motion, Part 1: Some Special Functionals (Lectures in Mathematics, ETH, Zürich)*. Birkhäuser Boston, 1992.
281. Moshe Zakai. On the optimal filtering of diffusion processes. *Z. Wahrscheinlichkeitstheorie und Verw. Gebiete*, 11:230–243, 1969.
282. O. Zeitouni. On the tightness of some error bounds for the nonlinear filtering problem. *IEEE Trans. Autom. Control*, 29(9):854–857, 1984.
283. O. Zeitouni and B. Z. Bobrovsky. On the reference probability approach to the equations of nonlinear filtering. *Stochastics*, 19(3):133–149, 1986.
284. Ofer Zeitouni. On the filtering of noise-contaminated signals observed via hard limiters. *IEEE Trans. Inform. Theory*, 34(5, part 1):1041–1048, 1988.

Author Name Index

A

Adams, R. A. 165, 166
Aggoun, L. 192
Allinger, D. F. 35

B

Baker, J. E. 280
Baras, J. S. 179
Bayro-Corrochano, E. 194
Beneš, V. E. 8, 142, 197–199
Bensoussan, A. 8, 9, 95, 104, 196, 356
Bharucha-Reid, A. T. 7
Bhatt, A. G. 126
Billingsley, P. 303
Blake, A. 286
Bobrovsky, B. Z. 196
Bourbaki, N. 27, 296
Breiman, L. 294
Brigo, D. 199, 202
Brockett, R. W. 8
Bucy, R. S. 6, 7, 192
Budhiraja, A. 9
Burkholder, D. L. 353

C

Carpenter, J. 230, 280
Chaleyat-Maurel, M. 8, 9
Chen, J. 8
Chiou, W. L. 8
Chopin, N. 280

Chung, K. L. 329, 330, 332, 338, 343, 355
Cipra, B. 6
Clark, J. M. C. 7, 8, 35, 129, 139, 348
Clifford, P. 230, 280
Cohen de Lara, M. 199
Crisan, D. 230, 249, 279, 281, 285, 286

D

Daniell, P. J. 301
Darling, R. W. R. 199
Daum, F. E. 8, 199
Davis, M. H. A. 7, 149, 250
Del Moral, P. 249, 250, 281, 286
Dellacherie, C. 307, 308, 312, 317, 319, 330
Dickinson, B. W. 8
Dieudonné, J. 32
Doob, J. L. 18, 58, 88, 301, 329
Doucet, A. 285
Duncan, T. E. 7, 9
Dynkin, E. B. 43

E

Efe, M. 194
Elliott, R. J. 9, 192
Ethier, S. N. 298, 303, 305, 330

F

Fearnhead, P. 230, 280
Fleming, W. H. 196

Author Name Index

Friedman, A. 101, 103
Frost, P. 7
Fujisaki, M. 7, 34, 45

G

Getoor, R. K. 27, 28
Gordon, N. J. 276, 286
Grigelionis, B. 9
Gyöngy, I. 9, 139, 209

H

Halmos, P. R. 32
Handschin, J. E. 286
Hazewinkel, M. 8, 9
Heunis, A. J. 9, 95, 113, 114, 126
Hu, G.-Q. 8

I

Isard, M. 286
Itô, K. 360

J

Jacod J. 249
Joseph, P. D. 192

K

Künsch, H. R. 279, 280
Kailath, T. 7
Kallianpur, G. 7, 8, 34, 35, 45, 57
Kalman, R. E. 6
Kao, J. 194
Karandikar, R. L. 8
Karatzas, I. 51, 88, 310, 330, 355
Kitagawa, G. 286
Kloeden, P. E. 251
Kolmogorov, A. N. 5, 13, 31, 32, 301
Krein, M. G. 5
Krylov, N. V. 7, 93, 139, 209, 355
Kunita, H. 7, 9, 34, 45, 182
Kuratowksi, K. 27
Kurtz, T. G. 9, 126, 165, 249, 298, 303, 305, 330
Kushner, H. J. 7, 9, 139, 202

L

Lévy, P. 344, 362
Le Gland, F. 9
Leung, C. W. 8
Liptser, R. S. 9
Lototsky, S. 202, 204
Lucic, V. M. 95, 113, 114, 126
Lyons, T. J. 230, 249, 250, 281, 285, 286

M

Mangold, M. 194
Marcus, S. I. 8
Maybank, S. J. 8
Mayne, D. Q. 286
McKean, H. P. 360
Meyer, P. A. 27, 45, 307, 308, 312, 317, 319, 330
Michel, D. 8, 9
Miclo, L. 250
Mikulevicius, R. 9, 202
Mitter, S. K. 8, 9, 35
Mortensen, R. E. 7

N

Newton, N. J. 9
Novikov, A. A. 52, 350
Nualart, D. 348

O

Ocone, D. L. 9, 126
Oleĭnik, O. A. 105
Ozbek, L. 194

P

Pardoux, E. 7, 9, 182, 193, 196
Picard, J. 9, 195, 196
Pitt, M. K. 285
Platen, E. 251
Prokhorov, Y. V. 45
Protter P. 330, 351

R

Radkevič, E. V. 105

Rigal, G. 286
Rogers, L. C. G. 17, 32, 58, 293, 296, 300, 301, 307, 308, 319, 321, 329, 339, 343, 348
Rozovskii, B. L. 7, 9, 93, 176, 177, 182, 202, 355
Rubin, D. B. 286
Runggaldier, W. J. 9

S

Salmond, D. J. 276, 286
Saloff-Coste, L. 166
Salut, G. 286
Schmidt, G. C. 199
Schwartz, C. A. I. 8
Sharpe, M. J. 332
Shephard, N. 285
Shiryaev, A. N. 7, 9
Shreve, S. E. 51, 88, 310, 330, 355
Smith, A. F. M. 276, 286
Stein, E. M. 166
Stratonovich, R. S. 7
Striebel, C. 8, 57
Stroock, D. W. 28, 298
Sussmann, H. J. 8
Szpirglas, J. 125

T

Tsirel'son, B. S. 35
Tulcea, I. 298

V

Varadhan, S. R. S. 298

Veretennikov, A. Y. 249

W

Watanabe, S. 35
Whitley, D. 230, 280
Whitt, W. 303
Wiener, N. 5
Williams, D. 17, 32, 43, 58, 293, 296, 300, 301, 307, 308, 319, 321, 329, 339, 343, 348, 362
Williams, R. J. 329, 330, 332, 338, 343, 355
Wonham, W. M. 7
Wu, X. 8

X

Xiong, J. 165, 249

Y

Yamada, T. 35
Yau, S.-T. 8
Yau, S. S.-T. 8
Yor, M. 28, 360

Z

Zakai, M. 7, 196
Zatezalo, A. 93
Zeitouni, O. 9

Subject Index

A

Announcing sequence   321
Atom   298
Augmented filtration   *see* Observation filtration
Averaging over the characteristics formula   182

B

Beneš condition   142, 196
Beneš filter   141, 146, 196
  the d-dimensional case   197
Bootstrap filter   276, 286
Borel space   301
Branching algorithm   278
Brownian motion   346
  exponential functional of   360, 361, 363, 365
  Fourier decomposition of   360
  Lévy's characterisation   344, 346
Burkholder–Davis–Gundy inequalities   246, 256, 353

C

Càdlàg path   303
Carathéodory extension theorem   300, 347
Change detection filter   *see* Change-detection problem
Change of measure method   49, 52
Change-detection problem   52, 69

Clark's robustness result   *see* Robust representation formula
Class
  $\mathcal{U}$ 96, 97, 100, 107, 109, 110, 113, 118
  $\bar{\mathcal{U}}$ 109, 110
  $\mathcal{U}'$ 110, 111, 113, 114, 116
  $\bar{\mathcal{U}}'$ 116
Condition
  U   97, 102, 107, 110
  U'   113, 114, 116
  U''   114
Conditional distribution
  of X_t 2–3, 191
    approximating sequence   265
    density of   174
    density of the   200
    recurrence formula   261, 264
    unnormalised   58, 173, 175
    regular   294
Conditional expectation   293
Conditional probability
  of a set   294
  regular   32, 294, 296, 347
Convergence determining set   323
Convergence in expectation   322, 324
Cubic sensor   201

D

Début theorem   307, 314, 339, 341
Daniell–Kolmogorov–Tulcea theorem   301, 302, 347
Density of ρ_t
  existence of   168

smoothness of   174
Dual previsible projection   332
Duncan–Mortensen–Zakai equation
  see Zakai equation

E

Empirical measure   210
Euler method   251
Evanescent set   319
Exponential projection filter   201
Extended Kalman filter   194

F

Feller property   267
Feynman–Kac formula   182
Filtering
  equations   4, 16, 72, 93, 125, 249, 308
    see Kushner–Stratonovich
    equation, Zakai equation
  for inhomogeneous test functions
    69
  problem   13, 48
    discrete time   258–259
    the correlated noise case   73–75,
      109
Finite difference scheme   207
Finite-dimensional filters   141, 146,
  154, 196–199
Fisher information matrix   199
Fokker–Planck equation   206
Fujisaki–Kallianpur–Kunita equation   see Kushner–Stratonovich
    equation

G

Generator of the process X 48, 50, 51,
  151, 168, 207, 221
  domain of the   47, 50–51
    maximal   51
Girsanov's theorem   345, 346
Gronwall's lemma   78, 79, 81, 88, 172,
  325

H

Hermite polynomials   203

I

Importance distribution   285
Importance sampling   273
Indistinguishable processes   319
Infinitesimal generator   see Generator
  of the process X
Innovation
  approach   7, 49, 70–73
  process   33–34
Itô integral   see Stochastic integral
Itô isometry   337, 338, 349
Itô's formula   343

K

Kallianpur–Striebel formula   57, 59,
  128
Kalman–Bucy filter   6, 148–154, 191,
  192, 199
  1D case   158
  as a Beneš filter   142, 148
Kushner–Stratonovich equation   68,
  71, 153
  correlated noise case   74
  finite-dimensional   66
  for inhomogeneous test functions
    69
  linear case   151
  strong form   179
  uniqueness of solution   110, 116

L

Likelihood function   260
Linear filter   see Kalman–Bucy filter
Local martingale   330, 344

M

Markov chain   257
Martingale   329
  representation theorem   348
  uniformly integrable   330, 346
Martingale convergence theorem   318,
  329, 345
Martingale problem   47
Martingale representation theorem   35,
  38, 44
Measurement noise   1

Monotone class theorem   29, 31, 293, 295, 311, 318, 336
Monte Carlo approximation   210, 216, 222, 230
  convergence of   213, 214, 217
  convergence rate   215, 216
Multinomial resampling   *see* Resampling procedure
Mutation step   273

N

Non-linear filtering   *see* Stochastic filtering
Non-linear filtering problem   *see* Filtering problem
Novikov's condition   52, 127, 131, 218, 222, 350

O

Observation
  filtration   13–17
    right continuity of the   17, 27, 33–40
    unaugmented   16
  process   1, 3, 16
    discrete time   258
  σ-algebra   *see* Observation filtration
Offspring distribution   224, 252, 274–281
  Bernoulli   280
  binomial   280
  minimal variance   225, 226, 228, 230, 279, 280
  multinomial   275–277
  obtained by residual sampling   277
  Poisson   280
Optional process   320
Optional projection of a process   17–19, 311–317, 338
  kernel for the   27
  without the usual conditions   321

P

Parabolic PDEs
  existence and uniqueness result   100
  maximum principle for   102
  systems of   102

uniformly   101, 121
Parseval's equality   204, 205
Particle filter   209, 222–224
  branching algorithm   225
  convergence rates   241, 244, 245, 248
  correction step   222, 230, 250
  discrete time   272–273
    convergence of   281–284
    prediction step   264
    updating step   264
  evolution equation   230
  implementation   250–252
    correction step   251, 252
    evolution step   251
  offspring distribution   *see* Offspring distribution
  path regularity   229
  resampling procedure   250, 252
Particle methods   *see* Particle filter
Path process   259
PDE Methods
  correction step   207
  prediction step   206
π
  the stochastic process   14, 27–32
  càdlàg version of   31
π_t *see* Conditional distribution of X_t
Polarization identity   342
Posterior distribution   259
Predictable σ-algebra   *see* Previsible σ-algebra
Predictable process   *see* Previsible process
Predicted conditional probability   259
Previsible σ-algebra   331
Previsible process   321, 331, 338
Previsible projection of a process   317, 321, 340, 341
Prior distribution   259
Projection bien measurable   *see* Optional projection of a process
Projection filter   199
Projective product   261

Q

Q-matrix   51
Quadratic variation   332, 335, 342

R

Reducing sequence   330
Regular grid   207
Regularisation method   167
Regularised measure   167
Resampling procedure   276
Residual sampling   277
ρ see Conditional distribution of X_t, unnormalised
   density of   173, 178
   dual of   165, 180–182, 233, 238
Riccati equation   152, 192
Ring of subsets   331
Robust representation formula   129, 137

S

Sampling with replacement method   see Resampling procedure
Selection step   274
Sensor function   4
Separable metric space   296
Sequential Monte Carlo methods   see Particle filter
Signal process   1, 3, 16, 47
   discrete time version   257
   filtration associated with the   47
   in discrete time   257
   particular cases   49–52
SIR algorithm   276
Skorohod topology   304–305
Sobolev
   embedding theorem   166
   space   166
Splitting-up algorithm   206
Stochastic differential equation
   strong solution   355
Stochastic filtering   1, 3, 6, 8, 9, see also Filtering problem
Stochastic Fubini's theorem   351
Stochastic integral   330–341
   limits of   358
   localization   343
   martingale property   337
Stochastic integration by parts   342
Stopping time   306
   announceable   321

T

TBBA   see Tree-based branching algorithms
Total sets in L^1 355, 357
Transition kernel   257
Tree-based branching algorithms   230, 279
Tulcea's theorem   298, 303, 347, 348

U

Uniqueness of solution   see Kushner–Stratonovich equation, uniqueness of solution, see Zakai equation, uniqueness of solution
Usual conditions   16, 319

W

Weak topology on $\mathcal{P}(\mathbb{S})$ 21–27
   metric for   26
Wick polynomials   203
Wiener filter   5–6

Z

Zakai equation   62, 69, 73, 154, 177
   correlated noise case   74, 111
   finite-dimensional   65
   for inhomogeneous test functions   69, 97
   strong form   67, 175–178, 202–203, 206
   uniqueness of solution   107, 109, 114, 182

Printed in the United States of America